THE ROUTLEDGE HANDBOOK OF RELIGIOUS NATURALISM

Ecological crisis is being widely discussed in society today and therefore, the subject of religious naturalism has emerged as a major topic in religion. *The Routledge Handbook of Religious Naturalism* is an outstanding reference source to the key topics, problems, and debates in this exciting subject and is the first collection of its kind. Comprising thirty-four chapters by a team of international contributors, the Handbook is divided into seven parts:

- Varieties of religious naturalism and its relations to other outlooks
- Some earlier religious naturalists
- Pantheism, materialism, and the value-ladenness of nature
- Ecology, humans, and politics in naturalistic perspective
- Religious naturalism and traditional religions
- Putting religious naturalism into practice
- Critical discussions of religious naturalism.

Within these sections central issues, debates, and problems are examined, including: defining religious naturalism; religious underpinnings of ecology; natural piety; the religious-aesthetic; ecstatic naturalism as deep pantheism; spiritual ecology; African-American religious naturalism; Christian religious naturalism; *Dao* and water; Confucianism; environmental action; and practices in religious naturalism.

The Routledge Handbook of Religious Naturalism is essential reading for students and researchers in religious studies, theology, and philosophy. The Handbook will also be useful for those in related fields, such as environmental ethics and ecology.

Donald A. Crosby is Professor of Philosophy Emeritus, Colorado State University, USA.

Jerome A. Stone is Professor of Philosophy Emeritus, William Rainey Harper College, USA.

ROUTLEDGE HANDBOOKS IN RELIGION

RECENTLY PUBLISHED:

THE ROUTLEDGE HANDBOOK OF MUSLIM–JEWISH RELATIONS
Edited by Josef Meri

THE ROUTLEDGE HANDBOOK OF RELIGIOUS NATURALISM
Edited by Donald A. Crosby and Jerome A. Stone

FORTHCOMING:

THE ROUTLEDGE HANDBOOK OF RELIGION, MEDICINE AND HEALTH
Edited by Dorothea Lüddeckens

THE ROUTLEDGE HANDBOOK OF RELIGION AND JOURNALISM
Edited by Xenia Zeiler and Kerstin Radde-Antweiler

THE ROUTLEDGE HANDBOOK OF ISLAM AND GENDER
Edited by Justine Howe

THE ROUTLEDGE HANDBOOK OF RELIGIOUS NATURALISM

Edited by Donald A. Crosby and Jerome A. Stone

Routledge
Taylor & Francis Group
LONDON AND NEW YORK

First published 2018
by Routledge
2 Park Square, Milton Park, Abingdon, Oxon OX14 4RN

and by Routledge
52 Vanderbilt Avenue, New York, NY 10017

First issued in paperback 2020

Routledge is an imprint of the Taylor & Francis Group, an informa business

British Library Cataloguing-in-Publication Data
A catalogue record for this book is available from the British Library

Library of Congress Cataloging-in-Publication Data
Names: Crosby, Donald A., editor.
Title: The Routledge handbook of religious naturalism / edited by Donald A. Crosby and Jerome A. Stone.
Description: 1 [edition]. | New York : Routledge, 2018. | Includes bibliographical references and index.
Identifiers: LCCN 2017040565| ISBN 9781138292079 (hardback) | ISBN 9781315228907 (ebook)
Subjects: LCSH: Naturalism—Religious aspects.
Classification: LCC BL183 .R68 2018 | DDC 201/.77—dc23
LC record available at https://lccn.loc.gov/2017040565

ISBN 13: 978-0-367-58089-6 (pbk)
ISBN 13: 978-1-138-29207-9 (hbk)

Typeset in Bembo
by Apex CoVantage, LLC

CONTENTS

CONTRIBUTORS

George Allan is Professor of Philosophy Emeritus at Dickinson College, where he taught for 33 years, the last 21 of which he was also the College's senior academic officer. Allan has published three books on the metaphysical foundations of social value: *The Importances of the Past: A Meditation on the Authority of Tradition*; *The Realizations of the Future: An Inquiry into the Authority of Praxis*; and *The Patterns of the Present: Interpreting the Authority of Form*. He has also published three books in philosophy of education: *Rethinking College Education*; *Higher Education in the Making: Pragmatism, Whitehead and the Canon*; and *Modes of Learning: Whitehead's Metaphysics and the Stages of Education*. Allan has published around 100 articles and book chapters in metaphysics, social philosophy, philosophy of history, philosophy of education, and issues in higher education, usually from a process/pragmatic perspective. His current projects include reinterpreting Whitehead's cosmology in a functionalist and secular manner, and exploring the relevance of Susanne Langer's ontology of the Act for an interpretation of the conditions of civilized existence.

Victor Anderson is the Oberlin Theological School Professor of Ethics and Society at the Divinity School. He is also the Professor in the Program in African American and Diaspora Studies and Religious Studies in the College of Arts and Sciences. He holds degrees from Calvin Theological Seminary including the Master of Divinity and Master of Theology in Philosophical and Moral Theology. He earned the M.A and Ph.D. in Religion from Princeton University in Religion, Ethics, and Politics (1991, 1992). Anderson has published three books: *Beyond Ontological Blackness: An Essay in African American Religious and Cultural Criticism* (1995), *Pragmatic Theology: Negotiating the Intersection of an American Philosophy of Religion and Public Theology* (1999), and *Creative Exchange: A Constructive Theology of African American Religious Experience* (2008). He teaches courses in philosophy of religion, philosophical, theological and social ethics, African American religious studies, and American philosophy and religious thought. He is currently working on a book, *Creative Conflict and Creative Exchange: A Christian's Social Witness to the Public and Its Problems*, which is planned for spring 2017.

Whitney Bauman is Associate Professor of Religious Studies at Florida International University in Miami. He teaches and lectures on science and religion, religion and nature, and religion and queer theory. His books include *Religion and Ecology: Developing a Planetary Ethic* (2014); *Theology Creation and Environmental Ethics* (2009); with Lucas Johnston, *Science and Religion: One*

Planet, Many Possibilities (2014) and with Kevin O'Brien and Richard Bohannon, *Grounding Religion: A Fieldguide to the Study of Religion and Ecology* (2010). He is currently working on a manuscript that examines the religious influences on Ernst Haeckel's understanding of the natural world.

Michael Cavanaugh is a Louisiana lawyer who retired early to concentrate on figuring out the relation between science and religion. That resulted in his deep involvement with The Institute on Religion in an Age of Science and with the journal *Zygon*. He is a co-founder of Religious Naturalist Association.

David E. Conner holds M. Div. and Th.D. degrees from the Iliff School of Theology. He also was a Ph.D. student for a year in the Claremont Graduate School Religion Department. His bachelor's degree is in chemistry. Conner has written articles on an empirical interpretation of Whitehead's thought, on the application of certain scientific and cosmological discoveries to theology and religion, and on other topics. He has been an ordained minister for over 40 years and is currently serving as co-pastor of the Wheat Ridge United Church of Christ in Colorado.

Robert S. Corrington is the Henry Anson Buttz Professor of Philosophical Theology in the Graduate Division of Religion at Drew University in Madison, NJ. He is the author of 12 books and around 80 articles. His most recent book is *Nature and Nothingness: An Essay in Ordinal Phenomenology*. In this and previous books and articles he has created his own philosophical perspective of ecstatic naturalism. He has an interest in nature religions, e.g., Neo-paganism, and nature "itself." He correlates nature with nothingness because the word "nature" has no referent and cannot be defined. Key to his ecstatic naturalism is Spinoza's distinction (with a medieval provenance) between *natura naturans* and *natura naturata*. This distinction enables one to talk of the unconscious of nature (i.e., nature naturing) and the innumerable orders of the world (i.e., nature natured). He weaves together ordinal psychoanalysis with ordinal phenomenology to get a more capacious understanding of the human process as it is imbedded in the one "nature" that there is.

Donald A. Crosby is Professor of Philosophy Emeritus of Colorado State University, where he taught for 36 years. He also taught for three years at Centre College of Kentucky. His Ph.D. in the field of Philosophy of Religion and Ethics is from the Joint Program in Religion at Union Theological Seminary and Columbia University in New York City. His current main research interests are in the areas of religious naturalism, metaphysics, American philosophy, and philosophy of religion. Crosby's latest published books are *Nature as Sacred Ground: A Metaphysics for Religious Naturalism* (2015) and *The Extraordinary in the Ordinary: Seven Miracles of Everyday Life* (2017). He has recently had accepted for publication a new book manuscript: *Consciousness and Freedom: The Inseparability of Thinking and Doing* (Lexington Books). Crosby uses the term *religion of nature* as the label for the version of religious naturalism he explicates and defends.

Pamela C. Crosby is a former high school English teacher, department chair, and academic center director. A National Milken Family Foundation Award Educator, she is the co-editor of the *Journal of College and Character* and chief editor of *JCC Connexions*, sponsored by NASPA – Student Affairs Administrators in Higher Education. Both publications examine the influences of the higher education setting on the moral development and civic behaviors of college students.

Her research interests focus on character education, college student development, and adult religious education as they relate to the moral implications of the works of William James and Justus Buchler. She earned her Ph.D. from Florida State University in Philosophy of Education and has advanced degrees in English Education and Philosophy.

Willem B. Drees serves as Dean of the Tilburg School of Humanities of Tilburg University, the Netherlands, and Professor of Philosophy of the Humanities, as well as the editor of *Zygon: Journal for Religion and Science.* Previously, he was Professor of Philosophy of Religion at Leiden University (2011–14) and president of ESSSAT, the European Society for the Study of Science And Theology (2002–08). He has earned doctorates in theology and in philosophy, and an advanced degree in theoretical physics. He is the author of *Religion, Science and Naturalism* (1996), *Beyond the Big Bang: Quantum Cosmologies and God* (1990), *Religion and Science in Context: A Guide to the Debates* (2010) and *Creation: From Nothing until Now* (2002). Edited volumes include *Is Nature Ever Evil? Religion, Science and Value* (2003), *The Study of Religion and the Training of Muslim Clergy in Europe: Academic and Religious Freedom in the 21st Century*, with Pieter Sjoerd van Koningsveld (2008), and over 20 other volumes. He twice received a Fulbright Grant, for periods of research in Berkeley, Chicago and Princeton. He has held fellowships and given invited lectures at various universities worldwide.

Jay N. Forrest, DMin, is the Education Director for the Spiritual Naturalist Society. He is the author of *Practical Buddhism* and *Secular Buddhism: An Introduction.* He is a certified meditation teacher and an ordained Humanist Celebrant.

Ursula Goodenough is Professor of Biology at Washington University where she teaches courses in cell biology and molecular evolution to undergraduates and graduate students and conducts research on unicellular eukaryotic algae. In 2000–10 she co-taught, with a physics and an earth-sciences professor, an undergraduate course for non-science majors called "The Epic of Evolution." Her book *The Sacred Depths of Nature* (1998) represents an early articulation of the orientation she called religious naturalism, and continues to sell regularly. She has also published numerous journal articles and blog entries on the topic; these can be accessed here: http://religious-naturalist-association.org/ursula-goodenough-2/. She currently serves as president of the Religious Naturalist Association. Her professional honors include election to the presidency of the American Society for Cell Biology and as a fellow of the American Academy of Arts and Sciences.

Walter B. Gulick is Professor Emeritus of Philosophy, Humanities, and Religious Studies at Montana State University Billings. He understands philosophy as ideally a synoptic discipline; he is especially interested in issues of meaning and value as played out in religious traditions and in everyday life as well as in the various academic disciplines. Michael Polanyi's thought has been a primary source of inspiration. Gulick has served as president of the Polanyi Society and for 25 years has been the book review editor of *Tradition and Discovery*, the Polanyi Society's journal. Twice a Fulbright Scholar, he has taught in seven countries.

Bruce M. Hannon is Jubilee Professor of the Liberal Arts and Sciences at the University of Illinois. He is an engineer by training and has modeled ecological and economic systems. He has been an active environmentalist for the past 50 years, living in Champaign County, Illinois, his entire life.

Philip Hefner is Professor Emeritus of the Lutheran School of Theology at Chicago and founding director of the Zygon Center for Religion and Science (1988–2004). He was editor-in-chief of *Zygon: Journal of Religion and Science* for 20 years (1988–2008). Although he has lectured and taught in universities, colleges, and seminaries in Asia, Europe, Africa, and the United States, his regular occupation has been teaching ministerial students in three Lutheran seminaries: Wittenberg University in Springfield, Ohio; Lutheran Theological Seminary in Gettysburg, Pennsylvania; and 35 years at the Lutheran School in Hyde Park, Chicago. In Hyde Park, he worked with Ralph Wendell Burhoe, developing his thinking in the field of religion-and-science. His best-known books are *The Human Factor* (1993), in which he develops the concept of humans as created co-creators, and *Technology and Human Becoming* (2003). He is presently at work on a book that expands the concept of created co-creator.

Cedric L. Heppler earned the following degrees: B.A., Stetson University, B.D. and Th.M., Southeastern Baptist Seminary, and M.S.L.S., University of North Carolina. He was Campus Minister for United Ministries in Higher Education at Humboldt State University and Eastern Washington University. He has served as Head Librarian, North Carolina Wesleyan College and Reference Librarian (Humanities and Social Sciences), North Carolina State University. Heppler was editor of Henry Nelson Wieman, *Seeking a Faith for a New Age* (1975), and editor of Henry Nelson Wieman, *The Organization of Interests* (Wieman's doctoral thesis at Harvard University, 1985). He was co-editor with William S. Minor of Henry Nelson Wieman's book *Man's Ultimate Commitment* (reprinted 1991). His essay "Creative Naturalism and Creative Interchange" was published in John Broyer and William S. Minor's, *Creative Interchange* (essays in memory of Henry Nelson Wieman 1986), and his essay "To End Is to Begin" appeared in Ramakrishna Puligandla and David Lee Miller, editors, *Buddhism and the Emerging World Civilization* (1996).

Michael S. Hogue, Ph.D., is Professor of Theology at Meadville Lombard Theological School (Unitarian Universalist) in Chicago, IL. He is also currently a fellow with the Enhancing Life Project, funded by the John Templeton Foundation and organized through the University of Chicago and Ruhr-Universität Bochum in Germany. He received his doctorate in theological ethics from the Divinity School at the University of Chicago in 2005. His writing and teaching explore the intersections between theology, ethics, and the environment. In 2008, he was awarded a John Templeton Prize for Theological Promise for his first book, *The Tangled Bank: Towards an Ecotheological Ethics of Responsible Participation*. He is also the author of *The Promise of Religious Naturalism* (2010). He has published articles in numerous journals and is the editor of *American Journal of Theology and Philosophy* and is on the editorial boards of *Zygon: Journal of Religion and Science* and *Science, Religion and Culture*. He is currently finishing a book on the contemporary theopolitical significance of the pragmatist, process, and naturalist lineages of American religious naturalism. Tentatively titled *American Immanence*, this book will be published in the Insurrections series with Columbia University Press in 2017.

Stephanie Kaza is Professor Emerita of Environmental Studies at the University of Vermont and former Director of the Environmental Program. Her courses include *Unlearning Consumerism*; *Religion and Ecology*; *Women, Health, and Environment*; and other values-based courses. She co-founded the Environmental Council at University of Vermont, a campus-wide consortium on sustainability, and served as the faculty director for the UVM Office of Sustainability and

Sustainability Faculty Fellows program. Her books include *Mindfully Green* (2008), *Hooked! Buddhist Writings on Greed, Desire, and the Urge to Consume* (2005), *Dharma Rain: Sources for Buddhist Environmentalism* (2000, co-edited with Kenneth Kraft), and *The Attentive Heart: Conversations with Trees* (1993).

Todd Macalister is Communications Director at the Religious Naturalist Association (RNA). He works as a medical writer, developing professional education programs. He attended the School of Theology at Boston University and is the author of *Einstein's God: A Way of Being Spiritual Without the Supernatural* (2008).

Thomas Norton-Smith, an enrolled member of the Piqua Sept Shawnee Tribe, Professor of Philosophy at Kent State University, earned his Ph.D. from the University of Illinois at Urbana-Champaign. His professional interests include American Indian philosophy and the philosophy of mathematics. He is also a contributor to the ethical debate on the use of American Indian sports team mascots and nicknames. Dr. Norton-Smith served as chair of the American Philosophical Association's Committee on the Status of Indigenous Philosophers. His publications include *The Dance of Person and Place: One Interpretation of American Indian Philosophy* (2010), "Indigenous Numerical Thought in Two American Indian Tribes," in Anne Waters (ed.) *American Indian Thought* (2004), and "What a Puzzle Teaches about Moral Justification For and Against the Use of American Indian Sports Team Imagery," *Ayaangwaamizin* 3:1 (2003).

Jea Sophia Oh is Assistant Professor of Philosophy at West Chester University of Pennsylvania. Her areas of specialization are comparative philosophy and religion and ecology. Her book, *A Postcolonial Theology of Life: Planetarity East and West* is the first approach to bridge postcolonialism and ecological theology with the use of Asian spirituality as the philosophical underpinning for the argument that all forms of *Life* are sacred. She is a member of Institute for Ecstatic Naturalism, formulated by Professor Robert S. Corrington at Drew University. Her paper at the Second International Congress on Ecstatic Naturalism (2012), "Nature's Spontaneity and Intentionality: Ecocracy, Doing Non-Doing Principle of Donghak [Eastern Learning]," is published in *A Philosophy of Sacred Nature: Prospects for Ecstatic Naturalism* (2014). Her paper, "Vulnerable Transcendence of Nature: A Transhuman Understanding of Hybridity," won "The Ralph Waldo Emerson Prize" for the Best Paper at the Sixth International Congress on Ecstatic Naturalism, Drew University (April 2016).

Karl E. Peters (M.Div. systematic theology, McCormick Seminary, Ph.D. philosophy of religion, Columbia University) is Professor Emeritus of Philosophy and Religion, Rollins College. He has been editor and co-editor of *Zygon: Journal of Religion and Science*, and the President of the Institute on Religion in an Age of Science and of the Center for Advanced Study in Religion and Science. Peters has published *Dancing with the Sacred: Evolution, Ecology, and God* (2002), *Spiritual Transformations: Science, Religion, and Human Becoming* (2008), and over 35 essays in science and religion. Currently he is developing a Christian Naturalism with papers on "Human Salvation in an Evolutionary World," *Zygon* (December 2012) and "A Christian Naturalism: Developing the Thinking of Gordon Kaufman," *Zygon* (September 2013). In 1999 Karl was married to UCC minister and pastoral counselor Marj Davis. With an M.S. neurology, Cornell U. and an M. Div. Yale, she is his intellectual, moral, and spiritual soul mate. They live in Granby, CT.

Varadaraja V. Raman, Emeritus Professor of Physics and Humanity at the Rochester Institute of Technology, is a philosopher, writer, and physicist. He holds a Ph.D. in Theoretical Physics from the University of Paris. He has authored a dozen books and scores of articles on the historical and philosophical aspects of science, as well as on aspects of bridge building among cultures, and between science and religions of all denominations. His most recent books are *Truth and Tension in Science and Religion* (2010), *Indic Visions in an Age of Science* (2011), *Sivapuranam: Commentaries on a Mystic Poem* (2012), *The Scientific Enterprise* (2011), *Remembering Great Scientists* (2016), and *The Bhagavad Gita: Non-Traditional and Cross-Cultural Perspectives* (2016). Raman is a senior fellow at the Metanexus Institute (Philadelphia), a fellow of the International Society for Science and Spirituality (Cambridge, UK), recipient of the Raja Rao Award from the Jawaharlal Nehru University, Delhi, and of the Academic Fellow Award of the Institute of Religion in an Age of Science (IRAS). He is an ex-president of IRAS, and is on the editorial board of *Zygon: International Journal for Science and Religion*, as well as of *Science and Theology*. He has served as president of the Rochester Interfaith Forum, and is a board member of the Center for Interfaith Studies and Dialogue, from which he is the recipient of an award. He has been an invited speaker in conferences at Madrid (Spain), Cambridge and Birmingham (UK), Budapest (Hungary), Luxembourg, Frankfurt am Oder (Germany), Porto (Portugal), New Delhi, Chennai, Bangalore (India), Seoul (South Korea), Cape Town (S. Africa), and several places in the United States.

Michael L. Raposa is Professor of Religion Studies and the E.W. Fairchild Professor of American Studies at Lehigh University, where he has been a member of the faculty since 1985. Previously (1981–85), he taught at Sacred Heart University in Fairfield, Connecticut. Raposa received his B.A. degree from Yale University (1977), an M.A.R. degree from the Yale Divinity School (1979), and the Ph.D. from the University of Pennsylvania (1987). He is the author of numerous articles and reviews devoted to topics related to pragmatism and American religious thought, as well as three books: *Peirce's Philosophy of Religion* (1989), *Boredom and the Religious Imagination* (1999), and *Meditation and the Martial Arts* (2003). A fourth book, *Theosemiotic: Religion, Reading and the Gift of Meaning*, is nearing completion and under contract with Fordham University Press. For five years, from 2010 through 2014, Raposa served as editor of the *American Journal of Theology and Philosophy*.

Loyal Rue is Emeritus Professor of Philosophy and Religion at Luther College in Decorah, Iowa. He has written six academic books and one novel. The recipient of two Templeton awards, Rue has held fellowships at Harvard and Durham University (UK). He continues to play Vintage Base Ball.

John R. Shook (Ph.D. Philosophy) is Research Associate in Philosophy, and Clinical Instructor of Science Education for the online Science and the Public Ed.M. program, of the University at Buffalo, New York. He is also a lecturer in philosophy at Bowie State University in Maryland. In recent years he has been a visiting fellow at the Institute for Philosophy and Public Policy at George Mason University in Virginia, and visiting fellow at the Center for Neurotechnology Studies of the Potomac Institute for Public Policy in Virginia. His research areas include philosophy of science, neurophilosophy, moral psychology, ethical and political theory, and science-religion dialogue. Among his recent books are *The Future of Naturalism* (co-edited, 2009), *The God Debates: A 21st Century Guide for Atheists, Believers (and Everyone in Between)* (authored, 2010), *The Essential William James* (edited, 2011), *American Philosophy and the Brain: Pragmatist Neurophilosophy, Old and New* (co-edited, 2014), *Dewey's Social Philosophy: Democracy as Education* (authored, 2014), and *The Oxford Handbook of Secularism* (co-edited, forthcoming).

Lisa H. Sideris is an associate professor of religious studies at Indiana University with research interests in environmental ethics and narratives at the intersection of science and religion. She is author of *Environmental Ethics, Ecological Theology, and Natural Selection* (2003) and co-editor of a collection of essays on the life and work of Rachel Carson, *Rachel Carson: Legacy and Challenge* (2008). Her recent work focuses on the role of wonder in environmental and science-religion discourse, and particularly on efforts to recast scientific narratives, including narratives of the Anthropocene, as sacred, shared stories for humanity. Her current project, *Consecrating Science: Wonder, Knowledge, and the Natural World*, will be published by the University of California Press. Sideris has been a fellow at the Center for the Study of Religion at Princeton University and the Rachel Carson Center for Environment and Society in Munich. She serves as associate editor for the *Journal for the Study of Religion, Nature, and Culture*.

Dan Solomon (Ph.D., Illinois Inst. of Tech.) is a retired computer scientist. He currently serves as Secretary of the Institute on Religion in an Age of Science (IRAS). He is a member of the Learning Associates at the Unitarian Church of Evanston, where he teaches Adult Education classes, often focusing on Jewish heritage. He previously taught a two-year course on Jewish Intellectual History at Congregation Beth Or, a Humanistic Jewish congregation in Deerfield, Illinois.

Leslie E. Sponsel earned a B.A. in Geology from Indiana University and an M.A. and Ph.D in Biological Anthropology from Cornell University. He has taught at seven universities in four countries, two as a Fulbright Fellow. He joined the Anthropology faculty at the University of Hawai'i in 1981 to develop and direct the Ecological Anthropology Program. Although retired as a Professor Emeritus in 2010, he still teaches one course a semester, including on Sacred Places, Spiritual Ecology, and Anthropology of Buddhism. The rest of his time is devoted to research and publications. Sponsel has published numerous journal articles, book chapters, and encyclopedia entries as well as four edited books. His recent monograph, *Spiritual Ecology: A Quiet Revolution*, won the science category of the Green Book Award in San Francisco in 2014. The companion website is: http://spiritualecology.info.

Eric Steinhart grew up on a farm in Pennsylvania. He received his BS in Computer Science from the Pennsylvania State University, after which he worked as a software designer for several years. Many of his algorithms have been patented. He earned an M.A. in Philosophy from Boston College and was awarded a Ph.D. in Philosophy from SUNY at Stony Brook. Since then, he has taught at Dartmouth College and in the Philosophy Department at William Paterson University. His books have concerned Nietzsche, the logic of metaphor, mathematics, and life after death. He has published several dozen academic articles. He has been interested in using new computational concepts to solve old philosophical problems. He is especially interested in new and emerging religions. He loves New England and the American West, and enjoys all types of hiking and biking, chess, and photography.

Jerome A. Stone is Professor of Philosophy Emeritus, William Rainey Harper College. He is a former member of the adjunct faculty at Meadville Lombard Theological School and was a United Church of Christ minister for 18 years. He is currently community minister of the Unitarian Church of Evanston, IL. Stone has a Ph.D. in philosophical theology from the University of Chicago. He is author of *The Minimalist Vision of Transcendence* and of *Religious Naturalism Today*, co-editor of *The Chicago School of Theology*, and guest editor of the Special Issue on Eco-justice and the Environment for the *American Journal of Theology & Philosophy*, XVIII: 1, 1997.

He is also Faculty adviser to The Ecocentrics, an environmental student organization at William Rainey Harper College.

Daniel T. Strain is Executive Director of the Spiritual Naturalist Society. He is a Humanist minister, speaker, and writer on the topics of ethics, spirituality, and ancient philosophy. His Humanist Contemplative group and blog inspired a similar mindfulness group at Harvard University. He is former president of the Humanists of Houston and has served in the Chapter Assembly of the American Humanist Association. Daniel writes for the *Houston Chronicle* belief page online and his work has appeared nationally in other magazines, on Houston PBS, and the journal *Essays in the Philosophy of Humanism*. He leads meditation and occasionally speaks at Jade Buddha Temple and other venues.

Mary Evelyn Tucker is co-director of the Forum on Religion and Ecology at Yale University where she teaches in an MA program between the School of Forestry and Environmental Studies and the Divinity School. With John Grim she organized 10 conferences on World Religions and Ecology at Harvard. They are series editors for the 10 resulting volumes from Harvard. She co-edited *Confucianism and Ecology, Buddhism and Ecology*, and *Hinduism and Ecology*. She has authored with John Grim, *Ecology and Religion* (2014). They also edited Thomas Berry's books including *Selected Writings* (2014). With Brian Swimme she wrote *Journey of the Universe* (2011) and is the executive director of the Emmy-award winning Journey film that aired on PBS. She served on the International Earth Charter Drafting Committee and was a member of the Earth Charter International Council.

Abigail T. Wernicki earned her Ph.D. in philosophical and theological studies from the Theological School at Drew University in 2015 under the direction of Robert Corrington. Her dissertation engages Arthur Schopenhauer's aesthetics and argues for a third category of the sublime that she calls the "ontological sublime." Dr. Wernicki regularly delivers papers at the annual Congress on Ecstatic Naturalism and contributed an entry on Indian Materialism to the Internet Encyclopedia of Philosophy. She currently serves as Director of Admissions and Enrollment Services at Reading Area Community College (Reading, PA) and adjunct instructor of philosophy at Alvernia University (Reading, PA). Her research interests include the philosophy of Arthur Schopenhauer, aesthetics, American naturalism, and Indian philosophy.

Nicholas J. Wernicki serves as Dean, Communication, Arts and Humanities at Delaware County Community College in Media, PA. Previously, he was Associate Professor of Humanities and founding Faculty Fellow for the Teaching and Learning Center at Peirce College in Philadelphia. Dr. Wernicki is a board member for the Institute for Ecstatic Naturalism and Acting Area Editor for Continental Philosophy for the Internet Encyclopedia of Philosophy (a peer reviewed academic resource hosted by the University of Tennessee, Martin). He is an active member of the North American Sartre Society and his research interests include philosophical theology, empathy studies, phenomenology and moral aesthetics. He earned a Ph.D. in Theological and Philosophical Studies from Drew University in Madison, New Jersey under the direction of Robert S. Corrington.

Demian Wheeler is Assistant Professor of Philosophical Theology and Religious Studies at United Theological Seminary of the Twin Cities. He holds a Ph.D. from Union Theological Seminary in the City of New York, where he specialized in American liberal theology. His research and scholarship focus on the "Chicago school" of theology and the streams of theological and philosophical thought that have flowed into and out of it: religious naturalism, pragmatic historicism, empirical theology, and process philosophy. He is the author of a number of scholarly articles and essays and is the co-editor (with David Conner) of *Conceiving an Alternative: Philosophical Resources for an Ecological Civilization*, which will be published by Process Century Press in early 2018.

Carol Wayne White is Professor of Philosophy of Religion at Bucknell University, and the author of *Poststructuralism, Feminism, and Religion: Triangulating Positions* (2002); *The Legacy of Anne Conway (1631–70): Reverberations from a Mystical Naturalism* (2009); and *Black Lives and Sacred Humanity: Toward an African American Religious Naturalism* (2016). She has published articles on process philosophy, religious naturalism, and critical theory. White has also received national awards and fellowships, including an Oxford University Fellowship in Religion and Science, a Science and Religion Course Award Program Development Grant (The John Templeton Foundation), and a NEH Fellowship. She is currently writing a book that explores the tenets of deep ecology and insights of religious naturalism expressed in contemporary American nature poets and writers.

INTRODUCTION

Donald A. Crosby and Jerome A. Stone

Religion is a broad term that covers many quite different outlooks, practices, and points of view. For people of the West, it is normally associated with belief in and commitment to some kind of personal, conscious, purposive deity seen as separate from nature and as the creator of all that is in nature. This connotation of the term, while commonplace, is too narrow and provincial. It does not work, for example, with Daoism, some forms of Hinduism (such as Advaita Vedanta), or some forms of Buddhism (such as Theravada or Zen) because in none of these traditions is the principal focus on a personal God. But these outlooks are entitled to recognition as religious in their basic stance and character.

The too narrow theistic connotation of the term *religion* also does not work with religious naturalism, where nature itself or some aspect of nature, not a separate personal God (or radically nature-transcending spirits, principles, powers, or presences of any kind), is regarded as the center and source of all that is holy or religiously profound and thus as deserving in its own right of fervent piety, reverence, commitment, and service. Religious naturalism is avowedly atheistic in the sense of *theism* described above, but it is far from being either anti-religious or non-religious. Moreover, it is not necessarily opposed to all possible conceptions of God so long as these are conceived as radically immanent or internal to nature, as is shown by some of the essays in this volume. To place exclusive or contentious emphasis on religious naturalism's atheistic character (in the traditionally dominant Western sense of this term) would be analogous to placing predominant stress on traditional theism's *adaoist*, *ahindu*, *abuddhist*, or *anaturalistic* character. It would fail to do justice to what conventional religious theism *is*, rather than what it is *not*. A similar point holds for religious naturalism.

With its positive focus on nature or aspects of nature as the ultimate focus of religious commitment, religious naturalism is entitled to recognition as a legitimate member of the family of distinctive religious outlooks in today's world. The task of *The Routledge Handbook of Religious Naturalism* is to give expression to some varieties of religious naturalism and perspectives on religious naturalism as these have been developed or are in process of being developed by proponents, expositors, and critics of this outlook on human beings and their world—an outlook that is coming increasingly into notice in our time. The authors featured in the volume provide such expressions in their essays. In doing so, they help to give substantive shape, character, and content to ways in which religious naturalism can be conceived, practiced, and appraised.

Some of the authors also suggest ways in which this newly awakened and growing religious vision can be compared and contrasted with, informed by, or shown to contribute illuminating contexts, emphases, and meanings to more established and readily recognized religious traditions and angles of vision. Nature is common to us all and of vital importance to us all. Thinking deeply about nature and our place as human beings in nature is an urgent and salutary activity for each of us and for the institutions of our societies, no matter what our personal religious or secular outlooks may be—especially in this time of rampant species endangerments, global climate change, and looming ecological crisis.

The editors' hope and expectation is that the essays in this volume will contribute in their own fashion to this much needed deep thinking and to the aspirations, practices, and policies for which it calls. The essays are but a sampling of what could have been said about religious naturalism in forms of it envisioned, adhered to, or presented by persons other than those in this book, to say nothing of what cannot now be known about its future forms and developments. But they can serve to indicate significant contours, traits, and varieties of the character and range of religious naturalism and some of its significant comparisons and contrasts with other more traditional and familiar types of religious thought, practice, and commitment.

The essays by Jerome A. Stone and Willem B. Drees in Part I help to bring into view different ways of characterizing religious naturalism as such and raising critical questions regarding some of its relationships with other closely related points of view. In Part II Whitney Bauman, Cedric L. Heppler, John Shook, Abigail T. Wernicki, and Nicholas J. Wernicki present versions of religious naturalism developed by certain influential thinkers from the nineteenth to the latter part of the twentieth century. In Part III Robert S. Corrington and Demian Wheeler present and defend pantheistic types of religious naturalism, while Donald A. Crosby endorses a religious naturalism committed to a radical form of materialism. George Allan reflects on the valuative inspirations, challenges, and opportunities that reside in the splendor of the natural world.

In Part IV Leslie E. Sponsel discusses relations of what he calls "spiritual ecology" to religious naturalism. Carol Wayne White focuses her discussion on religious naturalism's critique of White supremacy and other types of hierarchy and exclusion, its mandate for cherishing all forms of life, and its critical bearing on what it is to be a human being. Michael S. Hogue devotes his attention to the political outlook and responsibilities implicit in a religious naturalist outlook—a crucial aspect of religious naturalism's holistic vision of human life in a world threatened by widespread ecological devastation and danger. Victor Anderson develops the idea of what he calls "pragmatic naturalism" and shows how the metaphysics, valuative dimensions, and symbolic expressions of this outlook can guide public and planetary life.

In Part V Karl E. Peters and Loyal Rue highlight religious naturalism's relations to Christianity. Peters presents a case for a Christian form of religious naturalism. Rue talks about his early Lutheran Christian upbringing and his transition from it to the development and espousal of a type of religious naturalism. Jay Forrest and Stephanie Kaza explore some important connections of religious naturalism with Buddhism, showing how this Eastern religious outlook can contribute to an expansion of vision and an awakening of critical perspectives on traditional Western ways of viewing humans and their relations to the world. Thomas Norton-Smith emphasizes the sole reality, sacredness, and inherent value of the natural world in his own Shawnee Indian heritage and questions what he sees as the too-abstract character of many forms of religious naturalism.

Jea Sophia Oh explores the image of water in Daoism, showing how this image can help us to see Daoism as a highly relevant form of religious naturalism that can deepen our sense of intimate relatedness to nature and our urgent need to contribute in every way possible to its wellbeing. Mary Evelyn Tucker shows how Confucianism can also be seen as a significant and long-lasting form of religious naturalism. Varadaraja V. Raman indicates how aspects of the Hindu religious tradition show strong affinities with religious naturalism, both in conceptualization and in practice. Dan Solomon shows how resources of the Jewish religious tradition can be drawn on to develop a Jewish form of religious naturalism. In doing so, he also describes his personal vision of religious naturalism.

In Part VI Bruce M. Hannon describes his personal path of commitment to religious naturalism and places strong emphasis on the necessity of organizing groups dedicated to putting its ecological implications and responsibilities into practice in local areas. Eric Steinhart discusses numerous particular ways of expressing and celebrating religious naturalism, while Daniel Strain describes what he calls "naturalistic spirituality" as a practice. Pamela Castellaw Crosby explains the philosopher Justus Buchler's concept of *Query* and explores the central role it can have for educating adults in the outlooks and practices of religious naturalism. Ursula Goodenough, Michael Cavanaugh, and Todd Macalister describe the process of setting up a religious naturalist organization and bringing it to the internet, while Walter Gulick comments critically on the prospect and challenge of bringing religious naturalism to the fore as an engaging option and workable practice for peoples of a rapidly globalizing world in which there is growing ecological awareness.

In Part VII David Conner critically examines five types of religious naturalism and explores the relative degrees of "religious availability" of each of these types, as shown in terms of seven types of religious value. Michael Raposa mounts a critical perspective on religious naturalism by claiming that all of its forms involve a kind of nostalgia for the supernatural that lurks beneath their explicit rejections of supernaturalism. Lisa H. Sideris is critical of the notion that fervent and responsible environmental behaviors will be inspired by various aspects of the cosmogenesis story of contemporary science. She argues instead that the wonder inspired by this story lacks positive ethical and environmental dimensions. Philip Hefner argues that nature is a text requiring interpretation and suggests six basic approaches to this interpretation. He warns against viewing nature too exclusively in scientific terms, contends that much religious naturalism is inclined to do so, and argues that religious naturalism needs to draw on the richness of traditional religious approaches in its interpretations of nature.

The two editors of this Handbook are convinced that religious naturalism can be a significant religious option for persons unable to endorse the outlooks and commitments of religious traditions in which supernatural realms, personages, presences, powers, and events play the dominant role—but for whom the insistent lure of religious aspirations, values, and meanings is persistent and compelling. Religious naturalism offers an entirely this-worldly, here-and-now way of viewing the world and the place and responsibility of humans within it. There is no in-principle opposition between its view of the world and the findings of the natural sciences. Religious naturalism in its various forms provides a deeply spiritual and inspiring religious vision with levels of assurance, demand, and empowerment that can command response, consent, and commitment from many of those for whom traditional supernaturalistic types of religious faith have no convincingness or appeal.

Religious naturalism is a particularly relevant and demanding outlook on the world, in the judgment of the editors and most others writing in this volume, in a time of grave ecological crisis that threatens large numbers of the life forms of the earth—including humans beings—and the health and wellbeing of the earth as a whole. This outlook calls upon humans to be servants of nature rather than assuming themselves—as human beings often have in the past—to be the focus of nature and entitled to be served by nature. The centering is shifted from us humans to the multifarious community of earthly beings to which we belong as fellow creatures of nature. The new centering, when properly embraced and lived, can have radical saving effects for us all. This is the good news that lies in the heart of religious naturalism. It is news that calls for committed effort and response but also for grateful recognition of our privilege as human beings to be part of a nurturing earth and magnificent universe.

PART I

Varieties of religious naturalism and its relations to other outlooks

1
DEFINING AND DEFENDING RELIGIOUS NATURALISM

Jerome A. Stone

As a preliminary description, religious naturalism may be said to be a movement that asserts the possibility and desirability of a robust religious/spiritual life without recourse to the supernatural, whether deity, soul, or heaven. A number of people discussing naturalism have stressed that proponents of naturalism need to emphasize what naturalism stands *for*, rather than *against*. In line with that, naturalism "affirms that attention should be focused on the events and processes of this world to provide what degree of explanation and meaning are possible to this life" (Stone 2008: 1). However, many recent supernaturalists make a similar affirmation; hence the negative assertion of a denial of the supernatural seems appropriate.

Not all religious naturalists use that term to describe themselves, and sometimes there are boundary issues or vacillation. To go into these in detail would take us beyond the scope of an introductory essay. There is no central organization, institutional basis, or organ to disseminate the ideas and practices of religious naturalism, although there are a number of bonds of cohesion, some of which are described in the selection by Goodenough, et al.

The application of the term "religious naturalism" is something like the term "Impressionism" to describe certain French painters of the nineteenth century. The latter term was coined by a journalist to describe the paintings in the first exhibition organized by Impressionist artists and has been found useful by art historians and critics. The term "religious naturalism" was used frequently among some theological writers in America in the 1940s and 1950s, especially at the University of Chicago's Divinity School and its *Journal of Religion.* (Many issues in the present essay are treated more fully in Stone 2008.) Thus religious naturalism is a not a clearly delineated natural kind, like a solar system, nor a purely fictive construction, like a stellar constellation. It is much like a galaxy with loosely defined borders or perhaps like an ecosystem.

Naturalists have different views of nature. Contrast Spinoza (pro) and Crosby (con) on causal determinism. Many follow a variety of emergentist approaches. One may question whether religious naturalism is really religious at all. The justification for the term "religious" is that attitudes and beliefs among religious naturalists are sufficiently analogous to attitudes and beliefs among the paradigm cases of religion that they may be called religious. This approach creates a difficulty, since the choice of paradigm cases of religion will vary depending on the socio-historical context (see Stone 1992: 21–27).

Is religious naturalism a distraction from more pressing issues? Is it only for people who can afford it? Is religious naturalism for the elite? There are a number of replies to these questions.

(1) Religious naturalism can help sensitize people to the importance of the environment. Earth is the home, the life-support system of everyone. (2) Religious naturalism can help attune the victims of injustice to the healing and renewing resources of the non-human world. (3) Religious naturalism helps remove the oppressive aspects of some traditional theisms. Furthermore, it fosters a religious or spiritual approach that can be at least as fulfilling and liberating as much traditional theism. (4) Since religious naturalism is compatible with the methods and results of scientific inquiry, it is useful for challenging many conservative social and political movements. (5) Many religious naturalists have been especially focused on social justice and human empowerment. Finally (6), religious naturalism can motivate care for the non-human world as I try to show in *Sacred Nature* (Stone 2017).

The meanings of "naturalism"

There is no single, agreed-upon meaning of "naturalism." Religious naturalists espouse a variety of meanings of the term, most of which would satisfy all but the upholders of the strictest definition of it. The *Oxford English Dictionary* makes the suggestion that the philosophical meaning of the term goes back to the eighteenth century and meant a view of the world in which "only the operation of natural (as opposed to supernatural or spiritual) laws and forces is admitted or assumed" (Volume X: 245). The term is also said to mean the view that moral concepts are to be analyzed by means of concepts applicable to natural phenomena. This later meaning is not normally the focus in discussions of religious naturalism.

Owen Flanagan, in his article in *The Oxford Handbook of Religion and Science*, asserts that "the terms 'naturalism' and 'naturalist' lack a single determinate meaning" (Flanagan 2006: 432). Indeed Flanagan specifies 15 meanings or implications of the term naturalism and asserts that there are more (Flanagan 2006: 430–431). However, he notes that there is a common core to all the varieties of naturalism:

> Anti-supernaturalism forms the common core, the common tenet, of "naturalism" insofar as "naturalism" is anything like a coherent philosophical doctrine spanning the last four centuries.... [T]he objectionable form of "supernaturalism" is one according to which (i) there exists a "supernatural being or beings" or "power(s)" outside the natural world; (ii) this "being" or "power" has causal commerce with this world; (iii) the grounds for belief in *both* the "supernatural being" *and* its causal commerce with this world cannot be seen, discovered, or inferred by way of any known and reliable epistemic methods.
>
> (Flanagan 2006: 433)

In my characterization of religious naturalism I include the notion that it denies or at least does not employ the notion of an ontologically distinct and superior reality. See my comments on David Griffin under "Neighboring movements." In my judgment, Griffin is not a naturalist, although he aspires to be. Flanagan would have a harder time excluding Griffin from the fold. I apologize for the exclusivist language here, but some distinction needs to be made.

Flanagan goes on to make a familiar distinction between naturalism as a strong ontological claim about "what there is" and a weaker epistemological claim (often called a methodological claim) about what can be used in explaining things. Religious naturalists take a variety of positions on this distinction. Flanagan also makes a distinction between what he calls "imperialistic ontological naturalism" and a "non-imperialistic ontological naturalism." The former makes a strong claim that the supernatural does not exist, the latter a more modest claim that "*for all we know and can know, what there is, and all there is, is the natural world*" (Flanagan 2006: 437). Again, religious naturalists take a variety of positions on this distinction.

Arthur C. Danto, in *The Encyclopedia of Philosophy*, asserts that "naturalism is polemically defined as repudiating the view that there exists or could exist any entities or events which lie, in principle, beyond the scope of scientific explanation" (Danto 1967: 448). Alan Lacey in *The Oxford Companion to Philosophy* takes a similar position (Lacey 1995: 604). The necessity of scientific explanation in the *explicans* of naturalism is a common narrowing of the term. Again, religious naturalists take a variety of positions on this necessity.

Mario De Caro and David Macarthur have published two collections of essays, *Naturalism in Question* and *Naturalism and Normativity* (De Caro and Macarthur 2004; De Caro and Macarthur 2010). They define "scientific naturalism" as committed to an exclusively scientific conception of nature and conceive of philosophical inquiry as continuous with science (De Caro and Macarthur 2004: 1–6). Instead they nurture a "liberal" or "pluralistic" naturalism that challenges this viewpoint. According to them, liberal naturalism asserts that the claims of scientific naturalism extend beyond the limited scope of scientific assertions and also explores the weakness of scientific naturalism in dealing with the topics of mind, agency, and normativity, especially ethical and aesthetic normativity. Liberal or pluralistic naturalists share four features: (1) a shift in focus from nonhuman to human nature, conceived as a historically conditioned product of contingent forces, (2) a nonreductive attitude to normativity, (3) a view of philosophy as in some respects autonomous from scientific method, and (4) a pluralistic conception of the sciences, rejecting the ideal of the unity of the sciences as unrealizable and conceding that there is no clear demarcation of science from non-science. Religious naturalists are generally, though not necessarily, in sympathy with the liberal or pluralistic naturalism depicted by these authors. They tend to sidestep the concerns of "scientific naturalism" as described here, since religion or spirituality is a different way of engaging the world than science.

Neighboring movements

There are a number of movements related to or overlapping with religious naturalism. It is helpful to clarify their relationships. One such view is materialism or physicalism. As naturalists, most religious naturalists are materialists. However, the two positions are logically independent. It is possible to reject any supernatural realm and yet to hold that the world is not material but something else, perhaps mental or spiritual. Furthermore, the materialism of most religious naturalists is a generous materialism, often emergentist, which allows for the reality of much of what we call mind or values. Loyal Rue put this idea in picturesque language when he said we need to replace the "grunge" theory of matter with a "glitz" theory (Rue 2011: 52–53). When we realize what the material world has produced, we need not deprecate it, but appreciate its creativity. We need not think of matter as lifeless and inert. Some of it is, but some of it is animate, even capable of thinking. (For a vigorous exposition of religious naturalism in physicalist terms see Hardwick 1996.)

Any materialistic view should include the notion that information, relationships, and perhaps possibilities are part of the universe and are to be included among the fundamental aspects of the world. Patterns can be replicated in different times and places. Nevertheless, they seem always to have a physical basis when they are so replicated.

Another position overlapping with religious naturalism is humanism, especially religious humanism. I am referring to such clergy, starting in the 1920s, as John Dietrich, Cutis Reese, and Charles Francis Potter, and also to the signers of the various *Humanist Manifestos*. These humanists are clearly naturalists in that they focus on this world and reject such notions as God, soul, and heaven. They could also be called religious naturalists, especially those that call themselves religious humanists. This is because their devotion to human betterment and the value they place on the search for truth is analogous to the devotion of people we usually call religious. Indeed many of what William Murry calls the newer humanists have an openness "*to wonder and mystery and transcendence in a naturalistic framework*" (Murry 2006: 84). Thus, many humanists today belong even more clearly to the religious naturalists than the humanists of the 1920s and 1930s.

Another related movement is process theology. Many process thinkers refer to themselves as naturalists. However, there is an important difference between most of them and the religious naturalists treated in this *Handbook*. They refer to themselves as naturalists because their panentheism permits them to speak of God as immanent in the world and thus refer to themselves as naturalists. Robert Mesle developed "process naturalism" in Chapter 17 of his *Process Theology: A Basic Introduction* (Mesle 1993: 127–133). To add to the confusion, John Cobb, in a chapter written for Mesle's book, says that he sometimes calls himself a "naturalistic theist" or a "theistic naturalist" (Cobb 1993: 134).

For the process thinkers close to Charles Hartshorne there is one entity that is unique in being surpassable by no other entity except itself in a future state. This entity is supremely related and compassionate and often is thought of as conserving value. These characteristics make this one entity so ontologically distinct and supreme that it is a form of the supernatural. An entity surpassable by none except itself is not naturalist—immanentist yes, naturalist, no.

The process thinker David Griffin, in *Religion and Scientific Naturalism*, develops a "naturalistic theism" which does not belong with the religious naturalists. "Variable constitutive divine influences would be understood as *part* of the normal pattern of causes and effects, *not* an interruption of this pattern" (Griffin 2000: 40). Such a position, Griffin claims, is a form of naturalism because it rejects any supernatural interruptions. God so conceived is a supreme power, the only entity involved in the origination of every other event and giving to each of them its ideal aim. This God is ontologically distinct and supreme, and thus not really an aspect of naturalism as I am characterizing it.

Wesley J. Wildman has grouped a number of philosophical theologies together (e.g., Plato, Plotinus, Thomas Aquinas [in his God as Being Itself and Pure Act moments], G. W. Hegel, Paul Tillich, Robert Neville, and himself) under the heading of ground-of-being theologies (Wildman 2006). This group is characterized by "two important negations: they *deny that ultimate reality is a determinate entity*, and they *deny that the universe is ontologically self-explanatory*" (Wildman 2006: 612). This is an important family of theological writers, often overlooked by those outside of theological circles, and is to be contrasted with personalistic theism and process theology. Because they deny that ultimate reality is a determinate entity, both their theoretical formulations and their general outlook are often close to religious naturalism. However, the other negation, that the universe is ontologically self-explanatory, separates them from religious naturalists. [I find Wildman's claim that some positive formulations of ground-of-being theologies "are indistinguishable from religious naturalism" to be an overstatement (Wildman 2006: 612).]

Yet another term closely allied with religious naturalism is pantheism. It is probably best to think of these as intersecting concepts. Many pantheists are religious naturalists. However, some pantheists might have an animistic outlook, which places them outside of naturalism. And there are many religious naturalists who identify God with only part of the universe, such as the American philosopher of religion Henry Nelson Wieman, for whom God is the integrative process within the world. [See also Gordon Kaufman's *In Face of Mystery* and his smaller *In the Beginning … Creativity*; also *Dancing with the Sacred* by Karl Peters (Kaufman 1993; Kaufman 2004; Peters 2002).] Such religious naturalists are not pantheists.

While the meaning of the term "pantheism" is contestable, usually it refers to an identification of God (or the Divine) with the whole universe. It might be argued that for Spinoza God and Nature are interchangeable terms insofar as the universe is considered as a logically interconnected system. For Samuel Alexander, the early twentieth-century British metaphysician, the focus of religious sentiment is not towards the universe as a present whole, but to the universe insofar as it is growing towards a new and higher level. Edward Scribner Ames, an early twentieth-century psychologist of religion and theologian at Chicago, referred to God as the world in certain aspects and functions; namely, orderliness, love, and intelligence or order, beauty, and expansion (Ames 1929: 154, 157). A similar comment could be made about Chicago theologians George Burman Foster and Bernard Loomer. For Foster "God" designates "the universe in its ideal-achieving capacity" (Foster 1909: 108–110; Peden and Stone 1996, Volume I: 52). Loomer, in his rich essay *The Size of God*, asserts that "God as a wholeness can be identified with the concrete, interconnected totality of this struggling, imperfect, unfinished, and evolving societal web." By "societal" he is referring to the interconnected nature of the universe (Loomer 1987: 42). Paul Harrison, founder and president of the World Pantheist Movement, says that to say that the universe as a whole is divine does not mean that oil slicks, bits of chewing gum on the pavement, nuclear weapons, smokestacks, or mass murderers are divine (Harrison 1999: 71). Many recent naturalists distinguish between *natura naturans* (nature as creating) and *natura naturata* (the totality of the myriad created things) and think of the former as the appropriate object of religious orientation.

Perhaps the Roman Catholic distinction between worship (*latreia*) and veneration *(dulia)* might be of help here. In this tradition only God is worthy of worship, while the saints can be objects of veneration without committing idolatry. Analogously for a naturalist, the entire universe might be the proper object of reverence (divested of personalist overtones), while various things, events, and systems might be objects of something akin to veneration. These analogies might not be helpful, especially for religious naturalists who have a strong antipathy to the connotations of "worshiping" nature.

Many people think that pantheism involves absorption into the infinite ocean of being as the goal of the spiritual journey or as a prospect after death. However, Charles Milligan suggests that in the past century or so pantheists stress independence and autonomy of the human self (Milligan 1987). Perhaps a better way of speaking is to stress the interdependence of the human self with its total ecological nexus, physical, biological, historical, and familial.

There is also some similarity between the Gaia movement and those naturalists who refer to the entire universe in religious terms. It may be said that these naturalists usually use religious language of the entire universe(s) rather than just the planet Earth, as the Gaia movement usually does. The term "Gaia" is also sometimes used in an interesting but debatable scientific hypothesis about the self-corrective nature of the Earth's planetary processes. A helpful study of some types of naturalism is to be found in Bron Taylor's *Dark Green Religion*. However, this book covers several viewpoints that move beyond naturalism (Taylor 2010).

A case for religious naturalism

No conclusive argument or rigorous proof can be made for religious naturalism. A case can be made for it, but in the end it is a conjecture or insight that makes sense to its adherents. It should be noted that it is not a scientific theory or an empirical generalization, although it is a scientifically informed surmise. It is a philosophical position and way of living. Therefore it cannot be dismissed on the ground that it goes beyond the empirical evidence.

Part of the case for religious naturalism is that arguments for the existence or reality of a God conceived of as an ontological ultimate fail to be convincing. A complete critique of the arguments for a maximally conceived deity (as distinct from some varieties of a naturalistically conceived God) is not possible here. However, the ontological arguments and the one based on rejection of an infinite regress deserve brief mention.

The ontological arguments for the existence of God, which allegedly prove that since God is by definition a being that must necessarily exist, are frequently considered unconvincing. However, Charles Hartshorne and Schubert Ogden have made vigorous defenses of these arguments based on what is known as modal logic, thus allegedly invalidating religious naturalism. Using a criticism similar to that of John Hick, I have elsewhere challenged Hartshorne and Ogden by differentiating between the necessity of a proposition and the alleged necessary existence of an entity (Stone 1992: 170–181).

The cosmological arguments for the existence of God are given classic expression by Thomas Aquinas. In his arguments, he moves from the existence of the universe or an aspect of it to the existence of a cause for the universe or aspect. He makes three assumptions. First, that all finite beings are dependent upon other beings for their motion or existence. Second, an infinite regress in the chain of dependence is inexplicable if not supported by something non-contingent. Third, the anchor of this chain, the first cause or necessary being, is the proper object of religious devotion. Although there is some controversy about the third assumption, current discussion has centered on the second. Briefly put, naturalists question the tenability of the second assumption because for them the existence of the universe needs no explanation.

A more positive case can be made for accepting religious naturalism as a surmise. A naturalist can avoid many of the disadvantages of traditional theistic viewpoints. She does not have to wonder why God is allowing bad things to happen to her. She does not have to puzzle over many of the conflicts between religion and science. She does not have to go through the intellectual gymnastics of the standard theodicies. She can avoid the acute sense of guilt and outmoded ideas that often accompany traditional religions. She does not fan the flames of bigotry and religious wars. Naturalists need not fight against the sorrier aspects of organized religion as many liberal theists do. Further, living as a *religious* naturalist means that she does not live in a totally alien world. Although nature as a whole is indifferent to her, although the earthquake does not ask about her religion, nor does God help her to score a goal or her team to win, even so the universe has set the stage for the physical, biological, and historical evolutions that have produced her. Even more, living as a religious naturalist means that she can have many of the positive values of a religious or spiritual life, as many of the chapters in this *Handbook* illustrate.

There is an admitted downside to any of the varieties of religious naturalism. One does not have the solace and comfort of a governing super mind, of divine intervention, of ultimate redemption or immortality. There is no cosmic companion who understands, although many of us have close neighbors and friends. When an earthquake destroys Lisbon, when hurricane Katrina strikes, or when the depth of human evil is revealed in genocide, child abuse, or the horrors of slave ships, there is no God to cling to. We cannot invoke a deity to make sense of it all or to save us. It is our responsibility to strengthen the levees or restore the wetlands. It is our job to prepare for emergencies, to comfort the grief-stricken, to resist genocide, and to remember those who have perished. And surely God is not responsible for those who are not spared, as if their guilt was stronger or their faith or prayers not as effective as in the cases of those who are spared.

To live without God or the hope of immortality may require some mourning. However, mourning is a part of maturation. This does not mean that non-naturalists are immature, but that moving into a naturalistic framework may take some grief work.

Critics of religious naturalism (especially of my own version of it)

Some critics of naturalism use the term in an outmoded sense, are unaware of it, or have not taken seriously the possibility of a religious naturalism. Rudolf Otto, for example, thought of two types of naturalism: an enthusiastic, obscurantist type, typified by Goethe, which eventually ended in nature worship, and strict naturalism, to which he devoted his polemic (Otto 1907: 20). Otto characterized this latter type as cold and indifferent, seeking to simplify and reduce everything (Otto 1907: 30). He found three key areas where the religious and the strict naturalist outlooks conflict. Strict naturalism (1) denies purpose in the universe and thus rejects teleological explanations, (2) asserts that the cosmos is self-sufficing and self-governing, and (3) seeks to have everything clear and intelligible, rejecting mystery. Most naturalists of a religious variety agree with (2). However, some assert that there are aspects of purpose and meaning in the universe, and they vary concerning the place of mystery.

For Shailer Mathews there are two "logically tenable" worldviews: "the materialistic or naturalistic" and "the religious" (Mathews 1924: 4). According to him, in its older form naturalism involved a "dead matter" of which thought and emotion are outcomes. Mind and matter were regarded as opposites, the second alone being finally real. Although the recent naturalists regarded mind as an expression of matter as energy, he "asserts that the uniformity of nature is hostile to any freedom of personality," that "there is no evidence of purpose in the universe," and that "there is no form of existence other than those 'envisaged by physics and chemistry'." Further, "the materialist minimizes the human activities we call personal and reduces all knowledge to sensation. This is not science; it is philosophy or ... metaphysics" (Mathews 1924: 5). Thus Mathews ignored or was not aware of the "emergentist naturalists" such as Samuel Alexander and Jan Christian Smuts who sought to find a place for thought and personality with a naturalistic framework, even though Mathews's colleague at the University of Chicago's Divinity School, G. B. Smith, was touting their significance.

An astute criticism of my version of religious naturalism is Langdon Gilkey's essay "Response to Stone's 'The Viability of Religious Naturalism'" (Gilkey 1993: 42–48). This was a reply to my "The Viability of Religious Naturalism" (Stone 1993: 35–42). Gilkey had been my dissertation adviser at the Divinity School of the University of Chicago when I wrote on secular experiences of transcendence as seen by Bernard Meland, H. Richard Niebuhr, and Paul Tillich.

Gilkey's main point is that I, in representing religious naturalism, (a) have too much confidence in the presence of healing forces in life, (b) ignore the question of the relationship of these forces to ideals, and (c) ignore the question of the source of the ideal as a permanent, universal character to existence (Gilkey 1993: 48). According to him, I am too serene and too confident that these healing forces will always be there. At its center, according to Gilkey, religion is more about the question of whether these resources are actually there than about the confidence that they are there (Gilkey 1993: 43, 46).

In addition to these three points Gilkey affirms that naturalists such as myself and John Dewey are wrong to affirm that language about transcendence is dangerous, while the language about naturalism is benign. He claims that I have swallowed too much of Dewey here, that naturalists can be quite as dogmatic and dangerous as theists.

In "Concluding Reply by Stone" (Stone 1993: 49–50), my main point is that Gilkey's criticisms miss their mark. He finds naturalism incapable of seeing ambiguity. However, he does not take Wieman or Loomer seriously on this score. Probably he is so critical of the pretensions of modernity that he fails to discern the critical and prophetic power of religious naturalism. He finds only two options open to modern humans: overconfidence or despair. He may find me inconsistent because of the limits of his dichotomous categories.

Gilkey and I agree that transcendent resources and norms sometimes appear as overwhelming and seem "disclosed" rather than humanly created. However, I do not see that the overwhelming character of such experiences—which we both recognize—evinces a non-natural Ultimate. Also, Gilkey challenged my optimism that ideals and healing are always present. But by "continually" challenging ideals, I do not mean "constant" but "never-ending." Similarly I assert that healing resources are occasional and often sporadic. The healing herb, for example, is not always available. Hence the need for renunciation and grief work. I would say that religion concerns the absence as well as the presence of transcendence, but this absence is known only in contrast to its sporadic presence. As for the issue of the relationship between the healing resources and challenging norms, these experiences are often intertwined. But they are also often separate, and we should not speak too quickly of their unity. I agree with Gilkey when I say, "Modernism can be dogmatic. That is why I attack its closure and lack of a sense of ambiguity. On the other hand, Gilkey does not address my arguments for ontological restraint and avoids my challenge to his arguments" (Stone 1993: 50).

John Cobb spends little time in polemics with process naturalists. He prefers collaborative efforts on what both theistic and naturalistic process thinkers can contribute to public issues such as economic and environmental policy. However, he directly discusses the differences between theistic and naturalistic process thinking (Cobb 1993: 134–147). Although his tone is remarkably tentative and modest in this essay, it amounts to a sympathetic critique of religious naturalism.

Cobb starts by noting that conscious beliefs make some, albeit small, difference. For example, those who deny human freedom are less likely to accept responsibility for their own decisions as time goes on. Similarly, when conscious belief in God fades, there is a tendency to believe that life has no meaning or importance. Cobb's personal judgement is that process naturalism does not provide for a ground of meaning. God, when properly conceived, provides such a ground and also provides a sense of forgiveness and companionship.

An example of the difference that process theism makes to public policy concerns the issue of species extinction. There is a deep, spontaneous revulsion to this catastrophe by many people. In nontheistic circles this revulsion is expressed and justified by practical considerations concerning the medicinal or ecological value of these species. But many lost species, Cobb argues, have no medicinal value and the damage to the biosphere from the loss of most of these species is trivial. This revulsion, Cobb claims, comes from a sense that whether or not humans can appreciate the difference these species make, it makes a difference to reality as a whole. But that means "that reality as a whole is the sort of thing that can be impoverished—that is, that it has subjective qualities. Panentheism grounds and explains this judgment. For God, the variety of creatures provides the contrasts that enrich the divine experience" (Cobb 1993: 143). Cobb finally asserts that naturalism cannot provide a comparable vision.

A second implication of process theism for public policy, according to Cobb, concerns the role of the Pareto optimality principle when economists offer guidance on public policy. Employing this principle, economists offer no sense of minimal sufficiency. That is, economics as usually understood offers no guidance in the distribution of wealth. On the other hand, process theism

> affirms that there is the One who includes, and can, therefore, compare, the feelings of loss or gain of diverse peoples. God experiences the benefits to the poor as greater than the loss to the rich, and we, believing in this God, can propose appropriate policies.
> (Cobb 1993: 144)

A different type of criticism is to be found in Wentzel Van Huyssteen's sympathetic and careful treatment of this writer's *The Minimalist Vision of Transcendence* (Van Huyssteen 1999: 202–213; Stone 1992). Van Huyssteen accurately sketches the main positions in the book, including the phenomenology of concrete religious experiences, the transactional realism, and the fact that the outlook is presented as "a metaphysical position, that needs whatever justification a metaphysical position can get" (Stone 1992: 7). He then makes what, on my reading, are three major and three lesser criticisms. The first is that some of my "epistemic decisions and value judgments, because they are essentially shaped by [my] prior choice for naturalism, are made in advance" (Van Huyssteen 1999: 205). This criticism puzzles me, because he had already acknowledged that my naturalism is a philosophic system. I may not have made a case for it, but it seems hardly to the point to say that it is a prior philosophical commitment. His second point is that my minimal model of transcendence results in a "fairly innocuous, inoffensive, generic notion of the divine," naturalistically conceived of course (Van Huyssteen 1999: 206). This certainly may appear to be so, but the continuing challenges of the ideal aspect of the naturalistically conceived transcendent have revolutionary potential in the right context. I do not consider Chapter 3 of *The Minimalist Vision*, "The Ethics of Openness" to be innocuous but quite demanding. His third point is that the minimalist vision is not rooted in a living tradition, that it lacks historical and social context, and that it is remote and empty, "intellectually esoteric" and for "a selected intellectual few" (Van Huyssteen 1999: 206–207). When I first read this critique I felt that it had some bite. However, since then I have found a spiritual home in a religious community with a living historical context and where many people of seemingly wide divergence of intellectual ability find my ideas helpful and even inspiring on occasion. Thus I no longer take this third point so seriously.

Van Huyssteen has three other points he makes in passing. First, he questions whether my minimalist model of transcendence can be distinguished from psychological self-actualization (Van Huyssteen 1999: 206). I suspect that the boundary between my minimalist vision and self-actualization may be vague, but this is because there is a type of naturalistic (hence minimal) transcendence in self-actualization. Second, he suggests that belief in "the inadequacy of natural explanations to account for all our experiences may even be more invariant across cultures than the belief in any specific God" (Van Huyssteen 1999: 206). This hardly seems to be a significant criticism, since many beliefs once invariant across cultures, such as a belief in a flat earth, are left behind when there are reasons to challenge them. Finally he suggests that I do "not really show why maximalist theistic views fail, but only why highly restricted—and already problematical—arguments for maximalist positions of theism fail" (Van Huyssteen 1999: 207). I grant that my criticisms of the theistic arguments (see my earlier comments on the ontological and cosmological arguments) may seem irrelevant or unconvincing to many believers. However, I have found it necessary to develop these counter-arguments because a number of my critics have felt that they had triumphed because I had no adequate response to their theistic arguments!

Another prominent critic of naturalism is John Haught in his book *Is Nature Enough?* (Haught 2006). Haught argues that naturalism lacks both the intellectual and the spiritual resources to satisfy human longings. He has been answered extensively by Loyal Rue (Rue 2011: 116–122). Haught's argument for the intellectual deficiency of naturalism is based on his assertion that naturalists cannot give an adequate explanation for: (a) the existence of the natural order or for (b) the teleological or "anticipatory aspect" ("openness to new possibilities") of the world manifest in the goal-directed behavior of all living organisms, most clearly evident in the intentional quality of human experience and finally in the universe as a whole (Haught 2006: 179).

Rue's response is that many naturalists would challenge this last assertion that the universe itself is purposeful. As for the more localized teleological phenomena, the emergentist view is that "radically new systematic properties of nature can arise spontaneously from the modifications and amplifications of component relationships" (Rue 2011: 119). In other words, the emergentist view is that teleological or anticipatory phenomena within the cosmos (rather than the cosmos itself) can be explained naturalistically. Put simply "purposeful behavior can emerge spontaneously within a purposeless universe" (Rue 2016: 119). Rue asserts that Haught has no answer for the emergentist view. To put it in his colorful language, Haught is stuck with a "grunge" view of matter.

Haught also charges naturalism with conceptual incoherence. His argument starts with the notion that humans have a high regard for truth and their ability to pursue it (Haught 2006: 35). This is true certainly for naturalists, he goes on, for regard for the truth is the foundation of scientific inquiry. However, naturalism cannot account for this inherent regard for truth. Human abilities are only evolved to achieve *adaptive* beliefs, not *true* ones. In addition, the naturalistic account of human knowledge stresses the vulnerability of our cognitive systems to error and deception. Thus naturalism disposes us to doubt the imperatives of our mind.

> So if adaptive evolution, or accidents of nature, or social conditioning ... constitute the *ultimate* explanation of your own mental functioning, then why are you not suspicious right now that you may be deceiving me and yourself by claiming that naturalism is true?
>
> (Haught 2006: 115)

Rue's reply is that normally we trust the results of our cognitive endeavors, but that "we are well served by periodic moments of self-doubt. Belief and doubt are equally adaptive" (Rue 2011: 121).

Haught also argues that naturalism is spiritually deficient. Religion "is a *conscious appreciation of and response to the mystery that grounds, embraces and transcends both nature and ourselves*" (Haught 2006: 22). Thus, Haught can say that religious naturalism is a logical contradiction. Rue's response is that Haught's definition of religion amounts to condemning it to be a delusion, since we cannot get beyond nature.

Haught refers to religious naturalists as "sunny naturalists" who hold that there is enough in nature for spiritual contentment—enough beauty, exhilaration, and human love. He contrasts these sunny naturalists with the "sober naturalists" who affirm that nature is devoid of meaning and cannot satisfy our deepest longings. Sober naturalists are hopelessly tragic, but at least they are honest about the implications of naturalism, whereas the groundless optimism of religious naturalists is cowardly and deluded (Haught 2006: 194). Rue comments that we seem to be at an impasse, where both sides "hurl accusations of spiritual delusion" (Rue 2011: 122). My final comment goes back to my earlier point that the case for religious naturalism will be an inconclusive yet hopefully persuasive argument for it as a scientifically informed surmise.

References

Alexander, S. (1920, 1966) *Space, Time and Deity: The Gifford Lectures at Glasgow, 1916–1918*, Two volumes. London: Macmillan; reprinted New York, NY: Dover Publications.

Ames, E. (1929) *Religion*. New York: Henry Holt and Company.

Cobb, J. (1993) "Process Theism," in C. R. Mesle (ed.) *Process Theology: A Basic Introduction*. St. Louis, MO: Chalice Press.

Danto, A. (1967) "Naturalism," in *The Encyclopedia of Philosophy*, Paul Edwards, Editor-in-Chief. New York: The Macmillan Company.

De Caro, M. and D. Macarthur. (2004) *Naturalism in Question*. Cambridge, MA: Harvard University Press.

——. (2010) *Naturalism and Normativity*. New York: Columbia University Press.

Flanagan, O. (2006) "Varieties of Naturalism," in Philip Clayton and Zachary Simpson (eds.) *The Oxford Handbook of Religion and Science*. New York, NY: Oxford University Press.

Foster, G. (1909) *The Function of Religion in Man's Struggle for Existence*. Chicago, IL: University of Chicago Press.

Gilkey, L. (1993) "Response to Stone's 'The Viability of Religious Naturalism'," *American Journal of Theology and Philosophy* 14: 43–48.

Griffin, D. (2000) *Religion and Scientific Naturalism: Overcoming the Conflicts*. Albany, NY: State University of New York Press.

Hardwick, C. (1996) *Events of Grace: Naturalism, Existentialism and Theology*. Cambridge: Cambridge University Press.

Harrison, P. (1999) *The Elements of Pantheism: Understanding the Divinity in Nature and the Universe*. Shaftesbury, UK: Element Books Limited.

Haught, J. (2006) *Is Nature Enough?* Cambridge: Cambridge University Press.

Kaufman, G. (1993) *In Face of Mystery: A Constructive Theology*. Cambridge, MA: Harvard University Press.

——. (2004) *In the Beginning … Creativity*. Minneapolis, MN: Fortress Press.

Lacey, A. (1995) "Naturalism," in Ted Honderich (ed.) *The Oxford Companion to Philosophy*. New York, NY: Oxford University Press.

Loomer, B. (1987) *The Size of God: The Theology of Bernard Loomer in Context*, William Dean and Larry Axel (eds.) Macon, GA: Mercer University Press; also available in *American Journal of Theology and Philosophy* 8.

Mathews, S. (1924) *Contributions of Science to Religion*. D. Appleton and Company; reprinted 1970, Freeport, NY: Books for Libraries Press.

Mesle, R. (1993) *Process Theology: A Basic Introduction*. St. Louis, MO: Chalice Press.

Milligan, C. (1987) "The Pantheistic Motif in American Religious Thought," in Peter Freese (ed.) *Religion and Philosophy in the United States of America*, Volume 2. Essen, Germany: Die Blaue Eule.

Murry, W. (2006) *Reason and Reverence: Religious Humanism for the 21st Century*. Boston, MA: Skinner House Books.

Otto, R. (1907) *Naturalism and Religion*. J. Arthur Thomson and Margaret R. Thomson (trans.), W. D. Morrison (ed.) New York: G. P. Putnam's Sons.

Oxford English Dictionary, The (1989) 2nd edition. Oxford: Clarendon Press.

Peden, C. and J. Stone (1996) *The Chicago School of Theology—Pioneers in Religious Inquiry*, Volumes I and II. Lewiston, NY: The Edwin Mellen Press.

Peters, K. (2002) *Dancing with the Sacred: Evolution, Ecology, and God*. Harrisburg, PA: Trinity Press International.

Rue, L. (2011) *Nature Is Enough: Religious Naturalism and the Meaning of Life*. Albany, NY: State University of New York Press.

Smuts, J. (1926, 1961) *Holism and Evolution*. New York, NY: Macmillan; reprinted 1961 by New York: Viking Press.

Stone, J. (1992) *The Minimalist Vision of Transcendence: A Naturalist Philosophy of Religion*. Albany, NY: State University of New York Press.

——. (1993) "The Viability of Religious Naturalism," and "Concluding Reply by Stone," *American Journal of Theology and Philosophy*, 1993: 35–42, 49–50.

——. (2008) *Religious Naturalism Today: The Rebirth of a Forgotten Alternative*. Albany, NY: State University of New York Press.

——. (2017) *Sacred Nature: The Environmental Potential of Religious Naturalism*. London: Routledge.

Taylor, B. (2010). *Dark Green Religion: Nature Spirituality and the Planetary Future*. Berkeley, CA: University of California Press.

Van Huyssteen, W. (1999) *The Shaping of Rationality: Towards Interdisciplinarity in Theology and Science*. Grand Rapids, MI: William B. Eerdmans.

Wieman, H. (1946) *The Source of Human Good*. Carbondale, IL: Southern Illinois University Press.

——. (1958) "Naturalism," in M. Halverson and A. Cohen (eds.) *A Handbook of Christian Theology*. New York: Meridian Books.

Wildman, W. (2006) "Ground-of-Being Theologies," in P. Clayton (ed.) *The Oxford Handbook of Religion and Science*. New York: Oxford University Press.

Further reading

Stone, J. (2008) *Religious Naturalism Today: The Rebirth of a Forgotten Alternative.* Albany, NY: State University of New York Press. (Analyzes several dozen major philosophical and theological writings of religious naturalists starting with Samuel Alexander, George Santayana, John Dewey, and some Chicago theologians.)

——. (2017) *Sacred Nature: The Environmental Potential of Religious Naturalism.* London: Routledge. (Treats philosophical and religious topics including perception, theistic naturalism, relations to American Indians, and public theology.)

Taylor, B. (2010) *Dark Green Religion: Nature Spirituality and the Planetary Future.* Berkeley, CA: University of California Press. (Covers several recent religious and quasi-religious movements, including non-naturalist ones.)

2

RELIGIOUS NATURALISM AND ITS NEAR NEIGHBORS

Some live options

Willem B. Drees

I have sympathy for religious naturalism, but prefer not to identify as a follower (Drees 2006: 121). Why this reservation? As I will discuss here, I think that there are credible philosophical alternatives to naturalism and interesting religious alternatives to religious naturalism. Before considering these, let me offer two minor considerations: the American character of religious naturalism as a movement, and a professional reservation about being labelled, and thereby constrained.

Geographically, articulations of "religious naturalism" are mostly American; see the list of highly respected colleagues contributing to this volume. In the Netherlands and elsewhere in Western Europe, there does not seem to be a demand for this label, nor for new organizational structures when one no longer identifies with a religious position or denomination. As Robert Putnam and David Campbell showed in *American Grace*, religious life in the USA is mostly congregational. Many of those who do not to join a church, still desire to have a congregational home or at least a recognizable label, while in Europe most "nones" do without a congregational home or recognizable label.

As the group of "religious nones" is increasing, Putnam and Campbell conclude "that there is a potential constituency for a new form of religion within the contemporary United States" (2010: 163). So far, we have had good books, many by contributors to this volume, and websites (see Goodenough in this volume). In 2014, a community of intellectuals advocating religious naturalism formed the Religious Naturalist Association. Personally—and this may reflect my European bias—I have no interest in joining such a community, even though I find the issues discussed in those books genuinely interesting.

My individualism aligns well with a professional consideration. I am a philosopher, who served previously in a department for the academic study of religions. Professionally, the primary stance is to study ideas and practices, whether one agrees with them or not. Ideas one has sympathy for, still need to be studied with critical academic distance. Identifying with a substantial label might be premature. Unlike membership in professional societies, I find membership in a society advocating a particular position in my field of study makes me uneasy, professionally, and so too for adopting a particular label.

Let us return to the main line of this chapter. I will take my point of departure in the sciences, especially the natural sciences, as our main source of knowledge about the world we live in. After presenting "science-inspired naturalism" in the first section, we will turn to "philosophical naturalism" and "religious naturalism" in the next two sections, and hence to philosophical and religious alternatives to religious naturalism.

Science-inspired naturalism

In our time, any form of naturalism worthy of serious consideration involves, at least, that one accepts the scientific approach to reality as informative about the way things are. Naturalism is, at least, *naturalism about the world.* There are no exceptions; "no supernatural or spiritual realm distinct from the natural world shows up *within* our natural world, not even in the mental life of humans," as I wrote in *Religion, Science and Naturalism* (Drees 1996: 12).

All entities in the world seem to be made up of the same constituents, and these are best described by physics—atoms, and these consist of quarks, electrons, and other ingredients of the standard models of weak and strong interactions, and beyond those the ingredients of a deeper underlying theory. If exceptions seem to arise (e.g., the current search for "dark matter" and "dark energy" in cosmology), we may need a further development of physics; the basic assumption that physics is about the fundamental elements remains the same.

When we come to "higher" phenomena, we may need concepts that do not belong to the vocabulary of fundamental physics (emergence), even though these phenomena are based in physics (reduction). For living beings, evolutionary biology offers the best, functional explanations for the emergence of various traits in particular contexts.

We, humans, are no exception to the natural world—a world that includes cities, and other products of culture as well. It is as with mountains, in an image taken from John Dewey (1934: 3): "Mountain peaks do not flow unsupported; they do not even rest upon the earth. They *are* the earth in one of its manifest operations." Thus, our mental and cultural life, made possible by our brains and social environments, reveals possibilities of the natural world. This insight does not downgrade us; it shows the great potential of material existence—matter, properly organized, can be Isaac Newton or Albert Einstein, Henrik Ibsen, Wolfgang Amadeus Mozart or Rembrandt Harmenszoon van Rijn, Siddhārtha Gautama, Jesus of Nazareth, or Muḥammad ibn ʿAbdullāh, or any other human who contributed, significantly or modestly, to our intellectual, cultural, and religious history. Epithets such as the Buddha, the Christ, or the Prophet are understood as honorific titles, given by humans, and not as terms that set them apart from the natural, human community.

Non-naturalists may reject the prominent role of physics and chemistry in understanding the material world, for instance by including additional forces, auras, angels, or spirits in the inventory of the world. Among the non-naturalists are also those who reject the evolutionary view of the biological world, whether straightforward (e.g., young earth creationism) or partially, by adding interventions to natural processes (e.g., "intelligent design" advocates, though the additional interventions they claim imply that the initial design was not that intelligent after all). And it includes those thinkers who add a non-physical "soul" as an additional pseudo-material ingredient to human nature or in some other way exempt humans from material, biological existence. (Of course, speaking of a "soul" to speak of important characteristics of a person, is not excluded at all.) Among the non-naturalists may be some who emphasize "alternative realities," a tradition going back to alchemy, alongside some religious believers who take it that their religious faith must be at odds with scientific knowledge, understood in this more encompassing naturalistic way.

Disagreements between science-inspired naturalists and non-naturalists may be about substantial issues— how we see the world—but they may also relate to sources of knowledge.

For a science-inspired naturalism, reliable knowledge arises via experimentation, modelling and the like, and is discussed and evaluated by scientific colleagues; "the scientific forum." Intuitions may be heuristically of interest, but are not by themselves authoritative. So too for claims about special divine revelation or other appeals to authority, bypassing public justification. "Revelation" may be an affirmative title of honor, but it is not a short cut to knowledge.

Does a science-inspired naturalism, as sketched briefly here, imply philosophical naturalism and religious naturalism? Those two questions will be central to the next two sections.

Philosophical naturalism

If one accepts the science-inspired naturalism outlined previously, there are further questions and concerns that go beyond science, such as questions about assumptions and criteria involved in such a naturalistic project. Whether one should be a naturalist on those issues or whether a different philosophical approach is preferable, is debated by philosophers. I will present here my own line of thought, rather than review the extensive philosophical literature. I see two major issues for a science-inspired naturalism: issues about interpretations in science and the development of science, and questions that seem to be outside the scope of science.

Scientific theories are open to multiple interpretations. Quantum physics is an extremely successful theory, with very precise mathematical formulations. Its predictions have been confirmed with great accuracy. Technology shows the adequacy of such physics. However, how to interpret this theory as an informative theory about reality remains disputed. Various interpretations challenge common sense understandings of causality and reality. Quantum physics has a bewildering plurality of interpretations. Hence, even if the theory is accepted as true, we would not understand what reality is like. The theory under-determines our worldview. Other fundamental theories are also open to multiple interpretations. Einstein's theories of relativity raise issues about the understanding of time; should we understand our reality as evolving in time or as a four dimensional block of space-time, existing all at once?

Not only do current theories allow for multiple interpretations, but future theories may offer us radically different ways of understanding of reality. For instance, while Newtonian physics treated gravity as a force, working at a distance, Einstein's General Relativity Theory treated the same phenomena as consequences of the curvature of space-time; the force at a distance has disappeared. At a practical level, Newton's formulas are still useful for systems with modest velocity and mass, but these can no longer be taken to describe reality. And Einstein's theory need not be the final word on reality either. Einstein's theory and quantum physics are both extremely successful, but they cannot be combined in a coherent understanding of reality without making some changes to at least one of these theories. Some physicists have developed theories that treat matter as vibrating strings in a space with additional dimensions; this has resulted in complex mathematics, but not yet in a successful theory. More recently, the Dutch physicist Erik Verlinde (2016) has proposed to work with the concept of information in a holographic universe as fundamental, thus moving beyond Einstein's theory, and, in passing, doing away with the need for "dark matter" to explain current observations.

Whether it will be superstrings, information, or something else, such proposals should make us aware that current science does not deliver the fundamental ontology of reality. It is very successful, but also open to further development. From the perspective of future theories our current theories will remain as good approximations at the appropriate level of description, but future theories will lead to different possible views of the nature of reality. With a metaphor from *The Logic of Scientific Discovery* of Karl Popper (1992: 94), the building of knowledge rests on pillars driven into a swamp:

> Science does not rest upon solid bedrock. The bold structure of its theories rises, as it were, above a swamp. It is like a building erected on piles. The piles are driven down from above into the swamp, but not down to any natural or 'given' base; and if we stop driving the piles deeper, it is not because we have reached firm ground. We simply stop when we are satisfied that the piles are firm enough to carry the structure, at least for the time being.

We can build, and we can drive the pillars deeper, but there is no absolute foundation. Nonetheless, as the city of Amsterdam shows, one can have a great city built on soft foundations. In fundamental physics, we are driving for deeper fundamental theories, while at the same time in other branches of physics and in the life sciences, colleagues are developing insights drawing on regular chemistry, as current theories are good enough for their work.

Given the multitude of interpretations and the possibility of future theories that are continuous with current knowledge in practice but not in ontology, a problem for a science-inspired naturalist might be that we do not really know what nature is like, deep down, even though we have very secure knowledge at more mundane levels of existence.

Not only does a science-inspired naturalism fail to deliver answers to questions about the ultimate structure of reality, but it also seems unable to address certain important facets of existence. I'll consider three that I find interesting and important: mathematics, values, and ultimate origins.

Mathematics is odd, if one comes at it with an empiricist mind set. Within a science-inspired naturalistic ontology, pure circles, triangles, cubes and the like do not exist, nor do imaginary numbers, Lie groups, or Bessel functions. Nonetheless, we can make claims about their properties. We can even make mathematical existence claims such as that there is (or that there is not) an even integer greater than 2 that is not the sum of two primes (Goldbach's conjecture). The fact that we currently do not know whether there is such a number does not undermine the conviction that either there might be a smallest even number that is not the sum of two primes or not one. The existence of this number does not depend upon our preferences.

One interpretation of this feature of mathematics has been "Platonism" (used here without regard for historical accuracy), the view that mathematical realities exist "out there" in an objective but immaterial world. If so, mathematical truth can be understood as correspondence between our propositions and mathematical reality. Mathematicians make discoveries. Roger Penrose (1989) is a modern advocate of such a view. As an ontology this "Platonic" reality is so distinct from material reality that it is hard to envisage where it might be. If one dismisses this as a non-problem, given the categorical difference between material reality and this Platonic reality, a second problem arises: how do we material beings have access to those non-material lands? "Mathematical intuition," the possibility to make "observations" in this Platonic realm, would be a remarkable addition to the experiential, causally mediated repertoire we are supposed to have. Such an ontology of mathematics seems too remote to fit the epistemic challenge of how mathematical knowledge is acquired and developed (e.g., Kitcher 1984: 102). A different but somewhat related problem is how it might be possible that mathematics as an axiomatic-deductive system is useful for the physical world, if it deals with abstract entities rather than with material objects and natural processes.

A quite different view of the nature of mathematics is constructivist in kind. Mathematical objects are human creations, a conceptual world that is "up to us." But if it is just a construction, why do we not see much more variation? Why do mathematicians agree on mathematical insights, across cultural, linguistic, and ideological differences? One approach might take inspiration from Immanuel Kant, by treating mathematics not just as a construction, but as the construction that abstracts from human practices such as counting and measuring certain necessary, "transcendental" features of human reasoning.

The philosophy of mathematics is a scholarly profession in itself, which I will not delve into here. However, a Platonic view of mathematics is problematical for a naturalist, as it introduces a transcendent realm of its own. A constructivist view that involves transcendental arguments avoids such major ontological assumptions, but it moves beyond the repertoire of science-inspired naturalistic approaches to reality.

A similar argument might be made for *values* that have their origins in human social practices but at the same time are supposed to be valid for all. Upon a naturalistic perspective, pro-social behavior has its roots in enlightened self-interest, whether of the individual or of the "selfish" genes. Biologists and others have come up with evolutionary explanations for support given to children, nieces, nephews, and other kin (same genes); support given to one's partners (a shared project, though if one can mobilize the other and nonetheless do less, even better—cheating co-evolves with sociality); support given to neighbors (direct reciprocity); and to the larger community (indirect reciprocity, group selection). It all seems ultimately to come down to enlightened self-interest, unconsciously pursued. What about values in such a world of self-interest? Is there any place for reason in such a world where proximate causes (mechanisms that guide behavior) are selected according to "ultimate" evolutionary success? A naturalist who is a relativist about values might find this view enough. Morality evolved, and that seems intelligible; there is no further issue of validity or justification. However, if one takes values more seriously, perhaps one might argue as follows.

Evolution has delivered more than was ordered. Fingers did not evolve to play a piano, but they can be used to play the piano. Intelligence and communication, brains and language may have evolved because they were useful for the four Fs that are essential for survival as individuals and as a species: feeding, fighting, fleeing, and reproducing. Once intelligence and language evolved, our communicative and intellectual abilities may have been used in other tasks as well. This "more" that was delivered allows morality to be genuinely moral, for our intelligence allowed us to reconsider our own behavior. We may discover that we are "naturally" inclined to treat men and women differently. However, by becoming aware of this we can also act against that which comes "naturally." Communication may also contribute. Imagine that once an offended hominid asked a fellow hominid: "Why do you behave thus?" In the presence of others he or she was challenged to justify the behavior in question with arguments that would be recognizable and acceptable to the others, and thus, to formulate general principles justifying his or her behavior. In many incidents of this kind, natural behavior guided by enlightened self-interest may have become reconsidered, intentional behavior. The social context of our lives may have pushed towards universality and accountability, hallmarks of morality, and towards law rather than individual preference.

We are occasionally open to reasons, to argument. Since ideas spread faster than genes, culture may develop enormously. There is no reason to assume that the biological basis would always overrule the effects of culture. Thanks to the emergence of culture as a second type of heritage, alongside the genetic one, and thanks to the capacity for reflection and to the impulse to public justification, we are not victims of our evolutionary heritage. We are biological beings, but as these particular biological beings we have a moderate amount of freedom with respect to our genetic drives. We therefore also have responsibility.

We are practical beings, who measure and count, evaluate and judge. In doing so, we reach beyond the practical, and discover or create values and ideas that seem to be universal and timeless, transcendent relative to the world of facts. A strict science-inspired naturalist might have a problem here. Within such a naturalism, one can understand the emergence of a human *practice* of making normative evaluations, judging others to behave well or not. The emergence of moral discourse can be understood as a feature of our history, as presented by Philip Kitcher in the work *The Ethical Project*, and many others. To understand such moral discourse as having a higher standing, as appealing to universally valid norms, as justified, may be problematical for a science-inspired naturalist. Of course, one might opt for relativism; the norms are socially created, but have no further standing; they have no place in the fundamental inventory of reality. An alternative option is a transcendental constructivism, also with respect to morality. A recent defense by Sem de Maagt (2017) used a transcendental argument that roots the justification of values in conditions for agency, engaging the work of Alan Gewirth and Christine Korsgaard

When one considers the interpretation of science and non-things such as mathematical objects and moral values, one cannot just turn to science. More is needed—a question is whether that "more" still falls within the category of "naturalism."

One more issue seems to push us beyond a science-inspired naturalism: questions about *ultimate origins*. Why is there something rather than nothing? Why is the world the way it is? Any scientific explanation involves assumptions about that which exists and about the relevant laws. To predict the weather for tomorrow, we need to know the weather today as well as atmospheric physics. The Big Bang theory is a very successful theory about the evolution of the universe after an initial hot dense state, but it does not explain that state nor the laws of physics. A further theory is needed, a "quantum cosmology." However, even if a quantum cosmology would be successful by embedding Big Bang cosmology in a larger framework that integrates all known physics in a coherent and consistent whole, there would still be the contingency of existence, and hence the cosmological question: "Why is there something rather than nothing?"

Even if our current questions are answered by future theories, new questions will emerge in the context of those future theories. For instance, the inflationary model in cosmology solved many questions regarding the universe as observed, but it cannot explain why the universe is such that inflation happens; some assumptions are always made. The reach of explanation is impressive, but again and again, questions emerge at the limits of scientific understanding. They may be resolved, but the deeper understanding will trigger further questions, on elements assumed rather than explained.

Questions will remain even if one day physics and cosmology were to agree on a theory explaining all known phenomena in a unified, coherent way. Imagine, a single article, a single formula answering all our questions. But the article is on a piece of paper; the formula consists of symbols. Thus, there is no answer to the question: Why does reality behave as described here? It is as with a drawing of the Belgian artist René Magritte, a careful drawing of a pipe, as used for smoking tobacco. Underneath it, he has written "Ceçi n'est pas une pipe"—"This is not a pipe." And he is right. One cannot fill the image with tobacco. There is a difference between an image, how accurate it may be, and reality. This is also the case for a good scientific theory. However accurate the theory, the question remains why reality behaves as described in the theory. Even if the question is not about a first cause in a temporal sequence, there is a genuine philosophical question, namely why the reality described by the theories is actual. A theistic answer could be envisaged—but that too isn't much of an answer, as the same question might be posed with respect to God. The physicist

Charles Misner (1977: 97) expressed this well: "Saying that God created the universe does not explain either God or the Universe, but it keeps our consciousness alive to mysteries of awesome majesty that we might otherwise ignore, and that deserve our respect." Misner, a Roman Catholic, uses theistic language, but similar issues arise for a naturalist. The naturalist too has no answer to questions such as: Why is there a universe? Why is the universe as it is? We cannot escape the limitations of our concepts and ideas, except by relying on other concepts and ideas, and the limitations of our existence. We never see the universe "from outside," from the perspective of eternity, but always from within. This is the case for a theist, who speaks of something more, from within the universe. It is also a problem for a naturalist: we operate within natural reality, without resources to explain that reality itself.

The more we know, the more we may become aware of the limitations of our knowledge. *De docta ignorantia* (About learned ignorance) was the title of a book of Nicolas of Cusa, a cardinal in Europe in the fifteenth century. The scientific road to knowledge has shown itself to be very successful; we have learned more than Nicolas of Cusa and his contemporaries might ever have expected. But we are also confronted with fundamental questions concerning the nature and ground of our reality. Thunder is no longer a voice of the gods, nor a mystery. But that does not exclude wonder regarding the reality of which both we and the thunderstorms are part. In the end existence remains a mystery. This can inspire a naturalist, with awe and humility. But it also shows that as an intellectual project, a science-inspired naturalism is incomplete. If one accepts the science-inspired naturalism of the previous section, one need not be a philosophical naturalist in a more encompassing sense. Kantian constructivism and other philosophical approaches might be serious options.

We now turn towards "religious naturalism." If one accepts a science-inspired naturalism as the basic orientation with respect to science and knowledge, is "*religious* naturalism" the only "live option?" What is it that makes it "religious?"

Religious naturalism

Is naturalism not, by definition, excluding what is typical of a religious view? The answer depends not only on the understanding of naturalism, but also on one's concept of religion and the particular variety of religious or non-religious naturalism one espouses (see, for instance, Gregersen and Stenmark 2016; Stenmark 2016). If one accepts naturalism, under what conditions could one qualify it as religious? How does religious naturalism stand compared with naturalistic varieties of theism? And what about atheism? We'll begin at the theistic end of the spectrum and gradually come to more purely naturalistic positions. Before considering various positions, let me briefly consider what might make a philosophical or metaphysical position religious.

To approach the concept of religion, let us consider two dimensions of theism (Drees 2016: 196). First, in theism God is understood to be the creator of all that exists, a necessary being who is not dependent upon anything else for existence. To count as religious, we might ask of any alternative view whether it offers a frame to speak of the ultimate metaphysical question regarding *existence*.

Second, in theistic discourse God is the ultimate judge. From this image, I take on our exploration the ultimate "moral" question about the possibility of evaluating our behavior from an impartial perspective that *transcends* all human interests and biases. A religious view seems to offer language to speak of the axiological dimension of existence.

Last but not least, a key issue is whether these two dimensions—the cosmological and the axiological one—can be combined. I consider this combination of the two dimensions crucial for a vision to be worthy of being considered religious (rather than just metaphysical or meta-ethical). This is in line with Clifford Geertz's anthropological definition of religions as systems of symbols that shape moods and motivations (the axiological side of things) by presenting us with an understanding of reality (the cosmological dimension) that is taken as true, thus supporting those moods and motivations (Geertz 1966: 3; Drees 2010: 68, 76–82, 136–139; Drees 2016: 196)—thus, religions intertwine models *of* the world and models *for* the world. Let me give a theistic example: speaking of God as creator might integrate an idea about the source of existence with an attitude of gratitude, humility, and respect for fellow creatures.

We will now consider religious naturalism and some alternatives, beginning with *naturalistic theism* (Drees 2006: 116). There is a traditional way to articulate theism that avoids a confrontation with the natural sciences, and that is to emphasize the uniqueness of God's mode of being and activity (e.g., Stoeger 1995; Kaufman 1972; Wiles 1986). This is articulated in the notion of *creatio ex nihilo*, which does not apply to any natural causality. God creates and sustains all things as their primary cause; all natural causes are real, just as are all entities and events, but they are so because they have been created by God. Such real natural causes are "secondary causes." This distinction between primary and secondary causality was developed in the European Middle Ages, for instance by Thomas Aquinas, but its roots can be traced back at least to Augustine (Hebblethwaite and Henderson 1990; McMullin 1985, 1988; Burrell 1993). God creates everything—past, present, and future events—and God creates them not as an amorphous bag of events but with their temporal, spatial, and causal relations. The distinction between God and God's activity on the one hand, and creatures and creaturely activity on the other, often is articulated also as a difference with respect to time: creatures are temporal, whereas God, as conceived in this view, is not temporal. God's eternity is not everlastingness but timelessness. Accepting the whole natural world as the creation of a timeless transcendent God may be consistent with a naturalistic view of the world, since it accepts the world as understood by the natural sciences as God's creation. There is no need for particular gaps within the world.

A naturalistically-minded theist might argue that the sciences are explanatory within the world, but not explanatory of the world as such. This, in my opinion, is consistent with a science-inspired naturalism. Science offers explanations, but every explanation assumes an initial state and laws. Thus, science explains within a framework, but does not explain the framework as such; limit questions are persistent (Drees 1996: 266–269).

Responses to limit questions may be quite different for theists and naturalists. If naturalism is defined as including the assumption "that nature is necessary in the sense of requiring no sufficient reason beyond itself to account either for its origin or ontological ground" (Hardwick 1996: 5–6), such a naturalist by definition excludes a transcendent ground of reality. In my opinion, however, science-inspired naturalism need not imply the dismissal of a theistic response to limit questions regarding the existence, structure, and intelligibility of the world. I find *naturalistic theism* a genuine and attractive possible point of view.

Naturalistic theism has one major problem, as I see it. Once one accepts a naturalist understanding of created reality, "since there are no real 'gaps' to fill, we may be left without an argument for God's existence of the kind that would convince a science-minded generation" (McMullin 1988: 74).

Less dualistic is *theistic naturalism*, a label I use for the position of those who speak of God as Ground of Being (Wildman 2016; Drees 2006: 117). A major figure in the articulation of such a theological position has been Paul Tillich. This view has come to be formulated in pan-en-theistic terms—understanding the world to be in God, even though God surpasses the world (Clayton and Peacocke 2003). I found a most inspiring poetic expression among aphorisms in *The Aristos* of John Fowles (1980: 27): "The white paper that contains a drawing; the space that contains a building; the silence that contains a sonata; the passage of time that prevents a sensation or object continuing forever; all these are 'God.'"

Whatever the precise formulation, this position has more deeply ingrained naturalistic presuppositions by seeking to avoid the dualism of a transcendent God and the natural world, even though it maintains a concept of God as surpassing the world—and thus I still consider it a species of theism.

Most contributions in this handbook of religious naturalism develop or assume positions further removed from theism, as they do not understand God as an entity or as the all-encompassing reality, but at most use the term as a symbol to speak of our existence and the world we live in. Some such naturalists might prefer to speak of "the sacred" rather than of God. Are such positions still religious? Stone describes religious naturalism as

> a variety of naturalism whose beliefs and attitudes assume there are religious aspects of this world that can be appreciated within a naturalistic framework. Occasions within our experience elicit responses that are analogous enough to the paradigm cases of religion that they can appropriately be called religious.
>
> (Stone 2003: 89)

Speaking of *religious* naturalism may thus be justified if the attitudes and responses are sufficiently analogous. Stone's book offers various proposals. Let me briefly introduce a few varieties of such forms of religious naturalism, to illustrate some of the diversity within this sphere (Drees 2006: 117f.).

Gordon Kaufman interprets the Christian symbolism of God as a figure of speech to speak of an overwhelmingly significant characteristic of processes in the universe, namely their serendipitous creativity (1993, 2003). He connects such aspects with traditional understandings of God as creator and the moral call and vision as understood in the Christian tradition.

Charley Hardwick gives up on ontology, relating theological content to valuational aspects of our existence. "Although *God* does not refer (any more than rights, duties, values, or point masses need have ontological references), *God* or *God exists* can serve as a complex meta-expression for a form of life that is expressed as theistic seeing-as" (1996: 114). In his book *Events of Grace*, Hardwick reconstructs classical Christian conceptions, such as sin and grace. This is not accidental, but part of his understanding of the religious naturalistic agenda. He does not seek a religious Esperanto without roots in a particular tradition: "I am constantly reminded here of Santayana's dictum that 'the attempt to speak without speaking any particular language is not more hopeless than the attempt to have a religion that shall be no religion in particular'" (2003: 115).

Other religious naturalists are less engaged with a particular tradition. Perhaps "the evolutionary epic" might become *Everybody's Story* (Loyal Rue 1999). Such more universal inspirations also come from scientists with humanistic interests, such as Ursula Goodenough's *Sacred Depths of Nature* (1998).

Some religious naturalists dismiss all God-language. Donald Crosby speaks in this context of *naturism*:

> to distinguish it from conceptions of religious naturalism that make fundamental appeal to some idea of deity, deities, or the divine, however immanental, functional, nonontological, or purely valuational or existential such notions may be claimed to be. The focus of naturism is on nature itself as both metaphysically and religiously ultimate.
>
> (Crosby 2003: 117)

Each of these positions intertwines cosmological and axiological interests, or, to draw on Clifford Geertz again, models *of* the world and models *for* the world. Thus, they can all be considered religious, whether explicitly drawing on a tradition or not.

Such positions also share a fundamental philosophical problem with religions. When the key issue of religion is that facts and values, worldviews and ethos, are intertwined, one is at odds with the categorical distinction between values and facts, the is-ought distinction. This is not merely a philosophical issue; the "is-ought" distinction alerts us also against unjustified inferences that dismiss certain forms of behaviour as "unnatural" as if thereby they must be "immoral." The power *and* the danger of religious stories and visions, whether theistic or naturalistic, is in this combination of worldviews and values, of what is and what ought to be.

There are other alternatives to "religious naturalism," certainly if one includes approaches inspired by non-Western philosophical and religious traditions. However, let me consider one more within the Western context: *atheism*. Atheism seems to be the default alternative for theism. If one does not believe in God, one is an atheist. Of course, in a linguistic sense, this is trivial. However, the religious naturalist seeks to bring more nuance to the table—even if one does not refer to God as an entity "out there," there may be features of reality that deserve to be considered "sacred or holy." But how does atheism do on the key issues raised in this section: ultimate origins and the way values and worldview are intertwined?

In *The Oxford Handbook of Atheism*, Erik J. Wielenberg considers the perspectives for morality in the context of atheism. His conclusion is that "whether there are objective moral truths is independent of the existence or nonexistence of God" (2013: 89). He argues that theories that anchor morality in theism fail, basically because of a dilemma that goes back to Plato's dialogue *Euthyphro*: either God is beyond morality, and what is moral is decreed by God—a theological voluntarism that undermines morality—or God follows morality, but that would undermine God's primacy. Anyhow, without God, one has the discussion touched upon in the previous section, on philosophical strategies to envisage the justification of values.

Another question might be whether atheism has a motivating narrative, linking the worldview and the values. That might be a challenge, though a particular atheist might have a historical narrative that argues for modern values, rooted in the Enlightenment. Science might be valued, for instance as a resource to counter superstition, as Carl Sagan showed with his book *The Demon-Haunted World: Science as a Candle in the Dark*.

Atheism, as a position that accepts science as the main source of knowledge, could thus formally allow for a "religious" variant, in the terminology used here, by incorporating a value dimension (e.g., "scientism") and motivating historical Enlightenment narrative, but labelling "atheism" as religious seems to push the Geertzian use of this terminology too far. It seems to me more appropriate to consider the variety of *religious naturalisms* the "religious" position at this end of the spectrum, and understand atheism as a negative position, denying both a religious ontology and any religious narrative intertwining facts and values.

Religious naturalism and its near neighbors: does one need to choose?

There is more between heaven and earth than theism and atheism. Among those seem to be variants that are consistent with a science-inspired naturalism about the world. This seems to be a minimal requirement for someone who takes the sciences seriously, which I consider the appropriate stance in our time. This does not imply that science is the only source of knowledge. There is common sense knowledge—I know the names of my children—but I would also speak of knowledge in the context of mathematics, but that would not be scientific knowledge about the world. The section on philosophical naturalism considered some of the issues that go beyond science—mathematics, values, limit questions—that are genuine issues even though they are beyond the scope of science. Both with respect to scientific insights, which are open to multiple interpretations, and with respect to more encompassing philosophical issues, there is a multitude of worldviews possible. I appreciate positively this multitude of possible positions; reality and the conceptual universe are both so rich that none of our models is exclusive and final.

With respect to religious naturalisms, my main concern is that these tend to root values in "nature," the "nature" that is studied by science or experienced personally. Of course, functional human values are rooted in our biological and cultural existence, and thus in nature, but whether those values should be valued, seems to me to be a question that highlights the categorically distinct character of values—their actuality in our natural and cultural reality is not by itself enough evidence. Too often, humans have criticized others with an appeal to what is natural (white and male dominance), or what is counter-natural (same sex relationships), while later generations challenged those value judgements. In the end, I consider us to be as agnostic on ultimate values as we are on the ground of existence, though perhaps the other contributions in this handbook show me mistaken. As I see it, for all practical purposes one might take a science-inspired naturalistic stance in daily life (e.g., when needing medical assistance), consider Kantian constructivism an attractive strategy when it comes to philosophical justification of values, appreciate the motivating power of narratives that integrate ethos and worldview, while considering oneself an agnostic on matters of ultimate explanations and values.

References

Burrell, D.B. (1993) *Freedom and Creation in Three Traditions.* Notre Dame, IN: University of Notre Dame Press.

Clayton, P.H. and Peacocke, A. (eds.) (2003) *In Whom We Live and Move and Have Our Being: Panentheistic Reflections on God's Presence in a Scientific World.* Grand Rapids: Eerdmans.

Crosby, D. (2003) "Naturism as a Form of Religious Naturalism," *Zygon: Journal of Religion and Science* 38: 117–120.

De Maagt, S. (2017) *Constructing Morality: Transcendental Arguments in Ethics*, Ph.D. dissertation defended at Utrecht University, 27 January 2017. Questiones Infinitae, Volume 99, Utrecht: Utrecht University.

Dewey, J. (1934) *Art as Experience.* London: Allen and Unwin.

Drees, W.B. (1996) *Religion, Science and Naturalism.* Cambridge: Cambridge University Press.

—— (2006) "Religious Naturalism and Science," in P.H. Clayton and Z. Simpson (eds.) *The Oxford Handbook of Religion and Science*, Oxford: Oxford University Press, 108–123.

—— (2010) *Religion and Science in Context: A Guide to the Debates.* London: Routledge.

—— (2016) "The Divine as Ground of Existence and of Transcendental Values: An Exploration," in A.A. Buckareff and Y. Nagasawa (eds.) *Alternative Concepts of God: Essays on the Metaphysics of the Divine*, Oxford: Oxford University Press, 195–212.

Fowles, J. (1980) *The Aristos* (rev.ed.) Falmouth, Cornwall: Triad/Granada.

Geertz, C. (1966) "Religion as a Cultural System," in M. Banton (ed.) *Anthropological Approaches to the Study of Religion*, London: Tavistock, 1–46.

Goodenough, Ursula. (1998) *The Sacred Depths of Nature.* New York: Oxford University Press.

Gregersen, N.H., and Stenmark, M. (2016) "Naturalism and Beyond: Religion and the Varieties of Naturalism," in N.H. Gregersen and M. Stenmark (eds.) *Naturalism and Beyond: Religious Naturalism and Its Alternatives*, Studies in Philosophical Theology, Volume 59, Leuven: Peeters, 3–14.

Hardwick, C.D. (1996) *Events of Grace: Naturalism, Existentialism, and Theology.* Cambridge: Cambridge University Press.

—— (2003) "Religious Naturalism Today," *Zygon: Journal of Religion and Science* 38: 111–116.

Hebblethwaite, B. and Henderson, E. (eds.) (1990) *Divine Action: Studies Inspired by the Philosophical Theology of Austin Farrer.* Edinburgh: T&T Clark.

Kaufman, G.D. (1972) *God the Problem.* Cambridge, MA: Harvard University Press.

—— (1993) *In Face of Mystery. A Constructive Theology.* Cambridge, MA: Harvard University Press.

—— (2003) "Biohistorical Naturalism and the Symbol 'God'," *Zygon: Journal of Religion and Science* 38: 95–100.

Kitcher, P. (1984) *The Nature of Mathematical Knowledge.* Oxford: Oxford University Press.

—— (2011) *The Ethical Project*, Cambridge, MA: Harvard University Press.
McMullin, E. (1985) "Introduction: Evolution and Creation," in E. McMullin (ed.) *Evolution and Creation*. Notre Dame, IN: University of Notre Dame Press.
—— (1988) "Natural Science and Belief in a Creator: Historical Notes," in R.J. Russell, W.R. Stoeger, and G.V. Coyne (eds.) *Physics, Philosophy, and Theology: A Common Quest for Understanding*. Vatican City State: Vatican Observatory, 49–79.
Misner, C.W. (1977) "Cosmology and Theology," in W. Yourgrau and A.D. Breck (eds.) *Cosmology, History, and Theology*, New York: Plenum Press.
Penrose, R. (1989) *The Emperor's New Mind: Concerning Computers, Minds, and The Laws of Physics*. Oxford: Oxford University Press.
Popper, K. (1992) *The Logic of Scientific Discovery*. London: Routledge.
Putnam, R.D. and Campbell, D.E. (2010) *American Grace: How Religion Divides and Unites Us*. New York: Simon & Schuster.
Rue, L.R. (1999) *Everybody's Story: Wising Up to the Epic of Evolution*. Albany, NY: State University of New York Press.
Sagan, C. (1995) *The Demon-Haunted World: Science as a Candle in the Dark*. New York: Random House.
Stenmark, M. (2016) "Religious Naturalism and Its Rivals," in N.H. Gregersen and M. Stenmark (eds.) *Naturalism and Beyond: Religious Naturalism and Its Alternatives*, Studies in Philosophical Theology, Volume 59. Leuven: Peeters, 59–84.
Stoeger, W.R. (1995) "Describing God's Action in the World in Light of Scientific Knowledge of Reality," in Robert J. Russell, *et al.*, *Chaos and Complexity: Scientific Perspectives on Divine Action*. Vatican City State: Vatican Observatory and Berkeley: Center for Theology and the Natural Sciences, 239–261.
Stone, J.A. (2003) "Varieties of Religious Naturalism," *Zygon: Journal of Religion and Science* 38: 89–93.
Verlinde, E. (2016) "Emergent Gravity and the Dark Universe," preprint https://arxiv.org/pdf/1611.02269.pdf.
Wielenberg, E.J. (2013) "Atheism and Morality," in S. Bullivant and M. Ruse (eds.) *The Oxford Handbook of Atheism*. Oxford: Oxford University Press, 89–103.
Wildman, W.J. (2016) "Reframing Transcendence: Ground-of-Being Theism and Religious Naturalism," in N.H. Gregersen and M. Stenmark (eds.) *Naturalism and Beyond: Religious Naturalism and Its Alternatives*, Studies in Philosophical Theology, Volume 59. Leuven: Peeters, 123–150.
Wiles, M. (1986) *God's Action in the World*. London: SCM Press.

PART II

Some earlier religious naturalists

3

ERNST HAECKEL'S CREATION

Developing a non-reductive religious naturalism

Whitney Bauman

Triune monism: the problems with dualism, idealism, and materialism

> The whole marvelous panorama of life that spreads over the surface of our globe is, in the last analysis, transformed sunlight.
>
> (Haeckel 1900: 139)

I begin my reflections with this widely cited quote from Ernst Haeckel to help ward off accusations by many that Ernst Haeckel was merely a reductive materialist. Quite often philosophical monism gets lumped in with the reductive variety of monism, when in actuality there were many different types of monism ranging from materialistic to idealistic to pluralistic (Holt 1967). I would argue that Haeckel's monism is of the pluralistic variety in part because he outlines a triune structure of substance that includes material, energy, and experience "all the way down" (Haeckel 2008: 69). One of his final books was, after all, *Krystal Souls*, in which he argues that even the forms of crystals display some sort of continuity with biotic life (both plant and animal) (Haeckel 1917). In fact, the difference for Haeckel between biotic and abiotic life is that the internalization of "experience," which happens in the cell, leads to some sort of interiority that abiotic life does not have (Weber 2000a: 91). Though this may sound like vitalism or spiritualism at some level, Haeckel also fought against these notions as they were found in the works of Bergson and in Theosophy (Weber 2000b: 6–7). He thought such works turned everything that is real and of value into the ideal or spiritual. In other words, he neither wanted to move too far in the direction of reductive materialism or reductive idealism, but rather wanted to give full reality to all life: ideal/material, mind/body, energy/matter, culture/nature. At the same time, he wanted to avoid what he saw as the largest philosophical mistake in Western thought: ontological dualism. It was in articulating his position against dualism that Haeckel managed to construct a triune version of monism, which angered both scientists (mostly materialists) and theologians (mostly idealists or dualists). Here I want to focus on his construction of triune monism as form of ontological pluralism (which later would be articulated by the likes of American Pragmatists), a monism that was much more based in aesthetics than metaphysics. Then, I will move to some of the implications of what placing humans in an evolutionary context meant for Haeckel in terms of ethics and values. Finally, I will end with some lessons we might learn from Haeckel's version of Non-Reductive Religious Naturalism, particularly from within our planetary context in the twenty-first century.

For Haeckel the worst philosophical sin was that of assuming some form of ontological dualism, and he thought the church was the worst offender and propagator of this mistake. He couched his own work around evolutionary thinking and its implications in a long line of shifts toward immanent ways of thinking including those shifts made by Copernicus, Galileo, Spinoza, and later Goethe, and von Humboldt (Haeckel 1905: 53–57). As he understood it, all of human knowledge had been sidetracked as a result of dualistic thinking, whether in the form of God and the world, energy and matter, soul and body, or even self and other. It was Darwin's theory of evolution that finally enabled him to promote a form of philosophical monism in which all things could be made intelligible within an immanent, emergent framework. In doing so, he managed to attack dualism in any form that he found it, and Kantian thought was one of his primary targets.

For Haeckel there was much to be commended in Kant's thought, especially the very notion that we can never escape hermeneutics to reach some sort of "bare facts" about the world. In other words, Haeckel would agree that there are no un-interpreted facts, much to the dismay of many materialist scientists of his time (e.g. Haeckel 1905: 5–6). Kant, however, made two related wrong turns in which dualism, Haeckel thought, slipped back into his otherwise sound philosophy. First was Kant's assumption of the *ding an sich*. This assumption meant, according to Haeckel, that there was something outside of relationality and thus meant that there was some sort of dualism, or space of removal in which some sort of essence of the "other" could not be known. This essence, for Haeckel, was impossible from an evolutionary perspective in which biotic life emerges from abiotic life, animal life from plant life, and human life from animal life. There simply is no room for any type of essentialist understanding of self and other, and no hard separation between the different categories of life that scientists use.

Kant's second mistake, according to Haeckel, was his insistence on *a priori* statements in the form of categorical imperatives (Haeckel 1905: 10–11, 69). From Haeckel's point of view Kant was right in thinking that we could never fully have an objective view of the world, but not because there was a space from which that objectivity was possible. There is no hidden "outside" of experience that we do not have access to; rather, everything is radically interrelated and evolving, and thus we are always in the midst of life. There is no way we can have a bird's-eye view of "the way things really are." For Haeckel, an evolutionary perspective meant, among other things, that anything we might think of as *a priori* must in fact be an *a posteriori* idea/fact/concept whose bio-historical construction has just been covered over by time (Haeckel 1905: 25). In this sense, Haeckel comes very close to many postmodern thinkers that follow similar arguments found in Michel Foucault. For instance in *The Order of Things*, Foucault argues that what seems to be "given" or "natural" is, rather, historically constructed and shifts over time (Foucault 1970). Similarly, despite Haeckel's rhetoric in the *Riddle of the Universe*, which insists that one day science will solve all of life's riddles, the conclusions of his triune Monistic way of thinking do not bear out this possibility (Haeckel 1900). Haeckel argues, in my view correctly according to his assumptions about placing humans into an evolutionary framework, that there are no bare, un-interpreted facts. Such an assumption opens the door to the perspectivism and contextual assumptions that are at the heart of what Foucault (and other postmodern thinkers) are arguing.

At the risk of taking responsibility and agency away from Haeckel, which I certainly do not want to do, I do argue that his overconfidence in the scientific ability to one day overcome these epistemic uncertainties stems from his overzealousness in stamping out theological dualisms. In creating his "anti" position, he mirrors the structure of authority that he is fighting against.

Whereas priests and the church once had authority in metaphysics, science will now have authority in all things: from educational reform, to aesthetics, to "spiritual" matters and the ethics surrounding life and death (Haeckel 1895; Weber & Breidbach 2006: 161–174). His zeal for arguing against theology and his inflammatory comments about the God of Christianity, which Haeckel often publicly declared was nothing more than a "vertebrate gas bag," brought him scorn not just from the Church and some protestant theologians, but also from Darwin himself (Weber 2000b: 14; Krauße 1987: 116). In one letter from Darwin to Haeckel, Darwin begged Haeckel to tone down his rhetoric against theology so that people would listen to his evolutionary ideas more. In other words, Darwin knew that the radical implications of evolutionary theory could fall on deaf ears, so it was better to stick with the more strictly "scientific" side of things (Wuketits 2006: 19). Indeed it was largely the rhetoric of Haeckel and Huxley that led to huge rifts between proponents of evolutionary theory and those that positioned themselves against these theories who would later become known as creationists. This constructed rift between an evolutionary understanding of the world and a religious understanding of the world has had grave consequences for the world I do not need to rehash here. Instead, I turn now to some other implications of Haeckel's placing humans within an evolutionary context, that have consequences not only for later emergence theory and religious naturalisms, but also for post-modern thought. In doing so, I would argue that Haeckel represents much less of a "warfare" model between science and religion than he does a "dialogical" or "disputational" approach (Stenmark 2013). His religious naturalism is a hybrid, then, of scientific and religious ideas.

Human thought and history in an evolutionary context

My argument in this section is not that Haeckel himself or any other nineteenth-century scientist of his time was post-modern. Rather, my argument is that some of the tenets of post-modernity—all reality is interpreted, identities are non-essential and fluid, and values are co-constructed over time—depend in large part upon humans being placed within an evolutionary context. In other words, if human identity, values, and histories emerge out of the common planetary context as does all other life, then human knowledge of the world is also emergent. Haeckel understood this emergent nature of human identities, histories, and values in several ways.

First, and as I already mentioned, Haeckel the scientist argued that there are no bare facts in nature; rather, all understandings of nature involve interpretation. Put another way, all knowledge is *a posteriori* knowledge; hence, one of his major disagreements with Kant and neo-Kantians. For Haeckel, this conviction made developing a naturalistic worldview all the more important. He was flustered by the fragmentation of chemical, biological, and physical sciences in the nineteenth century and understood evolutionary theory as a way to finally bring these multiple perspectives on "nature" together into a single worldview. He understood this worldview as a form of monism, but again, a pluralistic form of monism. For Haeckel, multiple perspectives from the various sciences would give us a much better understanding of the world in which we live than any single science could (Haeckel 1895). Again, this "better" understanding for Haeckel was not based upon objective retrieval of bare facts, but upon the collaboration between perspectives that made better sense of the phenomena of life than any one science could provide.

In addition to the role of the sciences in understanding the world, nature aesthetics was equally important. Haeckel's *Kunstformen der Natur* and other sketches of abiotic and biotic life, in addition to his studies on embryos, looked for common patterns from crystals, to plants, to animals, and humans (Haeckel 1904; Haeckel 1917; Haeckel 1874). Indeed, he even made statements suggesting that human and dog embryos were virtually the same, but that the context (the ecology, biology, and even culture and education) leads to two quite different organisms

(Haeckel 1900: 65–66). His point was that knowledge of the world and our interrelatedness with it could only come about with the help of a nature aesthetic in which we could witness the evolution and emergence of forms, rather than assume that forms were imposed upon nature *ex nihilo* in some cosmic beginning.

Second, this loosing of forms from transcendent foundations (in a creator God or some other antecedent fixer of forms) meant that the world was much more diverse than humans could previously have imagined. Placing us in an evolutionary context was more than just re-thinking what it means to be human vis-à-vis other animals, or what this meant for one's religious understanding of the world, as important as these things were and are; it also meant tracing how such things as thought, ideas, and values, emerged from a long line of evolutionary developments. Rather than a "top down" approach to knowledge, this new way of thinking about the world and what it means to be human required a bottom-up way of thinking about the world, and with that an understanding that the forms we once held to be stable, actually change over time (Haeckel 1904: 9–18). What does this mean for "the human?" What does this mean for "male" and "female?" What does this mean for "beauty?" These are the types of question that an evolutionary context required new methods for answering, and Haeckel realized this. Part of his vehemence against theological dualisms, idealisms and reductive materialisms was that he understood these as old ways of thinking from within a pre-evolutionary context.

One example of this new way of thinking about the world and the proliferation of possibilities for what life could become was Haeckel's support of Magnus Hirschfeld in the late nineteenth and early twentieth century (Jacobson 2005: 208; Richards 2008: 275). Hirschfeld, a renowned sexologist of the time, founded the Institute for Sexual Research in Berlin in 1919. Prior to that time, he admired the work of and found support from Haeckel, particularly in his studies surrounding the evolution of forms in nature. Haeckel and he understood that once humans were placed into an evolutionary context, the idea of gender and sexual dimorphism, and even heteronormativity would give way to a diverse number of ways in which human beings could develop. Human sexuality, then, would naturally be just as diverse as the forms of sex and sexuality found throughout the natural world. The evolutionary context was essential to this type of scientific normalization of diversity in sex and sexuality.

Carl Linnaeus only a century earlier, for instance, had no real concept of the evolution of all of life, and this fact is apparent in the way he transposed human relationships to his description of the plant world. Plants were male and female, and had children, much like a human Victorian model of the family (Schiebinger 1993: 1–40). I argue that the lack of a strong form of evolutionary theory helped to enable Linnaeus to impose his cultural forms for thinking about sex and sexuality onto the world of plants, as if they were merely "natural." In other words, because he didn't have a concept of humans being placed within an evolutionary context (and thus human language, ideas, and forms), he was able to project his own understanding onto the world as if that was merely evident and natural. Of course, Haeckel too had his own cultural location that he imposed upon evolution in that he argued that European humans, and particularly Germans, were at the "top" of the evolutionary tree (Marks 2010). This idea fueled his more racist, nationalist, and anti-Semitic views of the human world. More will be said about this fact. However, his deep observation of the diversity of sexuality in nature and his placing humans into an evolutionary context meant that there could be no more understanding of God creating forms "in the beginning," nor an original creation of humans as male and female. Rather, the forms of organisms, including humans, evolve and emerge out of natural processes in which organisms interact with other organisms. The interactions of organisms within nature, according to Haeckel, fueled evolutionary changes. In other words the context or environment were very important for shaping the evolution of the organism and for this reason Haeckel coined the term "ecology" (Richards 2008: 409). So from the very beginning the idea of ecology is linked with diversity and change. Put into a contemporary context we might say the queerness of nature has everything to do with the queerness of human forms as well: evolutionarily, biologically, and psychologically, forms of life interact and shift over time to such an extent that naming one life form this or that is a heuristic device that scientists use to understand the world. The same would be true of Haeckel's understanding of aesthetics and ethics.

According to Haeckel, and again contrary to Kant, Beauty, Truth, and Goodness and all such ideals and forms emerge out of a continuous process of evolution. His work on the *Soul Life of Crystals*, *Art Forms of Nature*, Gustav Fechner's work on *The Soul Life of Plants*, and Wilhelm Bölsche's *Love Life in Nature*, were all nineteenth-century attempts to describe how some type of value or ensoulment was present in life "all the way down" (Haeckel 1917; Haeckel 1904; Fechner 1848; Bölsche 1926). It was not just that Beauty, Goodness, and Truth appeared on the scene with the first self-reflective *Homo sapiens*; rather, these things must have emerged from the evolution of earlier forms one can find throughout the natural world. For Haeckel, the source for such values in biotic life was to be found in the protoplasm of the cell, which he thought to be the unit of internalized "memory" of living cells. But even in abiotic life such as crystals, the world displayed forms of beauty and pattern in the ways that rocks and other things were shaped by the world around them. There is, in this sense, intelligence to life, but not of a teleological variety. Rather, it is an emergent form of intelligence, or an intelligence that is not predictive "from the beginning" of what and how the process of life will unfold, but rather may develop in many different directions that could not and cannot be anticipated. Nature, for Haeckel, was poesies and Art (Larson & Brauer 2009). Humans were not imposing these traits upon the world; they were "fine-tuned" over millions of years of evolution to make possible the perception of the beauty of the natural world.

According to Haeckel, religions (and philosophies) served an evolutionary purpose in as much as they helped human beings to become more aware of beauty, goodness, and truth. However, he argued, the arrival of the natural sciences and particularly the arrival of evolutionary theory, which would put into context all sciences and humanities, meant that old religions should be replaced by some form of what we might call religious naturalism, which he referred to as *monism*. Haeckel was certain that all peoples would benefit from close observation and study of the rest of the natural world from which they came, rather than reflecting merely on words, books, and ideas. It was this reason that he and the *Monistbund* in general (which he founded) argued vehemently for education reform that would lead to more study of the natural world, and why he fought so vehemently against what he perceived as the ignorance of theological and religious education. For Haeckel, promoting an evolutionary worldview was more than just a scientific necessity; it was also a religious duty. The religiosity with which Haeckel pursued a Monistic worldview in his arguments against dualists, idealists, and reductive materialists is evident in that his argumentation often is reminiscent of someone like Tertullian arguing against the Gnostics, or Augustine arguing against the Pagans. This religious fervor also might help explain some of his more short-sighted, nationalistic, and xenophobic understandings of the human world. I think we can learn from Haeckel's mistakes in this regard, and we should pay attention to his failures as a cautionary tale for anyone who seeks to develop a religious outlook based upon the evolving natural world of which we are a part. What Haeckel lacked, namely a bit of humility and an apophatic understanding of the natural world, I think we can embrace in our planetary context, a context in which we are globalized, understand ourselves as part of a larger community of evolving life, and recognize just how much scientific findings shift from one decade to the next.

Haeckel's non-reductive religious naturalism for a planetary context

As older generations can attest, there is really nothing certain except change. Scientific, political, social and economic realities are constantly changing, as are ideas about religion and the role religion plays in the lives of individuals, in politics, and in culture writ large. Yet, there still seem to be those who, in any given generation, lay claim to having arrived at a more stable version of the truth. I am reminded of thinkers such as Richard Dawkins, or any number of less-evangelical people who hold to "scientism" and truly believe the science of today gives us the correct understanding of every aspect of the world. This is not to deny that science may have an understanding of the world that takes into account many more embodied realities than other understandings of the world (past and present), but it cannot be the final word.

Haeckel also thought that the end of the nineteenth century was going to give way to the era of evolution, and that the natural sciences would solve all the riddles of the world (Haeckel 1900). This type of certainty and sense of arriving at the truth of things flies in the face of his own understanding of all ideas and values being *a posteriori*, and that there are no bare facts, but rather all facts must be interpreted as such. Such certainty also flies in the face of an evolutionary perspective of the world. If indeed we are part of an evolving planet, and our bodies and minds are evolving along with it, then our knowledge of the world will change along with the world. Finally, this type of certainty seems to go against the scientific method in which hypotheses are tested through experimentation and validated/invalidated in terms of how adequately they can explain the data. There are always "remainders" or data that fall outside of a theory or even a law, and these "remainders" may eventually shift us toward a different understanding with new lines of experimentation.

It seems to me that Haeckel's largest mistakes, and the largest mistakes of most scientists, come from not being able to recognize even their most powerful insights as coming from a particular perspective on the world. In other words, denying partiality and contextuality can lead to projecting one's own cultural-social-biological location onto nature. One example of this would be the way in which Haeckel projects his own ethnocentrism and racialized understanding of the world onto a hierarchical understanding of evolution that leads from the lowest and darkest peoples right up to the height of European (and even German) cultures and peoples (e.g. Haeckel 1905: 74–77). Though he thought these "lower" cultures could be "taught" through a "good" Monist oriented education that included a great deal of natural sciences and observation of nature, he nonetheless thought of European cultures as superior (Richards 2008: 245–276). He, like many others of his time, also read into the history of religions an evolutionary perspective on religious ideas. Like so many nineteenth-century religious scholars who were beginning to compare religions, he too thought some forms of "animism" were more primitive than the supposed highly developed forms of "monotheism" (Haeckel 1905: 53). Monism and his naturalistic worldview based upon evolutionary theory was for him just the next step in evolutionary progress; and this progress, according to Haeckel, called for us to take some responsibility for the future of humanity in the form of eugenics (Weir 2012: 6). This colonial so-called scientific approach to comparing differences was itself uncritically adopted from the preceding theological forms of colonialism: just as Christianity was bringing its alleged light to the assumed darkness found in the rest of the world, so now would science. These ways of thinking, whether by religious scholars, theologians, philosophers, or scientists, had and still have, as we should recognize, so many horrific consequences for humans and the rest of the natural world. One such horrific consequence during Haeckel's time was eugenics.

Haeckel was far from being the originator of eugenics, but he was a member of the Eugenics Society. He also questioned whether it was good to allow those who are born severely disabled to continue living and questioned whether it was immoral to allow someone who was severely depressed or mentally ill to end their own lives (Haeckel 1905: 101). These are questions that many still ponder today, and questions of *in utero* genetic testing, abortion, and designer babies bring all of this up. In some ways the conversation has not advanced beyond some of Haeckel's musings, but in other ways they have changed drastically.

First of all, the racial aspects of Haeckel's thought and the ethnocentrism of his thought would and should be roundly rejected by most scientists today (not to mention most educated people). The horrors of WWII revealed the worst of the technologies of eugenics and that line of thinking. Though I do not think Haeckel's ideas lead in a direct line to the Holocaust, as Daniel Gasman argues, I do think that placing humanity in an evolutionary context and then ranking different cultures in a hierarchy helped feed into that way of thinking (Gasman 1971).

Second of all, the Modern synthesis, the mechanism by which we now understand evolution to work, places the responsibility on genetic selection and variation, not on Haeckel's understanding of the protoplasm of the cell and recapitulation. For Haeckel, remember, information was "remembered" by a given cell's protoplasm. Within an organism, these cells made up societies, which were more democratic in things like plants and more monarchical in animals as you moved up the chain to human beings. His idea of recapitulation, that ontogeny recapitulates phylogeny, was that every individual organism went through the history of the whole course of evolutionary development. The developing organisms' environment—including the mother's womb, the environment/ecosystem, and education—led to the "higher" or "lower" developing of that given organism (Haeckel 1913; Brain 2009: 92–115; Haeckel 1898: 42). Thus, for Haeckel, science and medicine could intervene more easily and directly in the future of evolution than we would now think possible. The complexity of genetic selection, genetic variation, and evolutionary drift; our lack of understanding of the complexity of the genome and the role of epigenetics; and a non-hierarchical and non-progressive understanding of evolution today help to guard against the types of societal eugenic programs that were popular at the beginning of the twentieth century.

Third and perhaps foremost, we have had women's suffrage, civil rights movements, workers' rights movements, and a huge advance in communication and transportation technologies that allow us to see a pluralistic planet in a way that was not quite possible during Haeckel's time. None of what I say here is meant to excuse Haeckel from his racist and xenophobic interpretations of evolutionary history. Not all scientists of the time were ranking the races and cultures in terms of a progressive evolutionary scale. For instance, Virchow, one of the founders of ethnology argued vehemently against this approach to the point of rejecting the evolution of humans from primates. He maintained a separate and unified evolutionary history for human beings in order to combat the racist assumptions he saw being drawn from evolutionary theory. He argued for a monogenic origin of the human species, whereas Haeckel argued for a polygenic origin (Marks 2010). My question is why, at that time, "mono" and "sameness" were equated with equality, and "poly" and "difference" were equated with inequality? I argue that this has in part to do with the adoption of some lingering quality of monotheism into the very "scientific" worldview these scientists were beginning to construct. In other words, the monism proffered by Haeckel against dualism, idealism, and reductionism, even though it was a "triune" and what I might call a pluralistic monism, still had a strong understanding of the oneness and unity of ultimate truth.

Accordingly, differences had to be rank-ordered according to their value and/or veracity. This was as true for differences in hypotheses as it was for visible differences among cultures and races. Whereas Virchow rejected the evolution of *Homo sapiens* from primates in order to maintain the unity of (and thus not rank the differences among) *Homo sapiens* cultures, Haeckel assumed that if humans evolved from the rest of the natural world, there were probably different primates from which extant humans evolved (a polygenic origin), and thus these differences could be ranked.

The tension here is not really, in my mind, unity vs. diversity, but the way in which one values and accounts for differences. In a monotheistic view of truth—which was assumed by the Natural Sciences in terms of Natural Laws that could be discovered, objectively via Reason—multiple truths about a set of given facts can be problematic. Objectivity, under such a system, calls for conformity; otherwise, objectivity gives way to relativism. This false binary between objectivity and relativity has plagued Western religious, philosophical, and scientific thinking because of its equation of objectivity at some level with unity.

However, I would argue that such thinking goes against the grain of Haeckel's own understanding of a triune monism in which there are no "bare" facts, there is no "thing in itself" removed from relations, and in which all ideas and values are "a posteriori." In other words, there is no logical reason why differences must be ranked in the way that Haeckel ranked them. His ranking was simply a cultural bias that he was unable to see as such and which had racist and xenophobic implications. It is not uncommon for scientists breaking into new territory to allow their own biases to prevent them from following through with the conclusions of their own thinking: Darwin himself did not want to suggest that humans were evolved from primates, and even Einstein in the twentieth century did not want to admit that his mathematical conclusions suggested that the universe was expanding (he wanted it to remain constant). Though not all biases have immediate and negative social consequences, they are biases nonetheless.

We must learn from these mistakes. From where we stand in a pluralistic, globalized world, within a 4.5 billion-year-old understanding of planetary evolution, and a 13.8-billion-year-old understanding of universal expansion, our entangled reality can be entangled without being unified under one truth. In other words, perhaps because we are multiply embodied creatures, none of whom can escape that entangled embodiment and achieve a "god's eye" view of the world, the manifold, multiple perspectives on reality are what give us a sense of what Sandra Harding calls "strong objectivity" (Harding 1993). Strong objectivity simply means that if we are really going to take seriously our embodiment, the subjectivity of the observer must be taken into any accounting of the facts. Furthermore, it means that different embodiments will likely see different aspects of the world and have different accounts of the same phenomena or event. We gain "better" understandings of the world the more perspectives we can listen to and include. Rather than orthodoxy, we might be in a place in our planetary journey where polydoxy suits us much better. Polydoxy is, however, much messier and requires that we allow for some type of uncertainty in our own understanding of the world so that we can let other understandings in and seriously entertain them as well. This is not unlike what Haeckel and the Romantics were doing in their own time. Thus I end with a brief spiraling back to the nineteenth century in order to make sense of our own context.

Conclusion

Although there are many differences between nineteenth-century Germany and our own era, such as the fact that we are now living in a much more globalized and hyper-connected world, thanks to the speed of advances in communications, transportation, and production technologies, there is also much we might learn from that era. As stated earlier, in the nineteenth century there was not yet a decided upon, single methodological foundation and worldview for the natural sciences. Some argued for idealism or materialism, others dualism, and still others for some combination that was pluralistic and/or non-reductive. This methodological/epistemological/ontological pluralism is something that might benefit us today as we navigate the multiperspectival planet of which we are a part. One type of explanation does not fit all (despite the hope of most of these nineteenth-century scientists). For instance, despite Haeckel's focus on what I would call perspectivism (that there are no bare uninterpreted facts and that all ideas are a posteriori), he still concluded that the emerging natural sciences (and especially evolutionary theory) would one day solve all of the world's riddles. Where single explanations still ruled the imagination among nineteenth-century scientists, perhaps a hangover from a monotheistic worldview, multiple methods, explanations, and causality can and perhaps should be embraced within the sciences (broadly conceived) today.

It is also, finally, important to remember that many nineteenth-century scientists, especially of the romantic stripe such as Haeckel, were already beginning to struggle with the problems of the Industrial Revolution. Goethe, von Humboldt, Bölsche, Fechner, and Haeckel, among others, were in part seeking a naturalistic worldview that would take account of all of nature, in light of a world that was rapidly being changed through the instrumental assumptions about nature during the Industrial Revolution. Forests were disappearing, pollution was becoming a problem in cities, and already societies concerned about the preservation of animals were beginning to appear. It is not too much of a stretch to see the parallel with our own situation today, even though our changes are happening at a much more rapid pace.

The non-reductive models for understanding a naturalistic worldview, and thus for providing a non-reductive basis for the natural sciences, were all but swept away by the demands of the War Industry in the two World Wars in the twentieth century. Scientists were coopted by a technology transfer model (one in which science became reduced to technological outcomes), which is based upon an instrumental, reductive, and productive model: first for war technologies, then for agriculture in the so-called green revolution, and later for continued advances in transportation, communication, health, and production technologies in general. This "development" model for science, though it has brought humans many benefits, began to clearly buckle around the late 1960s/early 1970s with the publication of Rachel Carson's *Silent Spring*, and the beginning of the modern environmental movement (Carson 1962). In other words, science has again begun to question models that are reductive and begun to throw chaos, complexity, probability, and emergence into the analysis of natural phenomena and events. It may be time to critically retrieve some of these nineteenth-century models for the future of a planetary community: one which may come to see reductive materialism as a 40-year blip within the history of the natural sciences rather than as the normative basis for a naturalistic worldview.

References

Bölsche, W. (1926) *Love Life in Nature: The Story of the Evolution of Love*, 2 vols., New York: Albert and Charles Bonni.

Brain, R. M. (2009) "Protoplasmania: Huxley, Haeckel and the Vibratory Organism in Late Nineteenth-Century Science and Art," in B. Larson and F. Brauer, eds., *The Art of Evolution: Darwin, Darwinisms, and Visual Culture*, Hanover: Dartmouth College Press.

Carson, R. (1962) *Silent Spring*, New York: Houghton Mifflin.

Fechner, Gu. (1848) *Nanna oder über das Seelenleben der Pflanzen*, Leipzig: L. Voss Verlag.

Foucault, Michel. (1970) *The Order of Things: An Archeology of the Human Sciences*, New York: Random House.

Gasman, D. (1971) *The Scientific Origins of National Socialism*, Piscataway: Transaction Publishers.

Jacobson, E. P. (2005) *From Cosmology to Ecology: The Monist Worldview in Germany from 1770 to 1930*, Bern: Peter Lang.

Krauße, E. (1987) *Ernst Haeckel: Biographien hervorragender Naturwissenschaftler, Techniker und Mediziner*, Leipzig: Teubner Verlagsgesellschaft.

Haeckel, E. (2008 edition) *Gott-Natur (Theophysis), Kommentar Nachdruck von Olaf Breidbach und Uwe Hoßfeld*, Stuttgart: Franz Steiner Verlag.

—— (1917) *Kristallseelen: Studien über das Anorganische Leben*, Leipzig: Alfred Kroner Verlag.

—— (1913) *Die Natur als Künstlerin*, Berlin: Vita Verlag.

—— (1905) *The Wonders of Life: A Popular Study of Biological Philosophy*, translated by J. McCabe, New York: Harper and Brothers.

—— (1904; 2015 edition) *Kunstformen der Natur*, Leipzig: Verlag des Bibliographischen Instituts.

—— (1900) *Riddle of the Universe*, New York: Harper and Brothers.

—— (1898) *The Last Link: Our Present Knowledge of the Descent of Man*, London: Adam and Charles Black.

—— (1895) *Monism as Connecting Religion and Science: The Confession of Faith of a Man of Science*, translated by J. Gilchrist, London: Adam and Charles Black.

—— (1874) *Anthropogenie oder Entwickelungsgeschichte des Menschen*, Leipzig: Verlag von Wilhelm Englemann.

Harding, S. (1993) "Rethinking Standpoint Epistemology: What Is Strong Objectivity?" in L. Alcoff and E. Potter, eds., *Feminist Epistemologies*, New York: Routledge, 49–82.

Holt, R. N. (1967) *The Social and Political Ideas of the German Monist Movement*, New Haven, CT: Yale University PhD Dissertation.

Larson, B. and F. Brauer (2009) *The Art of Evolution: Darwin, Darwinisms, and Visual Culture*, Hanover: Dartmouth College Press.

Marks, J. (2010) "Why Were the First Anthropologists Creationists?" in *Evolutionary Anthropology*, 19: 222–226.

Richards, R. (2008) *The Tragic Sense of Life: Ernst Haeckel and the Struggle over Evolutionary Thought*, Chicago: University of Chicago Press.

Schiebinger, L. (1993) *Nature's Body: Gender in the Making of Modern Science*, Piscataway: Rutgers University Press.

Stenmark, L. (2013) *Religion, Science and Democracy: A Disputational Friendship*, Lanham: Lexington Press.

Weber, H. (2000a) "Der Monismus als Theorie einer einheitlichen Weltanschauung am Beispiel der Positionen von Ernst Haeckel und August Forel," in P. Ziche, ed., *Monismus um 1900: Wissenschaftskultur und Weltanschauung*, Berlin: VWB Verlag.

—— (2000b) *Monistiche und Antimonistiche Weltanschauung: Eine Auswahlbibliographie*, Berlin: VWB Verlag.

Weber, H. and Breidbach, O. (2006) "Der Deutsche Monistbund 1906 bis 1933," in A. E. Lenz und V. Mueller, eds., *Darwin, Haeckel und die Folgen: Monismus in Vergangenheit und Gegenwart*, Neustadt am Rübenberge: Angelika Lenz Verlag.

Weir, Todd (ed.) (2012) *Monism: Science, Philosophy, Religion and the History of a Worldview*, New York: Palgrave.

Wuketits, Franz (2006) "Mein Lieber Haeckel! Ernst Haeckel, Charles Darwin und der Darwinismus," in A. Lenz and V. R. Mueller, eds., *Darwin, Haeckel und die Folgen: Monismus in Vergangenheit und Gegenwart*, Neustadt am Rübenberge: Angelika Lenz Verlag.

4
THE RELIGIOUS NATURALISM OF HENRY NELSON WIEMAN

Cedric L. Heppler

The launching of a career

In his 1917 doctoral thesis for the Philosophy Department of Harvard University, Henry Nelson Wieman (1884–1975) wrote in three sentences what I would call the epitome of his entire career:

> The growth of consciousness reaches its acme when the antagonism of a great many interests forces the mind to take cognizance of that concrete field of time and space in which they all operate. All this is multiplied many times when the conflicting interests consist of those total systems which constitute human individuals or groups. There is no situation in which creative interest is so greatly stimulated and so abundantly satisfied as in those interactions between individuals which arise out of the necessity of adjusting themselves to one another.
>
> (Wieman 1985: 132)

By a familiarity of Wieman's career, one can see his concerns with ethics, knowledge, and metaphysics in these sentences. However, on deeper analysis, one can see that these three sentences are *in nuce* the beginnings of Wieman's ideas of God, creative interchange, science, and community, among others. Since these statements come at the beginning of his career, one would not expect of them the maturity of expression that characterized Wieman's thoughts on these matters even up to the time of his death in 1975. However, his doctoral thesis, *The Organization of Interests*, contains much that he came back to from time to time in his later writings. But more significantly, it was from within his dissertation that he launched his career. One idea he put forward in this thesis should be looked at further before turning to the specific discussion of Wieman's "religious naturalism." The phrase above, "the mind to take cognizance of the concrete field of time and space in which they all operate," indicates that Wieman saw human life and all that is important to its growth, survival, and betterment as cradled in and as a part of what is generally referred to as "nature," and he assumed that nothing outside of "nature" would intervene supernaturally to rectify or "save" human life from its destructive propensities.

The founding of creative interchange

Wieman announces his "thesis" in the opening paragraph of *The Organization of Interests*: "Our problem will be to discover that organization of human interests which is most conducive to their maximum fulfillment. The object of our quest is the greatest good. The principle of organization, which we propose, we shall call creativity" (Wieman 1985: 3). Wieman was to argue against a particular type of philosophical persuasion that proposed to lead to "the greatest good" by setting forth how individual humans and groups could overcome conflicts, animosities, and antagonisms by coordinating them into a harmony where none would have the advantage over the others. In his argument against this view of harmonization of conflicts that purports to lead to "the greatest good," Wieman wrote:

> With reference to this method of coordination we shall raise the question whether it is possible to reduce the inalienable interests of human nature to a harmony. Freudian psychology has demonstrated that there are ineradicable interests which cannot be simply ignored. It is not possible to cultivate indifference with respect to them. They must find expression in some form or other ... The constant pressure they exert, even though it be altogether subconscious, will eventually turn the beautiful harmony we have fashioned into gall and worm-wood. This is the inevitable working of suppressed tendencies.
>
> (Wieman 1985: 6)

When Wieman wrote these words the United States was just entering the "war to end all wars." It was arguably a bad time for "harmonious happiness" to gain many adherents, and in the lead-up to the beginning of World War I, humans in general seemed not to have exerted themselves with any dedication to adopt and follow the principles of harmony of any persuasion for any length of time. The history of warfare is an open book of evidence against such harmonious happiness. Incidentally, the two persons advocating the harmonious happiness against which Wieman established his thesis were two professors in the philosophy department of Harvard while Wieman was there from 1915 to 1917; namely Edwin Bissett Holt, in his book, *The Freudian Wish* (Holt 1915), and Ralph Barton Perry, in his book, *The Moral Economy* (Perry 1909). One could hardly have a more intimate relation with one's antagonists.

Wieman argued, given the fact that humans have trouble living together harmoniously, that a better option for life and any attainment of the greatest good would be to learn how to organize life so that its highest fulfillment would be possible in the midst of conflicts, animosities, and antagonisms. Wieman proposed an "organization of interests" wherein humans would take into account the "interests" of each other and learn from each other by coming to know the values of the other and to live in a creative interaction with each other. *The Organization of Interests* is an analysis of how this creative interaction may be aspired to, and Wieman uses the human dimensions of humor, play, work, art, friendship, and society at large to analyze the distinction between his "creative interaction" and "harmonious happiness." He writes:

> In humor, art and friendship it is the interaction of antagonistic tendencies which quickens the consciousness to a certain vivacity, emotional tension or process of idea-forming which is cherished by the person concerned as one of his most precious experiences. But it is the organized conflict of minds which is most highly valued and is most fruitful in developing all the ideas and emotions of which human nature is

> capable. Friction between persons is necessary in order that there be any mutual interpretation of minds. This friction is an integral part of the interpretation.
>
> (Wieman 1985: 7)

At this point Wieman speaks of what later, especially in his three books *The Source of Human Good* (Wieman 1946), *Man's Ultimate Commitment* (Wieman 1958), and *Religious Inquiry* (Wieman 1968), he called the "creative event" and "creative interchange":

> Through this interaction each individual comes first to understand the purpose of the other, and then forms his own purpose in relation to this other, either as a cooperator or antagonist. The other person reacts in the same way, and there ensues a mutual creation of mind out of which has arisen that marvelous growth of idea and emotion which we call culture.
>
> (Wieman 1985: 7)

Here are the four component parts of the "interaction" on which Wieman elaborated in various ways and with different symbols throughout his career and especially in his three books, namely, "the fourfold creative event" and "creative interchange."

When Wieman arrived at Harvard in 1915 Josiah Royce was living out his last year. The year before, Royce had published *War and Insurance*, his lamentations over the opening of The Great War in Europe (Royce 1914). Wieman's reaction to *War and Insurance* was critical, and he took the opportunity to elaborate further on his just-born concept of creative interaction: "In his book *War and Insurance* Royce proposes the banker, the judge and the insurance company as the great socializing and communizing agencies" (Wieman 1985: 101). Wieman points out that neither the banker, the judge, nor the insurance company can really deal with the conflicts, animosities, and antagonisms between human individuals or groups, unless through binding arbitration with the judgment of the arbitrator standing firm regardless of how close it comes to solving the original conflict. In such arbitration, neither party of the conflict would have had an interpersonal relationship with the other party. Perhaps persons of neither party would feel that the process of arbitration was especially creative, nor had they attained "maximum fulfillment" of their lives. Wieman concludes his criticism of Royce's "third party" arbitrator by saying,

> [a]s long as the bank intervenes, there is no mutual consciousness such as one can say: I know that you know that I know that I am serving you in this undertaking; I appropriate your interests and you assimilate mine; thus the consciousness of both of us is modified, its range expanded, its content further evolved toward a universal consciousness, and a genuine community established between us. As long as the bank persists nothing of the kind takes place. It inhibits all such creative interaction and establishes instead an intervening mechanism by which their activities are concatenated, like the cogs of a machine, so that certain results are produced without the need of personal intercourse and mutual understanding.
>
> (Wieman 1985: 101)

Establishing a view of nature

Wieman's "Introduction" to his initial book, *Religious Experience and Scientific Method*, gave a new definition of God that was both perplexing and applauded on the American theological scene in 1926. At this time Wieman was still using theistic language to try to expound on his earlier formulation of the concept of creative interchange (or interaction). However, even with the use of theistic language, Wieman tried to make sure that the "theism" of which he spoke was not the supernaturalism of systematic theologians of most main-line denominational seminaries and divinity schools of the late nineteenth and early twentieth century, such as Charles Hodge's *Systematic Theology* wherein God was omnipotent and omniscient, as well as being wholly transcendent to the world. By 1926 Wieman was bold enough to exclaim:

> Whatever else the word God may mean, it is a term used to designate that Something upon which human life is most dependent for its security, welfare and increasing abundance. That there is such a Something cannot be denied. The mere fact that human life happens, and continues to happen, proves that this Something, however unknown, does certainly exist.
>
> (Wieman 1926: 9)

It is important to note that Wieman says, "whatever else the *word* God may mean, it is a *term* to designate that Something." He would have said that this "Something" is a symbol; he tries to make clear that the word God is not "God Himself." The word *God* has always been a symbol and the symbol has to be elaborated upon within a community of humans who are trying to say what it is that they worship because of the security, welfare, and increasing abundance they feel has been bestowed upon them.

Wieman continues his claim about this Something as functioning in life for human security, welfare, and increasing abundance:

> [O]ne can say that there are innumerable conditions which converge to sustain human life ... But in that case either one of two things is true. Either the universe is a single individual organic unity, in which case it is the whole indivisible universe that has brought forth and now sustains human life; or else certain of these sustaining conditions are more critically, ultimately and constantly important for human welfare than are others.
>
> (Wieman 1926: 9)

The first example given by Wieman is pantheism, which he rejects. The second example explicates the notion that within the whole of the universe, and specifically with reference to human existence and life, there are conditions that function to bring about the security, welfare, and increasing abundance of humans. Wieman is making a fine distinction here. He has rejected traditional supernaturalism, and he has rejected pantheism because when the All is Everything, there are no distinctions. Wieman avoids Spinoza's "God or Nature." For Wieman, Nature is that field of influence from which we are born; it is the field of activity upon which we live, mature, and die; it is the field of evolutionary differentiation where the inorganic and organic come into and pass out of existence; it is the field upon which we humans have our environments in which we establish our own habits and develop the cultures peculiarly identifiable as human—we still have basic instincts but reason and language distinguish humans from other forms of organic life. For Wieman, that Something for which he is trying to find empirical evidence is a process within Nature, along with and over against other processes that either sustain or destroy human existence.

In 1930 Wieman wrote the essay "What's the World to Me" in an anthology of essays, *Ventures in Belief.*

> We cannot live without the support of the world. Also we cannot live without making the world different so that it will support us more constantly and adequately than it otherwise would ... We are a part of the world and the world works in us and upon us and round about us.... We as part of the world are sustained and made by that of which we are a part. This part which is ourselves reacts upon some other part to make it different; then this other part, thus changed, works upon us to make us still more different in ways that we never wholly foresee. The point is that in discussing the world we must not think of it as something outside ourselves and different from us, but as including us.
>
> (Wieman 1930: 79–80)

Wieman was not advocating that we humans can know the whole of the world, or the whole of nature, or the whole of the universe. We know only that which we change and which changes us. We can be aware of changing that part of the world that lends itself to us for the necessities of our livelihood, and we can be aware of our own lives being changed by the reaction of the world upon us. But we can never devise a "science" to explain everything: "Why can we not? Because there is no such thing as a science about the world. We have only the several different sciences and no one of them studies the world as a whole. No one of them even tries to study that feature or aspect of the world which most critically depends upon human welfare" (Wieman 1930: 81).

In line with Wieman's views on human dependence and interdependence with nature, there is an interesting, brief essay he wrote in 1945 for the cooperative volume edited by Vergilius Ferm, *An Encyclopedia of Religion* (Ferm 1945). Wieman was assigned to write eight short interpretive essays, and the one I turn to now is titled "Faith."

In this essay, Wieman assumes that humans live dependent on and interdependent from the various fields of influence within Nature. Thus Wieman does not consider faith to be based on "revelation ... or ... indubitable intuition or ... mystical experience or in some other way ... set beyond the bounds of human inquiry and testing" (Wieman 1945a: 270). He asserts, "If the reality to which one gives himself in faith is physiologically and psychologically inescapable, one does not need to cling to any belief." He goes on to say,

> Thus neither knowledge nor belief is the source of genuine faith. The knowledge which one gets is a consequence of the faith. The source of faith is in the human body. The body so reacts that one is coerced to be aware that one is sustained and that something-or-other is most important. Beliefs of the mind may distort, suppress, conceal or confuse this apprehension that emerges from the body. They may also clarify and inform the bodily apprehension with true knowledge. But beliefs of the mind cannot do this latter if they are themselves made the objects of faith. They can clarify and inform the faith only when used as means of reaching out after whatever may be most important however vaguely at first it may be sensed. Such a faith magnifies intelligent understanding. It also opens the way to all supreme fulfillments of life.
>
> (Wieman 1945a: 270–271)

Wieman also wrote the essay "Naturalism and Theology," for *An Encyclopedia of Religion* (Wieman 1945b). In it, he notes,

> The most marked development in the newer naturalism is the growing recognition that there are many levels and orders in nature. Values, ideals, personality, community, the sense of beauty, tragedy, heroism, religion and God are as readily acknowledged and upheld by some forms of the newer naturalism as the quantum theory and the physics of relativity ... Naturalism is based upon a certain method of inquiry ... Knowledge is achieved by discovering how events (happenings) are related to one another, or how they might possibly be related ... But when happening or event is analyzed it is seen to be necessarily temporal and spatial, whatever else it may also be. Therefore naturalism holds that all actual reality is necessarily temporal and spatial ... Nature is precisely the totality of all that is temporal and spatial together with whatever possibilities the temporal and spatial process may carry ... Therefore anything that we can ever experience must be some quality, form or movement pertaining to temporal and spatial reality. Since nature includes all temporal and spatial reality together with its possibilities, all that we can ever experience must be nature. This is the view of naturalism.
>
> (Wieman 1945b: 518–519)

In this quotation, Wieman speaks of "the most marked development of the newer naturalism." He had used the phrase "new(er) naturalism" previously, especially in the book he co-authored with Walter Marshall Horton, *The Growth of Religion* (Wieman and Horton 1938; hereafter this work is referred to as Wieman 1938, and pages listed will be from his part of the book). *The Growth of Religion* was a very important project for Wieman, because in it he confronted the established and rapidly growing influence in America of Karl Barth and Emil Brunner, as well as many American theologians who came to be identified as Neo-orthodox.

Wieman's opening chapter in *The Growth of Religion*, "New Growths in Religions" (Wieman 1938: 247–272), is a discussion of what he considered at the time to be five significant areas of religious activity: Literalism and Traditional Supernaturalism; Humanism; the New Supernaturalism; Theistic Naturalism; and the New Naturalism and the New Supernaturalism Compared. This "new supernaturalism" is what Wieman understood Barth and Brunner to be espousing in their classes and writings, and what their American sympathizers were supporting. In this chapter, Wieman loosely uses "theistic naturalism" and the "new naturalism" interchangeably. In a contrast between theistic naturalism and the new supernaturalism, Wieman observed,

> Theistic naturalism is the exact opposite of the new supernaturalism in respect to the use of reason and the empirical method. It asserts that all we know is mere guesswork unless it meets the tests of rationality and observation ... We know only by acting, but action gives us knowledge only when we are able to predict in some measure what the outcome of the action will be. First we act under physiological propulsion. In time we learn, by observation, to know what the outcomes of action will be when it is carried out according to certain patterns under certain conditions. When we reach that stage we have knowledge; and knowledge gained in this way is the only knowledge there is.
>
> (Wieman 1938: 258)

In Wieman's chapter, "Forms of Religious Apprehension" (Wieman 1938: 420–448), he explains how he thinks the new supernaturalist would answer the question, "How do we know?"

especially with regard to the knowledge of God. The following is Wieman's imagined question to the new supernaturalist and his imagined answer to the question.

> How, then, can we know this supreme God? By revelation and faith, by intuition and myth. God reveals himself directly to the heart and mind of the individual. Since reason falls into self-contradiction when it seeks to find the way to God, and since it cannot portray the nature of God, the human mind must make use of myth. When God is revealed to us we can set forth his being by means of myth, and there is no other way of doing it, since the abstract categories of reason fail us. How this is done we do not know. But the individual who receives this revelation and accepts it in faith, knows. He knows God. He cannot, by rational procedure, tell you what God is nor how he knows. But the knowledge is his. God and he are in living relation to one another. There is vital connection.
>
> (Wieman 1938: 429)

One could judge these latter assessments by Wieman as being mild but definitely partisan. One can understand that Wieman's philosophy of religion is naturalistically based: the religious object of faith, traditionally called "God," is that Something which is a part of nature and functions within nature; human knowledge grows out of the interchange of the human rational mind with the processes, events, qualities, and relations of other natural actualities exemplified by and within the same bounds of time and space that exemplify human life. Wieman's new naturalism is in direct opposition to the theology of Barth and his sympathizers.

During the second half of the nineteenth century, in America at least, the prevailing idea about the relationship between God and His world was immanence. God had not only created the world but He was also continually present in the world, providing guidance to His followers for their best lives and avoiding sin and evil. In 1884, Alexander V.G. Allen (1841–1908) published *The Continuity Of Christian Thought: A Study Of Modern Theology In The Light Of Its History.* This book was something like the handbook on the immanence of God. It went through 11 editions by 1895, and was still being published well into the 1920s. Allen was more concerned with the nature and character of God and His beneficial relationship with His followers than he was with making political or sociological statements about the human condition. However, the contemporaneous political and sociological assessments of the inevitable and continuous progress for the betterment of humanity by the continued expansions westward in the territories of the United States and its expansion of transportation upon land and sea, the founding of factories for the production of needed goods and services, the expansion of farming that lessened America's need for the dependence on costly trade with other countries, and the reestablishment of the Union in the aftermath of the Civil War, all could easily be judged to be a part of the work of God by His immanence in the world, and especially in the United States. Allen's book could easily have been turned to for "spiritual guidance" in this new era of bountiful production and consumption.

In 1918 Karl Barth, a Swiss pastor, published his *Der Römerbrief.* This book changed the Western theological worldview back to one in which God was transcendent, "wholly other" in relation to the world and humanity. An English translation of the 1928 sixth edition of *Der Römerbrief* was published in 1933 with the title, *The Epistle to the Romans* (Barth 1933). The following is a classic statement by Barth regarding this return of God's transcendence:

> The Gospel in not a religious message to inform mankind of their divinity or to tell them how they may become divine. The Gospel proclaims a God *utterly distinct* from men. Salvation comes to them from Him, because they are, as men, incapable of knowing

> Him, and because they have no right to claim anything from Him. The Gospel is not one thing in the midst of other things, to be directly apprehended and comprehended. The Gospel is the Word of the Primal Origin of all things, the Word which, since it is ever new, must ever be received with renewed fear and trembling. The Gospel is therefore not an event, nor an experience, nor an emotion—however delicate! Rather, it is the clear and objective perception of what eye hath not seen nor ear heard. Moreover, what it demands of men is more than notice, or understanding, or sympathy. It demands participation, comprehension, co-operation; for it is the communication which presumes faith in the living God, and which creates that which it presumes.
>
> (Barth 1933: 28; my emphasis)

This is the Neo-orthodoxy (the New Supernaturalism) that had hegemony in Western Christian thought from the 1920s till the announcement of the "death of God" in 1965.

The historical and theological significance of *The Source of Human Good*

The hegemony of Neo-orthodoxy in the United States (and of Barthianism in Europe) was one of the causes (with varying dimensions of conscious and unconscious motives and intentions on the part of Wieman) which inclined Wieman to write the content of *The Source of Human Good* the way he did. There were certain assumptions and declared statements of ultimate reality by the Neo-orthodox that Wieman, as a naturalist, thought it necessary to speak against and declare what he understood to be the misdirection and harmful tendencies of Neo-orthodoxy and its deprecation of reason in theological thought.

By consensus, *The Source of Human Good* is the book with which Wieman is most identified. This book comes the closest of all those Wieman wrote to being a "metaphysics" in the classic sense of that word. Wieman speaks directly to his "metaphysics" in two places in the book (Wieman 1946: 190–196; 298ff).

If people were to read *The Source of Human Good* and reach a point where they think they understand what Wieman had to say about "God" and his use of the new phrase "creative event," which he introduced at this time (Wieman 1946: 56ff), and then put the book aside, they would leave undiscovered most of the book and its significance for the American theological scene in 1946. Not unlike Edgar Allan Poe's "The Purloined Letter," Wieman's *The Source of Human Good* has an overarching focus that lies "hidden in plain view," but has rarely been elaborated on by any of his commentators who have mentioned it.

Before we turn to Wieman's overarching focus of *The Source of Human Good*, let us look at how he established his methodology of inquiry and his underlying assumptions. In Chapter 1, "The Way Good Increases" (Wieman 1946: 3–26), Wieman compares and contrasts six philosophies of value used by various schools in seeking to achieve the greatest good in human life. Of the six possible approaches to value, Wieman says,

> The sixth interpretation, in which value is identified with the total complex context, is the one that we shall develop in the pages that follow. As we have said, we take it not because it alone is true ... Both in practical action and in intellectual inquiry, we can do more with value when we take it this way ... Hence it is deliberately designed from the beginning for utility. But this usefulness must be demonstrated. That is largely what this book tries to do.
>
> (Wieman 1946: 6)

Immediately, Wieman elaborates on his philosophical assumptions for using the "contextualist" theory of value for his methodology. Under the heading "Naturalism," Wieman continues:

> This reply to the question "What is value?" is the answer of *the newer naturalism*, a movement or development in contemporary thought taking to, but to be sharply distinguished from, the older naturalism. It asserts that there is nothing in reality assessable to the human mind more basic than events and their qualities and relations ... No knowable cause or explanation for anything that happens can reach deeper than events and their structure and qualities ... We shall have no recourse to any "transcendental grounds, orders, causes or purposes" beyond events, their qualities, and relations. Naturalism bases this claim on thorough analysis of the method by which any knowledge whatsoever can be obtained.
>
> (Wieman 1946: 6–7; my emphasis)

Wieman titled Chapter 2 of *The Source of Human Good* "The Human Predicament." This chapter reads as if it were written by an existentialist, sitting on the left bank of the Seine while quaffing an espresso. In the opening paragraph, Wieman introduces his overarching focus that only a few interested in his thought have ever dealt with and none extensively: "Three features (to be indicated below) intrinsic to man's way of apprehending value render life perilous, *with the rising power of technology*" (Wieman 1946: 27; my emphasis).

Wieman saw the human predicament in 1946 as that of the dilemma of power, the power that comes from science and its technology. In essence, humanity has always had a science and its technology, as in when the first humans discovered how to chip stones and began making bows and arrows. As humanity evolved, a science and its technology were fashioned in relation to how humans strove to satisfy their desires and needs. In 1946, the dilemma had reached a crisis that could no longer be ignored or avoided. With the coming of the atom bomb, Wieman recognized that the *power* of science and its technology had forced a radical shift away from the human control of power to the constant threat of scientific and technological power as the potential destroyer of human life. Wieman asserts, "This peril can be escaped by redirecting human endeavor from service of good already created to service of the generating source of all good" (Wieman 1946: 27).

The "three features intrinsic to man's way of apprehending value render life perilous, with the rising power of technology" are (1) "the limited range of human appreciation," (2) the fact that humans are dominated by the sense of self-concern, self-centeredness, (3) and the resistance of humans to change "in the structure of appreciative consciousness" (Wieman 1946: 28). In the past when science and its technological power was limited to human lifting, pushing, running, riding horses, driving chariots and wagons, and building huge catapulting machines, the total threat to the of the human race was limited to small areas undergoing some sort of immediate turmoil. Wieman writes,

> Today modern technology has changed all that. Newly generated distinctions of good and evil peculiar to the new situation often carry the issues of life and death, because the power we wield today involves us with groups and cultures alien to our own, accelerates change, and deepens the levels at which it occurs.
>
> (Wieman 1946: 28–29)

Wieman saw the nature of the human predicament in 1946 as being in dire straits. He sensed what the governments had to offer for the resolution of the situation where the potential destructive power was in the hands of humans whose political guiding principles by and large were based on laws, rules, and regulations that had not been revised and updated since from before World War I. He also knew from his position as a theologian in the tradition of Western Christianity, and as one who attempted to keep pace with what other theologians and leading spokespersons of the Christian denominations were saying and doing, that, in general, Western Christianity was not taking seriously the challenge of science and its technology in the new era of atomic power. In fact, Wieman generalized the lack of an adequate response to this new threat and declared,

> The great religions ... have developed an ideology peculiarly fitted to the need of man in the days of his weakness ... Science must be directed to searching out the conditions demanded by the source of human good so that this creative power may produce the values of life for all; and technology must be applied to setting up those conditions which science discovers to be required ... But the creative source of human good is not now interpreted in such a way that science and technology can be applied to its service. The great religions portray it [the creative source of human good] as being the shaper of events, or the overruled of them, or somehow generating them, but not itself a structure of events ... Therefore, unless we can find the source of human good in the form of an order of events, we are doomed, if our analysis of the human predicament is correct.
> (Wieman 1946: 30, 31)

One doctrine of the Neo-orthodox Christianity of America in 1946 was that God is not "the form of an order events" but rather a God beyond human reason, not to be found naturally in time and space, and known only through a unique revelation originating at the pleasure of this transcendent God. "Knowledge of Him" could be received only as a supernatural gift from God. In telling exasperation, Wieman summarizes the difference between his rational-empirical inquiry and the non-rational and speculative theology of the Neo-orthodox: "This is not the problem of 'reconciling science and religion' ... That problem is twaddle compared to this. We are here discussing, not logical inconsistency, but life and death" (Wieman 1946: 31). He continues in this vein, saying, "We are here declaring is that the position [Neo-orthodoxy], held by so many today, is fatal to man in the days of his power" (Wieman 1946: 33).

The *Source of Human Good* has to be read in light of the threat of the power of the atomic age being unchallenged by governments and theologies structured on principles and doctrines of exclusiveness and transcendence. Given the nation states that have nuclear capabilities today, although there are only "a few," the primary ethical, moral, and theological principles of each of these nations are predicated on a worldview that is under the control of some transcendent being (God), which is radically transcendent above the affairs of humans. One can only wonder how the world would have been different if Wieman's perspective had gained more traction in the same affairs.

Wieman on religious naturalism

In 1971, in response to my question about his philosophical development, Wieman wrote:

> MY PHILOSOPHICAL DEVELOPMENT
>
> You refer to my present thought as that of naturalism. I am not happy with that characterization. When the term was first applied to my philosophy by Harold Bosley it was primarily negative by intent, meaning opposed to supernaturalism. Of course

> I am opposed to supernaturalism; but I think my work has been primarily affirmative than negative, constructive rather than destructive.
>
> On the other hand, when the word "naturalism" has been given its most conventional meaning, it has referred to the total process of nature below the human level. For this reason the humanists do not call themselves naturalists. They are as much opposed to supernaturalism as I but they are equally insistent on the sharp distinction between human existence and the total subhuman process of nature. In consequence they call themselves humanists and not naturalists. Their religious commitment is given to ideals beyond all process of time and space and therefore beyond all nature.
>
> There is still another meaning of "naturalism," different from all the above and also different from my own philosophy. It is that philosophy of religion which merges human life into the total cosmic process and interprets this cosmic process as dominated and directed by a cosmic mind. The Whiteheadian philosophies are of this sort represented by Charles Hartshorne, Paul Weiss, Schubert Ogden, John Cobb and others. My own philosophy of religion is clearly distinguished from these.
>
> On the other hand I am a naturalist in the sense (1) that I am opposed to supernaturalism, (2) that I am opposed to humanism because creativity is not merely an ideal but involves processes of nature and is a temporal and spatial, chemical and biological, psychological and social and historical process; (3) but this process of creativity distinguishing human existence is not a cosmic process, does not include all the diverse forms of process occurring in nature, yet is a process occurring in space and time with its own distinctive form of chemical, biological, psychological, social and historical process. This human form of process is a distinctively different process from the many other forms of process found throughout nature. If the word "naturalism" should obscure this difference it would be misleading when applied to my philosophy.
>
> I suppose the conclusion is that I am a naturalist but since "naturalism" has all these other meanings different from the meaning it has when applied to my philosophy, the word does not adequately distinguish my philosophy from others.
>
> I do not mean to repudiate the label. I do not know of a better name unless it be the religious philosophy of "Creativity." It might be called "Creative Naturalism" to distinguish it from other forms of naturalism and other theories about creativity. On the other hand, "empirical" philosophy of religion might be the best name to distinguish this interpretation of religion; but when one says that, he is immediately confronted with the many meanings of "empirical."
>
> (Wieman 1971)

Conclusion

I have tried to show throughout this essay that Henry Nelson Wieman had some far-reaching insights early in his career concerning the nature of human existence and the place of humans in their world and cultures. I hope I have shown that regardless of how many ways in which Wieman categorized, conceptualized, and symbolized these insights, he remained consistent in pursuing his insights about creativity, interpersonal interchange, and securing the means and conditions for humans to achieve the greatest quality of life across barriers of conflict and animosity. He committed his life to the principles of the ideal of the self-correcting methodology of science, and his worldview was structured on and guided by the assumptions of naturalism and empiricism. Wieman's position was never popular with traditional religious laity and officials, and he was never one to join the latest theological fad; a number of them came and passed away during his lifetime.

References

Allen, A.V.G. (1884) *The Continuity of Christian Thought*, Boston: Houghton, Mifflin and Company.
Barth, K. (1933) *The Epistle to the Romans*, trans. E.C. Hoskyns, London: Oxford University Press.
Holt, E.B. (1915) *The Freudian Wish*, New York: Henry Holt and Company.
Perry, R.B. (1909) *The Moral Economy*, New York: Charles Scribner's Sons.
Royce, J. (1914) *War and Insurance*, New York: The Macmillan Company.
Wieman, H.N. (1926) *Religious Experience and Scientific Method*, New York: The Macmillan Company.
____ (1930) "What's the World To Me," in H.P. Van Dusen (ed.), *Ventures in Belief*, New York: Charles Scribner's Sons.
____ (1938) *The Growth of Religion* (co-authored with W.M. Horton), Chicago: Willett, Clark and Company.
____ (1945a) "Faith," in V. Ferm (ed.), *An Encyclopedia of Religion*, New York: The Philosophical Library.
____ (1945b) "Naturalism and Theology," in V. Ferm (ed.), *An Encyclopedia of Religion*, New York: The Philosophical Library.
____ (1946) *The Source of Human Good*, Chicago: University of Chicago Press.
____ (1958) *Man's Ultimate Commitment*, Carbondale: Southern Illinois University Press.
____ (1968) *Religious Inquiry, Some Explorations*, Boston: Beacon Press.
____ (1971) "My Philosophical Development," Henry Nelson Wieman Papers, Special Collections, Morris Library, Southern Illinois University, Carbondale.
____ (1985 [1917]) *The Organization of Interests*, edited by C.L. Heppler, Lanham: University Press of America.

Further reading

One scholar to whom Wieman turned over the years for insights into the relationships between the body and the world is C. Judson Herrick, anatomist and neurologist, who wrote *An Introduction to Neurology* (Philadelphia: W.B. Saunders Company, 1915) (Wieman cites this book in *The Organization of Interests*); *Brains of Rats and Men* (Chicago: University of Chicago Press, 1927) (Wieman cites this book in *The Wrestle of Religion With Truth*, New York: The Macmillan Company, 1927); and *The Thinking Machine* (Chicago: University of Chicago Press, 1929). Two other scholars who corroborate Wieman's insights that faith is structured in our bodies (or is biologically embodied) are: Joseph Haroutunian, "Toward a Piety of Faith," in Philip J. Hefner (ed.) *The Scope of Grace, Essays in Honor of Joseph Sittler* (Philadelphia: Fortress Press, 1964); and Philip J. Hefner, "Towards a New Doctrine of Man: The Relationship of Man and Nature," in Bernard E. Meland (ed.) *The Future of Empirical Theology* (Chicago: University of Chicago Press, 1969). Some readers may be interested to read the six other essays Wieman wrote for V. Ferm (ed.), *An Encyclopedia of Religion*. They are: Christ, the living; God, modern conceptions of; Myth; Prayer; Symbol and Symbolism; and Worship.

5
A UNITY WITH THE UNIVERSE
Herder, Schelling, and Dewey on natural piety

John R. Shook

This chapter situates John Dewey's type of religious naturalism within the course of nineteenth-century philosophy of religion. His intellectual genealogy from Herder and Schelling through Goethe, Coleridge, and Trendelenburg infused his university years by way of the idealisms of James Marsh and George Morris. The ample continuities between his early ethical idealism and his late naturalistic humanism were central to his entire philosophy. Dewey's shocking 1934 announcement in *A Common Faith* that naturalism can accommodate a God was no surprise to himself.

Dewey's God and his view of the religious life remains one of the few viable options for religious naturalism in the twenty-first century. Pursuing the question, "Why did Dewey in his mid-seventies return to the religious philosophy he absorbed in his twenties?" discovers no retreat from naturalism, but the fulfilment of naturalism.

Nature and God

Religious naturalism, generally speaking, offers some sort of naturalism compatible with religiosity. Dewey's variety of naturalism (Shook 2011) can accommodate natural territories suitably hospitable for, and worthy of, religious attitudes and activities. Religiosity has a robustly ontological object, if it would accept both scientific and philosophical guidance. God's reality can pass Dewey's strictest empirical standard for proof, a pragmatist's proof:

> In reality, the only thing that can be said to be "proved" is the existence of some complex of conditions that have operated to effect an adjustment in life, an orientation, that brings with it a sense of security and peace. The particular interpretation given to this complex of conditions is not inherent in the experience itself.
>
> (*LW* 9: 10)[1]

Dewey's God is a special complex of natural conditions effecting adjustments and satisfactions in ethical living, so it is a candidate for this pragmatist sort of proof. This God is not reducible to anything merely subjective or projective. God does not subsist only within human experience, past or future; nor does God reduce to what humans ideally aspire to, presently or potentially. As this chapter explains, the reality of Dewey's God does not depend on our aspiration to pursue religious lives; our lives become religious when we aspire to pursuing intelligent lives.

Two competing conceptions of "God" as either a particular being or a set of ideals have dominated the imaginations of Western minds, as Dewey reminds us in *A Common Faith* (*LW* 9: 29). He then promptly rejects both conceptions—his God is neither just substantial, nor merely ideal. Instead, their integration earns Dewey's approval, because natural forces are necessary for the meaning and realization of ideals in life's activity. "It is this *active* relation between ideal and actual to which I would give the name 'God'" (*LW* 9: 34). Something natural that is undergoing active relations cannot merely be something humanly ideal, nor something transcendentally real. Very few philosophers have rightly understood Dewey's God, typically misled by mistaking his initial opposition of a theistic God to an idealistic God for the supposition of a forced dichotomy.[2]

For Dewey, our intelligent lives are within God and guided by God, and God is correspondingly changed by our living pursuits and achievements. Indeed, not only should this God be recognized for its naturally ontological status, Dewey's God also participates in many features we take to be characteristic of life itself. This God is organized and dynamic, and it displays coherence, self-maintenance, and growth. In a naturalistic sense, as we shall see, Dewey's God is a living God.

Dewey's naturalism is grounded in a distinctive philosophy of nature that had notable predecessors. Herder and Schelling advanced a *Naturphilosophie*—a Nature Philosophy—proposing that a World-God envelops, sustains, and advances Mind's knowledge of the world. *Naturphilosophie* dominated the early nineteenth-century aftermath of the Lessing-Jacobi "Pantheism Controversy" over Spinoza's controversial legacy (Yasukata 2002). G. E. Lessing had reportedly admitted his pantheism to F. H. Jacobi with the phrase, "*Hen kai Pan!* I know of nothing else." Herder did likewise with his own formulation, "We are on even ground on God's creation," also confided to Jacobi (Beiser 2009: 102, 160).

Dewey, like Herder and Schelling, developed a Nature Philosophy akin to pantheism, but even more indebted to the scientific worldview. None of them were Idealists in the usual sense of the term, for they denied that reality is fundamentally conscious and only exists for mind, and they refused to conceive God as a personal being. Herder, Schelling, and Dewey regarded their philosophies of nature as realistic and scientific. All three philosophers also denied that mechanistic atomism must be the final word from science, anticipating the integration of biology with chemistry without reducing one science's categories to the other's. Life and matter are only qualitatively distinct, distinguishable by their proper activities and contexts. Life is demarcatable where there is a logical and ontological priority of wholes over parts (organicism) and a dominance of ends over means (teleology).

Dewey's inclusive philosophy, both early and late, should be situated within a general worldview indebted to Herder, Schelling, and Hegel, and developed by later thinkers such as Coleridge and Trendelenburg.

The Sentient and the Sacred

The worldview Dewey inherited and developed amounts to a Nature Philosophy emphasizing these seven themes:

(a) Life is no accidental and irrelevant sheen upon nature, because life's essential powers are as abidingly and importantly natural as anything.
(b) This organicist view of nature guarantees that life and sentience cannot be separate from nature, but are expression of, and perpetuation of, nature's vital powers.

(c) The mentality of sentience especially expresses the purposive engagement of life with its environs, so it cannot be explanatorily reduced to static, formal, or mechanistic matters.
(d) As intelligence advances, it becomes ever more social and scientific, and communal intelligence creatively expands the complexities to life's engagements with environing natural conditions.
(e) Because creative intelligence is at home within nature, encultured piety towards nature's ways is not reliant on rationalistic categories framing experience (contra Kant) but directly expressive of intelligent conduct dealing with nature's resources.
(f) Conduct's enlarging engagement with nature correlates closely with due reverence and natural pieties, so valuing nature neither requires mystical absorption into nature nor "intrinsic" values for nature.
(g) The object of natural piety ("God") is not transcendent or aloof from humanity, but intrinsically interfused with humanity's own sentient energies in growing harmonious relationships.

For this Nature Philosophy, the Sentient and the Sacred are not divided by ontology or axiology. To be naturalistically religious is just intelligence's own realization of what it can accomplish and what it is ultimately for. Intelligence and piety have a common natural basis and a shared object of attention.

In *A Common Faith*, published in 1934, Dewey provided an encapsulation of natural piety and its proper natural object:

> Our successes are dependent upon the cooperation of nature. The sense of the dignity of human nature is as religious as is the sense of awe and reverence when it rests upon a sense of human nature as a cooperating part of a larger whole. Natural piety is not of necessity either a fatalistic acquiescence in natural happenings or a romantic idealization of the world. It may rest upon a just sense of nature as the whole of which we are parts, while it also recognizes that we are parts that are marked by intelligence and purpose, having the capacity to strive by their aid to bring conditions into greater consonance with what is humanly desirable. Such piety is an inherent of a just perspective in life. Understanding and knowledge also enter into a perspective that is religious in quality. Faith in the continued disclosing of truth through directed cooperative human endeavor is more religious in quality than is any faith in a completed revelation.
>
> (*LW* 9: 25–26)

Dewey then labels as "God" that cooperation of human goals and wider nature—that "union of ideal ends with actual conditions" (*LW* 9: 35).

Nothing about this God is hidden or hypothetical. Although God cannot be an object for scientific fields or the objective of science's theorizing, this God must the ground for every successful intellectual inquiry. God is revealed through any intelligence's work with the world, and that work transforming the world makes God real. This understanding of God was explored by Dewey over 40 years earlier (Shook and Good 2010). His 1893 essay "Christianity and Democracy" set out that understanding of God:

> God is essentially and only the self-revealing, and the revelation is complete only as men come to realize Him. ... The revelation is, and can be, only in intelligence... . Beyond all other means of appropriating truth, beyond all other organs of apprehension, is man's own action.
>
> (*LW* 4: 6, 7)

After this profound statement conjoining God as truth with humanity as knowers, Dewey fell silent about his own view of God for several decades. However, his elaboration of a Nature Philosophy continued. Major works during the 1920s, particularly *Human Nature and Conduct* and *Experience and Nature*, accounted for intelligence as a part of the natural world. Dewey never proposed a dual-aspect ontology or a property-dualistic ontology, he dismissed any Spinoza-style metaphysics, and he rejected metaphysical absolutism in all forms (including reductivist materialism). For Dewey, mind is thoroughly unified with nature, and that unity need not be a secret kept from mind. Not only can minds come to understand this unification with nature, intelligence can appreciate and pursue that natural unity.

The universe unified in intelligence

In *Experience and Nature* (1925), Dewey wrote, "[A] mind that has opened itself to experience and that has ripened through its discipline ... knows that its wishes and accomplishments are not the final measures of the universe." Yet mind's "power and achievement" still "implies a unity with the universe that is to be preserved" (*LW* 1: 313).

How can humanity preserve its unity with universe? "Fidelity to the nature to which we belong, as parts however weak, demands that we cherish our desires and ideals till we have converted them into intelligence, revised them in terms of the ways and means which nature makes possible" (*LW* 1: 314–315). Dewey regards intelligence as deeply intimate with nature, so that our cherished values can be realized through intelligent activity. To the extent that significant ends are objects of devout commitment, those ends must receive transmutation through intelligence to be intelligible—to be humanly realizable. The most realistic philosophy shall be the most idealistic, and the most idealistic philosophy shall be the most realistic.

Dewey's reconstruction of the realism-idealism dispute pre-dates his acquaintance with pragmatism. His worldview emerged from his undergraduate and graduate years, and it animated his writings during his first decade (1884–1894) as a professor. Science, not religion, guided this metaphysical reconstruction. His 1894 article "Reconstruction" proclaims that science has overturned the medieval picture of the world by depicting the world in terms of activities, causalities, forces, and energies occurring in space and time. No external metaphysical principle, power, or creator is needed. Everything that the cosmos requires, it has always possessed in abundance: "Now we see the universe as one all-comprehensive, interrelated scene of limitless life and motion. No bound can be put to it in imagination or in thought. No detail is so small that it is not a necessary part of the whole; no speck is apparently so fixed that it is not in reality a scene of energy" (*EW* 4: 102). On this vision, any existing thing has both particularity and interrelatedness: it is dynamic rather than static, and it is more like activity than rigidity. Vitality and life are no strangers to nature, and nature does not hide itself from mind. "Science has made real to us, and is bound to make still more real, the actual incarnation of truth in human experience and the necessity for giving heed to it" (*EW* 4: 103).

How does mind participate in this "incarnation of truth?" Dewey's understanding of the new empirical psychology convinced him that the field of mental life, the self, is just as dynamic and interactive as anything and everything else in the universe. His 1893 article titled "Self-Realization as the Moral Ideal" rejected conceiving the self as a schematic form. Dewey's substitute is "the self as always a concrete *specific* activity" and he stated that his goal is "to emphasize *the notion of a working or practical self* against that of a fixed or presupposed self" (*EW* 4: 43, 44, italics in original). Dewey never saw any reason why the nature of the self must be radically different from, or opposed to, the environing processes around one's self. From his 1887 book *Psychology* we read,

> Mind has not remained a passive spectator of the universe, but has produced and is producing certain results. These results are objective, can be studied as all objective historical facts may be, and are permanent. They are the most fixed, certain, and universal signs to us of the way in which mind works. Such objective manifestations of mind, are, in the realm of intelligence, phenomena like language and science; in that of will, social and political institutions; in that of feeling, art; in that of the whole self, religion.
>
> (*EW* 2: 15)

In *Psychology* Dewey also affirmed that

> It needs only to be recognized that every act of knowledge is an intuition of truth, and that the goal of all knowledge is the complete intuition of truth, and that this truth is the complete manifestation of the unifying and distinguishing activities of the intelligence... . It is the intuition of God as perfectly realized intelligence that forms the cognitive side of the religious consciousness.
>
> (*EW* 1: 212)

The ideal of perfectly realized intelligence can be nothing other than the ultimate harmonization of intelligence's work with the natural world.

Dewey emphasized in his writings early and late that the only sort of intelligence capable of pursuing that harmonizing end is social intelligence manifested in social (ethical) institutions. Therefore, the ideal of realized ethical institutions is the same thing as the ideal of God. In Dewey's *Psychology*, religion guides the growth of the free individual within God's processes. Dewey's *A Common Faith* offers the same understanding of religion.

As Dewey began to formulate his psychological and philosophical views in more naturalist and pragmatist terminology, his core Nature Philosophy was not abandoned. By 1888 Dewey had judged that philosophy cannot justify the assertion of a rationalistic Absolute, and no longer treated "God" as that Absolute. By then, Dewey was seeking a third alternative between God as a projector of selves (but this God is too real and a threat to individuality) and God as a projection of selves (but that God is unreal and unmotivating for individuals). By 1893 Dewey was prepared to expressly claim that a perfect and unchanging God can play no real role in human intelligence or morality (*EW* 4: 159–161).

If the growth of the human mind is not a poor duplication of God's perfection, then the idealized community (as intelligence unified) is the spiritual ideal for each individual (as intelligence growing). The growth of human intelligence is not about representing for ourselves some perfect order. Nevertheless, human intelligence is capable of appreciating nature's order, because intelligence is participating in its reorganization. In 1898, Dewey's "Evolution and Ethics" points to the needed integration of individuality with naturality, and demands the progressive harmonization of humanity's ends with nature's ways:

> [T]he laws and conditions of righteousness are implicated in the working processes of the universe; when it is found that man in his conscious struggles, in his doubts, temptations,

> and defeats, in his aspirations and successes, is moved on and buoyed up by the forces which have developed nature; and that in this moral struggle he acts not as a mere individual but as an organ in maintaining and carrying forward the universal process.
>
> (*EW* 5: 53)

Although "Evolution and Ethics" does not mention God, this passage presages the definition of the religious character of experience in *A Common Faith*, and it clearly invokes "the universal process." The ethical advancement of humanity cannot be independent from the ongoing ways of environing nature: their integration is what Dewey here denotes as the "universal process." In 1908 Dewey first uses the term "natural piety" as "the sense of the permanent and inevitable implication of man and nature in a common career and destiny" (*MW* 4: 176). Could this universal process be that *Hen kai Pan*, the "One and All," which Lessing and Herder proclaimed to Jacobi?

Dewey and organicism

During Dewey's early period, he repeatedly encountered and absorbed German organicism, originating in Herder and Schelling. His own accounts of his formative years (1877–1879) as an undergraduate at the University of Vermont, and as a graduate student (1882–1884) at Johns Hopkins University, tell the tale.

Dewey recalled T. H. Huxley's *Elements of Physiology and Hygiene* from a junior-level physiology course, describing how he had derived from that textbook

> a sense of interdependence and interrelated unity that gave form to intellectual stirrings ... Subconsciously, at least, I was led to a desire a world and a life that would have the same properties as had the human organism in the picture of it derived from study of Huxley's treatment.
>
> (*LW* 5: 147–148)

Huxley's own philosophical sympathies were displayed in 1869 when he was asked to provide the first article of the inaugural issue of *Nature*. Huxley's offering was his translation of selected aphorisms on nature by Johann Wolfgang Goethe, prefaced by a line of pantheistic poetry from William Wordsworth. The first aphorism of Goethe as translated by Huxley reads: "Nature! We are surrounded and embraced by her: powerless to separate ourselves from her, and powerless to penetrate beyond her." A finer encapsulation of Nature Philosophy could not be desired.

Dewey also read about Goethe for himself while at Vermont, when he borrowed Eckermann's *Conversations with Goethe* from the university library (Good 2005: 103). He would have read Goethe's views on God's intimacy with Nature and God's endless productivity of Life, in expressions such as these:

> Nature and we men are so penetrated by the Divinity, that it holds us; we live, weave, and are in it; that we, under eternal laws, suffer and enjoy; that we practise them, and they are practised on us, whether we recognize them or not.
>
> (Eckermann 1874: 525)

Although Dewey may not have read Schelling during his university years, Goethe's intellectual debts to Schelling are ample and well-documented. Behind the *Naturphilosophie* of Goethe and Schelling stands the towering figure of Herder, as Frederick Beiser recounts:

> The most powerful and influential voice behind the new vitalism in Germany was Herder. His 1778 *Vom Erkennen und Empfinden der menschlichen Seele* argued that the mental and physical are simply different degrees of organization and development of organic force; and his *Gott, Einige Gespräche* (1786) took his vitalism one giant leap forward by applying it to Spinoza's substance, which was no longer a static thing but a living power, "the force of all forces."
>
> (Beiser 2009: 367)

That cosmically supreme source of all force is God, the All. Herder's *God, Some Conversations* boldly approved this Spinoza-like One and All worldview:

> He, the Self-dependent, is Power in the highest and only sense of the word, that is, the primal Force of all forces, the Soul of all souls. Without Him none of them came into being, without Him none are active, and all in their innermost connection express in every limitation, form, and appearance, His self-dependent nature, through which they all exist and work.
>
> (Herder 1940: 104)

This organicist view of nature dramatically transformed psychology, and Beiser credits Herder again:

> [T]he mind is not a disembodied spirit, but only the highest degree of organization and development of the physical powers of the body. It is important to see that this theory does not define the mind by what kind of thing or entity it is, but only by its distinctive purpose or function, which is to integrate, control, and organize all the various functions of the body.
>
> (Beiser 2009: 146)

Beiser then situates Schelling within this organicism:

> As Schelling metaphorically summarized this view in his first work on *Naturphilosophie*: "Nature should be visible spirit, spirit should be invisible nature." The young idealists then reinterpreted Spinoza's dual-attribute doctrine in such vitalist terms. Unlike Spinoza, the mental and the physical are no longer simply different perspectives, or different forms of explanation, of a single substance, which themselves cannot interact with one another. Rather, the mental and the physical refer to only different appearances, manifestations, or embodiments of a single living force... . The mental is not simply the effect of the physical, then, but its realization or development; conversely, the physical is not merely the effect of the mental, but its embodiment or organization.
>
> (Beiser 2009: 368)

Dewey's early organicism and mature naturalism was destined to become an elaboration of this insight into the mental and the physical, outlined by that concluding sentence to Beiser's account of *Naturphilosophie*.

Dewey became further acquainted with this organic worldview and functional psychology in his senior year course on moral philosophy. Professor H. A. P. Torrey, who once admitted his preference for pantheism to Dewey (*LW* 5: 148), required students to read Marsh's edition of Coleridge's *Aids to Reflection* and *The Remains of the Rev. James Marsh*. Marsh was among the first

American scholars to read Kant, Herder, and Schelling in their original German, not far behind Samuel Taylor Coleridge's similar precedent in England. Coleridge's studies of Schelling's *Ideas for a Philosophy of Nature* and *On the World-Soul* did not tempt him into pantheism, but Schelling's view that nature and the self are composed of the same active powers allowed Coleridge to explain why practical reason could be entrusted with knowing nature (Hamilton 2007: 93–95; Hedley 2000).

Fifty-five years later, in 1941, Dewey recalled the crucial influence of Coleridge and Marsh. He particularly recalled how Marsh was conveying an Aristotelian view more than a Kantian view (*LW* 5: 185). Dewey's recollection is significant because that Aristotelian organicism was essential to German idealism's revolt against dualism, whether in the form of empiricism, materialism, or Kantianism. Marsh's collected essays in *The Remains* elaborate a sophisticated natural philosophy and proto-scientific psychology. Dewey appeals to Marsh in one of his earliest articles, "Soul and Body" (1886) to support his view that "soul" and "body" are related as "function and organ, as activity and instrument" (*EW* 1: 112), as Aristotle had proposed.

During his graduate studies at Johns Hopkins, he was the rapt student of another idealistic Aristotelian, George S. Morris. In 1917 Dewey had occasion to reminisce how Morris sought "a union of Aristotle, Fichte, and Hegel." Morris, Dewey recalled, only interpreted Hegel with "an abiding sense of what he was wont to term the organic relationship of subject and object, intelligence and the world" (*MW* 10: 11–1112). In 1889, after Morris's death, Dewey composed a study of his philosophy, emphasizing the abiding impact of Morris's studies in Germany under Friedrich Adolf Trendelenburg in Berlin (*EW* 3: 7).

Trendelenburg's influential Aristotelian revival, in the context of Germany's *Naturphilosophie* movement, was deeply indebted to Schelling (Beiser 2013: 32–33). Morris's own major work, *Philosophy and Christianity*, embraced Trendelenburg's main positions, starting from the Leibnizian principle that activity is the "essence of substantial existence," and affirming Aristotle's doctrine that activity—energy—constitutes being or reality. (This explains how Dewey came to accept Morris's invitation to write a book about Leibniz in 1886 for Morris's book series, Grigg's Philosophical Classics.) Like Trendelenburg, Morris repeatedly denied "a complete and essential mechanical separation between human and divine intelligence, or between 'human reason' and the divine mind" (Morris 1883: 263). He therefore flirted with pantheism just as the Vermont idealists did, by proposing that God's supreme spiritual activity is absolute being, of which any human person is a participating manifestation. However, like most seminary-trained intellectuals of his era, Morris never worked out a fully organicist metaphysics encompassing God.

Idealists of Dewey's generation, far less encumbered by theology, directly confronted the trilemma of either reducing personal selves to ineffective aspects of God, eliminating individuality by endorsing pantheism, or retreating to dualism to preserve human freedom. Dewey's organicism allowed him to deny key assumptions framing that idealist trilemma: (a) God possesses all powers; (b) God possesses all substantiality; and (c) causation is a necessary relation. If individual selves are related to each other and God as purposive organs of a body, then each self has particular ends and accomplishments not duplicated by others or the Godly whole. One can be one as a functional component of the All and sharing in the powers of the All while enjoying one's own existence and fulfilling limited ends not achievable by the All. One and All are co-related and co-dependent; neither could be what it really is without the other.

This organicism requires the reality of purpose, transformation, and growth at all levels, not excepting God. If God is too static, too mechanistic, or already comprises everything realizable, then component individuals cannot contribute any effective growth. Organic growth for persons implied God's growth, and all growth implies a relevant environs, for God as well as ourselves. Dewey had to reject pantheism, instead affirming that God is always less than the totality of nature. Religiosity is concerned with a God at work within the world.

A friend to rational piety

Thomas Cogan reviewed Herder's *Gott, Einige Gespräche* for England's *Monthly Review* in 1792, extolling Herder as "a friend to rational piety." Cogan's Unitarianism explains that warm reception. Liberal theology generally embraced Herder, despite his pantheistic views on God's dynamism and growth:

> In a world in which everything changes, every force is in eternal activity, and hence in eternal metamorphosis of its organs. For this transformation itself is the expression of its indestructible activity, replete with wisdom, goodness and beauty.
>
> (Herder 1940: 187)

> Every blind force is infused with light, every lawless power with reason and goodness. None of its operations, no activity in creation was in vain. Thus there must be progress, advance in the realm of God, since there can be no standstill, and still less a regress.
>
> (Herder 1940: 189)

However comforting and inspiring this worldview may be, is it fully religious? Can religion rest on reason alone?

No one could mistake Herder for a rationalist in theology. Religion, for Herder, utilizes our capacities for imagining ideals and understanding nature. In his treatise *Von Religion, Lehrmeinungen, Gebrauchen* (1798) Herder asserts that religion is a conviction that comes from "our innermost consciousness of what we are as parts of the world, of what we ought to be and must do as human beings; this religion neither borrows nor expects its insight or its efficacy from any mathematical demonstration" (translation in Crowe 2009: 270).

Dewey recreates this view of religiosity inherited from Herder. His 1928–29 Gifford Lectures, titled *The Quest for Certainty*, connect his nature philosophy with religious psychology. Appreciating how mind's pursuit of ends is intertwined with nature's potentialities should inspire the further insight that everything worth living for must depend on nature.

> Religious faith which attaches itself to the possibilities of nature and associated living would, with its devotion to the ideal, manifest piety toward the actual ... Respect and esteem would be given to that which is the means of realization of possibilities, and to that in which the ideal is embodied if it ever finds embodiment ... Nature may not be worshiped as divine even in the sense of the intellectual love of Spinoza. But nature, including humanity, with all its defects and imperfections, may evoke heartfelt piety as the source of ideals, of possibilities, of aspiration in their behalf, and as the eventual abode of all attained goods and excellencies.
>
> (*LW* 4: 244)

The religious attitude proceeds from this piety towards the actual, but intelligence and piety are not enough for religiosity, as Dewey understands the religious. Pious respect is not yet motivating conviction. In *A Common Faith* he writes, "Any activity pursued in behalf of an ideal end against obstacles and in spite of threats of personal loss because of conviction of its general and enduring value is religious in quality" (*LW* 9: 19).

For both Herder and Dewey, the intellectual (what we actually are) and the idealizable (what we really want) must be fused with the moral (who we try to be) to produce the fully religious life. That intellectual/idealizable fusion lies in the realm of the aesthetic; the intellectual/moral fusion lies in the realm of the ethical; the idealizable/moral fusion lies in the realm of the mythical. Only their complete three-fold integration results in a devoutly religious life. Being religious is not just about God, but it depends on God. Here is Dewey's definition for God:

> A clear and intense conception of a union of ideal ends with actual conditions is capable of arousing steady emotion ... Whether one gives the name "God" to this union, operative in thought and action, is a matter for individual decision. But the function of such a working union of the ideal and the actual seems to me to be identical with the force that has in fact been attached to the conception of God in all religions that have a spiritual content; and a clear idea of that function seems to me urgently needed at the present time.
>
> (*LW* 9: 35)

Let us next compare this conception of God with Herder's, as Alex Englander explains it. "Herder, especially in *God*, adopts a theological conception of forces in the shape of God as *Urkraft*. In *God*, Herder substitutes "divine Kraft" for Spinoza's "God." The unity of the many forces the natural sciences reveal as explanatorily basic is the world's essential force, God. His manifestations in individual forces make his infinite attributes known to us (Englander 2013: 908–909). For Herder, God is Nature: God is all power and productivity; God is impersonal; God is dynamic growth; life is impermanent; and there is no personal immortality.

The resemblances between the God of Herder and the God of Dewey are now manifest.

The key naturalistic adjustment made by Dewey to Herder's God is the limitation of God to relevant natural forces that only include life's endeavors and the conditioning energies enveloping them. Dewey is neither a pantheist nor a panentheist. God is not all reality, and nothing about God is quite independent from life. God pervades and energizes everything life is and does, but God does not underlie or ground what the universe is doing. Dewey's God cannot be identified with "the mysterious totality of being the imagination calls the universe" (*LW* 9: 56).

Dewey's God permits his nature philosophy to fulfill the realized unity of the Ideal and Actual. Conceiving of God is not just an exercise of idealization nor is it an overreach of naturalization.

> We are in the presence neither of ideals completely embodied in existence nor yet of ideals that are mere rootless ideals, fantasies, utopias. For there are forces in nature and society that generate and support the ideals. They are further unified by the action that gives them coherence and solidity. It is this active relation between ideal and actual to which I would give the name "God."
>
> (*LW* 9: 34)

For Dewey, God is both the necessary *Urkraft* and the relevant *Umwelt* to humanity's existence, past and future. Taken to its ultimate conceivable extent, if all life anywhere is taken together, then God is also the *Uberwelt* for Life, for as long as Life endures. God may die. No ecologies are guaranteed to last forever.

This conception of God does not inspire religious optimism or pessimism. Once again, Dewey's nature philosophy allows us to transcend religious finalities and embrace an openness to the future. This much we can know: the universe is often favorable to intelligent action. So asserts our natural philosophy. That intellectual knowledge suffices for what we should do: continue the bold enterprise of life with reverence for our natural home. So avows our natural piety. That feeling of piety in turn guides how we may hope: devoting our lives to ethical ends can help realize a more ideal world for living well. So affirms our natural religiosity. That avowal of religiosity is the culminating organic fusion of intellect, feeling, and will promised by Dewey's early ethical idealism and fulfilled by his late naturalistic humanism.[3]

Together, natural intelligence, piety, and religiosity yields an answer to the question, what is humanity, that we should be mindful of it? God, in Dewey's hands, cannot be omitted from that answer.

> A clear and intense conception of a union of ideal ends with actual conditions is capable of arousing steady emotion. It may be fed by every experience, no matter what its material. In a distracted age, the need for such an idea is urgent. It can unify interests and energies now dispersed; it can direct action and generate the heat of emotion and the light of intelligence. Whether one gives the name "God" to this union, operative in thought and action, is a matter for individual decision. But the function of such a working union of the ideal and the actual seems to me to be identical with the force that has in fact been attached to the conception of God in all religions that have a spiritual content; and a clear idea of that function seems to me urgently needed at the present time.
>
> (*LW* 9: 35)

This God is not just any regulative ideal or set of high ideals, or an ideal vision of unified ideals. Nor is Dewey's God just the relation between ideals on the one hand and natural conditions on the other, as if those two things are separable and independent. Dewey takes every opportunity to emphasize how forces in nature and society generate and support ideals, ideals further unified by effective actions in that world lending them coherence and solidity. God is that organic relation uniting the ideal and actual, supplying the natural basis for a reasonable faith.

A naturally religious faith

We expect a religious person to place faith in something beyond oneself. However, for Dewey, that place is not God, or nature, or any knowable entity (*LW* 9: 23). Religious faith proceeds from piety towards the natural, but one does not thereby place faith in the natural. In *The Quest for Certainty*, Dewey describes religious faith as attentive to, and pious towards, the natural.

> Religious faith which attaches itself to the possibilities of nature and associated living would, with its devotion to the ideal, manifest piety toward the actual ... Nature and society include within themselves projection of ideal possibilities and contain the operations by which they are actualized. Nature may not be worshipped as divine even in the sense of the intellectual love of Spinoza. But nature, including humanity, with all its defects and imperfections, may evoke heartfelt piety as the source of ideals, of possibilities, of aspiration in their behalf, and as the eventual abode of all attained goods and excellencies.
>
> (*LW* 4: 244)

Natural piety has a natural object, accessible to anyone capable of that imaginative overview of life. However, natural piety lacks one crucial feature that dominates religious faith: the personal commitment of individual persons to recreate themselves for the advancement of ethical ideals they deem worthy.

What is religious faith and where should it be directed? Dewey's answer is, "I should describe this faith as the unification of the self through allegiance to inclusive ideal ends, which imagination presents to us and to which the human will responds as worthy of controlling our desires and choices" (*LW* 9: 23). Faith has an object because it arises from a prior stage of natural piety, but religious faith per se is not about an object. Religious faith seeks an objective, a future objective that includes one's own self as a phase of its development.

A person of natural piety appreciates how there are many human ideals all potentially realizable, and many have been of benefit to one's own life. Yet that contemplative appreciation is not the same as making a personal ethical commitment to furthering any of those ideals into the future. Dewey clearly thinks that the cultivation of natural piety should develop a person towards the fully ethical life, and he holds that the religious life cannot be lived without natural piety. But he does not equate natural piety towards what is actual with ethical religiosity. Nor does he equate faith in some set of moral ideals with genuine religiosity. Taking piety about the actual or faith about the ideal as sufficient for religiosity returns to the dualism against which Dewey warns. Only their organic unification suffices for genuine religiosity, but that unification must occur in one place: the moral growth of the individual self who seeks that unification. Neither the natural nor the ideal will accomplish that for us.

In *A Common Faith* Dewey expects religious faith to be personally transformative by ethically unifying the self in addition to imaginatively unifying the world.

> What has been said does not imply that all moral faith in ideal ends is by virtue of that fact religious in quality. The religious is "morality touched by emotion" only when the ends of moral conviction arouse emotions that are not only intense but are actuated and supported by ends so inclusive that they unify the self. The inclusiveness of the end in relation to both self and the "universe" to which an inclusive self is related is indispensable.
>
> (*LW* 9: 16)

For Dewey, the ethical life is unifiable with the religious life only when and where people devote themselves to fully inclusive ends: ends which are simultaneously (a) realizable through intelligent living within cooperative nature and (b) realizable through self-transforming social action. Pursuing just one or the other falls short of the religious life. There are innumerable ways to enjoy one's life intelligently and piously without the ethical integrity that Dewey recommends, and there are innumerable ways to sacrifice one's life whole-heartedly and fanatically without the intelligent practicality that he also recommends. Neither the pious but dissolute life nor the fanatical but fruitless life can truly be the religious life, and Dewey had nothing but scorn for the third alternative, the life of the idealist passively entranced by unreal and unrealizable ends.

The object of natural piety alone is not God for Dewey. Conceiving God as "the" Uberwelt, as a pre-unified substantial being, and then attempting to identify one's ends with this Uberwelt's ends (as far as we can imagine what those might be), may be a type of religious naturalism but it cannot by itself engender the religious life. That sort of God may be growing with or without us, and it may be using us while not really needing us. Such divine independence might elevate it to a status worthy of a God in the eyes of many, but allowing our ethics to depend on this God is highly questionable. Dewey accordingly rejects the speculative postulation of God as an already unified set of pro-active powers, and spurns submission to such a God as unworthy of humanity.[4]

Dewey reverses the priority of ontology over axiology. Only our own ethical ideals, when we can wholly commit our lives to potentially realizable values, should control what counts as God-in-nature. Each individual is, insofar as s/he is religious, not just a component of God by living intelligently, but also a vital contribution to the growth of God-with-nature. This God is intellectually knowable, aesthetically congenial, and ethically admirable—and so is humanity by living religiously within this God.

Notes

1 References to the *Collected Works* of Dewey (1967–1990) are indicated by *EW* (*Early Works*), *MW* (*Middle Works*), or *LW* (*Later Works*) followed by volume and page numbers.
2 Rockefeller (1991: 518–520) accurately describes Dewey's naturalistic God. William Rowe (2007) accepts that theistic vs. idealistic dichotomy, and decides that Dewey confusingly offers two Gods, one ideal and one natural. Pihlström (2013) substitutes a Kantian dichotomy of constructivism vs. realism instead, deciding that Dewey's God is transcendentally "real" only within practices that are already religious.
3 In Dewey's words, "What humanism means to me is an expansion, not a contraction, of human life, an expansion in which nature and the science of nature are made the willing servants of human good" (*LW* 5: 266).
4 See for example Dewey's criticisms of Henry Wieman's proposed God (*LW* 9: 200).

References

Beiser, F. (2009) *The Fate of Reason: German Philosophy from Kant to Fichte*, Cambridge, MA: Harvard University Press.
Beiser, F. (2013) *Late German Idealism: Trendelenburg and Lotze*, Oxford: Oxford University Press.
Crowe, B. (2009) "Beyond Theological Rationalism: The Contemporary Relevance of Herder's Psychology of Religion," *Method and Theory in the Study of Religion* 21, 249–273.
Dewey, J. (1967–90) *The Collected Works of John Dewey*, ed. J. A. Boydston, Carbondale: Southern Illinois University Press.
Eckermann, J. P. (1874) *Conversations of Goethe with Eckermann and Soret*, new edition, trans. J. Oxenford, London: George Bell and Sons.
Englander, A. (2013) "Herder's Expressivist Metaphysics and the Origins of German Idealism," *British Journal for the History of Philosophy* 21, 902–924.
Good, J. A. (2005) *A Search for Unity in Diversity: The 'Permanent Hegelian Deposit' in the Philosophy of John Dewey*, Lanham, MD: Lexington Books.
Hamilton, P. (2007) *Coleridge and German Philosophy: The Poet in the Land of Logic*, London: Continuum.
Hedley, D. (2000) *Coleridge, Philosophy and Religion: Aids to Reflection and the Mirror of the Spirit*, Cambridge: Cambridge University Press.
Herder, J. G. (1940) *God, Some Conversations*, trans. F. Burkhardt, Indianapolis: Bobbs-Merrill.
Morris, G. S. (1883) *Philosophy and Christianity*, New York: Robert Carter and Bros.
Pihlström, S. (2013) "Deweyan Pragmatic Religious Naturalism," in *Pragmatic Pluralism and the Problem of God,* New York: Fordham University Press, 2013, pp. 47–72.
Rockefeller, S. (1991) *John Dewey: Religious Faith and Democratic Humanism*, New York: Columbia University Press.
Rowe, W. (2007) "Religion within the Bounds of Naturalism: Dewey and Wieman," in *William L. Rowe on Philosophy of Religion: Selected Writings*, ed. N. Trakakis, Aldershot, UK: Ashgate.
Shook, J. R. (2011) "Varieties of Twentieth Century American Naturalism," *The Pluralist* 6 (2), 1–17.
Shook, J. R., and J. A. Good. (2010) *John Dewey's Philosophy of Spirit, with Dewey's 1897 Lectures on Hegel*, New York: Fordham University Press.
Yasukata, T. (2002) *Lessing's Philosophy of Religion and the German Enlightenment*, Oxford: Oxford University Press.

6

THE SUBLIME AS SACRED

Reading Schopenhauer as a religious naturalist

Abigail T. Wernicki

In *Religious Naturalism Today: The Rebirth of a Forgotten Alternative*, Jerome Stone proposes a typology of religious naturalism that organizes its various iterations into three broadly construed groups. The first two incorporate a concept of God but in distinct ways: one as creative process (I view this as a kind of panentheism) and the other as equivalent to the whole of nature (I view this as pantheism). Naturalisms belonging to the third type do not employ a concept of God but are characterized as religious for one reason or another. To be sure, there is plenty of variation and discord among the philosophical perspectives that fall within this third group, which includes a broad range of thinkers from Donald Crosby to Ursula Goodenough to Stone himself. Some who fall into this category assign special value and meaning to the whole of nature itself. Crosby, for example, argues that nature is "religiously ultimate" (Crosby 2002). Others avoid ascribing ultimate value to nature, but instead identify certain processes or experiences within nature that contain religious traits such as sacredness or grace. I propose that Arthur Schopenhauer's philosophical perspective belongs within the latter niche. While Schopenhauer does not profess an explicit theory of religious naturalism, his metaphysical outlook combined with his theory of aesthetic value form something like a nascent version of religious naturalism. The initial aim of this essay is therefore to expose Schopenhauer's worldview as wholly naturalistic through an analysis of his metaphysics of will. I will then establish the religious aspects of Schopenhauer's system by offering a critical reading of his aesthetic theory. The latter will include a consideration of Schopenhauer's concepts of aesthetic contemplation and the sublime through the lens of Donald Crosby's theoretical criteria for religious ultimacy.

The suggestion that Arthur Schopenhauer's philosophy resembles a kind of naturalism is neither original nor radical (Gardner 1999: 403–5; Corrington 2013: 38–41). Yet Schopenhauer remains on the margins of most discussions of historical antecedents to contemporary philosophical naturalism, although Robert Corrington has done his part to integrate Schopenhauer into the discourse (Corrington 2013). In fact, his Ecstatic Naturalism explicitly embraces Schopenhauer's aesthetic theory; additionally, Corrington's discussion of the perennial fissuring of nature into *nature naturing* and *nature natured* closely parallels, by Corrington's own admission, Schopenhauer's conception of the world as both will-in-itself and objectified will.

Historical background

Born in 1788, Arthur Schopenhauer was a child when Immanuel Kant published his three groundbreaking *Critiques.* Schopenhauer would come to view Kant as the most important philosopher of the modern age and, perhaps even more so than any of his contemporaries, his philosophical system would be heavily influenced by Kant's transcendental idealism as well as his aesthetics (Cartwright 2010). This is especially evident in Schopenhauer's seminal two-volume work, *World as Will and Presentation*, which can generally be understood as a response to some of the problems Kant left unresolved in *The Critique of Pure Reason.* Schopenhauer's high regard for Kantian philosophy was in contrast to his low opinion of much of the "professional" philosophy that came out of Germany leading up to and during his lifetime. His disregard for what he considered to be the "sophistry" of neo-Kantian idealists such as Fichte and Schelling was anything but subtle, but the most explicit object of his contempt was Hegel. Although Schopenhauer's distaste for Hegel stemmed in part from his association with the kind of self-important spirit of German nationalism that prevailed in the early part of the nineteenth century, he also claims in several of his publications that Hegel offered nothing of value to the field of philosophy (Schopenhauer 2014; Schopenhauer 2010). He calls Hegel a charlatan (more than once) and expresses his astonishment that despite his philosophy amounting to "folly and nonsense," Hegel manages to garner the admiration and allegiance of an impressionable generation of philosophers and theologians who would go on to spread his "pseudo-philosophy" (Schopenhauer 2014: 89). Although some scholars have speculated that Schopenhauer's distaste for Hegel was largely due to jealousy over his great success in the academy (Wicks 2008: 161), it is undeniable that Schopenhauer also finds Hegel's fundamental philosophical outlook utterly wrongheaded. Yet, despite their vast philosophical and personal differences, Hegel and Schopenhauer agree on one crucial point concerning metaphysical knowledge. In short, they both believe it is attainable. As Robert Wicks observes:

> Against Kant, both argue that, since the same being that constitutes the rest of the universe constitutes us as well, when we know ourselves we simultaneously know the world's essence. Schopenhauer and Hegel drift apart because their introspections reveal different foundations at the core of human being: Schopenhauer discerns an ever-striving, blind, unintelligent, and meaningless Will; Hegel discerns the dialectical, rational, and reconciliatory structure of self-consciousness.
>
> (Wicks 2008: 163)

When understood in the confines of Kant's epistemological system, the thing-in-itself (*Ding an sich*) would always remain in the noumenal realm, beyond the grasp of human cognition. Metaphysical knowledge was therefore ultimately unattainable for Kant. As the Kantian logic goes, we can know *that* the thing-in-itself exists; we just cannot know anything about *what* the thing-in-itself is (Kant 1999). As Wicks notes above, although Schopenhauer and Hegel both develop theories affirming the possibility of metaphysical knowledge—that is, knowledge of the thing-in-itself—their vastly different epistemologies give way to opposing conclusions. As I will argue in the following section, Schopenhauer's conclusion places him firmly within the realm of philosophical naturalism.

Schopenhauer's metaphysical standpoint as philosophical naturalism

To what extent does Schopenhauer's philosophy of will provide the foundation for a naturalistic worldview? To adequately answer this question, one must consider the historical roots of Schopenhauer's metaphysical orientation, which go back to his first publication, ambitiously titled *The Fourfold Root of the Principle of Sufficient Reason*. It is here that Schopenhauer first wrestles with Kant's epistemological theory of transcendental idealism. Kant's view, simply put, is that the fundamental components of understanding are comprised of the forms of sensible intuition (space and time) and the pure concepts of understanding (quantity, quality, relation, and modality). The pure concepts of understanding can generally be understood as modes of causality and all fall within the umbrella of the principle of sufficient reason (PSR). The PSR, as a broad philosophical concept, is the notion that everything that occurs in history has a reasonable explanation. In other words, everything that comes into existence and everything that happens is the next step in a chain of causation. According to this principle, no event that has ever occurred could have unfolded differently than it did. Furthermore, every particular fact of the universe—every event, entity, and physical law—is *necessary* in the sense that it could not have happened any other way.

Kant's epistemological position is that the application of the PSR to sensible intuitions provides us with cognitive content. In simpler terms, we apply the fundamental laws of causality to what we perceive, and this act results in knowledge. Kant's epistemological synthesis of rationalism and empiricism led to the realization that one cannot obtain knowledge from sense perception or reason alone. In *The Fourfold Root*, Schopenhauer probes deeper into the origin of human cognition than Kant previously had in *The Critique of Pure Reason* by establishing that the foundation for experience (and therefore knowledge) is not comprised *solely* of the faculties for perception and understanding, but is first rooted in the fundamental distinction between subject and object. In Schopenhauer's elucidation of this idea, he also exposes the paradox that to exist is to be both subject and object. Schopenhauer's revision of Kantian epistemology rests on the basic premise that the necessary and sufficient condition for an experience, and therefore cognition, is the presence of both an apprehending subject and an object to apprehend.

Schopenhauer refers to the PSR as a "principle of individuation" because it allows us to distinguish objects from one another, thus making them perceivable by consciousness in phenomenal reality. But Schopenhauer, like Kant, is most concerned with what Kant calls the noumenon, or that which eludes human reason. In other words, he wants to obtain knowledge that is beyond the confines of what is knowable through the PSR. As Wicks observes, *The World as Will and Presentation*[1] reveals the

> tension between what Schopenhauer recognizes as knowable and expressible within the constraints of the principle of sufficient reason and what he wishes to indicate as the reality that underlies and to some extent transcends, everything that can be expressed within the principle of sufficient reason's scope.
>
> (Wicks 2008: 36)

One might wonder, at this point in Schopenhauer's reasoning, whether he might employ a supernatural explanation for the realm of reality that escapes the grasp of human cognition. The following passage in Book Two of the first volume of *World as Will and Presentation* raises some ambiguity surrounding this question:

> What is now impelling us to inquire, however, is precisely that it does not satisfy us to know that we have presentations, that they are such and such, and that they are interconnected in accordance with these or those laws whose general expression is in every case the Principle of Sufficient Ground. We want to know the meaning of those presentations: we are asking whether this world is nothing more than presentation—in which case it would have to be passing before us like a dream with no essence, or a

> ghostly vision, unworthy of our regard—*or whether it is something else besides, something else beyond that, and what it might then be.*
>
> (Schopenhauer 2008: 135, italics added)

Schopenhauer continues this line of inquiry in the following passage, which seems to suggest that he will soon be leaving the realm of nature in order to provide an answer to the question of what lies beyond the knowable realm of phenomenal existence:

> This much is certain at once: that this something after which we are asking must be utterly and in its entire essence fundamentally distinct from presentations, to which even the latter's forms and their laws must be thus utterly foreign; thus we cannot attain to it starting from presentation, under the direction of laws that only connect objects, presentations, with one another. Such are the modes of the Principle of Sufficient Ground ...We already see here that the essence of things can never be approached *from outside* ...And yet this is the path that all philosophers before me have walked.
>
> (Schopenhauer 2008: 135–6)

The reader soon learns, however, that Schopenhauer does not wish to abandon nature in order to fully grasp the basic essence of reality. Instead, his departure is an epistemological one. In the passage above, we see Schopenhauer's turn away from an objective point of view toward a subjective one in his pursuit of something "utterly and in its entire essence fundamentally distinct from presentations." His revelation is grounded in his discovery that the PSR is necessitated by the subject-object distinction. He reasons: *if I can never know an object in itself, then I must seek ultimate reality not in objects, but elsewhere.* The elsewhere is not a god or transcendence but is the *subject* in itself.

Following this perspectival revelation, Schopenhauer reflects on the ways in which we know the self as both subject and object. He observes that there are two distinct ways in which the body is "given" to the self. As object, the body is given as presentation, which is mediated through the subject's perception, as are all other empirical objects it encounters. As subject, the body is known *immediately* as will. Therefore, although the body is always represented as an object to the subjective self, we have on the one hand knowledge of the body that is mediated through our faculties of perception and understanding, and on the other hand *unmediated* knowledge of the body *as will* (Schopenhauer 1974: 137).

Schopenhauer's "discovery" of the will reveals to him that the core of existence is a kind of blind principle of striving and desire, as well as repulsion and hostility. The only thing we can know about the will, Schopenhauer argues, is that it is aimed toward life. There is no grand teleology in Schopenhauer's theory of the "will-to-life," nor is there any transcendent logic or ethic. This is not to say, however, that Schopenhauer does not promote his own theory of value of human existence *in relation* to the will, which becomes apparent in his aesthetics.

Schopenhauer's epistemological turn therefore leads him to some grim metaphysical realizations, but it also lays the foundation for reading Schopenhauer as a philosophical naturalist. Recall the previously cited passage, in which Schopenhauer asserts that one cannot approach "the essence of things *from outside*," precisely because it implies that the structure of the encounter would necessarily be subject-encountering-object and therefore within the confines of the PSR. In other words, when we encounter an object *as such*, the thing-in-itself remains veiled. Schopenhauer accomplishes a novel way of establishing the possibility for metaphysical knowledge without assuming a supernatural source of reality.

To bolster the case for reading Schopenhauer's philosophy as wholly naturalistic, one might also consider how Schopenhauer viewed his own philosophy relative to the discipline of philosophy.

Several years after publishing Volume Two of *World as Will and Presentation*, Schopenhauer reflects on his philosophy of will in *Parerga and Paralipomena* and articulates it in terms of the metaphysical categories *naturing nature* and *natured nature*. In the section titled "Fragments for the history of philosophy," in which Schopenhauer scolds his contemporaries for resorting to theological arguments in their philosophy, he writes that "naturing nature is still far from being God," and that instead "its concept contains the realization that behind the ever so fleeting and restlessly changing appearances of natured nature an everlasting and indefatigable power must be hidden" (Schopenhauer 2014: 104). He goes on to say that it is through metaphysics that we seek the source of this power and ultimately arrive at the realization that "we ourselves too belong to nature and, therefore, possess in ourselves not only the closest and most distinct specimen of natured nature as well as naturing nature, but also the only one accessible from within" (Schopenhauer 2014: 104). In other words, Schopenhauer remains convinced that we are in fact made up of both the noumenal and phenomenal elements of reality, in the Kantian sense, but unlike Kant, Schopenhauer believes we can indeed know the noumenal realm through an exploration of ourselves from within.

Schopenhauer goes on to say more about whether the split between *naturing nature* and *natured nature* calls for a kind of theistic explanation, an idea he strictly rejects (Schopenhauer 2014: 105). He also clarifies for his reader that his philosophy of will ought not be conflated with any version of pantheism, for this would require that one equate God with the world itself and therefore assign properties such as "benevolence, wisdom" and "blessedness" to the world (Schopenhauer 2014: 105). Given Schopenhauer's assessment of the world as quite a hostile place and totally indifferent to all living beings, he finds this proposition utterly ridiculous. Schopenhauer seems to revel in the idea that his philosophical view avoids practically any categorization during his lifetime. Perhaps this remains the most accurate way to view his philosophical legacy, as a totally unique contribution to the discipline so unlike anything offered by a Western thinker in the Modern age that it defies classification. It is noteworthy, however, that Schopenhauer proposes a name for his worldview in the section of *Perarga and Paralipomena* titled "Some remarks on my own philosophy," that supports the argument that his philosophy belongs under the umbrella of naturalism. He writes, "[o]ne could call my system an *immanent dogmatism*, for its theorems are indeed dogmatic, yet do not go beyond the world given in experience, rather they explain *what this world is* by analyzing it into its ultimate components" (Schopenhauer 2014: 119). Later in the same section, he compares his philosophy to traditional theism and pantheism, concluding that he offers something utterly distinct from both.

> Theism also has the world proceed from a *will* and has the planets guided by a will in their orbits and a nature generated by it on their surface. It is just that theism naively shifts this will to the outside and lets it affect things only in a mediate way, that it, by having cognition and matter come in between, in the human manner, while with me the will acts not so much on the things as within them; indeed the things are nothing but the will's very visibleness ... *Pantheism* calls the will that acts in things a God, the absurdity of which I have often and strongly reprimanded. I call it the *will to life*, because this expresses what can ultimately be known in it.
>
> (Schopenhauer 2014: 121)

Two things should be clear at this point. First, Schopenhauer derives his metaphysical view from a wholly naturalistic foundation. He establishes an ultimate principle of reality that can be known through nature or, more specifically, our own bodies. Second, Schopenhauer's discovery of this ultimate principle, the "will-to-life," leaves him with a rather morose philosophical worldview: The core of existence is a blind life force that presses on without regard for the extent to which living things suffer as a result.

"Religious ultimacy" in Schopenhauer's naturalism

Thus far, I have developed an argument for why Schopenhauer's philosophy should be included under the umbrella of philosophical naturalism, but I have said very little about why he might be counted among the philosophers whose thought represents a kind of historical antecedent to *religious* naturalism. In what follows, I will argue that evidence for this proposition can be found in Schopenhauer's aesthetic theory and in particular upon a reconsideration of his concepts of aesthetic contemplation and the sublime. A careful examination of these central concepts of Schopenhauer's philosophy reveals at least a resemblance to the kinds of "religiously ultimate" concepts that are central to the religious naturalism of Donald Crosby.[2] However, unlike Crosby's religious naturalism, Schopenhauer's philosophy does not imply the religious ultimacy of *nature itself* or, as Schopenhauer would say, the world given in experience. Rather, for Schopenhauer, nature itself can be seen as the necessary result of the objectification of the will. My position is therefore that religious ultimacy, if possible in Schopenhauer's system, emerges as a mode of attunement to the world by which we can achieve metaphysical knowledge (enlightenment) and liberation from the constant demands and inevitable suffering that result from existing as an object among objects in the phenomenal world.

In Book Three of *World as Will and Presentation*, Schopenhauer offers a comprehensive account of his aesthetic theory, which follows directly from his theory of the will. A key element of his aesthetics is his distinction between aesthetic cognition and conceptual cognition. He proposes that aesthetic cognition is the unique way in which we can apprehend objects that reside within the realm of pure objectivization (i.e. Platonic Ideas), while conceptual cognition is how we come to know the manifold indirect objectivizations of the will (everything else). While Schopenhauer concedes in Book One that all cognition is subject to the laws of causality, he explains in Book Three that aesthetic cognition of Platonic Ideas is only constrained by the form of objectivization and *not* by space, time or causality. This is because Platonic Ideas are abstract concepts that exist outside of the confines of space, time, and causality. Conceptual cognition, on the other hand, is the mode by which we apprehend all things that inhabit the realm of space, time, and causality. Aesthetic cognition is closer to what might be considered an intuitive act of cognition, whereas multiple layers of representation mediate conceptual cognition.

Another sense in which Schopenhauer's concept of aesthetic cognition is distinct from conceptual cognition is that it facilitates one's detachment from the desires of the will. Although Schopenhauer suggests in *World as Will and Presentation* that cognition in general is always in "service of the will," he allows for an exception with respect to aesthetic cognition (Schopenhauer 1974: 219). For example, in the case of conceptual knowledge, the cognizing subject views the body as objectified will, and therefore seeks to know things as they relate to the body's interaction with the world as presentation. In other words, throughout the course of our ordinary existence, we are concerned with individual things, i.e. "objects" insofar "as they exist at this time, in this place, under these circumstances, through these causes, with these effects: in a word, as individual things" (Schopenhauer 2008: 220). For example, when I get out of the shower, I am aware that my feet are wet and will make the floor slippery when I step onto it. Therefore, I modify my posture to brace myself for this possibility. We are all familiar with this sort of existence in the world. A less trivial example would be someone who prepares for the loss of a loved one. Her attitude about the world, her interaction with her colleagues, her appetite for food and her ability to simply enjoy life is hindered by her awareness of the impending threat to her relationship with this person. As existing things, Schopenhauer argues, we are rarely free from these kinds of concerns, however mundane or heartbreaking. Yet, through aesthetic cognition, we are able to temporarily free ourselves from these concerns, and we become focused on the Ideas instead. We move away from the pressures and pain that result from concern over our particular existence to contemplation of existence in general. The transition from conceptual cognition to aesthetic cognition is therefore an act of liberation. Schopenhauer writes:

> The possible passage—but, as has been stated, it is to be considered only an exception—from ordinary cognizance of individual things to cognizance of Ideas occurs suddenly, with cognizance tearing itself away from the service of the will. Just by that fact the subject ceases to be merely individual and is now the pure, will-less subject of cognition, which no longer pursues relations according to the Principle of Sufficient Ground, but rests in constant contemplation of the given object beyond its interconnection with any others, and gets absorbed therein.
>
> (Schopenhauer 1974: 221)

The language Schopenhauer employs here evokes the mood of peaceful detachment; in a liberated state, the self is free from the particular demands of the phenomenal world, free from obligations to and dependence upon other individuals in the world, and is filled with "restful contemplation" of the Ideas.

At this point I want to pause and reflect on the ways in which Schopenhauer's concept of aesthetic contemplation might be considered religious in any respect. As a guideline, I find it useful to employ Crosby's six "role-functional" categories of religious objects, which include Uniqueness, Primacy, Pervasiveness, Rightness, Permanence, and Hiddenness. Together, they offer a set of criteria by which one might determine the religious ultimacy of an object. As Crosby explains, each one is "intended to identify an aspect of the distinctly religious function performed, or role played, by religious objects in the life of the religious person and in the cosmos as the religious person views it" (Crosby 2002: 118). By analyzing some of Schopenhauer's philosophical concepts using these criteria, I am not suggesting that Crosby's is the best or only way to determine whether an object has religious status. Rather, my aim is to reveal a strong resemblance between Schopenhauer's ideas and those of a prominent contemporary religious naturalist in order to underscore my position that Schopenhauer's philosophy is closely affiliated with religious naturalism as a school of thought.

It seems clear enough that Schopenhauer's concept of aesthetic cognition satisfies three of Crosby's six role-functional categories for religious ultimacy. These include Uniqueness, Primacy, and Hiddenness. As for Uniqueness, Schopenhauer's position is unambiguous: aesthetic contemplation has the unique ability to enable contemplation of the Ideas. This is in direct contrast to the way in which we experience cognition of all other objects. Aesthetic cognition can be understood as satisfying the category of Primacy insofar as human beings strive to achieve aesthetic contemplation as a superior mode of attunement to the world, one that is not entirely veiled by phenomenal existence. Hiddenness, Crosby observes, "brings to mind the overpowering sense of mystery and awe" that we experience when we engage with the religious object (Crosby 2002: 120). When aesthetic contemplation is experienced, one comes face to face with the Platonic Ideas, which in their eternal nature defy reason and inspire wonder.

The categories of Pervasiveness, Rightness, and Permanence are less obviously present in aesthetic contemplation. With regard to Pervasiveness, Schopenhauer certainly does not claim that all individuals can or will ever experience aesthetic contemplation. Although, I would argue that this speaks to its deep Hiddenness as a mode of attunement and does not necessarily suggest the property of exclusivity. Perhaps it is sufficient that aesthetic contemplation in theory exists as a possibility for all individuals. There is little to be said about the category of Permanence with regard to aesthetic contemplation. This largely stems from the fact that I am attempting to apply categories that were established as criteria for religious *objects* to aesthetic cognition, which is not really an object but rather a way of being in the world. Aesthetic contemplation, by definition, is impermanent. However, the Platonic Ideas, which become accessible through aesthetic contemplation, are indeed permanent and unchanging.

The final role-functional category proposed by Crosby is Rightness, which "makes explicit the valuative function of the religious object" and in so doing reveals the purpose of each individual's life along with the broader purpose of human life itself as being both relevant to and at home in the world (Crosby 2002: 119). The distinction between the two aspects of the category of Rightness—the personal and the cosmic—is important when considering whether or not Schopenhauer's concept of aesthetic contemplation functions in this way. Let's consider the personal aspect of Rightness first. For Crosby, this is the aspect in which the religious object helps to facilitate personal development by defining "the goal of human existence, laying out a path of spiritual progress toward that goal" (Crosby 2002: 119). It is true that Schopenhauer strictly denies any kind of teleological view of human existence. The notion of "spiritual progress toward a goal" simply does not fit into his system. It would be inaccurate to suggest that Rightness, in this sense, could be ascribed to the experience of aesthetic contemplation. However, Crosby goes on to say that the religious object, in its Rightness, functions as the "most profound healing, transforming, saving force in the life of the religious person," which more closely resembles the role that aesthetic contemplation plays in Schopenhauer's system (Crosby 2002: 120). Aesthetic experience transforms our perspective and liberates us, albeit temporarily, from suffering. I would offer the following response to this: while the concept of aesthetic contemplation does not imply the possibility for spiritual growth, it leads to metaphysical knowledge and in so doing, also relieves us from the inevitable suffering of our existence. Aesthetic contemplation is therefore uniquely equipped to diminish the suffering of human existence, if not able to provide the framework for spiritual development. It can therefore be understood as a mode of attunement by which the value of human life is enhanced. Thus, there is a sense in which the personal aspect of Rightness can be seen in Schopenhauer's concept of aesthetic contemplation, although it does not entirely match Crosby's definition.

The cosmic aspect of Crosby's role-functional category of Rightness inspires hope in the individual by, in Crosby's words,

> pointing to a goodness or fitness that the religious person regards as lying at the heart of the world. It means that human beings are not simply left to their own resources but reside in a universe that, due to cosmic Rightness (as well as Primacy) of the religious object, is in its depth responsive—not indifferent or inimical—to their yearnings and strivings for the triumph of the salvific ideal in themselves and the world.
>
> (Crosby 2002: 120)

Here, it becomes increasingly difficult to find the resonance between cosmic Rightness and the role that aesthetic contemplation plays in Schopenhauer's system. While aesthetic experience reveals certain truths to the individual subject, it does not suggest that a fundamental "goodness" or "fitness" lies at the core of things. Rather, it allows us to feel a familiarity—indeed, a kind of formal sameness—with the world, but not within a valuative construct.

Based on the above analysis, I would offer the following assessment: Schopenhauer's concept of aesthetic contemplation meets five of Crosby's six categories that are required in order for religious ultimacy to be established, Rightness being the sixth and not entirely satisfied category. Aesthetic contemplation, to a certain extent, carries out the personal aspect of the role-functional category of Rightness, but cosmic Rightness is not present in Schopenhauer's concept of aesthetic experience.

Schopenhauer's theory of aesthetic cognition is integrally tied to the aesthetic categories, namely the beautiful and the sublime. The beautiful, for Schopenhauer, has primarily to do with the Platonic Ideas. The beautiful becomes manifest in objects when they *accommodate* the Ideas (Schopenhauer 1974: 246). For Schopenhauer, recognition of the Ideas in objects, i.e. aesthetic experience of the beautiful, transports consciousness into a state of "pure perception," thus freeing it from service of the will (Schopenhauer 1974: 246). On the subject of the sublime, Schopenhauer agrees with Kant that it has primarily to do with the confrontation of one's insignificance relative to the world or the whole of nature. For both philosophers, the *feeling* of the sublime is a precarious balance between the pleasure of peaceful detachment and the pain of a particularly threatening environment or situation. In contrast to the feeling of the beautiful, Schopenhauer writes that the feeling of the sublime is aroused when the objects of perception embody the Ideas but also "stand in hostile relation to human will in general" (Schopenhauer 1974: 246). The feeling of the sublime consists of a paradoxical combination of pleasure and pain—simultaneous ecstasy and fear of annihilation.

Schopenhauer's view is that certain kinds of manifestations of will in nature evoke the feeling of the sublime in us more than others. He notes that whereas light or lightness can be correlated to the beautiful, darkness and emptiness are often characteristics of the sublime. Darkness and emptiness, he observes, are associated with a kind of lacking or mystery, which, instead of accommodating the will, act out of hostility for the will by withholding what it desires: knowledge and enlightenment. Similarly, extreme stillness or quietude in nature can lead to the feeling of the sublime because one feels alienated from one's surroundings. Schopenhauer alludes to this mood in the following passage, in which he suggests that one element of the feeling of the sublime can be the feeling of being utterly alone in the world:

> Let us transport ourselves into a most lonely region, with unlimited horizon, under utterly cloudless skies, trees and plants in entirely motionless air, no animals, no people, no moving waters, the deepest stillness—then such surroundings are like a summons to seriousness, to contemplation, together with a tearing of oneself away from all willing and its neediness.
>
> (Schopenhauer 1974: 249)

More threatening still, Schopenhauer argues, is a region like the one described above except it contains no plants or trees at all, and has been stripped of any object that might be pleasing to the individual will, or strive to relate to it. Thus, we can see a kind of hierarchy forming of lesser to greater degrees to which the feeling of the sublime can be achieved. Moving up the hierarchy, Schopenhauer argues that an even more powerful feeling of the sublime can result from one's experience of "nature in stormy movement; *chiaroscuro* produced by threatening black thunderclouds; monstrous, naked, overhanging cliffs that block one's view with their folds"; thus, we begin to see a kind of movement in nature that is parallel to Schopenhauer's theory of the sublime (Schopenhauer 1974: 249). First, nature begins to withdraw from the individual will, so as to leave the subject utterly alone, and then it moves through increasingly more threatening states of hostility to the will. Schopenhauer's argument culminates in examples of what causes the most intense feeling of the sublime in nature: a great, rushing waterfall, the raging sea during a torrential storm. The key to experiencing the sublime in these settings, instead of fear, is to remain in pure subjectivity, rather than in service to the will (Schopenhauer 1974: 250).

I will now return to Crosby's six role-functional categories for religious ultimacy and pose the following question: does Schopenhauer's concept of the sublime meet enough of Crosby's criteria for religious ultimacy so as to deem it a religiously ultimate mode of experience (while not exactly a religious *object* of engagement)? An affirmative answer would further support my position that Schopenhauer's perspective can be read as a kind of religious naturalism. Similar to my analysis of the concept of aesthetic contemplation, the sublime initially seems to satisfy some, if not all, of Crosby's categories for religious ultimacy.

I will begin with Pervasiveness, for that seems to be the most obvious one captured by the feeling of the sublime. While Schopenhauer cites specific examples in the passages above of types of experiences that can give way to the feeling of the sublime, he is also careful to say that the feeling of sublimity is applicable to any object of perception. Sublimity is not limited to particular things in nature, but instead is a mode of experience that can come about at any time, so long as the right cognitive conditions are met. Thus, sublimity is not exclusive to any one thing, and in this sense, it pervades all things. The feeling of the sublime is without a doubt uniquely equipped to evoke a paradoxical balance of pleasure and pain in the individual who experiences it. There is no other kind of experience that enables this special mix of humility and significance, fear and courage, or powerlessness and potency. In this sense, it can be understood as not only having the quality of Uniqueness, but also Primacy, at least in Schopenhauer's personal view, because he places the sublime above even the beautiful for its ability to facilitate our understanding of our paradoxical relation to the world as will. We are at once utterly dependent on the world as will and subject to the laws of causality—we could be annihilated at any moment—but, insofar as we are also in part the will in itself, we have an immediate connection to its magnitude. The Primacy of the feeling of the sublime is therefore evident in that through the sublime we intuit the fundamental fact that we are at once the source of the world and utterly dependent upon it. Furthermore, like aesthetic contemplation, the feeling of the sublime is both mysterious and rare and, ultimately, words fail to adequately describe it. Thus, it would be fair to say that the sublime fulfills the requirement of Crosby's category of Hiddenness.

The final two categories in Crosby's schematic are Rightness and Permanence. Again, as with my analysis of aesthetic contemplation, it is most difficult to find resonance with these in Schopenhauer's concept of the sublime, primarily because the sublime itself is not an object and therefore a consideration of its permanency (or lack thereof) is almost nonsensical. Furthermore, any stretch beyond this view would be a misconstrual of Schopenhauer's philosophy. It is also dubious that Rightness is applicable to the concept of the sublime, at least in the sense that Crosby defines it. While the feeling of the sublime, like aesthetic contemplation, reveals specific cognitive content to the individual, it does not promote any overarching purpose of human existence in the sense of "spiritual progress." There is a sense in which the experience of the sublime influences character development in Schopenhauer's philosophy that is similar to a kind of spiritual progress, but even this does not suggest that there is a kind of goodness in the world with which the individual ought to seek alignment. However, since Schopenhauer's philosophical perspective is just that: a *perspective* and not so much a robustly positive metaphysical view, I would suggest that one focus on the way in which the feeling of the sublime affects the individual's quality of existence.

Concluding remarks

While Schopenhauer's metaphysical outlook is somewhat grim, his philosophy expresses a passionate reverence for the character of the world, which is sometimes manifest in sheer terror, sometimes in ecstasy, and other times in peaceful detachment. Despite this erratic relationship to the world, Schopenhauer never resorts to total indifference, nor does he adopt a merely pragmatic attitude about things. Throughout his philosophical project, he is committed to "getting to the bottom of things," not in order to prove his system logically consistent or *right*, but rather out of an earnest desire to know and understand the meaning of the world and our place in it (Schopenhauer 2014: 120). That Schopenhauer characterizes the conditions of the world as largely hostile to existence does not make his worldview anti-religious. If the pursuit of ultimate truth coupled with the practice of cultivating meaningful experiences can broadly be construed as a kind of religious project, then a close examination of Schopenhauer's philosophy can, at least, be understood as having a religious flavor. Moreover, a generous reading of Schopenhauer's aesthetic theory in relation to Crosby's concept of religious ultimacy provides a possible foundation for an interpretation of Schopenhauer's philosophy as not merely vaguely religious, but also as belonging to a school of religious naturalism that incorporates religious concepts without reliance upon theism of any sort.

Notes

1 Alternative translations of Schopenhaur's *Welt als Wille und Vorstellung* translate "*Vorstellung*" to representation. I think Richard E. Aquila's translation of the word to *presentation* is a purer interpretation of Schopenhauer's intent. As he comments, "the case for 'presentation'" vs. representation "goes hand in hand with the need to avoid the sense of *possession* generally attaching to possessive pronouns. More positively, the point is to promote what we take to be the central intention in Schopenhauer's use of the term: not possession by, but presentation of objects *to*, a cognizant subject" (Aquila's Introduction in Schopenhauer 2008, xiii).

2 See Crosby's *A Religion of Nature*, especially pages 159–170.

Bibliography

Atwell, J. (1996) "Art as Liberation: A Central Theme in Schopenhauer's Philosophy," in D. Jacquette (ed.) *Schopenhauer, Philosophy, and the Arts*, Cambridge: Cambridge University Press.

Cartwright, D. (2010) *Schopenhauer: A Biography*, Cambridge: Cambridge University Press.

Corrington, R. (1992) *Nature and Spirit: An Essay in Ecstatic Naturalism*, New York: Fordham University Press.

—— (2013) *Nature's Sublime: An Essay in Aesthetic Naturalism*, Lanham: Lexington Books.

Crosby, D. (2002) *A Religion of Nature*, Albany, NY: State University of New York Press.

Gardner, S. (1999) "Schopenhauer, Will, and the Unconscious," in C. Janaway (ed.) *The Cambridge Companion to Schopenhauer*, Cambridge: Cambridge University Press.

Kant, I. (1999) *Critique of Pure Reason*, trans. P. Guyer and A. W. Wood, Cambridge: Cambridge University Press.

—— (2000) *Critique of the Power of Judgment*, trans. P. Guyer and E. Matthews, Cambridge: Cambridge University Press.

Schopenhauer, A. (1974) *On the Fourfold Root of the Principle of Sufficient Reason*, trans. E. F. J. Payne, La Salle: Open Court.

—— (2008) *The World as Will and Presentation*, vol. 1, trans. R. E. Aquila in collaboration with D. Carus, New York: Pearson Longman.

—— (2010) *The Two Fundamental Problems of Ethics*, trans. D. Cartwright and E. Erdmann, Oxford: Oxford University Press.

—— (2011) *The World as Will and Presentation*, vol. 2 trans. D. Carus and R. Aquila, Boston: Prentice Hall.

—— (2014) *Parerga and Paralipomena: Short Philosophical Essays*, vol. 1, trans. S. Roehr and C. Janaway, Cambridge: Cambridge University Press.
Shapshay, S. (2009) "Poetic Intuition and the Bounds of Sense: Metaphor and Metonymy in Schopenhauer's Philosophy," in *Better Consciousness: Schopenhauer's Philosophy of Value*, Malden: Wiley-Blackwell.
Stone, J. (2008) *Religious Naturalism Today: The Rebirth of a Forgotten Alternative*, Albany, NY: State University of New York Press.
Wicks, R. (2008) *Schopenhauer*, Malden: Blackwell Publishing.

7

JASPERS'S PHILOSOPHICAL FAITH

Toward a form of religious naturalism

Nicholas J. Wernicki

At first glance, Karl Jaspers seems out of place on a list of early contributors to contemporary religious naturalism. Current scholarship on Jaspers's work is generally associated with his early psychopathology and his post-war political philosophy, both of which still maintain some relevance in niche academic circles today. Beyond these threads Jaspers is, in my view, widely under read and subsequently underappreciated by the academy at large, and it could be argued that he never had "his moment" of philosophical fame in comparison to his existentialist counterparts of twentieth-century France and Germany. Motivated by the raw and tumultuous socio-political climate of post-war Europe, Jaspers found himself as both philosopher and public intellectual responding to the role his homeland played in the Second World War.

His public persona was befitting in view of his overarching perspective that philosophy, or what Jaspers considered the active process of *philosophizing* "exists wherever thought brings men to an awareness of their existence" (Jaspers 2000: 125). This simple conception of philosophizing becomes the cornerstone of Jaspers's philosophical theology, leading to what Jaspers calls *philosophical faith*, which develops in stark contrast to and, ultimately, as a rejection of, faith in revelation. For Jaspers, philosophical faith is both an epistemology and a philosophy of communication grounded in the possibility of transcendence.[1] Through the concept of philosophical faith, Jaspers illuminates the limits of empirical judgments while acknowledging the transcendent quality of that which lies beyond them. Jaspers writes in his *Philosophical Autobiography*:

> The fundamental philosophical operation at all times is, more or less consciously, to transcend towards that out of which the objective as well as the thinking of the subject intending the objective arises. What is neither object nor act of thinking (subject), but contains both within itself I have called the Encompassing. This latter does not speak for itself either through the object or through the subject, but through both in one as that which is the Transcendence at one and the same time of consciousness as well as of Being.
> (Jaspers 1981: 73)

Put simply, philosophical faith retains an existential posture, concerned with the complexities of the self, while honoring the possibility of transcendence as ultimacy. In Jaspers's words: "Philosophical faith is the indispensable source of all genuine philosophizing. From it comes the striving of individuals in the world to experience and investigate the appearance of reality with the aim of attaining the reality of Transcendence ever more clearly" (Jaspers 1971: 89).

It is important to note that when Jaspers refers to the reality of transcendence he is not presupposing, or offering a proof of, something that is outside of nature. Rather, he is addressing a kind of epistemology of transcendence that differentiates philosophical faith from religious revelation. Jaspers uses the concept of hope as one of the many distinguishing features separating the two, which may help define philosophical faith as epistemological. For Jaspers, hope in the context of revealed religious faith is grounded in the objective actuality of a savior or a revealed promise, whereas philosophical faith is based on a truth grounded in philosophical reason. Paradoxically, I arrive at hope through the process of philosophical faith by rationally comprehending the truth and by discovering that the truth will never fully show itself. Yet, through the hope associated with philosophical faith, I am compelled to reach beyond the objectively knowable toward the ultimacy of my situation.

Jerome A. Stone points out in his text *Religious Naturalism Today: The Rebirth of a Forgotten Alternative* that religious naturalism has been experiencing a revival over the last 30 years. Along with this revival comes a renewed interest in some of the thinkers that make up the intellectual history of religious naturalism today. The broad genealogy of religious naturalism represents an amalgam of philosophers and theologians ranging from the American Transcendentalism movement and American Pragmatism to the Chicago theologians and process thinkers as well as some strains coming out of continental thought. Theories of nature can be traced back to the Ancients and the Pre-Socratic thinkers and through the Vedas of Hinduism.

However, the seedling of contemporary religious naturalism is most clearly recognized through Spinoza's philosophical treatment of *Natura Naturans* (Nature Naturing) and *Natura Naturata* (Nature Natured).[2] It is worth noting that Spinoza was the focus of one of Jaspers's volumes in his *The Great Philosophers* series and that in his "Philosophical Autobiography" he reminisces, "At the age of seventeen I read Spinoza. He became my philosopher" (Jaspers 1981: 7). In addition to the influence that Spinoza had on Jaspers's thinking, it is important to recognize that Jaspers's metaphysics is generally recognized as a reordering of Kant's transcendental idealism. This is most clearly represented in Jaspers's seven modes of the encompassing (*Existenz*, Transcendence, *Dasein*, Consciousness-as-such, spirit, world, and reason).

It is important to pay close attention to Jaspers's primary dialogue partners (Spinoza, Kant, Kierkegaard, and Nietzsche) in order to properly situate him in the lineage of early religious naturalists. Jaspers remains grounded in metaphysics while maintaining that philosophizing is a personal, existential pursuit, which for some seems incongruent at best. The challenge here is that, for some, metaphysics should remain separate from our historical situation. Jaspers overcomes this objection through what he calls *Existenz-philosophy*. He tasks himself with illuminating the encompassing by putting science, theology, and reason in productive tension with one another, thus honoring metaphysics in the context of our historical, existential situation. The encompassing for Jaspers is reality that shows itself always as a limit; it gives notice of itself and recedes at once. It is a subject for both metaphysical and existential inquiry. For Jaspers:

> The Encompassing is that in which all Being is for us. Said another way, it is the very condition under which Being becomes Being for us. It is not everything in the sense of the sum total of Being, but rather the whole – which remains open for us – as the ground of Being.
>
> (Jaspers 2000: 27)

Jaspers's use of the term "ground of Being" may raise some eyebrows for religious naturalists because he remains somewhat ambiguous about what this term implies. This usage might be especially problematic if Jaspers were to categorize this concept as ontologically distinct from all other realities, which he does not. Furthermore, if we maintain that Jaspers's mode of inquiry remains a limit epistemology, he is merely implying that philosophizing requires that inquiry remain in an open system. He is not objectifying the ground of Being in a way that presupposes a reality outside of or ontologically distinct from nature. In doing so, Jaspers attempts to move beyond ontology, which he recognizes as limited for probing into what objectively is. Jaspers extends traditional ontology to what he terms *periechontology*, which is a cognitive process of attempting to apprehend the unknowable by transcending the merely objectifiable.[3] For Jaspers traditional ontology "demonstrates what, in immanent thought, is *virtually* visible" while periechontology "makes palpable what, in transcending thought, is hit upon indirectly" (Jaspers 2000: 199). The palpability or indefinable awareness of possible transcendence is the primary focus of Jaspers's inquiry. It is the examination of the resulting non-knowledge, which opens us to the possibility of transcendence. This examination is the task of philosophical faith.

With the exception of a few contemporary thinkers, Robert S. Corrington for example, existential thinkers such as Jaspers, Heidegger and Tillich remain, at best, on the margins of current conversations on religious naturalism. My aim is to bring Karl Jaspers's *Existenz-Philosophy* more fully into the fold of contemporary religious naturalism through the engagement of his critically important and lesser-known text *Philosophical Faith and Revelation* and his overarching theory of the encompassing.

When I refer to naturalism I am referring to the assumption that – put simply – nature is all that there is.[4] This position rejects any metaphysical dualism that leaves the door open for a complex that is over, above, prior to, or outside of the limitlessness confines of nature; and, in my connotation, naturalism cannot support a reality subject to any kind of ontological priority.

I do however remain open to the possibility of transcendence through Jaspers's treatment of the encompassing. For Jaspers, transcendence is in relation to the encompassing of all encompassing while standing at the same time in rejection of *supranatura*. I am especially friendly to what Donald Crosby refers to as epistemological or noetic transcendence, which not only plays a powerful role in my conception of naturalism; it delineates my conception as a *religious* naturalism (Crosby 2003: 253).

In an effort to define what constitutes a religious disposition from those human and historical constructs that are merely important or relevant Crosby offers some "role functional categories" of religious naturalism. They are: Uniqueness, Primacy, Pervasiveness, Rightness, Permanence, and Hiddenness, each having an interrelated personal and cosmic side. Jaspers's *Existenz-philosophy* falls into a number of Crosby's religious categories but Jaspers's thinking is most active in the categories of Primacy and Hiddenness. Jaspers thinking is also grounded in the interrelatedness of the personal and the cosmic represented in various modes of the encompassing (*Existenz* and World).

When Crosby is referring to the personal side of Primacy, for example, he is referring to the idea that an object of interest is "not just one concern among many others but the ultimate concern of that person" (Crosby 2002: 119). Crosby, following Paul Tillich's lead, firmly posits the religious in the realm of the existential on the personal side of Primacy while also recognizing the cosmic side as ultimacy. Crosby's use of the term *object* here is used in the most generic sense to denote a focus of concern and not an object in the empirical sense. Crosby goes on to describe the personal side of Hiddenness as the individual confrontation with the object as being overpowered with mystery and awe much like the sublime. He describes the cosmic side this way:

> Hiddenness as a cosmic function points to the religious object as the most mysterious of all realities, impervious to full exploration because it is the secret wellspring of all that is, the primordial source of existence, meaning, and value for the cosmos as a whole.
>
> (Crosby 2002: 121)

Crosby's categories help frame Jaspers as a religious thinker. Jaspers's *Existenz*, the mode of the encompassing that I am, demonstrates its ultimate concern for its situation through its will to communicate and, in turn, become possible *Existenzen* in community. Furthermore, Jaspers's entire system of *Existenz-philosophy* is predicated on Crosby's treatment of his term Hiddenness. For Jaspers, the most mysterious of all realities lies beyond the boundaries of cognition of possible transcendence (the encompassing of all encompassing). Jaspers's seven modes of the encompassing are totalizing, interrelated, and representational of Crosby's categories of the cosmic and the personal. *Existenz*, Transcendence, *Dasein*, Consciousness-as-such, spirit, world, and reason are equally real although distinct in their varied features of immanence and transcendence. Most importantly they do not fall outside of the limitlessness of nature.

This position is highly contentious, and part of my aim is to argue that when Jaspers talks about the encompassing, he does so in a totalizing way that leaves room for an epistemological transcendence while remaining true to the premise that nature is all that there is. Jaspers's move beyond ontology to his method of periechontology is an attempt to illuminate that which falls outside or beyond what ontology can confirm by objective clarification, thus exposing the limits of ontology. Jaspers argues that periechontology is a mode of inquiry that asks about the objectively unknowable. I am not convinced that Jaspers would include a space for inquiry that is open to a reality other than nature, because this would require the objectification of the supernatural, which he could not reach through periechontology.

To be sure, if Jaspers is a naturalist his notion of the encompassing cannot surpass the limitlessness of nature. However, periechontology also restricts Jaspers from embracing nature through scientific dogmatism; his methodology requires an open system. According to Jaspers: "Nature ... becomes encompassing, unobjectifiable existence" (Jaspers 1967: 180). What he attempts to achieve through periechontology is a methodology that inquires into the possibility of the transcendent other. Periechontology makes no objective claims nor does it construct an edifice about the determinacy of Being. Further, it does not make explicit that the transcendent other is other than nature or some supernatural construct of transcendence.

It is important to note that *Existenz*, which constitutes one of the four modes of the encompassing that we are (*Existenz, Dasein*/existence, consciousness as such, and spirit/Geist), should not be merely conflated with our objective psychophysical existence or *Dasein*; it cannot be discovered as an object of knowledge and thus remains unknowable to us in totality. Rather, it is a mode of Being "in possibility" always qua possible *Existenz*. It is always in relation to and striving toward communication with other *Existenzen* over the meaning of Being-in-the-world and ultimately the possibility of transcendence. Through the process of philosophical faith, authentic *Existenz* unavoidably encounters a limit situation whereby reason fails and ciphers withdraw, leaving Existenz in an unsettled state of foundering or shipwreck. For Jaspers, to encounter possible *Existenz* is to experience a limit situation. Crosby offers a similar interpretation of what he calls cosmic hiddenness:

> it is exemplified by nature's sheer givenness; we cannot account for the fact of nature's existence or creative ongoingness. We can only explain particular aspects of the universe in terms of their relation to one another. The existence of nature must remain utterly inexplicable.
>
> (Crosby 2002: 125)

Is it possible that Crosby finds himself in a Jaspersian limit situation when confronted with the utter inexplicability of nature's existence? Crosby's characterization of nature signals the possibility that religious naturalism can be a kind of philosophical faith. Crosby's religious naturalism is a periechontologcal inquiry rather than a purely ontological one. It extends to the limitless boundaries of nature while making no claims about the possibility of the supernatural. The recognition of the unknowable does not necessitate an objective Other that is outside of nature. In fact, by creating metaphysics of something as undefinable as nature Jaspers would betray the very spirit of "hiddenness" upon which his periechontology is predicated. This leaves readers of Jaspers to come to their own conclusions about the relationship between the encompassing and nature by asking: how, beyond the objectifiable and immanent actualities of nature can we account for its sheer givenness? What can the process of cognition reveal about the fact that we can only comprehend certain aspects of the universe in terms of their relation to one other, as Crosby asks? These questions articulate that when confronted by the utterly inexplicable features of nature, foundering becomes an inescapable mode of being-before-nature.

For Jaspers, *Existenz* is always ensnared in a particular situation; it is a historical way of being-in-the-world. It is always situated in nature while confronted by the reality that nature, in its totality, is always also epistemologically out of reach. This fact implies that *Existenz* is always in a situation whereby nature is paradoxically present and other, illuminated and hidden at the same time. *Existenz* and nature are constantly and eternally at play in a game of periechontolocal "hide and go seek." Authentic *Existenz* through communication with other *Existenzen* is compelled to seek the hidden other even though there is no determined object to be found. It is only through possible transcendence that there is an awareness of that which is eternally and epistemological out of the grasp of *Existenz*.

Crosby's thinking helps delineate between two distinct yet interrelated aspects of nature, *experienced* nature and *conceptualized* nature. According to Crosby:

> Therefore our epistemologies and metaphysics must deal with two sorts or aspects of nature. One is *experienced* nature, and the other is *conceptualized* nature, the nature portrayed in postulated models, analogies, metaphors, and theories.
>
> (Crosby 2002: 25)

Both Crosby's and Jaspers's thinking is far too nuanced to make a blunt comparison and it would be plainly and problematically reductive to relate Crosby's theory of nature to Jaspers's theory of the encompassing although I hope that I have made some congruencies recognizable. Crosby is right to recognize a conceptualized category of nature because the hidden features of nature to which *Existenz* becomes potentially aware remain as limit situations. There would be no need for analogy and metaphor if all of nature were plainly knowable. Nature would just be plainly and objectively available in its totality.

Jaspers's conception of reality as the encompassing is bifurcated into the categories of the-being-that-we-are (Existenz, consciousness-as-such, spirit, reason, and Dasein) and the-being-that-ensnares-us (world, reason, and transcendence). Reason is noted in both categories because it is the bond between all modes of the encompassing, which further affirms the possibility of a naturalistic philosophy coming out of his *Existenz*-philosophy. However, world and transcendence can never be the objects of our inquiry. World is a totality through which we may encounter objectivity yet always extending beyond the objective. If we consider world analogous to Crosby's treatment of nature there may be some common ground on which Crosby and Jaspers both stand. For Jaspers, *Existenz* finds itself experientially in the horizon of the limitlessness of world or, to use Crosby's words, as experienced and conceptualized nature. According to Jaspers:

> When we cognize objectively the Encompassing that we are as human beings, it too becomes something different for us from the things in the world. Insofar as we are capable of investigating ourselves, we are ourselves taken up into this being-world which for us is the incomprehensive Other, i.e. nature.
>
> (Jaspers 2000:142)

The incomprehensible Other that is nature throws authentic *Existenz* into a state of foundering. In an effort to flee from the anxiety of the unknowable and the state of epistemological shipwreck *Dasein* is tempted to objectify nature through the conceptual means of metaphor and analogy. *I cannot grasp or encounter this or that reality of nature however I know it must be like this or that thing that can be objectifiable.* On the contrary, when *Existenz* is in full existential communication with other *Existenzen* it remains open to moments of possible epistemological transcendence through philosophical faith. *Existenz* does not objectify that which is encountered through foundering but also does not fall prey to revelation as a source of truth about the inaccessible reality that lies in transcendence.

Jaspers can be read as a naturalist in that he rejects revelation and any form of religious dogmatics, which are the primary sources of supernaturalism. It is easy to jump to the conclusion that this position is threatened by the very same polemic on which it is based. In other words it would seem that if he outright rejects supernaturalism as a possibility he betrays the very character of the limit situation that is embedded in *Existenz* by closing off possible transcendence to that which is unknowable. It is important to note that foundering does not imply an epistemology that is simply open to all possible truth claims about reality. Reason bonds the modes of the encompassing and fosters possible *Existenz*. Foundering does not lead to possible *Existenz* willy-nilly. Formulating an argument against that which is other than nature requires a metaphysics in which Jaspers does not participate. There are merely various encounters with the encompassing and the one totality that is encompassing of the encompassing. For Jaspers:

> The encompassing of Being-itself encompassed by the encompassing that we are is called world and Transcendence...We have no way to objectify the encompassing of Being. In the world we go in all directions, finding things we can know, ad infinitum. The world as a whole is neither comprehensible nor adequately conceivable; it is not an object of our knowledge, only an idea challenging our research. As for Transcendence, we do not explore this at all. We are touched by it, metaphorically speaking, and we touch it in turn – as the Other, the Encompassing of all encompassing.
>
> (Jaspers 1967: 69)

It is reasonable to read *Existenz*-philosophy as a rejection of supernaturalism if the supernatural is categorized as the objectifiable "thingness" that revealed religion makes it out to be. *Existenz* in relation to the limitlessness of nature does not necessitate the possibility that what lies beyond the limits of cognition could be supernatural. The encompassing of the encompassing, that which encompasses all other modes of the encompassing could not reconcile the notion that nature has an edge or contour that demarcates one reality and begins a supernatural Other without relying on a revealed epistemology. Foundering merely opens *Existenz* to the possibility of transcendence.

If Crosby is, in fact, a limit philosopher then his inquiry about nature and its ultimacy always results in a Jasperian shipwreck. For Crosby, it would seem, the confrontation with the utterly inexplicable character of nature is more personal and more meaningful than a mere metaphysical conundrum and foundering is an ever-present mode of philosophizing for him. This position locates the Jaspers's overarching system of periechontology in the realm of Crosby's religious naturalism. Crosby calls nature the ultimate and, as such, the ultimate hiddenness. This is reminiscent of Jaspers's earlier thinking from Volume II of *Philosophie*. He writes:

> Thus we react meaningfully to limit situations not by planning and calculation in order to overcome them, but by an entirely different sort of activity: namely, by *becoming the Existenz possible within us*. We become we ourselves by entering into the limit situations with eyes open. To experience limit situations and to be *Existenz* are one and the same. (Jaspers 2000: 97)

For the religious naturalist, metaphysics becomes religious naturalism when the encounter with nature becomes a limit situation. The result is existentialism working in concert with metaphysics rather than being estranged from it, which distinguishes Jaspers from most other thinkers coming out of the existentialist movement. Jaspers's *Existenz*-philosophy is rooted in this very marriage of existentialism and metaphysics, which supports the idea that the activity of religious naturalism is predicated on becoming the *Existenz* that we potentially are.

In 1962, Jaspers published *Der philosophische Glaube angesichts der Offenbarung*, later translated into English as *Philosophical Faith and Revelation*. In it he offers a rejection of revelational faith and forwards the methodology of philosophical faith to which I have been referring. In this text he grounds philosophical faith in the process through which *Existenz* approaches possible transcendence (the encompassing of all encompassing). Jaspers devotes a small but important section of *Philosophical Faith and Revelation* titled "The Modern Tripartition of Science, Philosophy and Theology" to an explanation of the limits of science and revelational faith in an attempt to clarify what he means by the act of philosophizing. This section also demonstrates how philosophy (philosophical faith), the "scientific attitude" (rational cognition), and theology (faith in revelation) each contribute to the process of philosophical faith.

Jaspers regards the objective facts revealed through scientific discovery as vital to philosophical faith; a move to which most religious naturalists would be friendly. However, he is quick to point out that the practice of philosophical faith illuminates the limits of science in that science is bound to objective verification. It meets a boundary situation that denies the possibility for knowledge beyond the scientifically verifiable and rejects the transcendent. The claim that "nature is all that there is" should not presume that the objectively verifiable is all that there is. Positivism, for Jaspers, is an ontological state of fleeing possible *Existenz*. It is the fleeting acknowledgement of shipwreck and the immediate retreat from possible transcendence. Theology, on the other hand, brings the idea or notion of transcendence to philosophical faith. However, theology is guilty of objectifying transcendence and, in certain circumstances, conflating it with supernaturalism through the authoritative dogmatism of supernatural revelation, which religious naturalism rejects outright.

If Jaspers's philosophical faith maintains the naturalistic tenets of contemporary religious naturalism while remaining open to the possibility of an epistemological transcendence, he may find a philosophical home among the early religious naturalists. There are many objections to framing Jaspers's *Existenz*-philosophy this way and the contemporary critiques of Jaspers's work in relation to religious naturalism are formidable. First and foremost Jaspers never forwards a theory of nature per se, and the concept of transcendence makes even the most forgiving naturalist suspicious. However, Jaspers's concept of philosophical faith might serve the contemporary religious naturalist well as an epistemology or even a practice of religious naturalism in that it honors scientific discovery and offers a language for the confrontation with the mysteries of nature's hiddenness.

I recognize *Existenz*-philosophy as a possible practice for religious naturalism because it is grounded in the dialectic between possible *Existenzen* in communication vis-à-vis community. Philosophical faith is a uniquely personal pursuit of that incomprehensible truth, which lives beyond the intelligible limits of cognition goading Dasein toward possible *Existenz*. Jaspers argues:

> Truth is that which brings about community. Furthermore, religion and philosophy agree that the merely intelligible creates only pseudo-communities based on what is known objectively. The intelligible, it is true, is the vehicle of community within the realm of the incomprehensible which the intelligible brings into an unending process of clarification.
>
> (Jaspers 2000: 75)

Through philosophical faith, Jaspers's intention is to probe beyond the intelligible into the realm of the incomprehensible, beyond the immanent toward possible transcendence.

This inquiry, he argues, should be personal and communal, and it should draw the empirical realities of metaphysics out of the sterile historical realm into the existential realm. *Existenz*-philosophy thus opens the self to the cipher script of transcendence. This philosophical disposition can be traced back to Jaspers's early psychopathology.

In his first major text *Allgemeine Psychopathologie* (1913), translated as *General Psychopathology*, he offers a nascent treatment of the encompassing in the context of psychopathology, which also sets the stage for his later theory of communication. Jaspers describes the phenomenological encounter with the *Existenzen* as an act of *Einfühlen, Verstehen* (empathy of understanding). I recognize this concept of comprehension as the groundwork for his theory of communication for his *Existenz*-philosophy. The concept of *Einfühlen, Verstehen* is an accurate description of communication between *Existenzen*, in community, specifically for *Existenzen* in a community of philosophical faith. Fritz Kauffman correctly points out:

> There are, thus, different shades of meaning, and qualitative leaps involved in the use of the word 'communication' as we ascend from nature to the personal level. But the inner connection of these different meanings can also be seen in religious experience ... Is it only our natural being – as Jaspers has it – that communicates quasi bodily with nature at large and is – by Einsfühlung connected with the very ground of all things?
>
> (Kauffman 1981: 248)

Interestingly, the concept of *Einfühlung* (empathy or vicarious introspection) and *Einsfühlung* (feeling one with the objective other) are derived from German aesthetics. Possible *Existenzen* yearn for the communication of other *Existenzen* through the reciprocal features of *Einfühlung*, thus bringing about religious community in the context of philosophical faith. When confronted with the aesthetic other that nature can be, it is *Einsfühlung* that draws possible *Existenz* into communication with nature vis-à-vis philosophical faith. In this context, philosophical faith is also an aesthetic philosophy. Under special conditions, possible *Existenz* finds itself in relation to nature through an encounter with the sublime.[5] What makes the sublime so unique in character is that it involves two opposing feelings at once. It seduces *Existenz* into utter shipwreck by exposing the limits of cognition while drawing *Existenz* closer to transcendence than in any other mode of the encompassing. For the religious naturalist the sublime may serve as a cipher of transcendence.

Broadly conceived, communication over the aesthetic sublime of nature is a source of foundering for *Existenzen*, thus becoming a religious experience. It seems this might be particularly true for the religious naturalist in philosophical faith. With this in mind it is appropriate to ask: is Jaspers's philosophical faith a religious naturalism?

The resurgence of religious naturalism may signal a revival in Jaspers's studies in a most unlikely corner of his thought due, in no small part, to the attention paid to him in the evolving school of Ecstatic Naturalism founded by Robert S. Corrington. In his early text on Ecstatic Naturalism, *Nature and Spirit*, Corrington relies on Jaspers to explain his concept of worldhood as "that side of nature that its most directly available to the human process ... nature is the potency that enables worldhood to prevail at all" (Corrington 1992: 22). Corrington's use of Jaspers here builds an important bridge between nature and worldhood, which I read as one of the modes of the encompassing in Jaspers's thinking. Corrington's Ecstatic Naturalism breathed life into the study of Karl Jaspers for many of the same reasons he has been overlooked for so long. In some areas of his work Corrington takes the modes of the encompassing seriously and remains open to the general spirit of transcendence, with the qualification that there is no possibility of overreach toward the supernatural. I see this position particularly relevant to Corrington's later Aesthetic Naturalism.

Corrington's work is, without exception, the clearest example of Jaspers's thinking reflected in religious naturalism today. While Corrington does not claim philosophical faith as a methodology, he honors the tension between finitude and transcendence as part of the human process. In this capacity he is, like Jaspers and Crosby, a limit philosopher. Corrington parts ways with Jaspers in that Corrington features a strict and thoroughgoing description of Nature, a description that has evolved along with his development of his system of Ecstatic Naturalism. However, the similarities between Jaspers and Corrington run deep and include the recognition that there can be a productive relationship between psychoanalysis, metaphysics, and phenomenology in the context of religious naturalism.

In addition, both thinkers demonstrate a healthy admiration for Spinoza and, for Corrington more than Jaspers, a deep appreciation of Schopenhauer. This is particularly evident in Corrington's aesthetic naturalism. In it he argues that community members (for Jaspers possible *Existenzen*) become ciphers of transcendence through what Schopenhauer refers to as *Mitleid*.[6] Corrington rightly points out that *Mitleid* is an essential feature for communication and, I would argue, a key social construct of all possible *Existenzen*. As noted, Jaspers recognizes the value of Schopenhauer's thinking, albeit to a lesser degree than Corrington does. In the closing pages of *Philosophical Faith and Revelation* Jaspers quotes Schopenhauer in response to the question of whether or not revelational faith and philosophical faith must inherently reject one another: "No one who is religious comes to philosophy," Schopenhauer writes; "he does not need it. No one who really philosophizes is religious; he walks without leading strings, dangerously, but in freedom" (Jaspers 1967: 360). In this passage Jaspers calls on Schopenhauer to make plain that revelational faith and philosophical faith can never engage in productive dialogue due to the way in which each is grounded in an uncompromising epistemology, revealed truth on one side and philosophizing on the other. Corrington would be friendly to this line of thinking not only because his system of thinking rejects supernatural revelation due to its obvious contradictions with his theory of nature. From a social philosophy perspective Corrington is also increasingly sensitive to the inevitable reality of tribalism that comes with putatively revealed truth. Corrington's reliance on Jaspers waxes and wanes throughout Corrington's oeuvre but is never too far out of sight.

Ecstatic Naturalism first developed in Corrington's 1992 book *Nature and Spirit* – grounded in the naturalism of Justus Buchler. Following Spinoza, Corrington produces a schema through which he categorizes two aspects of divine nature grounded in Spinoza's *Natura Naturata* and *Natura Naturans*. Since then, Corrington has published a flurry of articles and texts that advance his naturalism and that, for most, fit the criteria of religious naturalism. These texts include *Nature's Sublime: An Essay in Aesthetic Naturalism* (2013), *Deep Pantheism: Toward a New Transcendentalism* (2016), and, most recently, *Nature and Nothingness: An Essay in Ordinal Phenomenology*. The latter represents his most intimate dialogue with Jaspers on the encompassing. Jaspers is one of the few thinkers (along with Justus Buchler, Paul Tillich, and John Dewey) that remain steadfast as Corrington's sustained dialogue partners. This is specifically evident through Corrington's reliance on Jaspers's notions of the encompassing and transcendence.

There have been some important critiques of Corrington's work published within the last few years, some of which take him to task for his use of Jaspers in a naturalistic system. Examples of contemporary scholarship on Corrington's use of Jaspers include Nam T. Nguyen's *Nature's Primal Self: Peirce, Jaspers, Corrington*, Martin O. Yalcin's *Naturalism's Philosophy of the Sacred: Justus Buchler, Karl Jaspers and George Santayana*, and a collection of essays titled *A Philosophy of Sacred Nature: Prospects for Ecstatic Naturalism*, edited by Leon Niemoczynski and Nam T. Nguyen.

Nguyen offers an important criticism of Jaspers's use of *Existenz* due to the special status assigned to it in his *Existenz*-philosophy. Nguyen's primary challenge to Jaspers's thinking is that features of *Existenz* are so novel and so metaphysically unique that they may not fit into a naturalistic framework. Nguyen argues,

> from the standpoint of ecstatic naturalism's semiotic ontology, Jaspers's periechontology errs in privileging *Existenz* over nature. From the principle of ontological priority, Jaspers's human selfhood is more 'real' than nature; in other words, Jaspers, at true humanist, has elevated his *Existenz* to the highest metaphysical status.
>
> (Nguyen 2011: 192–193)

When Nguyen, referring back to Justus Buchler, refers to ontological priority, he is referring to a presumed binary ontological principle of ontological priority and ontological parity. Ontological priority suggests that some complexes can be more or less real than others, whereas ontological parity claims that all complexes are equally real. Nguyen is making the argument that Jaspers's *Existenz*-philosophy so honors the process of possible *Existenz* that it must be assigned a metaphysical status of the highest degree, thus falling prey to the metaphysical trappings of ontological priority. If this is true, Jaspers may well fall outside of the parameters of naturalism proper. Nguyen is correct in that Jaspers's contributions to psychopathology, existentialism, and philosophical theology are humanistic in scope, and in that he privileges *Existenz* as a primary mode of inquiry in terms of understanding what is true. In this context, as I have argued, Jaspers aims to transcend the sterility of metaphysical inquiry in favor of a philosophical faith grounded in communication between possible *Existenzen*.

Does Jaspers "care" more deeply about an epistemology that leads to possible transcendence than a metaphysics of nature? Nguyen argues that Jaspers privileges "human meaning horizons" over what Nguyen refers to as nature's primal self. Jaspers is indeed guilty of both taking greater care to develop an epistemology (philosophical faith) instead of a metaphysics of nature and of privileging human meaning horizons. In addition, Nguyen argues that "*Existenz* is a distinct form or mode of the encompassing; it has its own fundamental origin out of which it thinks and acts.... Existenz is elevated to the highest metaphysical status; it cannot be defined or illuminated" (Nguyen 2011: 218). This claim is far more precarious. Jaspers does, unsurprisingly, privilege *Existenz*, but I am not convicted that *Existenz*, as one of the modes of the encompassing, has any metaphysical priority whatsoever. More relevant or more relational is not analogous to more real.

There is, according to Jaspers, a multiplicity encountered within the modes of the encompassing; however, assigned degrees of relevance and importance are different than metaphysical status. It is important not to confuse the unique features of *Existenz* with ontological priority. *Existenz* remains a mode of the encompassing different from the other modes but likewise enveloped in the encompassing of the encompassing, which I read as one encompassing natural reality. It is through the reading of ciphers that possible *Existenz* is approached through the communication with other *Existenzen* (*Einsfühlung*) and through community with nature itself (*Einsfühlung*). Jaspers describes the meaning of ciphers in the following way. "Words come to have two meanings. Nature, sex, race, drives, etc. are facts; but at the same time words are now used differently, indefinably as indices. 'Nature', then, becomes encompassing unobjectifiable existence" (Jaspers 1967: 97). I read this passage as Jaspers affirming that, through the reading of ciphers, *Existenz* can only be open to those realities that make themselves available. His posture as a limit philosopher closes him off to the possibility of making a metaphysical claim about the objectifiable certainty of an ontological status in the way that Nguyen frames the problem. Jaspers can only refer to the encompassing of the encompassing and the various unique modes it envelops.

Referring back to my earlier claim that for the religious naturalist "nature is all that there is," can Jaspers be placed in the category of religious naturalism in light of Nguyen's critique? Nguyen's critique is critically important for understanding how we might treat *Existenz* in a system of naturalism. I remain optimistic that Jaspers's notion of the encompassing of the encompassing does not extend beyond the limitlessness of nature. Yalcin's text takes a friendlier position to Jaspers as a religious naturalist. He offers an important reminder that Jaspers dispenses with any truth grounded in revelation. In doing so, he greatly narrows the possibility of any supernatural reality in Jaspers's thinking. In the same breath Yalcin solidifies Jaspers as a religious thinker. He correctly points out

> For Jaspers philosophizing is a religious attitude toward life, an orientation that arises in the hidden depths of the self. Philosophy is not ontology – it does not provide a ready-made metaphysics about the nature of ultimate reality as found in positive or revealed religions.
>
> (Yalcin 2013: 67)

Yalcin makes the case here that Jaspers's *Existenz*-philosophy is not ontology, so there should be no expectation for a comprehensive theory of nature.

However, in order for Jaspers to be counted as an early contributor to contemporary religious naturalism he must, at least in spirit, adhere to the premise that nature is all that there is and thus reject supernaturalism outright. Yalcin offers an important retort to Nguyen's claim that Jaspers fails to maintain a line of thinking that remains within the limits of ontological parity. I am in agreement with Yalcin when he argues:

> Using Buchler's terminology, one can see that Jaspers employs ontological parity and ordinality to his advantage. Striking similarities between Jaspers and Buchler abound.

> For Jaspers, nothing in the world is more or less real, has more or less existence, or is more or less ultimate.
>
> (Yalcin 2013: 51)

Yalcin's reading of Jaspers posits him squarely in the current of early religious naturalists. Perhaps what is most important to glean from Yalcin's commentary is that what Jaspers was hoping to accomplish was an orientation about the hidden depths of the self and perhaps of nature as the one encompassing reality. Nguyen's warnings should be heeded. However, it is critical to keep Jaspers's overarching goals in mind when reading him through the lens of religious naturalism. What Jaspers has contributed to religious naturalism far outweighs the ambiguities present in his incomplete theory of nature. Jaspers's philosophical faith is an important contribution to the genealogy of contemporary religious naturalism in that it bridges the gap between Spinoza and Schopenhauer and contemporary religious naturalists such as Crosby and Corrington.

Notes

1 For a comprehensive treatment of Transcendence see Jaspers's *Philosophy Volume 3*. In relation to what Jaspers calls "world-being" transcendence is the absolute other. It is what Jaspers will later call the encompassing of all encompassing that lies beyond our limits of cognition. Jaspers uses the word Transcendence as a technical term that extends beyond the standard definition. He capitalizes it in his writing, specifically when he relates it to the encompassing, so the term will remain capitalized here.

2 See Robert Corrington's *Nature and Spirit* (1992) for a treatment of natura naturata (Nature Natured) and natura naturans (Nature Natured) in the context of religious naturalism.

3 For a helpful juxtaposition of traditional ontology and periechontology see part one of Jaspers's *Von Der Warheit*.

4 This concept of nature is derived from Robert S. Corrington's Ecstatic Naturalism.

5 Abigail T. Wernicki's chapter, *Will, Aesthetic Contemplation and the Sublime: Reading Schopenhauer as a Religious Naturalist*, offers an important treatment of the sublime in relation to religious naturalism.

6 *Mitleid* is best translated as compassion but I read it in close relation to the aforementioned treatment of empathy.

References

Corrington, R.S. (1992) *Nature and Spirit An Essay in Ecstatic Naturalism*, New York: Fordham University Press.

—— (2013) *Nature's Sublime: An Essay in Aesthetic Naturalism*, Lahnam, MD: Lexington Books.

Crosby, D.A. (2002) *A Religion of Nature*, Albany, NY: State University of New York Press.

—— (2003) "Transcendence and Immanence in a Religion of Nature," *American Journal of Theology & Philosophy*, 24(3), pp. 245–259.

Jaspers, K. (1962) *General Psychopathology*, trans. J. Hoenig and M.W. Hamilton, Manchester: Manchester University Press.

—— (1967) *Philosophical Faith and Revelation*, edited by R.N. Anshen, New York: Harper & Row Publishers.

—— (1970) *Philosophy Volume 2*, trans. E.B. Ashton, Chicago: University of Chicago Press.

—— (1971) *Philosophy of Existence*, trans. R.F. Grabau, Philadelphia, PA: University of Pennsylvania Press.

—— (1971) *Philosophy Volume 3*, trans E.B. Ashton, Chicago: University of Chicago Press.

—— (1981) "Philosophical Autobiography," in P.A. Schilpp (ed.) *The Philosophy of Karl Jaspers*, LaSalle, IL: Open Court Press, pp. 5–94.

—— (2000) *Karl Jaspers: Basic Philosophical Writings*, edited and trans. E. Ehrlich, L. Ehrlich and G.B. Pepper, Amherst, MA: Humanity Books.

Kauffman, F. (1981) "Karl Jaspers and A Philosophy of Communication," in P.A. Schilpp (ed.) *The Philosophy of Karl Jaspers*, LaSalle, IL: Open Court Press, pp. 210–296.

Nguyen, N.T. (2011) *Nature's Primal Self: Peirce, Jaspers and Corrington*, Lahnam, MD: Lexington Books.
Stone, J. A. (2008) *Religious Naturalism Today: The Rebirth of a Forgotten Alternative*, Albany, NY: State University Press of New York.
Thyssen, J. (1981) "The Concept of Foundering in Jaspers' Philosophy," in P.A. Schilpp (ed.) *The Philosophy of Karl Jaspers*, LaSalle, IL: Open Court Press, pp. 297–337.
Yalcin, M.O. (2013) *Naturalism's Philosophy of the Sacred: Justus Buchler, Karl Jaspers and George Santayana*, Lanham, MD: Lexington Books.

Further reading

R.S. Corrington, *Nature and Nothingness: An Essay in Ordinal Phenomenology* (Lanham MD: Lexington Books, 2017) is an extension of Corrington's ongoing system of Ecstatic Naturalism. Chapter 6, "Encompassing Nothingness" is an excellent example of contemporary religious naturalism in dialogue with Jaspers. L. Ehrlich, *Karl Jaspers: Philosophy as Faith* (Amherst, MA: University of Massachusetts Press, 1976) is a comprehensive explanation of Jaspers's notion of faith in the context of his philosophy. P.A. Schilpp (ed.) *The Philosophy of Karl Jaspers* (LaSalle, IL: Open Court Press, 1981) is part of the Library of Living Philosophers series. This is the complete collection of scholarship spanning all of Jaspers's philosophy, including chapters on the encompassing, foundering, transcendence, and the concept of ultimate situations. H. Wautischer, A.M. Olson, and G. Walters (eds) *Philosophical Faith and the Future of Humanity* (Media, BV: Springer Press, 2012) includes multiple chapters on philosophical faith including: "Reflections on Philosophical Faith in the 21st Century." H. Wautischer is the editor-in-chief of *"Existenz": An International Journal in Philosophy, Religion, Politics, and the Arts* devoted to the philosophy of Karl Jaspers broadly defined.

PART III

Pantheism, materialism, and the value-ladenness of nature

8

ECSTATIC NATURALISM AS DEEP PANTHEISM

Robert S. Corrington

Ecstatic naturalism is a form of religious naturalism that has multiple roots. It is a form of *naturalism* because it affirms that nature is all that there is. It is a form of *religious* naturalism because it affirms the place of the sacred *in* nature, but does not see nature per se as sacred, only certain orders and in certain respects. Its initial roots lie in my discovery of Ralph Waldo Emerson in my junior year of High School (1967), which was the same time that I discovered Advaita Vedanta Hinduism as found above all in the *Upanishads*. In each it became clear to me that one could be both a monist (nature is all that there is) and a pluralist (there are archetypes that project innumerable gods and goddesses). For Emerson and the unknown writers of the *Upanishads* there is a sense of wholeness combined with a sense of multiplicity. Thus, following William James we can say that there are many ones and many pluralistic centers.

The second encounter with a different kind of naturalism occurred during my undergraduate days as a philosophy major at Temple University in a class taught by the Charles Sanders Peirce and John Dewey scholar Philip Wiener. There I encountered the major pragmatists and was immediately drawn to various forms of naturalism. Dewey became important to me, as he continues to be, through his metaphysics that places the human process in and of nature along with his rejection of supernaturalism. Peirce's three categories of "firstness," "secondness," and "thirdness," applied both cosmologically and phenomenologically (his phaneroscopy), appealed to my sense of capaciousness *and* precision. From Dewey, I gained a sense that naturalism need not be reduced to simple causal materialism but could envision a much wider and more multi-layered nature. Further, his insistence that naturalism goes with a form of democratic socialism and creative forms of education satisfied my hunger for social justice. My continuing appreciation of Peirce led me to write a book on him in 1993.

The third encounter with naturalism, more loosely defined, occurred when I discovered the Martin Heidegger of *Sein und Zeit*. His transformation of Husserl's transcendental phenomenology into hermeneutic phenomenology paved the way for his sense of our thrownness into the world. By speaking of Being-in-the-world, he paralleled Dewey's 1925 *Experience and Nature*, as others have pointed out. While Heidegger had not studied Dewey, his special kind of naturalism augmented Dewey's slightly more austere version. Heidegger's later writings unfolded his own form of ecstatic naturalism, albeit with some serious moral problems in his life and writing.

The fourth source for my kind of naturalism came from an in-depth encounter with the Columbia naturalist Justus Buchler, whose 1966 *Metaphysics of Natural Complexes* ranks with Alfred North Whitehead's 1929 *Process and Reality* as one of the two great metaphysical systems in the Euro-American traditions of the twentieth century. While I believe that Buchler's system is much the superior and that Whitehead's panentheism is more like science fiction, I appreciate the boldness and sweep of both perspectives. I had the privilege of working with Buchler personally as an unofficial student and learned more about the craft of philosophizing than I had learned from anybody else, except perhaps Karl Jaspers. Buchler's concepts of ordinality, ontological parity, natural complexes, strong and weak relevance, and a host of others have become central to my delineations of the innumerable orders of nature. While Buchler did not focus on religion, I believe that his categories can help in the search for a more robust religious naturalism.

These sources became enriched by further readings in Jaspers and new readings in Buddhism. From Jaspers, the concept of the "Encompassing" (*das Umgreifende*) became for me an analogue to the idea that nature envelopes or encompasses all that there is. I continue to use this concept, especially in my twelfth book, *Nature and Nothingness: An Essay in Ordinal Phenomenology* (2017). From Buddhism, especially the Mahayana tradition as embodied in Nagarjuna, I learned of the power of nothingness *in* nature and as its "surround." I will delineate the four forms of nothingness in nature at the end of the essay. In all these influences, from Emerson, to Hinduism, to Dewey, to Jaspers, to Peirce, to Heidegger, and to Buddhism, two themes have been constant: nature is all that there is and it has sacred potencies *within* it.

The scientific backbone of ecstatic naturalism is the Neo-Darwinian synthesis and its elimination of purpose or teleology from the innumerable orders of nature. This gives a somber tone to my form of naturalism, but that is offset by a sense that ecstasies prevail (self-transcendence) in nature as well. Note that these forms of self-transcendence still take place on the plane of immanence and do not entail a supernatural escape hatch. An ecstatic moment means a standing outside of one's previous self if only for an instant. This will connect the self to nature in a deeper way. As will be seen, this notion requires an expanded idea of the unconscious that has a triad: the personal unconscious, the collective unconscious, and the unconscious of nature. The notion of an *unconscious of nature* is perhaps the most criticized of my concepts and I can see why its acceptance requires a great sea change in understanding, as it did for me.

My version of ecstatic naturalism is a project that first appeared in print in my 1987 book, *The Community of Interpreters* and has been followed by other books and some 80 articles articulating ecstatic naturalism. However, there are many antecedents to my form of this kind of naturalism, just as there are many other forms of naturalism. I divide the forms of naturalism into four varieties: 1) the descriptive, 2) the honorific, 3) the process, 4) and the ecstatic forms. The descriptive form emphasizes a more scientific flavor that still has a great place for nature that is all that there is. It has slight tendency toward materialism and efficient causality but may have a place for formal causality. Among its chief exemplars are Dewey, George Santayana, and Buchler. Note that its concept of "matter" is usually an expanded one. Honorific forms of naturalism place a stress on the spirit or creativity within nature, while still rejecting supernaturalism. Among its chief exemplars are Friedrich Schelling, Emerson, and Heidegger. Process forms of naturalism, if they truly are naturalistic, stress plural centers of awareness that are connected to an alleged web of internal relations. Time and creativity are elevated in status and there is usually a more optimistic tone and a slight softening of Darwinism. Among its chief exemplars are Whitehead, Teilhard de Chardin, Charles Hartshorne, and Robert Neville. Finally, ecstatic naturalism has deep roots in history, and my work is deeply indebted to them. As noted, it stresses the potencies within nature that bring ecstasies when they are encountered. Among its chief exemplars are Peirce, Paul Tillich, Ernst Bloch, Carl Jung, and Julia Kristeva. Of course, no form of naturalism has absolute boundaries and they often entwine with each other, requiring strenuous phenomenological efforts to untangle the threads. Thus, ecstatic naturalism has strands from the other forms of naturalism and is enriched and challenged by them.

Nature is all that there is

The central themes that run through all forms of naturalism are that nature is all that there is and there is no supernatural realm. Thus, nothing exists outside of nature that could have created it out of nothingness. Nature has always been and always will be, both prior to the Big Bang and after, whatever the fate of the universe of astro-physics turns out to be. Nature was neither created nor can it be destroyed. There is no time when it was not and no time beyond time when it will cease to prevail. The categories of "time" and "space" are intra-natural rather than being ultimate and super-creative in some process sense. There is no one trait that is found in every complex. Thus, ecstatic naturalism rejects the imperial notion that somehow each order of nature must boil down to some *one* thing. Chief candidates for this imperial move have been: matter, energy, monads, actual occasions, spirit, will, substance, internal relations, divine immanence (as in *previous* forms of pantheism), will-to-power, sense data, atoms, objects, proper names, forms, biomorphic energy, orgone, and a rich variety besides. This view insists that the world would not make sense if there were no single trait found in *every* order. For the aesthetic imagination, a plurality of traits, with no one thing in common, is hard to envision. We long to have nature tidy and explainable, without ambiguity and with limits that we can fathom to preserve our sense that we are somehow the center of it all. The above-mentioned concept of "ontological parity" insists that nothing is more real than anything else, just differently real. A building is neither more nor less real than a passing idea in my head, just real in a different way. This entails that there is no one trait that is somehow more real than any other trait.

For ecstatic naturalism, nature has no outer contour or center but prevails as innumerable orders in innumerable ways. All container analogies must be rejected as they compel thought to reduce nature to a shadow of itself. There is no ultimate order nor is nature a super organism. Any attempt to circumscribe it fails to grasp, and be grasped by, the infinity of nature. Nor is it like Peirce's continuum of all continua, which denies the real tears and breaks within nature. That is, there are non-relations as well as relations between and among orders (natural complexes). For some, especially those who believe that nature is a vast web of internal relations, the very concept of non-relation is difficult to accept and weave into one's metaphysics. But naturalism, religious or otherwise, must let go of the longing for total inter-connectedness.

To put even more pressure on our notions of nature, we must assert that there is no actual *referent* to the term nature. This may sound strange, but the following reflections might help in showing why it is so. For Aristotle, definitions entail locating a specific difference under a genus that is more encompassing in scope. For example, Socrates is an individual member of the genus humankind. He is located under the larger genus and, in a sense, receives his being from the genus. If you add all the genera together you still do not get to the indefinite infinity of nature, which can never be the ultimate genus or have a specific difference from something else. Ironically, to *define* nature, as *per impossibile*, entails positing a supernatural realm "outside" of nature, under which the specific difference of nature can be located. Thus, for naturalism, nature cannot be defined. I suspect, however, that this argument already appeals to naturalists, religious or otherwise. Nature's ever-unfolding capaciousness eludes all efforts to define it.

Following a strategy of Heidegger, we can cross out the reference idea all together. In his continuing efforts to crack open the Being problematic he crossed out the word *Sein* (and *Seyn*) to show that language cannot refer to the other side of the ontological difference between a thing in being and Being. Following this strategy, we can cross out the referent ~~nature~~ to indicate that it is "nothing" to which we could possibly refer. Instead of crossing out the word ~~nature~~ each time, we will put double quotes around it to signal that we are using the word in a strictly non-referential sense. We will not do this throughout, but the special use must always be kept in mind. If there is no referent, then, as we shall see, the concept of nothingness must be introduced into the problematic. This brings the religious aspect into naturalism and is a special dimension of ecstatic naturalism. Thus, in the end, as in the beginning, "nature" is no thing, i.e., nothingness.

From the above arguments, ecstatic naturalism, with its *deep* pantheism, rules out any form of process panentheism. A brief discussion is in order since so many religious naturalists feel comfortable listing themselves as process thinkers. Ecstatic naturalism moves in an opposite direction from panentheism and its surprising non-naturalism. My claims and further arguments will certainly not dishonor panentheism of the process variety, but will attempt to show some categorial weakness in the system, noting that process panentheism can take many forms.

There are several problems with Whitehead's system from an ecstatic naturalist perspective. We have argued that there can be no one trait found in or as each order of the innumerable orders of nature. To posit actual occasions/actual entities as the universal trait of nature is to impose an imperial construct *onto* nature. An actual occasion (and its society) exists in the web of internal relations that my perspective rejects. The model is well known; namely, that each occasion feels (prehends) every other just dead actual occasion in the universe either positively or negatively. The other occasions need to be just dead (concresced) so that they are fully what they chose to be. No moving subjectivity can remain. Thus, all perception is memory of the just past, not an immediate prehension into a self-transforming actual occasion. A positive prehension allows a trait to enter the actual occasion's subjective form and subjective aim. A negative prehension is a rejection of any form of ingression into the actual occasion. There is an initial aim from God who holds out a lure for each occasion leading it to a richer realm of experience and creativity.

So far we have two critiques of process panentheism: 1) the positing of one trait for the indefinitely complex orders of nature, and 2) the untenable belief in an unbroken webwork of internal relations. A relation is internal when it enters the subjectivity of the relata, while an external relation merely affects some trait in nature by various forms of causality, especially material and efficient. The God problematic further heightens the differences between process panentheism and ecstatic naturalism (as a form of religious naturalism). In the process version God exists as a primordial mind just on the nether side of nature. This mind is constituted by eternal and essential forms or essences, which have internal causal efficacy through their luring effect on actual occasions (and their societies). This introduces final causality into the system that puts pressure on the Neo-Darwinian synthesis. Also, the realm of essences constitutes a sphere of formal causality. The latter form of causality is being rescued in subdued versions as it can help explain the drive for form in organisms and the human process. It has also proven to be especially useful in Gestalt theories and frameworks, whether applied to art or to the self-in-process.

The consequent dimension of God physically prehends all solidified (dead) actual occasions and remembers them for eternity via superjection. Thus, we have two more critiques of process panentheism: 3) the privileging of final causality and 4) the positing of two modes of Deity, one of which transcends nature—a tenet of pan*en*theism, which affirms that God is both in and beyond nature.

There is at least one more critique of process panentheism. It was created to avoid the extremes of patriarchal theism (especially in the Western monotheisms) and a pantheism that simply equates God with nature. Clearly, theism and a simplistic pantheism are not philosophically viable: the former because it not only violates the principle of ontological parity insofar as its God is held to be more real than any subaltern orders, but it also privileges culturally shaped male traits. The latter is not viable because, as Arthur Schopenhauer argued, if you have two equivalent terms, then one is irrelevant and should be dropped; thus one must use either nature or God, but not both. We conclude with a final critique: 5) process panentheism fails to articulate a possibility between the two extremes it is avoiding; namely, that pantheism can be resurrected in a radical new form that does not equate God with nature but sees the sacred within nature in certain orders, but not all. I call this new form "deep pantheism" insofar as it uses a very different categorial structure that places a strong emphasis on the unconscious of nature.

The two methods of ecstatic naturalism: ordinal phenomenology and ordinal psychoanalysis

For Peirce, the concept and use of method was an all-consuming concern. He developed his own phenomenology to probe into the realities of perception and of the larger world. Further, he developed a unique logic of relatives to augment and transform traditional syllogistic logic. Ecstatic naturalism has developed two specific kinds of method to probe into the realms of "nature" and into the deepest momenta and structures of the human process. In both cases the concept of *ordinality* is basic to the structure and power that transform these two traditional methods. The first method radically shifts the focus of the Edmund Husserl/Martin Heidegger trajectory in phenomenology toward a multi-layered form of query that insists that phenomenology must go all the way down and that it must allow the world to grasp *us* as we struggle to grasp *it*. The second method transforms traditional psychoanalysis, moving it toward a more robust encounter between the self-in-process and its sheer locatedness in the innumerable orders of the world—what I call the Selving process. It is decidedly post-Freudian and works hard to remove the patriarchal elements from the early psychoanalytic movement centered in Vienna. Two key dialogue partners in this process are C.G. Jung and Otto Rank. Jung is important for his discovery of the archetypes and the collective unconscious, which has roots in the unconscious of nature, while Rank is important for his post-patriarchal focus on the mother/child relationship and the life-long problems of the birth trauma. Further, like Jung, Rank has a central role for art and the creative process wherein the genius (cultural creative) both creates great works and creates a great personality in that process.

Ordinal phenomenology locates all objects (natural complexes) within the infinite momenta and structures of the world, and probes, as far as is possible, into the depth structures of nature. For Husserl, for example, we take an object, usually spatial/temporal, and rotate it through as many perspectives as possible, transforming the hidden co-given into the unhidden given. It is a process of shadowing that rotates around the phenomenon and paints a portrait of it that can be presented to other phenomenologists. Husserl combines this idea with a near obsession with certainty as he seeks the Evident. This quest for certainty and an alleged pure and unambiguous showing of the phenomenon is firmly rejected by ordinal phenomenology, which goes beyond the spatial/temporal phenomenon and acknowledges both our finitude as working phenomenologists and that ontological ambiguity goes all the way down.

For ordinal phenomenology, one must probe into as many orders of the phenomenon as is humanly possible, given time and energy. This is a communal process insisting on interpretive input from a team of phenomenologists working on the same project. The goal is to render as many unhidden traits of the various (innumerable) ordinal locations of the phenomenon as possible. Here we use the word "phenomenon" rather than natural complex because the former term has played such a key role in the history of phenomenology, but the richer concept of "natural complex" must be held in mind. The exact procedure is as follows: one must begin the process of slowly rotating the phenomenon through both its spatial/temporal traits (if they apply) and then move onto its larger relationality. An example will help. Take the instance of a scholarly monograph that has recently been published. Husserl would stop at its given and co-given traits within a narrowly located band of the world. The ordinal approach is different.

First, the book is indeed a spatial/temporal phenomenon that has easily accessible traits such as size, weight, coloring, type fonts, front and rear covers, and a binding of varying degrees of quality. But one must not stop here. The book is multiply located in orders that may or may not be spatial/temporal. The following *abbreviated* list is of other genuine ordinal locations that can be probed by ordinal phenomenology: 1) the book exists in the minds of the writer and editor; 2) in the mind of the reviewers; 3) in the perhaps fragile economy of the publisher; 4) in the ecosystem from which its paper material was harvested; 5) in the sphere of trucking and whether its driver was union or non-union; 6) in the economy and spatial arrangements of the warehouse; 7) as a felt sense of satisfaction (or not) in the mind of the author; 8) in the minds and lives of its readers, perhaps changing them in dramatic or important ways; 9) as a social/political event within the communities for which it was written; and 10) in the history of its genre. Obviously, this is an abbreviated list that could be expanded indefinitely, but it gives some idea of the power and scope of ordinal phenomenology. It can probe into spatial/temporal phenomena or non-spatial/temporal traits within the same phenomenon, or into phenomena that don't have spatial/temporal traits, such as the rules of games.

Ordinal *psychoanalysis* is slightly different in that it confines itself to the ordinal locations, conscious and unconscious, of the human process. Yet, as noted, the human process is fully rooted in nature from whence it has come and into which it will return; it cannot be outside of nature in any sense during its life trajectory, or, perhaps beyond. This new form of psychoanalysis does not privilege the conscious or the unconscious. Both are equally real in the ways that they are real. Further, their interaction can be partially examined and rendered unhidden, but not through a rigid topology such as that of id, ego, and superego. There is such fluidity in the conscious/unconscious dialectic that any proposed topology much be used in a loose hermeneutic fashion rather than as a container structure that claims to know all things about the psyche. The psyche has an uncanny way of overthrowing our topologies, no matter how refined.

The method of ordinal psychoanalysis goes beyond the rigid patriarchal systems of the past to embrace the pre-Oedipal connection between the mother and child and its correlation to attachment theory. Further, it probes into the maternal per se, from the biological birthing mother to the archetype of the Great Mother, who represents the oldest naturalist religion in human history, going back to the Paleolithic era. In fact, ordinal psychoanalysis is linked to theories about the world's great myths and with how they have enhanced or demoted religious naturalism, ecstatic or otherwise.

The work of Jung and Rank is augmented, and sometimes challenged, by objects relation theorists (e.g., Melanie Klein, Heinz Kohut, and Donald Winnicott) who continue to stress the maternal/child infant relation but drop the birth trauma theory, which I consider to be a mistake. Yet they have pushed beyond earlier reflections on bonding with the maternal. Kohut stresses that there are "self objects" that can substitute for the mother/infant relationship, especially if the attachment between them was weak. The failure of the caregiver to provide deep connection and affirmation can have dire consequences in the future trajectory of that infant. I believe that self objects, like transitional objects (teddy bears, blankets, etc.) are manifestations of the ongoing effects of the birth trauma and provide some minimal relief for the young child and the adult. They stand in for the biological and cosmic wombs, both interconnected. Any object has the potential to be a self object, such as a person, a religion, a work of art, a historical event (however mythologized), gardening, special rocks and crystals, and so on.

Ordinal phenomenology and ordinal psychoanalysis work together to give the most complete picture of the way of the orders of nature and of that order known as the human process. The methods require each other in a deep way. The reason is that ordinal psychoanalysis cannot attempt to locate the human process within the one nature that there is without a larger phenomenological probe into the innumerable orders of the world that surround and ground the self-in-process. In the other direction, ordinal psychoanalysis opens up the unconscious of the self and of nature in such a way that they can be further probed phenomenologically and, as always, be probed by phenomenology.

Nature naturing and nature natured

Perhaps no distinction is as basic to ecstatic (religious) naturalism as that between the two modes of self-fissuring *within* the one "nature" that there is. These terms have a rich provenance from the medieval period to the present. While Spinoza has been given the credit for the use of these co-implicated terms, many others have used them to great effect. Ecstatic naturalism uses these terms in new ways. Nature naturing (*natura naturans*) is herein defined as: "Nature perennially creating itself out of itself alone." The use of the word "perennially" is used to signal that there is no notion of eternity outside of time and the idea that nature continually creates itself out of itself rules out any notion of creation out of nothingness. Nature natured (*natura naturata*) is herein defined as "The innumerable orders of the World without any collective contour." Note that the Latin term "natura" is feminine, which ties in with our themes from ordinal psychoanalysis.

Clearly, the notion of nature naturing is more elusive to examine than are the realms of nature natured. The later term denotes any order within nature that is both co-hidden and hidden but is on the surface, as it were. Thus, we can communally probe into the orders of the World but, as noted, can find no contour or outside for them. There are just innumerable orders without any exclusive web of internal relations. No order of the World connects with all other orders, thus entailing, as noted, that there are genuine breaks in continua. There are also strong semiotic reasons why this is so, as I demonstrate in my 2000 book, *A Semiotic Theory of Theology and Philosophy*. There, among other things, I describe four types of semiosis that exist in the World and the limits and roles of each.

Nature naturing is more radical in momenta and structures. It represents and is the unconscious of nature and is manifest from out of the dark depths of "nature." Here we go to the very edges of ordinal phenomenology and ordinal psychoanalysis. Insofar as nature naturing is largely hidden, it is far less available to probing by finite mortals, even with communal support. However, all is not lost, as there are strong and weak manifestations of the unconscious of nature, especially as it erupts into the personal and collective unconscious. As best as I have uncovered thus far, nature naturing is manifest by innumerable pulsations that may or may not have a collective integrity. Such pulsations are prior to the distinctions between good and evil (or Nietzsche's good and bad). These pulsations seem to have two modes of manifestation. One is the eruptions that generate archetypes that straddle the divide between nature naturing and nature natured, while the other is a gentler manifestation of an ongoing series of emanations upon emanations in Emerson's sense. These emanations have neither beginning nor end and do not come from a Plotinian One. They are smaller, as it were, than what Neo-Platonism describes.

The ejections from nature naturing are thus the focus of more attention as they come from deeper down in "nature" and have a greater force and velocity than emanations. For some, of course, such talk of ejecting archetypes from an alleged unconsciousness of nature seems far from any reasonable form of religious naturalism. Many religious naturalists have a deep suspicion of the use of psychoanalysis in any case, let alone as one of two methods to probe into and be probed by the uncanny depths of nature naturing. But as I will attempt to show, naturalism cannot be truly religious if it doesn't take account of the seemingly bizarre fissuring at the heart of "nature." Again, there are not two separate natures but one nature in two modes. Aquinas was tempted to simply equate nature naturing with God, while deep pantheism refuses to do so. Without an unconscious of nature in the picture, one is at a loss to explain the emergence of archetypes (forms that combine power and meaning) or the myriad, but never closed, endless series of emanations that assure nature's fecundity. Any use of Occam's razor here (a bloody and clumsy "method" at best) would violate the phenomenological and psychoanalytic evidence and reduce nature to a bland caricature of itself. Nature is complexity and fecundity all the way down and all the way up, and this endless complexity (and ambiguity) must be dealt with, no matter how much energy it requires—for query is unending. Further, as argued by Dewey, the very quest for certainty has been a curse throughout human history, especially as allied with an obsession with simple and clear-cut categorial structures.

Archetypes are not only found in the collective unconscious of the human process, but are manifest throughout nature. For Joseph Campbell, they are manifest as both instinct and ideation. As instinct, the archetype is a triggering potency that enables innate release mechanisms to function, while as ideation, the archetype is strongly manifest in human dreams, world and local religions, global art, and psychopathology, which often parallels global ideations, that is, has the same roots. Hence, for example, the delusions and hallucinations in manic-depression, schizophrenia, and schizoaffective disorder, are directly parallel to major themes in global culture and are intertwined with them. On the other side, the manifest archetypal images in culture are a form of collective dreaming or even collective psychosis, as in fascist states. The dream of Armageddon, for example, is a pathological twisting of the archetype of growth and transformation. Yet this bizarre fantasy has grasped millions and threatens to undo democratic structures that use amelioration and care rather than some quick and dirty apocalypse. This communal fact shows that the archetypes must be wrestled with and integrated with care.

Nature naturing, then is partly the birthing ground for those archetypal determinants that shape both "nature" and its specific order—the human process. If archetypes are not taken seriously, then havoc could result. Note again that nature naturing and its "products" exists/prevails prior to the distinction between good and evil. It is solely up to human individuals and communities to struggle relentlessly to transfigure the potential evil into the good, however ambiguous that good may be. While intense suffering exists in nature, evil only exists in the human process. But such evil cannot be eliminated, as *per impossibile* by use of reason or argument, but only by fearlessly entering the powers and potencies as they are, not as we would wish them to be.

Fourfold nothingness

Several times we have noted/argued that "nature" has no outer contour or shape-of-shapes that would somehow limit "it" and contrast it to something other. There is nothing outside of nature; indeed, we have hinted that the "surround" for nature is nothingness. The concept of nothingness (*sunyata*) has strong family resemblances to Jaspers's concept of the Encompassing (*das Umgreifende*), also translated as the Enveloping. I suspect that many religious naturalists are at this point ready to jump ship, but I will argue that the concept of nothingness is entailed in naturalism per se. For if the word "nature" has no referent, then just "what" is "it?" It can only be nothingness, albeit in a fourfold sense. The underlying concept here is that of "nihilation," in the robust sense of nothingness "doing" something. Heidegger's brilliant use of his often ridiculed phrase: "The nothing nothings" (*das Nicht nichtet*), is phenomenologically astute and psychoanalytically rich. It points out that nothingness is an event that nothings or nihilates. It cannot be a thing or Being-itself. It neither prevails nor is hostile to what prevails. But its nihilating events make nature naturing and nature natured possible.

I have suggested that the ordinal phenomenological evidence points to four forms of nihilating nothingness "in" nature. Of course, they are all part of the "one" nothingness and not separate realms, even if each realm can be delineated and partially described on "its" own. As we shall see, some realms call forth a human response that exists on a spectrum from total rejection to mindful acceptance. The axiological goal of ecstatic naturalism (and deep pantheism) is to help the Selving (individuation) process move from the former to the latter mode of relation.

The first modality of nihilating nothingness is what I call "holes in nature," that are more finite moments of emptiness that punctuate both nature and the Selving process. For example, these holes can appear as death, loss, anxiety per se, blinding self-discovery of one's shadow dimension, sexual failure, political conflict, natural disaster, and mental illness, to name just a few instances. These holes, as like miniature black holes in astronomy, draw metaphorical light into them, thus deflecting the light-filled Selving process into what seems a yawning abyss. More often, one's reaction to encountering modes of nihilation is to experience anxiety, fear of fragmentation, and loss of identity structure. This generalized ontological anxiety can recoil back on the self as it seeks reconstituted identity. The problem with this type of recoil is that it negates the negation by forming a rigid and highly armored self that believes it is immune from nihilation per se.

The extreme example of this negation of the negation is found in born again experiences wherein the so-called "old" self is firmly negated/abjected along with its openness to nothingness. This produces innumerable problems for psychoanalysis and the democratic social systems. In the end, it devolves into a naked will-to-power that wishes to conquer anything that might remind one of the "old" self and its associated memories. The rigid fundamentalist communities that come out of this abjection and a continual negation of the negation are among the more dangerous forces in communal and cultural life. Herein lies the origin of barbarism within the self and within the community.

This abjection of the old self in the face of the holes in nature immediately translates into an abjection of nature. This requires for its continuance a patriarchal theistic system that makes nature "less real" than the self or the divinity, thus violating the principle of ontological parity. The abjection of nature (*natura naturata* and *natura naturans*) can only portend tragedy for the ecology of the planet, as its dramatically lessened ontological status makes it of little concern for immediate corrective action on global warming and its human disasters.

However, at the opposite extreme and the one less often attempted, one can face a hole in nature with a mode of mindfulness that lets the nihilating event open out and richly transform the self. Instead of attempting to negate the negation by a manic and armored "personality," one lets the event of nihilation clear away armoring and any detritus that stands in the way of the Selving process. Mindfulness, as attentiveness to the fecundity of nature and the depths of the self, gives the Selving process its method for plumbing the depths of nature, while being fully engaged in the spiritual practice of ontological parity. For the self of mindfulness, instead of the one of armoring, nature is resplendent and fully real in the myriad ways in which it is real.

The second mode of nihilation is what I call "totalizing nothingness" that is best experienced in moments of horror. For the Selving process, this encounter can become one step removed in literature or cinema so that cathartic possibilities remain at the end of the tale or film. Totalizing nothingness is experienced when horror puts the entire world, and its normative structures, into question. This negation is absolute and seemingly inescapable. Horror's main effect/function is to show the absolute indifference of nature to its foundlings. In the confrontation with mighty powers that dwarf the human or the power of religion, the Selving process experiences its total littleness in the face of the seeming nihilating void that ensnares the self in gloom and terror.

As before, the Selving process has two choices (on a continuum). Either it can remain ensnared in the utter abyss of the uncanny or it can learn to accept the indifference of nature and embrace it in a freeing way. In the latter case, there is something liberating about letting nature be the non-telic reality and potency that it is. This moment of a higher mindfulness can be called sublime, to replace religious language with aesthetic nomenclature. The sublime occurs on the edges of "nature" through its infinitizing potency to shatter all local meaning horizons. The two final forms of nothingness are also encountered as the sublime, which goes beyond and envelopes beauty, but without negating or rejecting it.

The third mode of nihilation is that of "naturing nothingness." Here the ordinal phenomenological evidence points to the correlation of the infinite unconscious of "nature," i.e., *natura naturans*, with the mobile and expanding horizonality of the Selving process. The collective unconscious of the Selving process embraces the unconscious of nature. The archetypes become activated in earnest as the overwhelming and vast powers that they are. As noted, one cannot encounter an archetype *an sich*, but only its manifest images and instincts. The archetypes are the ultimate determinates of nature and the human process. Above all, they are encountered in art and grand scientific theories. This mode of nihilating nothingness is especially correlated to the mobile and seemingly chaotic unconscious of nature, which has roots that go deeper down than ordinal phenomenology can go. At this point, one needs to introduce an abduction (retroduction) in Peirce's sense; namely, a logical/methodical process in which one makes a conceptual leap from the conditions of the observed (using deduction and induction) to the posited necessary conditions "behind" the observed. Peirce called this a method of "guessing," but it is impossible to function without such procedures in building a conceptual architectonic. This produces an a priori, but not in Kant's rigid transcendental sense. Rather, the abduction generates a "pragmatic a priori" that is self-adjustable as the deductive and inductive structures shift.

Finally, the fourth form of nihilation is the "encompassing nothingness" that has best been articulated in Mahayana Buddhism and by Jaspers. Hints have been made about this ultimate manifestation of nothingness (*sunyata*). It is the non-container "surround" for nature and is the final goal of ecstatic naturalism as a form of religious naturalism. Very little can be said about it, if anything at all, but as language users we struggle to do so in a way analogous to any "discussion" of Peirce's "firstness." This goes directly back to our argument that the word nature has no referent and is thus, by implication, nothingness. "Nature" and nothingness are the Same. The encompassing nothingness, which "contains" the other modes of nihilation," is the pre-linguistic fore-structure for anything and everything in whatever way or mode that they "are."

References

Buchler, J. (1966) *Metaphysics of Natural Complexes*, New York: Columbia.
Campbell, J. (1949) *A Hero with a Thousand Faces*, Princeton, NJ: Bollingen.
Corrington, R.S. (1987) *The Community of Interpreters*, Macon, GA: Mercer.
—— (1992) *Nature and Spirit: An Essay in Ecstatic Naturalism*, New York: Fordham.

—— (1993) *An Introduction to C.S. Peirce*, Lanham, MD: Rowman & Littlefield.
—— (1994) *Ecstatic Naturalism: Signs of the World*, Bloomington, IN: Indiana University Press.
—— (1996) *Nature's Self: Our Journey from Origin to Spirit*, Lanham, MD: Rowman & Littlefield.
—— (1997) *Nature's Religion*, Lanham, MD: Rowman & Littlefield.
—— (2000) *A Semiotic Theory of Theology and Philosophy*, Cambridge: Cambridge University Press.
—— (2003) *Wilhelm Reich: Psychoanalyst and Radical Naturalist*, New York: FS&G.
—— (2003) *Riding the Windhorse*, Lanham, MD: Hamilton.
—— (2013) *Nature's Sublime: An Essay in Aesthetic Naturalism*, Lanham, MD: Lexington.
—— (2016) *Deep Pantheism: Toward a New Transcendentalism*, Lanham, MD: Lexington.
—— (2017) *Nature and Nothingness: An Essay in Ordinal Phenomenology*, Lanham, MD: Lexington.
Dewey, J. (1925) *Experience and Nature*, Carbondale, IL: Southern Illinois Press.
Emerson, R.W. (1981) *Essays and Lectures*, New York: Library of America.
Hartshorne, C. (1984) *Omnipotence and Other Theological Mistakes*, Albany, NY: State University of New York Press.
Heidegger, M. (1958) *The Question of Being*, New Haven, CT: College and University Press.
—— (2010) Being and Time, Albany, NY: State University of New York Press.
Jaspers, K. (1986) *Basic Philosophical Writings*, Athens, OH: Ohio University Press.
Jung, C.G. (2013) ed. Storr, A. *The Essential Jung*, Princeton, NJ: Princeton University Press.
Kohut, H. (1977) *The Restoration of the Self*, Chicago: University of Chicago Press.
Kristeva, J. (1991) *Revolution in Poetic Language*, New York: Columbia University Press.
Neville, R. (2013) *Ultimates*, Albany, NY: State University of New York Press.
Peirce, C.S. (1998) *The Essential Peirce, Vols. I & II*, Bloomington: Indiana University Press.
Rank, O. (1968) *Art and Artist*, New York: W.W. Norton.
—— (1993) *The Trauma of Birth*, New York: Dover Publications.
—— (1998) *Psychology and the Soul*, Baltimore: Johns Hopkins.
Schopenhauer, A. (2008) *The World as Will and Presentation*, New York: Pearson Longman.
Whitehead, A.N. (1978) *Process and Reality: Corrected Edition*, New York: Free Press.

Further reading

Crosby, D. (2015) *Nature as Sacred Ground*, Albany, NY: State University of New York Press.
Goodenough, U. (2000) *The Sacred Depths of Nature*, Oxford: Oxford University Press.
Nguyen, N. (2011) *Nature's Primal Self*, Lanham, MD: Lexington.
Niemoczynski, L. and Nguyen, N. eds. (2015) *A Philosophy of Sacred Nature*, Lanham, MD: Lexington.
Nishitani, K. (1982) *Religion and Nothingness*, Berkeley, CA: University of California Press.
Yalcin, M.O. (2013) *Naturalism's Philosophy of the Sacred*, Lanham, MD: Lexington.

9

DEUS SIVE NATURA

Pantheism as a variety of religious naturalism

Demian Wheeler

In his controversial bestseller, *The God Delusion*, evolutionary biologist Richard Dawkins wryly observes that pantheism is just "sexed-up atheism" (Dawkins 2006: 40). The burden of the present essay is to demonstrate that pantheism can also be meaningfully construed as a variety of *religious naturalism*.

Defining religious naturalism

Simply defined, "religious naturalism" refers to a broad family of philosophical and theological perspectives that regards nature as both exhaustive of reality and worthy of deep reverence and devotion. Religious naturalism is, on the one hand, a form of naturalism, which stipulates that nature is "metaphysically ultimate" (Crosby 2002). What there is, and all there is, is nature; there is nothing above, behind, or beyond the natural. As monists, naturalists are polemically antisupernaturalistic and antisupranaturalistic (Stone 2008: 2; Wildman 2009: 20–25), excluding all extranatural realms and entities (including disembodied divine agents) from their "ontological inventory" (Wildman 2014a: 41–43).

On the other hand, religious naturalists venture that living a religiously fulfilling existence on a fully naturalistic basis—i.e., without supernaturalism or supranaturalism—is both possible and desirable (Stone 2018). Essentially, religious naturalism is a post-Enlightenment, but spiritually potent, worldview that enables one to respond—religiously, ethically, and theologically—to the perplexity, splendor, and power of *this* world. That is, religious naturalism recognizes not only the metaphysical ultimacy of nature but also the *religious* ultimacy of nature. To deem something religiously ultimate is to hold it *sacred*—i.e., vitally and centrally important and, therefore, deserving of our utmost loyalty and faith (Crosby 2002: 114; Rue 2011: 110–111). The religious naturalist speaks of the "sacred depths" of nature, which, according to Wildman, lies in nature's "self-transcendent potential" as well as "in its beauty, terror, scale, stochasticity, emergent complexity, and evolutionary development" (Wildman 2014a: 41). Clearly, then, the rejection of supernaturalism "does not entail the ... dismissal of the religious categories of the divine, sacred, or transcendence" (Hogue 2010: 203); such categories can rightly be applied to nature itself.

In a word, for the religious naturalist, *nature is enough*—enough to elicit our unwavering fidelity and commitment, enough to arouse spiritual affections of wonder and gratitude, awe and humility, enough to find meaning and value in our lives (Crosby 2002: 169; Rue 2011).

Pantheism and its place in the family of religious naturalism

Crosby, Rue, Stone, Hogue, and others tend to define religious naturalism *nontheistically*. In actuality, however, the choice to retain or repudiate God-language is one of two crucial issues that divide religious naturalists from each other. As we shall discover, certain classes of theism *are* compatible with naturalism, and many religious naturalists opt to reinterpret "God" naturalistically rather than jettison the symbol altogether. The other major area of division within religious naturalism has to do with whether the religious ultimate is *within* the world or *is* the world. That is, should the sacred be identified with a *particular aspect* of nature or with the *ambiguous totality* of nature?

Figure 9.1 pictures these dividing lines as bisecting axes; the vertical axis represents "the God question," while the horizontal axis represents "the religious ultimacy question." These axes form a grid, with each quadrant signifying one of four major groupings within the family of religious naturalism.

If the natural exhausts what is real, then the sacred must be identified either with a *part* or with the *whole* of nature. The religious naturalisms on the left side of the grid take the former option: although the ambiguous entirety of nature is *metaphysically* ultimate, only certain elements of it are *religiously* ultimate. To illustrate, the empirical theologians of the "Chicago School" of theology famously put forward variations of finite (or naturalistic) theism, which conceives of God as *one kind of process included within nature*. Shailer Mathews equated God specifically with "the personality evolving activities of the universe" (Mathews 1931: 192–234). A decade and a half later, Mathews' junior colleague at the University of Chicago, Henry Nelson Wieman, insisted that God is none other than the "creative event" in nature that augments "qualitative meaning" or, simply, "the good"—especially the human good (Wieman 1995: 7, 56).

The varieties of religious naturalism depicted in the lower left quadrant overlap with naturalistic/finite theism on the religious ultimacy question but part ways on the God question. To illustrate, for religious humanism, the primary focus of religious concern is *humanity and human ideals and values*—e.g., the demand on human beings to take responsibility for themselves and the planet (Murry 2007: 1–11). However, the vast majority of religious humanists, both historically and currently, are nontheists (see Reese 1927; Pinn 2012).

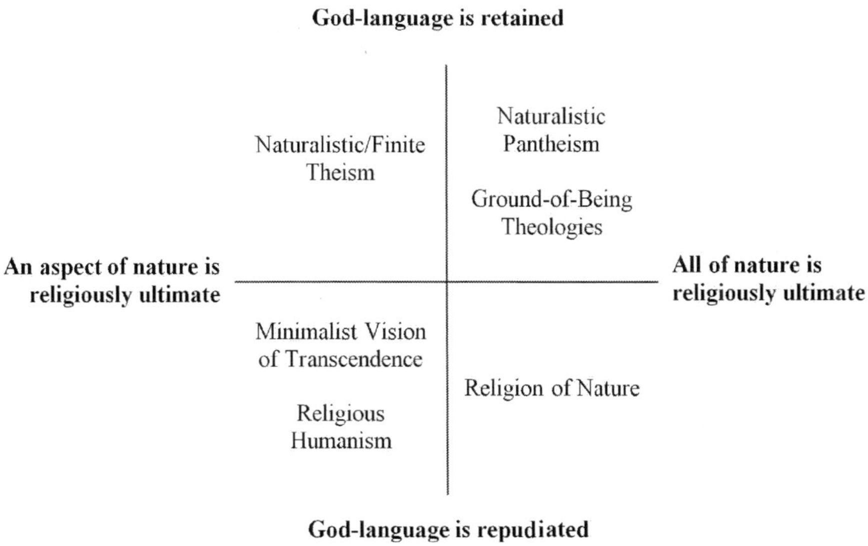

Figure 9.1 Varieties of religious naturalism

Jerome Stone is also a nontheist who argues that the "object of religious orientation" is "axiologically determinate." In his variation of religious naturalism, the sacred is limited to those "norms" and "powers" of nature that are "creative of the good," so that only "*some* things, like justice and human dignity, and the creativity of the natural world, are sacred" (Stone 2008: 194–195). Here, Stone moves beyond the more narrow scope of religious humanism, exhibiting "a greater sense that we are not masters of our fate, that we need to recognize the worth of, to nurture and be nurtured by, this-worldly grace and judgment" (Stone 1993: 35). But Stone, as I noted above, shares religious humanism's nontheism. Actually, his religious naturalism is a middle path between naturalistic theism and religious humanism, affirming a "minimalist vision of transcendence" (see Stone 1992; 2008: 143–145) without reinstating God-language.

Crosby, whose "religion of nature" is portrayed in the lower right quadrant of the grid, also avoids God-talk. Echoing Stone and nontheistic humanists, Crosby argues that "spirituality is contained within nature and provided by nature quite apart from the existence of a deity of any kind" (Crosby 2014). However, even though it is not a theistic faith, a religion of nature is far from irreligious or non-religious. In Crosby's view, the whole of nature is *both* metaphysically and religiously ultimate; it is (1) uncreated, everlasting, and self-sustaining, *and* (2) deserving of the same kind of "reverence, awe, love, and devotion we in the West have formerly reserved for God" (Crosby 2002: xi). On this latter score, Crosby departs from finite theism as well as from religious humanism and Stone's minimalist vision. In Crosby's version of religious naturalism, the most appropriate focus of our ultimate concerns and commitments is the *totality* of nature in all its vastness, ambiguity, diversity, mystery, and creative-destructive power—not just those parts of the world that "conduce to human good" or serve the aspirations and needs of humankind (Crosby 2007: 490–491, 494).

Crosby, though, is quick to distinguish a religion of nature from *pantheism*, which also hails the ambiguous totality of nature as religiously ultimate (see the upper right quadrant of the grid). Etymologically, pantheism simply means the whole (*pan*) is divine (*theos*). A religion of nature, by contrast, is *a*theistic, not *pan*theistic, *replacing* rather than *equating* God with nature. For Crosby, there is no divine spirit that guides and supports the world, nor is there any satisfying rationale for referring to the world as God.

That being said, the remainder of this essay will make a modest case for pantheism as a legitimate, distinctive, and compelling variety of religious naturalism. Crosby assumes that a materialist metaphysics rules out a pantheistic worldview, which (allegedly) sees nature as "suffused by a kind of divine presence" (Crosby 2018). What Crosby overlooks, however, is that there *are* thoroughly naturalistic varieties of pantheism; the pantheists canvassed below, for instance, do not view God as a disembodied or determinate entity and actually join Crosby in negating a divinely ensouled nature or any free-floating consciousness or being that presides over the universe (see Crosby 2002: 146). The chief difference is that Crosby rejects theistic imagery as "hopelessly anthropomorphic" (Crosby 2002: 9), whereas pantheists, much like finite theists, hazard that "God" is a complex, malleable, ever-evolving, and historically rich concept, capable of being (re)interpreted along non-anthropomorphic and naturalistic lines. After all, the idea of "God" is, as Gordon Kaufman persuasively demonstrates, a "human imaginative construction" that needs to continually undergo critical interrogation and reconstruction. A religious naturalist himself, Kaufman realizes that agential and supernaturalistic models of God are no longer intelligible in the modern era. Nevertheless, atheism is not the only alternative to classical theism, and by his late career, Kaufman began to reimagine the divine as the "creativity" manifest throughout the cosmos (see Kaufman 1993: 264–280; Kaufman 2004). Of course, Kaufman's attachment to God-language could easily be attributed to his self-interests as a Christian theologian. But, in actual fact, Kaufman held onto (and naturalized) the concept of God because he was convinced that it still conveys the utter mystery of things and continues to function, perhaps more powerfully than any other religious symbol in the West, as "an ultimate point of reference" that orients our lives and yet conditions everything human and finite (Kaufman 1993: ix–xv, 3–17, 32–44, 301–340).

I concur with Kaufman entirely, which is why I believe the pantheistic move is justified. Notwithstanding its conceptual limitations and historical baggage, "God" is a living metaphor in the Western imagination, one that has always expressly signified that which "relativizes" and "humanizes" our existence, that which humans take to be metaphysically and religiously ultimate. And if *nature* is truly ultimate in both of these senses, then the question becomes: why *not* speak of nature as God? Why *not* speak of "God or nature" (*deus sive natura*), as Benedict (Baruch) de Spinoza boldly did in the seventeenth century (see the following section)? My worry is that dispensing with God-talk risks losing one of the most potent and culturally recognizable concepts we have for symbolically engaging and worshipfully beholding the ultimacy and the sacred depths of nature. Thus, I would contend that pantheists are wise to retain the symbol "God," so long as it is duly de-anthropomorphized.

The fundamental theological instinct of pantheism is that the sacred is *coincidental with* nature in its mysterious, ambiguous, and infinite wholeness (Frankenberry 1993: 29). In this respect, pantheism runs counter both to finite theism, according to which the sacred is *less than* the whole of nature (e.g., the creative event), and to pan*en*theism, according to which the sacred is *more than* the whole of nature (i.e., a personal deity whose being contains and yet transcends the world). For the pantheist, God is not included within nature, nor is nature included within God. Rather, God *is* nature (although this statement will need to be nuanced and qualified below). As Milligan observes, the emphasis in pantheism is not on God as merely *in* nature, but on "God *as* nature, and Nature—in the richest, fullest, comprehensive sense—as God" (Milligan 1987: 584, emphasis added). By contrast, panentheistic perspectives (e.g., process theology), insofar as they posit a divine agent beyond natural processes, are not truly naturalistic. On the contrary, they are simply immanentalist iterations of supernatural theism—or to play on the title of Stone's text, *maximalist visions of immanence.*

Yet pantheism is (or at least can be) much more philosophically sophisticated than it appears at first glance; it need not be a simplistic conflation of God and the world. To be a pantheist is to be awakened to the holiness, indeed the divinity, of the entirety of reality. Be that as it may, pantheism does *not* mean that every single *part* of nature (e.g., nuclear weapons and student loan payments) is God or equally worthy of reverence and worship (Harrison 1999: 70–71). Citing the nineteenth-century pantheist, Allanson Picton, Milligan notes that more philosophically subtle variants of pantheism divinize "the all that is" not in "the sense of an aggregate," but in the sense of the "totality" and "oneness of things" as "phenomenal manifestation[s] of the energy of an infinite Life" (Milligan 1996: 237–238). The version of pantheism championed in the last section of this essay—which I dub "apophatic pantheism"—affirms just that. This model of pantheism even finds a completely naturalistic way of preserving the ancient monotheistic distinction between creator and creation, carefully distinguishing "nature naturing" from "nature natured." At this juncture pantheism interestingly intersects with certain modes of ground-of-being theology (see, again, the upper right quadrant of Figure 9.1)—a point to which we will return later.

Three naturalistic pantheists

Although the word "pantheism" was not coined until the early eighteenth century (by John Toland), pantheistic impulses and insights stretch back into antiquity. As Paul Harrison observes, pantheism is a "perennial heresy" that is present in nearly every religious and philosophical tradition of the world, both East and West, and is shared by a host of great thinkers throughout the ages (Harrison 1999: 13–38). For that reason, pantheism, as I intimated above, comes in many varieties, several of which are not congruent with naturalism.

But some are, and in what follows, I will look at pantheisms that are *naturalistic* in their metaphysics. I will focus on the thought of three pantheists in particular: Spinoza (an early-modern Jewish thinker), Bernard Loomer (a mid-twentieth-century process theologian), and Robert Corrington (a contemporary philosopher). Their naturalistic iterations of pantheism obviously stand apart from idealistic and dualistic ones as well as from other non-naturalistic perspectives such as panentheism (Milligan 1996: 235–236; 1987: 583–584, 593–594).

Admittedly, many present-day religious naturalists would balk at some of the key components of Spinoza's philosophy (e.g., his pre-Darwinian conception of a fixed, static, and deterministic nature). Nonetheless, Stone (rightly) claims Spinoza as a "forerunner," and perhaps even the "grandfather," of contemporary religious naturalism (Stone 2008: 18).

For one thing, Spinoza was an unblinking naturalist. He held that nothing exists outside of nature; that the world is devoid of teleology; and that humans are wholly natural beings, differing from the rest of reality only in degree, not in kind (see Spinoza 1996: 10, 27–29, 114, 146). He also refused to attribute personality, intentionality, and conscious awareness to God (see Spinoza 1996: 10, 34)—one of the factors that earned him a reputation as a heretic and led to his excommunication from the synagogue in Amsterdam. In his masterwork, the *Ethics* (1677), as well as in other important writings, Spinoza made quick work of the supernatural, anthropomorphic God of orthodox Judaism and Christianity and posited, instead, a thoroughly naturalized, non-personal conception of divinity (Nadler 2006: x–xi, 113–115; Crosby 2002: 148–149; Donagan 1996: 343, 354–355; Bennett 1984: 35–37; Capetz 2003: 115; Scruton 2002: 1, 9).

Traditionally, Spinoza's naturalistic theology has been classified as a type of pantheism. Part One of the *Ethics* lays out a monistic metaphysics in which the universe is conceptualized as a single, infinite, eternal, necessarily existing substance. (In)famously, Spinoza described this *substantia* as "God or nature"—*deus sive natura* (see Spinoza 1996: 114, 118–119). The Latin conjunction *sive* clearly signifies equivalence (Nadler 2006: 81; Curley 1988: 36). That is, the terms "*deus*" and "*natura*" have the same referent; God is none other than the one and only reality/substance that there is—i.e., nature (Spinoza 1996: 9–10). As B.A. Gerrish clarifies: "For Spinoza, God is not a being but being itself; nature and God constitute an indivisible unity, *deus sive natura*" (Gerrish 1993: 115).

Spinoza's interpreters disagree over the precise meaning of *deus sive natura*. For some, Spinoza was espousing straightforward pantheism, equating God with the whole of reality (see Bennett 1984: 32–35). However, other scholars, such as Steven Nadler, Edwin Curley, and Paul Capetz, point out that Spinoza differentiated between active and passive aspects of nature, between *natura naturans* (nature naturing) and *natura naturata* (nature natured). *Natura naturans* connotes the generative power within nature by which all things come into being and are preserved in being, while *natura naturata* simply refers to whatever has been "natured" or brought into existence, namely, the empirical world. In Part One of the *Ethics*, Spinoza unmistakably identifies God solely with nature naturing, not with the entirety of nature (see Spinoza 1996: 20–21). On this reading, to say that "whatever is, is in God" (Spinoza 1996: 10) is to assert that the relationship between God and "whatever is" is one of causation and dependence; the things of this world are not so much properties of God as they are effects of God. As nature naturing, God is the ground or efficient cause of being, that which produces and sustains—in a word, "natures"—all things. Regardless—and this is a critical point—nature naturing is the *immanent*, not the *transitive*, cause of all things—i.e., a cause whose effects are a part of, not other than, itself (Spinoza 1996: 16, 18–19). Nature derives its existence *from itself*, not from something (or someone) beyond it (Donagan 1996: 343). Put differently, the universe and its furniture are not created *ex nihilo* by a supernaturally transcendent deity, but are generated from and perpetually dependent on the laws, activities, causal principles, and creative powers of nature itself. The laws, activities, causal principles, and creative powers of nature itself—i.e., nature naturing—are precisely what is meant by "God" in Spinoza's system (Nadler 2006: 52–53, 72–83, 108; Curley 1988: 36–39, 42–45; Capetz 2003: 116).

Since the word "God" is reserved for the "naturing" dimension of nature, Curley questions whether Spinoza is a bona fide pantheist (Curley 1988: 149–150, n. 152). My own position is that Spinoza's theology is *philosophically* pantheistic, even if not *etymologically* pantheistic. Curley, Nadler, and Capetz are correct: the Spinozist God is not everything—i.e., is not identical with the totality of finite things. Even so, God literally *is* nature naturing, according to Spinoza, and the *totality* of finite things exists within—i.e., emanates from, depends on, and belongs to—God or nature. More significantly, *pace* Jewish and Christian theism, where God the Creator is wholly other than the creation, Spinoza's monism regards that which natures and that which is natured as ontologically inseparable aspects of the one nature that there is (Nadler 2006: 79, 83, 108, 113). This is pantheism, in my view, albeit a much subtler form of pantheism. It certainly is not pan*en*theism, given that the God-nature relation is not one of containment. As Nadler explains, "God is not 'in' Nature in such a way that nature contains, in addition to its natural contents, a distinct divine and supernatural content" (Nadler 2006: 118).

According to Nadler's analysis, however, Spinoza was effectively an *atheist*. After all, Spinoza renounced the existence of anything beyond the natural realm, including the personal God of the Judeo-Christian tradition, and reduced the divine to nature. The only reason he spoke of "*deus sive natura*" to begin with was because the basic characteristics of nature—eternity, necessity, infinity—are those historically ascribed to God (Nadler 2006: 112–121).

Ironically, Spinoza was accused of atheism by his own contemporaries—*an accusation that he resented and denied*! Bennett keenly observes that Spinoza "apparently did not think of himself as an atheist ... [H]e thought of himself as discovering things about God rather than as revealing that there is no God" (Bennett 1984: 35). Of course, as we have already seen, Spinoza *was* atheistic with respect to the supernatural, determinate-entity God of classical theism. Still, he proceeded to give nature the name "God." The question is *why*? Bennett concurs with Nadler that Spinoza divinized nature because nature was the most "Godlike" reality he knew—it is infinite, eternal, and self-caused, the ultimate source and explanation of everything that exists (Bennett 1984: 32–35). But Bennett believes that Spinoza also spoke of *deus sive natura* for another reason: "namely his view of Nature as a fit object for reverence, awe, and humble love, i.e., for the attitude traditionally reserved for God" (Bennett 1984: 34).

Spinoza was, in brief, a *religious* naturalist, a *pantheistic* religious naturalist! To quote Bennett again: "Spinoza did accept pantheism as a kind of religion" (Bennett 1984: 35). Spinoza was not exactly "drunk with God," as the German Romantic poet Novalis once effused. But nor was he a crass atheist. Rather, he was a proponent of the most radical kind of natural theology, naturalizing divinity and divinizing nature (see Donagan 1996: 343–357).

Loomer's brand of pantheism was much more philosophically clear-cut than Spinoza's, although no less naturalistic. Like Mathews and Wieman, Loomer was associated with the Chicago School of theology. More particularly, he belonged to a later generation of Chicago Schoolers who stood within the "empirical tradition" of Whiteheadian thought. One of the distinguishing characteristics of the empirical process theologians was their outright denial of the panentheistic God of conventional Whiteheadian theism. As naturalists, Wieman, Loomer, and the other process empiricists maintained that nature is the only reality there is, and if there is nothing in addition to or even slightly transcendent of nature, then "God" or "the sacred" must be either included within or identified with the natural world itself (Loomer 1987: 22–23).

Wieman, for example, completely blurred the Whiteheadian contrast between God and creativity. As Nancy Frankenberry helpfully explains, Wieman argued not "that wherever God is manifest, there is creative transformation, but precisely the opposite—wherever one finds creative transformation, *there* one finds what has been meant by 'God'" (Frankenberry 1987: 124). With Wieman, Loomer associated God not with a personal being that is distinct from nature but with the processes of nature itself. But against Wieman, the later Loomer even divinized nature's less-than-creative processes. The concrete world, Loomer contended, is utterly *ambiguous*. Becoming is metaphysically ultimate, but "the creative advance" is not so much an "adventure toward perfection" as it is a "struggle toward greater stature" or what he later dubbed "size," namely, the capacity to take in and sustain intense relationships, contrasts, tensions, and ambiguities (Loomer 1974; 1987: 42, 51). Moreover, the interrelated web of life comprises "a diversity of forces, many of which are either noncreative or destructive" (Loomer 1987: 40).

Loomer's God was commensurate with his realist cosmology. In the twilight of his career, he began to experiment with a kind of *process pantheism*, equating God with the *ambiguous totality* of the natural realm. God is none other than "the organic restlessness of the whole body of creation." As such, God must embody all the ambiguity actually found therein, "all the evil, wastes, destructiveness, regressions, ugliness, horror, disorder, complacency, dullness, and meaninglessness, as well as their opposites" (Loomer 1987: 40–43).

Loomer took issue with Whiteheadian efforts to circumvent the ambiguity that characterizes nature and to dissociate the divine from evil. Whitehead himself did this by "ontologically separating God and creativity" and imagining "an aesthetic form of persuasiveness that is pitted against the coercive and inertial powers of the world." Loomer harshly judged that this "unambiguous structure or character can be derived only by a complex abstractive process, the end result of which has no counterpart in reality" (Loomer 1987: 50, 38). Wieman improved upon Whitehead by urging that "the being of God is not other than the being of the world." But Wieman identified God with only "one aspect of the world or one kind of process," namely, the part of nature that is generative of good (Wieman 1995: 54–83). Thus, no less than Whitehead's deity, Wieman's deity is defined by pure goodness and is too clean and perfect to be concretely real; it is a bloodless, unempirical abstraction from a cosmos that is inescapably ambiguous (Loomer 1987: 21, 40, 48–50). Loomer, by contrast, held that God's activities are "not wholly or even primarily identified with the persuasive and permissive lure of a final cause or a relevant and novel idea ... God is also a physical, efficient cause that may be either creative or inertial in its effects." If "the size of God" embraces nothing less than nature in its ambiguous wholeness, then the divine lure "may exemplify itself as an expansive urge toward greater good" *or* a "passion for greater evil" (Loomer 1987: 41).

Corrington shares quite a bit in common with Loomer and other American religious empiricists (see Wheeler 2014). However, the metaphysical cornerstone of Corrington's philosophy of "ecstatic naturalism" is the Spinozist distinction between *natura naturans* and *natura naturata* (Corrington 2016: 7). This most primal of distinctions, according to Corrington, "is as old as thought itself" (Corrington 1997: 4). But Spinoza provided the "initial categorial framework" in which to probe the cleft *within* the natural sphere between the sheer generative power of nature naturing and its "natured" products—or to use Corrington's preferred terminology, between "nature perennially creating itself out of itself alone" and "the innumerable orders of the world" (Corrington 2014: 141–142; 2016: x, 1, 7–9).

Corrington, though, puts an *ecstatic* spin on Spinoza's distinction, portraying the natural difference as the "primal abyss" that is perpetually fissuring within the one nature that there is. From Corrington's point of view, Spinoza failed to discern "the uncanny underside" of *natura naturans*. What Corrington is alluding to here is nature's "underconscious"—i.e., the "unruly ground" or the "unconscious depths" from whence all natural complexes arise and the mysterious void to which they shall return. Corrington gestures toward the "fulsome" but "dark heart" of nature naturing, namely, the potencies that manically and indifferently churn away in the bosom of *natura naturans* and give birth to the innumerable orders of the world—including, as we will see, divinities and other "sacred folds" (Corrington 2016: ix–x, 1–18, 95).

"Deep pantheism," Corrington declares, is the theological perspective that is harbored within ecstatic naturalism. Akin to Spinoza, Loomer, and every naturalistic pantheist, Corrington repudiates determinate-entity theism: "we have providingness but no provider, natural grace but no bestower of grace, sheer availability but no intentionality, and a seed bed for consciousness with no consciousness in the seed bed" (Corrington 1997: 103). Corrington takes aim at "classical asymmetrical theism," which postulates a divine creator that is *other* to what has been created, as well as the "halfway house" of panentheism, which envisions an "in and above" relationship between God and the world. For panentheists (e.g., process theologians), "the divine retains its ontological perch" by living both within and beyond nature. Put differently, panentheism's nature is too small, while its deity is too big (Corrington 2016: xi, 1–2; 2007: 505; 1998: 169–171). Deep pantheism asserts the exact opposite: "Nature is the genus of which the sacred is a species" (Corrington 1997: 2). In other words, nature transcends and eclipses the sacred rather than vice versa. Baldly put, sacred folds are "natural complexes." Similar to all natural complexes, gods, goddesses, spirits, etc. are emergent from the underconscious of nature and are "ordinally located" within the orders of the world.

What makes Corrington's pantheism "deep" is its openness to the chaotic, ambiguous, hidden, and sometimes demonic underside of nature out of which the actualities and possibilities of the world emerge. With a nod to Spinoza, Corrington distances himself from a more "philosophically lazy" sort of pantheism in which God is casually equated with the universe. For one thing, the sacred is one natural complex among countless others, as is the universe itself; they are both "in and of" nature (Corrington 1997: 10; 2016: 53). From the standpoint of Corrington's "ordinal phenomenology," *no* natural complex is preeminently real or universally related to all other orders. Thus, as natural complexes, God and gods are not more relevant, encompassing, or ultimate than any other complex. Indeed, "there are innumerable nonsacred orders that lie outside of the holy" (Corrington 1997: 10). This is why deifying the whole of reality is problematic, for Corrington. Nor does Corrington identify the divine with nature naturing (contra Spinoza). "God-ing energies" are certainly vast in scope, manifest in the gap between *natura naturans* and *natura naturata*. Even so, the divine is a product, not a synonym, of nature naturing, an order, not the creator, of the cosmos (Corrington 2014: 151–152; 2016: xi, xxii, 1–18, 42, 44, 79–87).

Toward an apophatic pantheism

I want to conclude by laying out (in a preliminary fashion) my own pantheistic variety of religious naturalism, a position I will term "apophatic pantheism." Apophatic pantheism appreciatively (yet critically) draws on the three naturalistic pantheists analyzed in the previous section and also finds striking affinities with ground-of-being theologians such as Paul Tillich and Wesley Wildman.

Apophatic pantheism is *pantheistic* inasmuch as it is religiously and theologically oriented to "the all"—i.e., to reality in its entirety. The adjective *apophatic*, however, is intended to qualify pantheism in two important senses.

First, an apophatic pantheism is a naturalistic form of *mysticism*, pointing to the *absolute mystery* of nature or God. A truly *pan*theistic vision implies that there is, to borrow a lovely phrase from Michael Hogue, "no 'outside' of revelation—the whole cosmos rings with it, from the subatomic to the interstellar, from the unicellular to the civilizational" (Hogue 2014: 3). Yet Gordon Kaufman is right: "it is ultimately mystery within which we live" (Kaufman 1993: xii). Of course, apophatic pantheism *naturalizes* mystery; it is *nature* that is largely hidden behind a cloud of unknowing. Be that as it may, apophatic pantheists stand alongside contemplative mystics and negative theologians in every tradition, whose primary intuition is that whatever is divine or ultimate is inscrutable, incomprehensible, and ineffable. *Deus sive natura* discloses itself to us through multiple modes of inquiry but, in the end, finally surpasses human understanding. Its mysterious ways—its emergent properties and axiological possibilities, its exuding creativity and mindless destruction, its ordered regularity and unpredictable contingencies, its plural particulars and relational webs, its unimaginable vastness and sheer isness—evoke fascination, terror, wonderment, silence.

Indeed, the utter mysteriousness of existence is what invites a pantheistic interpretation of reality in the first place. Loomer realized that one of the principal justifications for divinizing the natural world is that nature contains yet enshrouds "a transcendent and inexhaustible meaning that forever eludes our grasp," while the word "God" adequately "symbolizes this incredible mystery" (Loomer 1987: 42). Corrington takes it a step further, suggesting that we can only speak of innumerable natural complexes; "nature *per se*" or "*the* nature" does not exist. And if there is no "what" to which the term corresponds, then "nature" can only be addressed "through a kind of unrelenting via negativa"—in a word, *apophasis*. This is why Corrington often insists on crossing out ~~nature~~. Such a tactic signifies that ~~God or nature~~ is of an entirely different order, defying easy description and transcending all genera. As such, it should only be used "elliptically and under erasure" (Corrington 2016: 4–5, 29; 2014: 142, 152).

Second, and perhaps less obviously, an apophatic pantheism is a naturalistic form of *ground-of-being theology*, underscoring the *infinite self-transcendence* of nature. It is here that the distinction both Spinoza and Corrington make between *natura naturans* and *natura naturata* becomes absolutely critical. Above all, the "natural difference" reveals that "nature" refers not only to the empirical world and everything therein (i.e., nature natured) but also to the ground and source of all existence, that is, to the creative-destructive processes that everlastingly bring new realities (even new universes) into and out of being (i.e., nature naturing). As naturalists, apophatic pantheists deny the existence of the *super*natural, of domains or entities that transcend nature. But they do stand with Corrington in affirming a kind of "intra-natural transcendence," a transcendent dimension *within* the one nature that there is (Corrington 2016: 80, xiii.). Put differently, nature is *self-transcendent*, infinitely transcending (and hence relativizing) whatever is natured, including the present "cosmic epoch." In that sense, apophatic pantheism is completely compatible with the various multiverse theories currently on offer within physics as well as with contemporary cosmology and its picture of a finite, historically contingent universe. Far from exhausting nature, the cosmos is a part of creation, generated and sustained—created—by the relentless powers of *natura naturans*. This world, the world subsequent to the Big Bang, is what Corrington terms "a subaltern world," one order among others within the innumerable orders of *natura naturata* (Corrington 2002: 142; 2016: 53).

Yet apophatic pantheism is consistent not only with contemporary science but also with the "radical monotheism" that lies at the heart of the Western faiths. The differentiation of nature naturing from nature natured is a naturalization of the distinction in classical theology between creator (or, less anthropomorphically, creativity) and creation. This is precisely where apophatic pantheism and ground-of-being theology converge, in my estimation.

To be sure, Tillich, the leading advocate of ground-of-being theology in the twentieth century, resisted pantheistic varieties of naturalism on account of their facile (and idolatrous) equation of God with the totality of the world. That is, pantheists, in Tillich's mind, discount the infinite distance between the whole of finite things and their divine ground, "with the consequence that the term 'God' becomes interchangeable with the term 'universe' and therefore is semantically superfluous" (Tillich 1957: 7).

Tillich's concerns are certainly not without warrant. Even Loomer, while helpfully collapsing the false Whiteheadian dichotomy between *creativity and God*, failed to make a sharp enough distinction between *creativity and the world*. Still, this essay has identified and elucidated trajectories of naturalistic pantheism that not only eschew the reductionistic tendencies that alarmed Tillich but also embrace the central convictions of his ground-of-being theology. Tillich himself seemed to acknowledge this with respect to Spinoza: "The phrase *deus sive natura*, used by people like ... Spinoza, does not say that God is identical with nature but that he is identical with the *natura naturans*, the creative nature, the creative ground of all natural objects" (Tillich 1957: 6). Tillich is spot-on here. But his point, which has been recently endorsed by Wildman (see Wildman 2014b: 14), is that Spinoza was not really a pantheist.

My reading differs from theirs. I would argue that there *is* a type of pantheism—call it apophatic pantheism—that recognizes the infinite self-transcendence of nature itself. I would further argue that Spinoza's theological naturalism, along with the ecstatic naturalism of Corrington and ground-of-being varieties of religious naturalism—are expressions of this non-superficial, non-reductive sort of pantheism, whether they employ the label or not (even Loomer's process naturalism, with the appropriate clarifications and qualifications, could be pushed in this direction). First and foremost, all of these perspectives *maintain yet naturalize* the infinite qualitative distinction between the world and its ontological ground. As Corrington explains, the ontological difference prevails *within and as* the one nature that there is; it does not comprise two separate realities. Again, nature is *self*-transcending; it perennially transcends itself, recreating itself from out of its own resources (Corrington 2016: x, 1–2, 8–12).

Wildman comes to a similar conclusion. He references Corrington's contention that nature *can* and *does* include "infinite transcendence of Tillich's sort," which makes Tillich a religious/ecstatic naturalist—in substance, if not in name (Wildman 2014a: 40). If there is a "God" in this type of naturalistic worldview, "it is the deep principles of nature understood as the conditions for the possibility of the natural world" (Wildman 2014b: 14). As a Tillichian, Wildman speaks of these conditions and deep principles as the "abysmal ground of being," which he associates with "the creative and fecund power source in the depths of nature" (Wildman 2010: 216–217). But Wildman allows that there are other apt theological symbols for the ground of all determinate things, including "Spinoza's *natura naturans*" (see Wildman 2014b: 14, 17)—and I would add Corrington's "underconscious of nature" and Loomer's "creative advance."

Yet the question remains: *is this pantheism*? After all, instead of divinizing everything (the literal meaning of "pantheism"), Spinoza (reasonably, in my view) reserved the word "God" for *natura naturans* (i.e., the ground of being). I want to venture that apophatic pantheism—or, perhaps, *self-transcendent* pantheism—*is* pantheistic by virtue of the fact that it attributes religious/theological ultimacy to the whole of nature rather than an aspect of nature (as with finite/naturalistic theism) or a person-like, benevolent divine agent that is somehow in addition to and better than nature (as with panentheism). To repeat, *natura naturans* and *natura naturata*—process and product, ground and consequent—are not dualistically other or ontologically bifurcated. Rather, they "belong together in the fullness that is nature in its encompassing" (Corrington 2016: x). Or to make the same point in Spinozist terms, nature naturing is the *immanent*, not the *transitive*, cause of nature natured—i.e., a cause whose effects are inseparable from itself. As

Corrington, commenting on Spinoza, astutely observes, there is, in the end, only one infinite divine substance: *deus sive natura*. Insofar as they are emanations from the one infinite divine substance, the orders of *natura naturata* are, necessarily, "part of the very being of god" (Corrington 2016: 9). To quote Tillich: "God ... is nearer to them than they are to themselves" (Tillich 1957: 7). Thus, when all is said and done, the apophatic pantheist *is* able to affirm the sacredness, even the divinity, of nature in its ambiguous entirety, even while retaining the distinction between the whole of finite objects and their infinite ground, creatures and creativity-destructivity itself. For apophatic pantheism, the ambiguous totality of existence—the all—is sacred by virtue of its emergence from and participation in *deus sive natura*—i.e., nature naturing, being itself, the ground of becoming, the mysterious power source in the depths of reality that natures all things and infinitely transcends whatever has been natured, whether gods or worlds. Undoubtedly, apophatic pantheism resembles panentheism in its retention of transcendence. However, it is pantheistic rather than panentheistic because the transcendent is a depth dimension in and of nature—not a divine being in and beyond nature.

References

Bennett, J. (1984) *A Study of Spinoza's Ethics*, Indianapolis, IN: Hackett.

Capetz, P. E. (2003) *God: A Brief History*, Minneapolis, MN: Fortress.

Corrington, R. S. (1997) *Nature's Religion*, Lanham, MD: Rowman and Littlefield.

——— (1998) "Empirical Theology and Its Divergence from Process Thought." *In:* Badham, R. A. (ed.) *Introduction to Christian Theology: Contemporary North American Perspectives*. Louisville, KY: Westminster John Knox.

——— (2002) "My Passage from Panentheism to Pantheism." *American Journal of Theology and Philosophy*, 23, 129–153.

——— (2007) "Deep Pantheism." *Journal for the Study of Religion, Nature and Culture*, 1, 503–507.

——— (2014) "Guest Editor's Introduction: Ecstatic Naturalism and Deep Pantheism." *Journal for the Study of Religion, Nature and Culture*, 8.

——— (2016) *Deep Pantheism: Toward a New Transcendentalism*, Lanham, MD: Lexington.

Crosby, D. A. (2002) *A Religion of Nature*, Albany, NY: State University of New York Press.

——— (2007) "A Case for Religion of Nature." *Journal for the Study of Religion, Nature and Culture*, 1, 489–502.

——— (2014) "Religious Naturalism and Its Place in the Family of Religions." *The Fourth R*, 27.

——— (2018) "Matter, Mind, and Meaning in Religion of Nature." *In:* Crosby, D. A. & Stone, J. A. (eds.) *The Routledge Handbook of Religious Naturalism*. London: Routledge.

Curley, E. (1988) *Behind the Geometrical Method: A Reading of Spinoza's Ethics*, Princeton, NJ: Princeton University Press.

Dawkins, R. (2006) *The God Delusion*, Boston, MA: Mariner.

Donagan, A. (1996) "Spinoza's Theology." *In:* Garrett, D. (ed.) *The Cambridge Companion to Spinoza*. Cambridge: Cambridge University Press.

Frankenberry, N. (1987) *Religion and Radical Empiricism*, Albany, NY: State University of New York Press.

——— (1993) "Classical Theism, Panentheism, and Pantheism: On the Relation Between God Construction and Gender Construction." *Zygon*, 28, 29–46.

Gerrish, B. A. (1993) *Continuing the Reformation: Essays on Modern Religious Thought*, Chicago, IL: University of Chicago Press.

Harrison, P. (1999) *Elements of Pantheism: Understanding the Divinity in Nature and the Universe*, Boston, MA: Element.

Hogue, M. S. (2010) *The Promise of Religious Naturalism*, Lanham, MD: Rowman and Littlefield.

——— (2014) "Religion without God: The Way of Religious Naturalism." *The Fourth R*, 27, 3–6, 15–16.

Kaufman, G. D. (1993) *In Face of Mystery: A Constructive Theology*, Cambridge, MA: Harvard University Press.

——— (2004) *In the Beginning ... Creativity*, Minneapolis, MN: Fortress.

Loomer, B. (1974) "S-I-Z-E." *Criterion*, 13.

——— (1987) "The Size of God." *In:* Dean, W. & Axel, L. E. (eds.) *The Size of God: The Theology of Bernard Loomer in Context*. Macon, GA: Mercer University Press.

Mathews, S. (1931) *The Growth of the Idea of God*, New York, NY: Macmillan.
Milligan, C. S. (1987) "The Pantheistic Motif in American Religious Thought." *In:* Freese, P. (ed.) *Religion and Philosophy in the United States of America*. Essen: Die Blaue Eule.
——— (1996) "The Eco-Religious Case for Naturalistic Pantheism." *In:* Crosby, D. A. & Hardwick, C. D. (eds.) *Religious Experience and Ecological Responsibility*. New York, NY: Peter Lang.
Murry, W. R. (2007) *Reason and Reverence: Religious Humanism for the 21st Century*, Boston, MA: Skinner House.
Nadler, S. (2006) *Spinoza's Ethics: An Introduction*, Cambridge: Cambridge University Press.
Pinn, A. (2012) *The End of God-Talk: An African American Humanist Theology*, New York, NY: Oxford University Press.
Reese, C. (1927) *Humanist Sermons*, Chicago, IL: Open Court.
Rue, L. (2011) *Nature is Enough: Religious Naturalism and the Meaning of Life*, Albany, NY: State University of New York Press.
Scruton, R. (2002) *Spinoza: A Very Short Introduction*, Oxford: Oxford University Press.
Spinoza, B. (1996) *Ethics*, London: Penguin.
Stone, J. A. (1992) *The Minimalist Vision of Transcendence: A Naturalist Philosophy of Religion*, Albany, NY: State University of New York Press.
——— (1993) "The Viability of Religious Naturalism." *American Journal of Theology and Philosophy*, 14.
——— (2008) *Religious Naturalism Today: The Rebirth of a Forgotten Alternative*, Albany, NY: State University of New York Press.
——— (2018) "Sacra Natura: Discerning the Immanent Sacred." *In:* Wheeler, D. & Conner, D. E. (eds.) *Conceiving an Alternative: Philosophical Resources for an Ecological Civilization*. Claremont, CA: Process Century.
Tillich, P. (1957) *Systematic Theology*, Chicago, IL: University of Chicago Press.
Wheeler, D. (2014) "American Religious Empiricism and the Possibility of an Ecstatic Naturalist Process Metaphysics." *Journal for the Study of Religion, Nature and Culture*, 8, 156–181.
Wieman, H. N. (1995) *The Source of Human Good*, Atlanta, GA: Scholars Press.
Wildman, W. J. (2009) *Science and Religious Anthropology: A Spiritually Evocative Naturalist Interpretation of Human Life*, Farnham: Ashgate.
——— (2010) "Narnia's Aslan, Earth's Darwin, and Heaven's God." *Dialogue: A Journal of Morman Thought*, 43, 210–217.
——— (2014a) "Religious Naturalism: What It Can Be, and What It Need Not Be." *Philosophy, Theology and the Sciences*, 1, 36–58.
——— (2014b) "Tillich's *Systematic Theology* as a Template for the Encounter of Christian Theology and Religious Naturalism." *Bulletin of the North American Paul Tillich Society*, 40, 13–18.

10

MATTER, MIND, AND MEANING

Donald A. Crosby

Matter, mind, and *meaning*: these three terms are bound inextricably together because the third requires recognition and interpretation by the second, and both of these are necessarily dependent on the first. Matter exists in great profusion in the universe and on our home planet, and it is generally devoid of mind and of mind's awareness of meaning. But there are also organizations of matter here on earth and probably elsewhere in the universe that are able to function as minds. There is no meaning without mind, and mind requires meaning for its effective operations. This statement is true of sensate meaning, conceptual meaning, artistic meaning, and existential meaning. Mind and meaning, in their turn, are functions of matter. Meaning lies in the relations of matter and mind and of mind with mind. Material existence gives rise to mind. Mind, in its relations to matter—the matter of its own embodiment and the matter of the world external to and interacting with its embodiment—becomes a field of meaning. The world external to itself with which a particular mind has relations includes material beings that, like itself, have varying degrees of mental capacity. With such internal and external relationships, *matter becomes meaning through the mediation of mind.*

There is a wide spectrum of degrees of both mind and meaning, from the most primitive forms of sentience and end-directed activity in relatively simple biological organisms to such extraordinary human intellectual and artistic feats as the invention of the alphabet in the second millennium BCE; the oral reciting and later writing of the Homeric epics *Iliad* and *Odyssey* toward the end of the eighth century BCE; the composition of the *Mahabharata* epic culminating about the beginning of the fifth century CE but with much earlier roots; the planning and building of the great Amiens cathedral in the thirteenth century CE; and the discovery of the DNA molecule and rise of molecular biology in the twentieth century CE.

Emergent matter, mind, and meaning

The version of religious naturalism I call *Religion of Nature* is based on a materialistic metaphysics or view of reality that regards all existence as diverse forms and functions of matter.[1] Its view of matter is neither Newtonian nor reductionistic. It is meant to take into account all that we have learned (and much we still have to learn) in physics about the nature and capabilities of matter since Newton's time, but it does not restrict its conception of matter to what can be described or explained by the discipline of physics. Rather than being reductionistic in this sense, it is emergentist or expansionist in its character. Approaches to a proper understanding of matter require the resources of all fields of thought, from physics, to chemistry, to biology, to psychology, to sociology, to philosophy, to art, to religion, and to the experiences of everyday life.

Matter *is* what matter has shown itself capable of *doing* and *becoming* from the origin of this universe about 13.8 billion years ago to the present, and that includes the myriad life forms on earth and us humans as one species of these life forms. With material life, material mind has come into being and, with that, material interactions, strivings, determinations, and purposes that crystallize into felt and intended meanings of many different sorts. Just as the phrase *material life* is not an oxymoron, neither are the phrases *material mind* and *material meaning*, when these are viewed in accordance with the metaphysical perspective to which Religion of Nature is committed.

Does the material universe as a whole have overarching purpose or meaning? Many religious thinkers hold that the answer to this question has to be positive; otherwise, human life would be deprived of the necessary basis for its immanent purpose and meaning. There can be no meaningful human life, they insist, in a purposeless universe. Such persons go on to reason that this overarching purpose or meaning must be bestowed on the universe by some kind of superconscious mind or spirit external to it or lying deeply within it. In other words, a primordial, purposive, and purpose-giving consciousness or mind must be regarded as the source and basis of the universe and the ultimate explanation for everything in it, including matter and finite minds and their meanings.

In my view, on the contrary, the universe has no overarching purpose or meaning conferred on it from without or within. It has come over time to contain many emergent purposes and meanings but does not itself have a comprehensive purpose or meaning. The material universe is for me self-explanatory and self-surpassing. It is self-explanatory in the sense that its existence, in all of its forms, is a given. It does not depend for its existence on something other than itself, unless that something is conceived as earlier versions of itself. Moreover, the face of the universe is not static in the way that Aristotle conceived it to be. Instead, it is dynamic and creative, surpassing itself with new creations and new kinds of existence over eons of time. The inherent dynamism brings new things into being even as it causes old things to cease to be. Its creations and destructions go hand-in-hand.

This analysis can be applied to an endless succession of universes, earlier ones undergoing eventual destruction, and the ashes of their destructions providing materials for the emergence of new ones. We should not think that a presumed fact of primordial infinite time entails the eventual emergence of some sort of infinite perfection over the course of that time, any more than it requires an infinitely powerful spiritual being to create, preside over, and give continuing support to the universe or a succession of universes. All existence, even over infinite time, can be presumed to be finite existence, with the ambiguities, uncertainties, perils, possibilities, opportunities, and joys of finite sentient existence—should this type of existence be present.

By my reckoning, all existence, including the various forms of mental existence, has and will have at its basis some form of matter at some stage of matter's irrepressible expansion and emergence. There is no reliable, publicly testable evidence for the existence of free-floating, non-material phenomena. Lewis Carroll's Alice, in his famous story *Alice in Wonderland*, remarks about her experience of the Cheshire cat that she has frequently seen cats without grins, but that she has never before seen the grin without the cat! Similarly, I have encountered many material things devoid of mind, but I have yet to encounter a mind separate from a material body. There is no reliable, publicly testable evidence for the existence of disembodied beings or for free-floating, non-material phenomena.

This observation is one of the reasons I do not accept a pantheistic view of nature, a view in which nature is seen as somehow God-like in character or suffused by a kind of divine presence.

Mine is a materialistic monism that allows for an extremely wide range of diverse phenomena. Over eons of historical change matter has done an astounding amount of different things. It is inherently protean and fecund in its materiality, something quite different from the externally related unchanging material atoms envisioned in Newtonian physics. My outlook is a kind of genuine pluralism consistent with a materialistic and monistic metaphysics. The universe is a ceaseless dance of nature natured with nature naturing. No external, underlying, or supplemental presence or power is needed to account for its ongoing dynamism and present character. Nature is unreservedly natural, through and through.

Within the present universe, which may well be only one member of a succession of universes stretching into the infinite past, new developments are the actualizations of possibilities brought to the fore by the more recent past, and these developments pose new possibilities for future actualities. This whole process of change and development is enabled by the creative capabilities of matter, ranging in the present universe from its earliest, simplest forms and functions (for example, plasmas, fields, forces, pulses of energy, particles, waves) to its ever more complex levels of organization and relationship as these emerge over vast reaches of time. Given a sufficiently high level of organization, life becomes possible. And with even higher levels, conscious minds and meanings in all of their manifestations become possible—first in minimal ways and later in the ways that enable me to write this essay and to transcribe it onto the glowing screen of a consciously designed laptop computer.

Purpose and meaning in human existence

Thus, even though the universe or nature has no purpose conferred on it from without or by some kind of superconscious purpose-giving presence or power within it there is abundant purposive activity and opportunity among sentient natural beings of various kinds, including us humans. There is no purpose of the universe, but it is replete with its own emergent purposive meanings and strivings toward meaning. According to Religion of Nature, we can find all the purpose we need for the living of our lives as we respond to, contemplate, and experience the innumerable challenges, opportunities, and meanings afforded by the natural world. This claim includes the existential purposes and meanings so fundamental to religious outlooks and religious faith.

Religion of Nature's answer to the question, "What is the meaning of or lives?" is that we humans can live our lives in order to express and seek to fulfil our maximum possibilities for creativity and good as natural beings—within ourselves, in our relations to other humans, and in our relations to nonhuman life forms and their earthly environments. The meaning of life is the meaning of natural, fully-realized, properly directed human lives, and the purpose of life is to strive ceaselessly as conscious, intentional beings for realization of this maximal natural meaning in everything we think, plan, and do.

This meaning includes such things as aesthetic appreciation and creation, advances in understanding the world, felt and practiced empathetic concern for others including nonhuman others, deepening sensitivity to all that is sublime and holy in the world, the envisioning, building, and maintaining of stable and just human societies, and joyous celebration and gratitude for the precious gift of life. It also includes growing ability to acknowledge and courageously cope with the ambiguities of nature, its inevitable sufferings and sorrows, diseases and disabilities, accidents and disasters, deprivations and losses, and inequitable distributions of gifts and misfortunes to its human and nonhuman creatures.

Environmental ethicist J. Baird Callicott wisely asks, "Why should we have a preordained *telos* to give meaning to our existence and a *raison d'etre*? Isn't it enough that we exist at all? Shouldn't we simply accept the mystery of our existence and pay it the homage it deserves by giving it a meaning of our own making?" (2013: 27). While applauding the spirit of his observation, I would prefer the phrase "meaning of our own *discovering*," because I do not think that we merely construct or invent all of the basic meanings that give point and purpose to our lives. We *find* many things to be intrinsically meaningful in our world; they are not just *made* meaningful by us or by our devising.

The most fundamental purpose and meaning of our lives, from the standpoint of religious naturalism, is to strive with all of our human capabilities to be grateful and responsible citizens of our earthly community, which includes nonhuman life forms as well as human ones and the whole earthly environment of which we are a part. To serve nature, our fellow creatures of nature, and the wellbeing of the planet earth, and not just to expect to be served by nature, is the ultimate meaning of life in a nutshell for Religion of Nature. In this religious outlook, nature as it exists both within us and outside of us—and nothing other than nature in its human and nonhuman, living and nonliving forms—is the appropriate focus of our religious devotion, commitment, and concern. Nature is sacred ground for us emergent beings. It deserves and demands our ardent recognition and response to it as sacred ground. Nature gives birth to us, nurtures us, enriches us, challenges us, empowers us, and sustains us through the course of our lives. Nature surrounds us with mysteries and wonders that constantly awaken astonishment and amazement, especially for minds that discipline and train themselves to be alert, expectant, and receptive. We can be thankful for these wonders even as we stand in awe of them.

Nature's wonders are obviously not always benign. Nature can shock us with sudden or incremental incursions or outcomes that remind us of the fact that we humans are not the sole or even the primary focus of nature, and that its contingencies, processes, and laws can sometimes harm us rather than help us. Nature can have for us as well as for its other creatures a bleak and threatening side as well as a benevolent and supportive side. The ambiguity of nature is the result, at least in part, of the ongoing creations and destructions that give nature its dynamic character. It stems from the numerous conflicts of goods that pervade nature and mark the lives of finite beings. The nourishment of the predator is bought at the price of the death of the prey, for example. The satisfaction of one creature's desires can mean denying or interfering with the satisfaction of another's desires. The best that can be hoped for or worked toward is a relative but significant balancing of needs, desires, and aspirations among the many creatures of earth.

There is considerable ambiguity, moreover, in the choices and behaviors of the human creatures of nature, ranging from deeds of great courage, compassion, and mercy to unbelievable acts of individual and social cruelty and evil. Ambiguity is not a unique fault or defect of nature, but it is an undeniable fact of nature. Formidable and frightening ambiguity also attaches to putative religious ultimates other than nature itself such as God, Shiva, or the Dao, at least to the extent that these relate in meaningful ways to the experienced world. Coping with nature's ambiguity is never an easy task, and in times of great trial, suffering, or loss can be an extremely demanding one. Grief and pain are real and cannot be brushed away by an airy romanticism.

The daunting ambiguity of nature should not deter us from reverencing, serving, and loving it in every way of which we humans are capable. We are fortunate to be conscious participants in the wondrous processes of nature, but with that good fortune also come unavoidable vulnerabilities and demanding responsibilities. We are not helpless in the face of nature's awesome powers, but we should not discount them or foolishly endeavor to lord over them. The nature spread over a vast universe does not depend on what we humans do or do not do. But we have the ability to contribute in lasting and significant ways to the good or ill of nature's creatures and their habitats here on earth.

Our present ecological crisis, with the prospects for thoughtful, effortful, and effective remediation it still affords, gives evidence of this fact. In this fact should reside a substantial portion of the challenge and meaning of our lives as human beings living in the twenty-first century of anthropogenic global climate change, worldwide species endangerment, and rampant habitat destruction. Our lives can have important and appreciable meaning to the extent that we live passionately for nature and do not just idly exist in nature.

The above is not the whole story of a meaningful life, but it is a significant part of it for us natural beings. Drawing on the boundless resources of nature and the capabilities of our own human nature can give joy and delight to our lives and provide incentive and strength for us to explore nature's depth and richness. It can give rejuvenation of spirit, and respite and rest in times of trial and adversity. The meaning of our lives as humans consists in nature's assurance and inspiration, not just in nature's demands. A meditative day in a quiet glen, in the mountains, or in the depths of a dark forest can enable us to return to our daily tasks with new motivation, insight, and hope. A reflective stroll in one's city or contemplation of the ultimately natural sources of everything in one's house or apartment can have this salutary effect, as can a tussle with one's frisky dog or a scratching session with one's itchy cat; and this is to say nothing of quiet attention to the astounding organization and functioning of one's own body. We can look around us and within us and exclaim, "I am part of all this, and I have a responsible role to play in relation to it." Coming fully to comprehend what it means to be a creature of nature and the beneficiary of nature's blessings, as well as being subject to its challenges and susceptible to its ambiguities and dangers—including those posed by actions or inactions of its human creatures—is the prospect and task of a lifetime.

The impersonalism of nature

I can imagine someone objecting at this point that nature by itself does not have sufficient resources to provide a fully meaningful human life. It does not, so the objection goes, because nature is impersonal. It cannot relate to us as what Alfred North Whitehead calls "the great companion-—the fellow sufferer who understands" (1978: 351). There is suffering aplenty in nature, but nature itself does not suffer. It cannot empathize with our sufferings or uphold us with personal help in time of need. We cannot pray to nature or enter into interpersonal communion with nature. Nature cannot personally guide us in the living of our lives. It cannot give us love or forgiveness in the way that a personal God is believed to do.

The Medieval and Renaissance scholar, essayist, story-teller, and Christian apologist Clive Staples Lewis tried at one point in his life to find in the Hegelian idea of the Absolute or in Berkeley's Idealism sufficient religious succor and meaning for his life. The Absolute or Divine of these two perspectives is certainly magnificent, all-encompassing, and awesome in its majesty and might. We can give our love and devotion to it. But Lewis remarks that despite our fervent *eros* reaching up toward it, there is no answering *agape* darting down from it. It is unable to share in our sufferings and delights and is, in fact, indifferent to them and to our lives as a whole. Since nature is like this too, he would claim, it can hardly be viewed as sufficient for a truly meaningful and sustaining religious life. It could be argued to offer us only what Lewis calls the "one-way street" of a "quasi-religion," not the reciprocal relationship and shared communion with a personal God of a fully adequate religious faith (1955: 210, 222–223).

Is a relationship with a personal God really *essential* to a meaningful and fulfilling religious life? I have no doubt that it is deeply meaningful to many religious people, but I do not think that it is necessary for an entirely adequate and sustaining religious outlook on the world or experience of the world. I think that religious naturalism can be such, despite its denial of the existence of such a God. We may desperately want to believe in God, especially in times of grave crisis. But *wanting* to believe and being *justified* in believing are two very different things.

I do not find it credible to believe in a being similar in many ways to us humans that is presumed to exist beyond and to support the whole vast universe. To me, this looks suspiciously like the wishful projection of a human face onto nature and raising it to the "nth" (and incomprehensible) degree, a kind of cosmic humanism or unconscious anthropocentricism. I regard conscious beings like us humans or other sentient beings as emergent from a material nature, not as giving rise to nature, sustaining nature, or existing outside of or along with nature. The notion of a personal, human-like God is to my mind far too small to fill the extraordinary bill assigned to it by theists. For Religion of Nature, the highest form of existence is that which is able to produce and sustain all other forms of being, personal and non-personal alike. This for Religion of Nature is nature itself. I do not set forth these brief remarks as knock-down arguments. To do so would be arrogant and foolish. I offer them mainly as the confessional conclusions of someone who has thought about these matters over the years.

To be without God is far from being bereft of relations with personal beings. Nature is full of "thous," both human and nonhuman. It is not just a collection of impersonal "its." We can commune with one another as human beings and find sources of help, encouragement, and empathy in one another. We can share our joys and sorrows with one another. We can look to one another for guidance. We can experience love in one another's company. We can forgive one another. We can find solace, help, and purpose in our relations with *nonhuman* forms of life. Nature is not bereft of love. There is love to be found within it and our own love to be given to it. And love can be given expression here on earth in individual and institutional forms of social and ecological justice. Our educational, political, economic, religious and other types of institution are of particular importance in this regard. They need in numerous ways to be radically reshaped, reformed, and redirected.

A nature without a personal God can still be suffused with personal relationships, institutional responsibility and care, and the spirit and experience of saving love. In the absence of God, we can turn our attention more firmly toward the needs of the world in our religious lives. There are no guarantees. There is no divine being to pick up the pieces after us or to compensate for the effects of our sins. We are responsible, and we can share this responsibility with others who, like ourselves, care for the wellbeing of the earth and all of its creatures, human and nonhuman

Religion of Nature and death

It is a pretty safe bet that within 150 years after their reading of this sentence every person doing so will have died— some sooner than others, but all of us over that span of time. An issue of great importance for most religious traditions is the inescapable fact of death. How does Religion of Nature propose that we interpret it or cope with it? Can life be affirmed in the face of the inevitability of death? Does not death bring our keenest hopes and aspirations to an abrupt end? Does it not bring crashing down all of the accumulated experience and wisdom, to say nothing of the continuing comforting presence, of the one who dies? And does not the frequent fact of premature death, death not in the fullness of time, call into serious question the meaningfulness of human life as a whole?

If we live only to die, and there is no kind of new life awaiting us beyond the grave, how can our lives here and now be said to be worthwhile? This is an especially pressing and poignant question for those separated by death from persons they earnestly love, persons who have meant most to them in the living of their own lives. If there is no respite from death, its inevitability for ourselves and our loved ones is bound to be a persistently haunting fact, one that casts the shadow of lingering uncertainty and dread—if not threatening despair—over even the brightest of our days. At the very least, this shadow lurks in our unconscious even when we are not fully aware of its menace. We know in the depths of our being that this hour, this day, this week, this year could be the last one for ourselves or for those we love. Death might come slowly or swiftly, but we know that it will come to each of us in due time.

All things material are subject to change, and the forms of material existence inevitably come into being and pass out of being. Even the most soaring, adamantine, majestic mountain face will eventually be eroded away and become part of the plains below. We humans, like all organisms, are material beings. We are born and eventually we will die. We share in this inevitable fact with all earthly creatures. Thanks to such factors as better nutrition, better sanitation, better living conditions, better understanding of diseases, and better medical care, we tend to live longer today—at least in peaceful, prosperous places—than did those of earlier generations. But we still must someday die and sometimes not in the fullness of time.

So-called near death experiences, when they occur, are more likely to be delusions than dependable revelations of a life beyond the grave. Religious claims to resurrection of the body or continued existence of the soul after bodily death are based more on the authority of long-ago teachers or alleged firsthand witnesses to a resurrection than on today's publicly accessible evidence. Philosophical or theological arguments for life after death are generally flimsy and tend, on careful examination, to be unconvincing. If all of this is true, where can we find the courage to live in the face of our own death and the deaths of those we love? How does Religion of Nature approach or propose that we deal with this question?

It does so, in the first place, by questioning the assumption of some persons that life can be fully meaningful and worthwhile in a religious sense only if it lasts forever. The logic of this assumption is not at all clear. Equally unclear is the idea that the loss of a loved one somehow cancels out or makes moot the inestimable value of the loved one or experience with the loved one while he or she lives. Our days as individuals and our time with one another are limited, but that is no reason for us to conclude that they must be utterly tragic and absurd.

This is not to deny the wrenching grief and profound sense of loss when someone near and dear to us dies. The grief and loss are real, and they will continue to ache in our hearts as long as we live. Naturalist, poet, and prose writer Helen MacDonald reflects on their inevitability in each human life:

> There is a time in life when you expect the world to be always full of new things. And then comes a day when you realise that is not how it will be at all. You see that life will become a thing made of holes. Absences. Losses. Things that were there and are no longer. And you realise, too, that you have to grow around and between the gaps, though you can put your hand out to where things were and feel that tense, shining dullness of the space where the memories are.
>
> (2014: 171)

I cannot think of the death of my parents without lamenting their deaths and missing their presence in my life. I experience surges of shock to this day when I must acknowledge once again that I will never be able to greet them, sit comfortably with them, laugh with them, reminisce with them, eat at their table, seek their advice, or bask in their love! But I do not resent the brute fact that my beloved parents had someday to die. This is an inescapable part of what it means to be a natural being, and all of us humans are natural beings. Our natural situation is not an either-or. There is room for both celebration and grief, and neither excludes the other.

We can affirm and be thankful for our relatively brief time as participants in the community of humans and other natural beings and for our conscious experience of the sustaining powers and wonders of nature, even while accepting the fact that the price of our lives here on earth is our inevitable deaths. I remember as a child not wanting to eat the chocolate bunny given to me at Easter because it would destroy the bunny. But I also very much wanted to enjoy the delicious chocolate! Similarly, having lives to enjoy—our own or the lives of those we love—requires that we accept and somehow come to terms with life's inevitable end. We cannot have the challenges and joys of life without the eventual destruction of life by death. And we cannot have them without the shadow of death and of the uncertain time of death for ourselves and those we love as their constant accompaniment.

I remarked earlier that nature's creations and destructions go hand-in-hand. One thing must be left behind in order that another thing can come into being. This is true even moment-by-moment. The past must be left behind in order that the present can come into being. The old must yield place to the new. Even a repetition of the past in the present is not quite that, because there must be a *new instance* of the past in the present. The old instance is left behind in order that the new one can occur. There could be no dynamism or change in nature without destructions, and nature is a dynamic system, as are all of its components.

Think of what the world would be like without routine deaths. There could presently be no lives, because each life form would have had to struggle with all of the others, and none of them would be able to die and thus to leave room for the others to continue to live. In our own case as humans, our population would long ago have become so horrendously large as soon to wipe us out as a species. And if we did not wipe ourselves out in this way—the way of having to live forever once we were born—voracious immortal insects would do so long before we have had sufficient time to bring about our collective demise. If these ruminations are not enough to remind us of how necessary death is to the continuance of life on earth, we have only to reflect on the history of biological evolution. In this history, close to 99 percent of past species have become extinct, and had they not done so, many of the present forms of life, including our own, would not now exist. Ongoing deaths are essential to the ongoingness of life.

Maybe humans are exceptions to this necessary connection of life with death. Perhaps we are special creatures destined for endless life in some other realm. This has long been the standard belief in religions such as Christianity and Islam. But this belief has no credence in Religion of Nature, because nature is the religious ultimate and human beings are creatures of nature. As such, they, like all the others of earth's creatures, are products of biological evolution and subject both to being born and to dying. Tasmanian writer on ecological sustainability Aidan Davison wisely acknowledges that throughout nature "death is the culmination, not the negation, of life" (2001: 210). There is a time to live and a time to die. Religion of Nature and other forms of religious naturalism can help to make us not only keenly aware of this inexorable fact but also able positively to affirm it.

This observation leads to a third way in which Religion of Nature addresses the issue of death, namely, that when we die something of our contribution can live on in the lives of the younger persons who come after us. They, in the freshness of their relatively young age, can go on to transform our contributions into new achievements of their own. In this way, progress is made possible from the older lives to the newer ones. The young are not weighed down by the established and often stubbornly entrenched beliefs and assumptions of the old. The former can dare new thoughts and hazard new paths of investigation. They can build on the contributions of the old without being blinded, inhibited, or engulfed by them. Death in this way of thinking is an instrument of needed progress, change, and refreshment through the generations of human beings. Individual human deaths make room for and can be a necessary stimulus to innovative futures in human history, culture, and life. A fond hope of those who must die is that they can leave something of use behind them, something to be reflected on, improved upon, or set aside. This hope can help to motivate and encourage them while they live. It can help to give meaning to their lives, mortal though they are.

These three lines of thought could be supplemented with others. But perhaps they are sufficient in this brief essay to provide insight into how Religion of Nature can address the fact of death and the pall it threatens to cast over the whole of life. These intellectual considerations do not dissolve or make unimportant the emotional side of the fearsome uncertainty of the impending time of death or the sad experience of the death of others, especially the others close to our own lives. Pervasive death as part of the pervasive life of this earth is indication of the unavoidable ambiguity of the world and life in the world. Frank acknowledgment of this fact may make it easier to cope with death, but it does not eliminate the existential agony and uncertainty of death. Somehow, even after a time of terrible and seemingly unquenchable grief, nature usually enables us finally to live with new confidence and hope when loved ones are lost. Somehow nature or our nature as human beings gives us the courage to live in the fact of death. There is mystery and miracle in the persistence and resolve of human lives when confronted with tragedy and grief that no amount of intellectual analysis can decipher or explain away.

A striking example of such courage and equanimity is this statement of neurophysiologist Oliver Sacks (2015), writing about his own impending death after learning that he had incurable metastatic cancer:

> And now, weak, short of breath, my once firm muscles melted away by cancer, I find my thoughts, increasingly, not on the supernatural or spiritual, but on what is meant by living a good and worthwhile life—achieving peace within oneself. I find my thoughts drifting to the Sabbath, the day of rest, the seventh day of the week, and perhaps the seventh day of one's life as well, when one can feel that one's work is done, and one may, in good conscience, rest.

Sack's words, written while confronted with the inescapability of his own fast approaching death, are much in the spirit of Religion of Nature and of religious naturalism in general. They show clearly how it is possible to live a meaningful life and to leave that life as one's distinctive heritage and gift after one's death. We can strive to accomplish this with our lives in one way or another, and we are indebted to all who have done so in their own lives and are now gone before us. Their "work is done," but its influence continues to live on and have valuable effects in the world.

Sacks lived a rich and full life spanning many years. Not all are as fortunate as he. Some live lives of persistent pain, desperation, sorrow, and disappointment. These lives are sadly unfulfilled. The lives of those who die prematurely or in childhood are undeniably so as well. We should not pretend otherwise. But the latter and even the former—brief or unfulfilled though both types of life are—may contribute in important and cherished ways to instruction and deepening of the lives of those who continue after them. This consideration is not compensation for joyless lives or early deaths, but it may indicate that their subjects have not lived or died entirely in vain.

Mere words can do no justice to such things as the haunting mystery of the approaching death of each one of us, the terrible sorrow evoked by the deaths of loved ones, or the premature, accidental, violent, or suffering deaths of innocent persons. But meaning can be found in the face of the inevitability of death. Each life within the limits of its birth and death contains the possibility of making distinct contributions to the wellbeing of nature and of one's fellow humans as part of nature. The fact of death with no promise of further life beyond does not, or need not, make all of life absurd. The limited field of thought and action this fact affords can serve to give to life urgency and intensity of meaning it might otherwise lack.

Practicing Religion of Nature

This essay has been devoted mainly to the task of understanding how Religion of Nature as a species of religious naturalism can be thought of or conceived. But the question is bound to arise, "How should it be practiced?" How does one go about living as a religious naturalist in today's world? I do not have space here to do justice to this question, but here are a few brief suggestions.

One should be constantly aware of and resistant to the rampant consumerism that is so much a part of the culture of the United States and some other parts of the world today. One should fight against unregulated capitalism and an unconstrained free market ideology that benefit the few at the expense of the many. One should take with utmost seriousness the global climate change and sad endangerments of species that direly threaten the earth, and seek ways to work for their amelioration. One should not base one's career choice on the amount of money to be earned or power and success to be attained but on the best contributions one can make to the earth and its creatures. And one should be generous in the giving of one's time and money to deserving agencies and organizations devoted to social, economic, and ecological justice.

One should support political candidates, policies, and programs that are ecologically alert, well informed, and seriously responsible—and that demonstrate a keen sense of the urgency of finding ways to deal with the severe ecological threats of our day. One should develop a personal diet and a diet for one's family that is respectful of the creatures of nature and their habitats. One should seek to understand the science of one's day, and especially ecological science. One should educate one's children in this science and help to promote educational institutions that give due attention to pressing ecological realities and issues. One should engage in carefully focused meditative practices that help to attune oneself to the marvels of nature and to one's privileges and responsibilities as a devotee of nature.

In the final analysis, the issue of how best to practice religious naturalism in one's own life is a matter to be weighed and considered by each individual in his or her own ways and in light of his or her passions, interests, and strengths. But it is important to recognize that Religion of Nature and religious naturalism in general as religious worldviews are not just matters for thought and reflection but inspirations and incentives for the whole of life and action. Such outlooks are intended to be ways of living and acting, not just ways of thinking.

Even as a general way of thinking, much work remains to be done in refining and developing a plausible, relevant, engaging, and coherent religious naturalist view of the world. In all of these respects, intellectual as well as practical, religious naturalism is an ongoing program and work in progress. It has no experts or final authorities, and it is not immune to probing questions and criticisms. In fact, these are critical to its ongoing development as a convincing and commanding religious way of life.

I write this essay, not in the spirit of trying to proselytize, but in the spirit of sharing what has become religiously plausible, meaningful, and important to me in the living of my life. Religion of Nature may not be the most appropriate and helpful religious outlook for many others, but it is for me at this stage of my life. I respect those with a theistic or pantheistic outlook incorporating some kind of deity, for example, even though I cannot concur in this outlook. Such outlooks can do and have done much good in the world. Ours is a time when all of the religions of the world should give heavy emphasis to thinking about, motivating, and working for the health and wellbeing of the earth and its creatures. All of them have ample resources for doing so.

Note

1 Or, more properly, matter-energy, and I mean to include in this conception of matter radiant energy and anti-matter as well as so-called dark matter and dark energy.

References

Callicott, J. B. (2013) *Thinking Like a Planet: The Land Ethic and the Earth Ethic*, New York: Oxford University Press.

Davison, A. (2001) *Technology and the Contested Meanings of Sustainability*, Albany, NY: State University of New York Press.

Lewis, C. S. (1955) *Surprised by Joy: The Shape of My Early Life*, New York: Harcourt Brace Jovanovich.

Macdonald, H. (2014) *H Is for Hawk*, New York: Grove Press.

Sacks, O. (2015) "Sabbath," *The New York Times*, August 16. http://mwr.nytimes.com/2015/08/16/opinion/sunday/oliver-sacks-sabbath.html?r=0.

Whitehead, A. N. (1978) *Process and Reality: Corrected Edition*, ed. David Ray Griffin and Donald Sherburne, New York: Free Press.

11
THE SOLEMNITY OF THE WORLD

George Allan

In attempting to develop a metaphysical interpretation of the natural world as sacred, I begin, as Aristotle says metaphysicians must, in wonder. Not wonder in the abstract, wondering why there is anything at all or whether God exists. Not wondering in general whether all moral imperatives are categorical or if beliefs that are not empirically verifiable can be meaningful. But rather, concrete wonder, wondering about the world as I find it, the buzzing blooming, coming going, hurting helping menagerie of things and events that I cannot avoid experiencing and therefore remembering and imagining. Here I must begin, from here my generalizations can grow and my abstractions spring.

Necessary contingency

A buzzing mosquito circles nearby and eventually settles on my arm, ready to insert her proboscis and draw out some blood. I short-circuit that intent, however, by swinging my other arm around and swatting her dead. Poor mosquito, her life normally so short, barely hatched and out for the first time in search of food when a force of which she was not aware until too late brought that life to a premature end. She would not have died so soon had she not been in search of food, and she would not have risked such a dangerous food source as myself had what she needed been ready at hand. But it wasn't, so she was left with no choice, for she would perish were she not to eat; and yet she has perished instead in the very quest for things to eat. She needs food, but she also needs protection from becoming food. Both needs, however, are not easily satisfied, both often ending for mosquitoes in momentary failure and after a few brief days in permanent failure. So it is with creatures all around me, whether prematurely or after a normal lifespan, or living to ripe old age: they die. Not just mosquitoes but also spiders and squirrels and purple finches, eagles and elephants and chimpanzees.

I include myself and others of my species among these creatures, understanding that we all go through the same kind of cycle, a journey from origination through a constant struggle for survival to eventual demise. Though my own cycle is not at its end, I understand that all too soon it will be. Thus, by abstracting and generalizing from my mosquito experience, I arrive at the metaphysical hypothesis that all living entities are essentially contingent.

Essential contingency is not an oxymoron, although in one sense it obviously is. Many of the events comprising the life of an organism could have been otherwise without changing what the organism is. Such contingences are accidental to its existence, as are also the time and place and manner of its coming to exist and ceasing to exist. I might have been distracted when the mosquito landed on my arm, and it would have enjoyed a satisfying meal that afternoon, although it might have soon thereafter become a meal for a passing bird or dragonfly.

Organisms are contingent in another sense, however, the sense in which they are necessarily so. They are systems that survive by means of processes the success of which are always at risk, a risk that increases the more complex the system, the more ways in which its functional components can falter and need repair or break down and need replacement. The risk of system failure also increases as the system ages and its processes grow less effective. Organisms are fragile systems that can forestall breakdown but are not able to prevent it. It is a contingent matter that any one of them arises, and a contingent matter how long it will survive, but it is necessary that once begun it will eventually end.

The reason why any individual life must have an end is that it is finite. It is something fashioned from limited resources by some already existing finite thing that has the power to bring it into existence. To exist is necessarily to be another's accomplishment. But these are fragile accomplishments because finite: made from limited resources by makers with limited powers. Although built to endure, they can be unbuilt instead, and if all organic things are made and can be unmade, then the makers of such things must also be themselves fragile achievements. The makers of contingent organisms must be themselves contingent. The power of the makers is contingent in two further senses as well. Although they can try to make or unmake something, they need not succeed in doing so. Nor do they have the power to create something that cannot be destroyed, for it cannot be the case that what is made cannot be unmade. Both the maker and what it makes necessarily can be undone and so necessarily will be.

What is made, given the risks of a contingent process, has worth. It is an achievement that might not have taken place, an integration of available resources into a new reality, but a fragile one. Precisely for this reason, its worth is intrinsic, a fundamental quality of what it is, apart from any additional value it might have because of its pragmatic usefulness to others or its moral importance or aesthetic import. It is precious, its value beyond price, because available only once and only for a while. To exist is to have worth doubly: as an achievement of making, and as a victory, albeit temporary, over the threat of its unmaking.

Fleeing contingency

Humans like all other organisms live out their brief lives attempting to overcome threats to their continued existence, seeking food and shelter, and avoiding or overcoming dangerous situations both inanimate and animate. In addition, however, humans have the capacity to understand that this is so, to grasp the fact that their lives are fragile and finite, and to imagine pathways toward more secure lives. Key to this capacity is language, the ability to use sounds or marks to stand for ideas, for abstract generalizations that refer to the world around them.

So language makes reflective consciousness possible. It expands the world as we know it to encompass not only what can be tasted, touched, smelled, seen, or heard, but also what can be remembered, imagined, abstracted, and generalized. It gives us a world in which we are aware of ourselves as having experienced aspects or features of it once before as well as now, and thereby are able to image how it could have been different or yet might be. The curse of language, however, is that we not only are fragile and finite but know we are, and know as well that the meaning we find in life is also fragile and finite. We come to believe that we need more than pantries stocked with food and shelters covered with mosquito netting. We need to be rescued from our fragility and finitude. We need our lives to be assuredly meaningful, and this they cannot be if at any moment they can abruptly end. We also need the meanings we embrace to transcend us, to last long after we have died, but this cannot be if those meanings are no less at risk that we are. "Out, out, brief candle" (Shakespeare "Macbeth": Act 5 Scene 5) cannot be the final word, for then what's the point of living?

The traditional answer to Macbeth's cry of despair has come in many forms, each in its special way protecting human worth by rescuing it from the threat of ultimate loss. The rescue requires introducing a kind of reality not in need of protection. One form of rescue looks to the past, arguing that the world, and we as a part of it, have been created by a power not dependent on some predecessor power that made it possible, a power that makes all other things possible. Another form of rescue looks to the future, to endings rather than origins, arguing that things happen as they do because they are shaped by a finality that has no successor, a fullness that needs no replenishment. A third form is oriented downward, arguing that things must have a basis, a foundation that has no foundation, on which all else can rest. A fourth form argues that the source of meaningful order lies not beneath but above, at the apex of a hierarchy of kinds of things, running from transient contingencies of little worth to timeless necessities that approach but never quite attain the complete perfection of a superior that has no superior.

All of these ways of rescuing the worth of what we do and are from the world's contingencies, of rescuing the world from meaninglessness, rely on an ontological exception. Even though all things are contingent, it is claimed that they cannot exist except by virtue of what is not contingent. Whether it be a First or Last reality, a Base or a Summit, it is the Other of the things composing the world as we find it. We are temporal entities that come to be and perish, yet what we are essentially is not temporal but timeless. The lives we live, and the things we make by our actions alone or in groups, are finite achievements, but they gain everlasting significance through their relation to an ultimate source. Alas, however, the value of these immortal things is taken as trumping the value of mortal things, which are judged inferior or even illusory. The world as we find it is thereby lost. Better then that we abandon the quest for ontological exceptions, accept instead the ontological necessity of contingency, and set about exploring the ways in which the implications of that ontology offer us meanings worth living by.

Bounded ambients

The world as we find it, because everywhere contingent, needs boundaries. Boundaries order things by setting them within a frame that orients us with respect to what is fundamental. They mark a definitive horizon and so identify an objective center, a midpoint equidistant at all points from that horizon. A boundary locates us within an intelligible system able to distinguish here from there, near from far, now from earlier and later. As this frame fills in, as the order expresses its particular implications, a world becomes manifest, a coherent structure spread out in space and time to its uttermost limits. That part of the world that serves as the primary resource for what an organism is able to accomplish, a resource that it can shape to suit its ends but that reciprocally shapes it, is its ambient (Langer 1967: 282–83). This ambient is meaningful if structured sufficiently so that particulars can be grouped into kinds, arranged sequentially, assessed by degree of worth, across which purposes can be formed and ends pursued, where failures can have reasons from which to learn how better to achieve an end or why a different end should be sought. I will call this kind of world a "bounded ambient."

A bounded ambient provides systemic stability and hence stability of meaning. Therefore the primary task of humans in attempting to make their lives more meaningful is to grasp the formal features of their ambient, the functional connections that organize things, and the causes of those features. Fundamental to them is replication, the components of systems repeating themselves, making it possible for things not merely to occur but to recur. Recurrences are how stability arises even though whatever occurs is unique and therefore fleeting. What recurs is the form, repeated across a content composed of sequences of novel behaviors of an organism, or across sequences of recurrent subordinate forms, or across sequences of organisms. It is the recurrence of the form that turns cacophony into patterns, piles into hierarchies, meanders into trajectories. That turns wilderness into civilization.

These forms are manifold: constitutional laws, governmental statutes, civic rituals, performative scripts, habits, rules of thumb. Across this vast range of forms, the function is always the same: to establish a way of securing similar behavior in different individuals in order to achieve and sustain something meaningful. We fashion a clearing in the wilderness and plant a garden, surrounding it with a stockade tall enough and sturdy enough to keep out wild animals and barbarous strangers. When technology improves, the stockade becomes a castle wall, perhaps complete with moat. After a while, walls may no longer be needed as the community is able to control its boundaries by the distinctive authority of its cultural tradition, perhaps supplemented by strategically placed military garrisons.

The serpent in our ambient garden is the temptation to deny the contingency of the most fundamental of these forms, those laws that assure all we hold important, that ground our sense of who we are and why we exist. Surely, it is thought, these cannot merely be stable. They must be permanent, self-evident in the sense of being unarguably obvious, rational in the sense of providing coherent and adequate conditions of existence fully sufficient for living meaningfully. In the absence of supernatural ultimates to provide such permanent principles of order, we attempt to fashion within the flux of history a social system that approximates them: a well-ordered stable society marked by a well-defined stable perimeter. Eventually, we come to forget that our clearing is an achievement, the result of a struggle to overcome the wild, to impose a system of common laws and the institutions that embody them on a lawless landscape. We come to think of our ambient not as a temporary approximation to some ideal of permanence, but as itself permanent. We are The People who have existed from the beginning of time, holding high the ideal toward which all the world's peoples should strive. Ours is a bounded ambient but an unthreatened one, ancient beyond imagining and as everlasting as the surrounding mountains. We take our good fortune for granted because we are not aware there is anything on which it depends and so nothing we need fear. The old ways and beliefs are self-justifying: they have always worked in the past and so they can be counted on to work in the future.

Thus a bounded ambient becomes another kind of flight from the radical contingency of the world. We imagine ourselves living in a world in no danger of destabilization because stabilized by an ultimate reality, or in a world where destabilization is taken as inconceivable because at odds with the range of possibilities its presumed stability affords. Both interpretations are blind to the dangers their ambient order has avoided, by a bit of good luck and a lot of hard work on the part of its citizens. One interpretation insists on the reality of ultimate ontological conditions and agencies, denying that these realities are of human origin, crafted as a way to achieve very human ends and so only temporary. The other interpretation insists that what is not known to have happened cannot happen, insisting that the ordered conditions giving meaning to life must therefore be permanent and so can be ignored because they are never in danger and hence never at issue.

The fate of such blindness to the necessity of contingency is to perish prematurely. In a world always changing, a hard carapace is at best a short-term adaptive strategy. Without flexibility, an organism or a group or a civilization will eventually face what it does not know how to avoid and will be shattered by unexpected blows for which it was not prepared. Living in a world where all things, including all makers, are vulnerable, we need constantly to do new things in new ways in order to cushion our vulnerability. Meaningful boundaries need to be established, functionally useful procedures put in place, but they cannot be taken as permanent. They must be recognized as having been made for reasons that may at any time no longer be relevant. The wild, in all its unknown power and mystery, must be recognized as an inescapable threat to the pleasant garden of our established ways.

Knowing we remain at risk, however, that our flight from risk has failed, is not enough. We need to use contingency positively, as a way to achieve a meaningful future by transforming our closed ambient into an open one.

Open ambients

In a world that has no ultimates, in which contingency is an inescapable fact, stability can never be guaranteed. It is always a temporary solution to processes that wear away and work around whatever well-wrought structures we might erect, whatever boundary conditions for such structures we might establish. How do we continue to live meaningfully if we abandon belief in a guarantee of the conditions for that meaning? How do we realize that there are limits to the ways by which we can secure our world against the destruction of its protections? How can we discover alternatives to the world as we know it, alternatives that it harbors but hides? Let me suggest three ways by which to break out of the constraints of our bounded ambients in order to further the survival of meaningful conditions for how we might live.

First, there are the strangers we encounter, their strange ways shattering our confidence in the self-evidence of our well-established beliefs and practices. However odd their dress and speech, their attitudes and mannerisms, they are living breathing humans whose ambient, like ours, has boundaries, but ones clearly not ours. Yet these are human-fashioned boundaries, so not unintelligible. To discover a world other than our own by encountering people who embody it is to realize concretely that our world's boundaries can be transgressed. An abstract claim about the virtue of boundary transgression permits us to wonder in leisure about its truth or falsity. But strangers are a living presence that forces our acknowledgment of the viability of their worlds and hints at alternative ways of being human that we might explore—and should; or must.

Art offers us another way by which the boundaries of our ambient are revealed as transgressible. It does so by presenting truths that cannot be said, not by asserting that something is true but by revealing it as true. One fundamental boundary of a society is that created by the limits of its language, for all is silence beyond what can be said. No theory can point our way into that silence, no set of facts pave a path, no rational extrapolation parachute us to a destination of which we are ignorant. It's no wonder that beyond what we can say is thought to be beyond what is real. Linguistic faltering is not due to bumping up against an ultimate boundary, however, but rather a cultural one, the outer limit of a coherent world order thought to be adequate but disclosed by the work of art to be inadequate. Beyond the boundary of our linguistically established world is more reality: things and events, and possibilities of things and events, that our world has left out. Any coherent system, any world, is necessarily finite, for reality is too complex to be contained within a coherent harmonic whole. Art can offer us a glimpse into what our established boundaries have excluded, let us hear what they have silenced, touch what they have denied as real. This beyond that art discloses is more than what we know, not other. It's a wider reality, not a different one. A beyond that is a function of our limited familiar world, and that will change if the boundaries of our world change.

In contrast, metaphors are a use of language that undermines the stability of our linguistic framework and hence the protections of the established boundaries of what can meaningfully be said, thought, or even imagined. A metaphor does this by taking the framework from one world of discourse and using it in place of the appropriate framework for the content of another world. It involves a shift of standpoint from a context where it is appropriate to another where it is not, thereby committing the signal sin of literalism: a category mistake. Except that a metaphor is not a mistake but rather a transformation. Romeo, seeing Juliet on the balcony above him, exclaims:

> But, soft! what light through yonder window breaks?
> It is the east, and Juliet is the sun.
>
> (Shakespeare "Romeo and Juliet": Act 2, Scene 1)

Now obviously, Juliet is by no means the sun; she's a young Capulet girl with whom Romeo has fallen in love. It is daybreak, Juliet's balcony faces west, and Romeo notes quietly to himself, and quite accurately, that the sun is just rising above the horizon behind it. But then he whispers a blatant falsehood, referring to her as the sun even though she is only human. The literal meaning is foolish, but Romeo is not speaking nonsense. He's in love, and lovers know how to use language in ways other than scientists use it, but with no less truth. Romeo takes the astronomical framework offered by the blaze of light announcing the sun's appearance, and uses it to describe his love: Juliet is the sun. A bevy of lovely implications suggest themselves: Romeo is a planet pulled into her orbit, captured by her, nurtured by her presence, dazzled by her radiance, unable to live without her. Negative implications hover at the edges, however: she is dangerous to approach too closely, her grandeur diminishing his to triviality.

An analogy works within a bounded ambient. Juliet is said to be like the sun with respect to something specific, her radiance for example. The framework is the standardly accepted one, within which a comparison is pointed out: there is a radiance to Juliet that is similar to the sun's, but of course different in kind and intensity. Romeo, however, is not making a comparison; Juliet *is* the sun. What this means is fundamentally unclear, the extent to which the astronomical framework alters the Italian content left vague, offering indefinite possibilities for what is meant. His metaphor creates an unresolvable tension between two standpoints, one of them astronomical and one cultural. And yet it is precisely because the metaphor holds these irreconcilables together, holds together a framework with boundaries that have no place for Juliet and another framework for which suns and moons and planets are of no concern, that our taken-for-granted view of things, our familiar way of understanding what goes on, is shaken and new possibilities able to breach its walls. Romeo's claim about Juliet is not factually true, but neither is it merely a rhetorical confection. It is his realization, and ours, of a deep truth about himself: the transforming intensity with which he loves that girl on the balcony. It is a change, but not like his earlier recognition of her genetic heritage. When Romeo realized that Juliet is a Capulet, his existing perspective was affirmed, changing Juliet from an attractive new acquaintance into an enemy he obviously must disdain. But now, realizing that Juliet is for him the sun, Romeo's earlier perspective is replaced by another, the implications of which are unclear, unsettling, and can be true only if taken as meaning that he has discovered that she is for him more valuable in a way fundamentally different in kind and importance than what he had thought.

So boundaries, although necessary, are never fully adequate. They can be breached; there is always more than what is contained in any of our philosophies. Strangers, works of art, and metaphorical expressions are important examples of the kinds of occasion in which we are carried beyond the limits of our familiar ambient, shoved dramatically or lured step by quiet step into the presence of more than we thought actual or possible. But this more is relative to a finite system; there is no absolute beyond.

Boundaries are necessary to meaning. Systems, to function well within the confines of their applicability, must be coherent and adequate. Order is essential to both inorganic and organic existence, to natural processes independent of human control and to human processes, including languages and cultures. But not just any order. Ordered processes need to be dynamic, able to adjust to changes capable of destabilizing them. Human order, to be effective, must be adaptively purposeful, its processes able to be redirected from the pursuit of initially determined outcomes toward ones found to be more likely to succeed. When our efforts at stabilization fail, when the boundary walls crucial to the endurance of our world system are breached, we can simply repair the walls, or buttress them so they do the old job better, or redesign them. All these options, however, result in reaffirming the established reason for the wall being there. What is to be done if the problem is that very reason, and therefore the wall it justifies? What if the fissures in the wall have revealed a beyond that cannot be ignored, that makes it clear the problem is the inadequacy of the system of which the broken walls are boundaries? How do we make the transition from a closed to an open ambient?

To begin with, we need to realize that the beyond of the culturally given is not a destructive agency that will leave us and our familiar purposes and practices bereft of meaning. The unknown is not unknowable. It is an unexplored land, a wilderness into which we can journey and from which we can return, because it is a continuation of the world that we know. It is what our world could have included but did not, a part of our world but without our world's pathways and waypoints, and because knowable, it might be relevant to our needs. In short, the boundless is a resource for reshaping the bounded. There can be no final achievements, no absolute systems, because any bounded system is finite. It leaves things out, with respect not only to its conditions for membership and its design of how member elements should be organized, but also to the shifting features of its ambient. Boundary-making is a journey with waypoints but no fixed destination, because goals in an ever-changing ambient are initially vague and can be clarified only by assuming a stability that is always at risk. There can be no fixed waypoints either, since as the destination changes so does the pathway and hence the waypoints.

So a meaningful open ambient is a creation always in the making. But the power of agents to create what is meaningful and the ability of their ambient to provide the conditions that make this possible must be constantly remade. Both agent and ambient must change so that they can both endure. Together they can marshal the power to invent on the fly solutions that suffice. Organisms interacting with other organisms are able to fashion mutual ambients that complicate these solution-making processes, but in doing so they also make possible new forms of sustainability and the emergence of higher orders of meaning, in particular human civilizations. The power to create a reality that makes it possible to amend that reality, in endless symbiosis, is how there can be a meaningful world in which humans live meaningful lives that increase the likelihood they will be able—for a while—to continue doing so.

Solemnity

I have characterized the world as composed exclusively of necessarily finite things—matters-of-fact of all kinds—but focusing on those with which we are most familiar: ourselves. I have characterized us as complex systems intrinsically related to other systems, oriented toward outcomes that perpetuate ourselves as individuals and as members of groups, creating novel systems that make our individual and collective lives more secure and meaningful. I have questioned approaches for achieving these aims that rely on supposed ultimate non-finite realities, arguing that we have only ourselves on which to rely, selves profoundly influencing and influenced by others: by other humans, other organisms, all the matters-of-fact that comprise our ambient. I have argued that we need to recognize the limitations of closed-boundary finite systems and affirm the adaptive power of those with open boundaries.

This understanding is "religious" in the sense that it affirms and seeks to express in daily life what is of ultimate significance: the making and securing of contingent matters-of-fact by the contingent efforts of contingent creatures in their endless struggle to create and sustain the only intrinsic values there can be. What is ultimate, worthy of our allegiance and source of our meaning, is a world of comings to be and perishings, a limited process with limited products, but of unlimited worth. To explain myself, I will draw out some ideas from Alfred North Whitehead's last two books.

Whitehead begins *Modes of Thought* by suggesting that among the ultimate notions with which philosophers should concern themselves two are fundamental: importance and matter-of-fact (1938: 4). Anything created "expresses its nature as being *this*, and not *that* ... [I]t expresses exclusion; and exclusion means finitude." But this finitude is "combined with the sense of modes of infinitude stretching beyond each finite fact" (78). What is merely finite is trivial; what is merely infinite is vacuous. "Importance arises from this fusion of the finite and the infinite" (79), an importance best expressed in the phrase "Have a care, here is something that matters!" (116).

The experience of actualities as mattering has on Whitehead's analysis three aspects, no one of them more significant than the other two: Intensity, Externality, and Totality. I sense my own intrinsic importance, my here-and-now factuality as a unique person with my unique story of achievements and failures, striving as best I can to live as well as possible. I'm also aware of other matters-of-fact, ones of which I am right now aware, but also those previously encountered or heard about or imagined as possible, friends and enemies, animal, vegetable, mineral, each with its own intrinsic importance. I am aware of myself and these others comprising a whole world stretching indefinitely in all directions, modes, and manners, a boundless totality of intrinsically important finite achievements.

My recognition of matters-of-fact as important in this threefold way is how I come to sense the "full solemnity of the world" (78). A sense of solemnity is not an experience of the sublime, of standing in awe at the immensity of the world, its vast measureless totality. It is not a spectacle that I behold but a reality of which I am a part. For I and my ambient are interdependent, and my sense of the totality of the world is my sense of the extent of that ambient. It is my awareness of the "necessary relevance" of my concrete here-and-now self, of this that I have made of my life, to what lies "beyond its own limitations." This sense of relevance beyond my finite matter-of-fact self is my "perspective of the universe" (79), a perspective that is necessary, for "mere matter-of-fact refuses to be deprived of its relevance to potentialities beyond its own actuality of realization" (83). My perspective, when it includes an appreciation, however dim, however far short of specific knowledge, of the perspectives of all other matters-of-fact, is the sense of totality. Totality is therefore not a reality other than myself. It is my importance, that of what I am that reaches beyond me, and that reaches beyond every other matter-of-fact, because it is our common creation.

"Actuality," says Whitehead, "is the self-enjoyment of importance. But this self-enjoyment has the character of the self-enjoyment of others melting into the enjoyment of the one self" (117–18). He immediately offers as "the most explicit example" of what he means the fusing of the remembered various self-enjoyments of the sequence of our recently past selves with our current self-enjoyment (118). The "one self," in other words, is the unique individual, myself for instance, whose present achieved actuality includes a sense of other individuals with their differing actualities, including those the actuality of which is only possible, and is at the same time a sense of myself and all those others selves as a boundless ongoing interdependent whole.

Crucially, the value of each individual matter-of-fact is intrinsic, including in its value the value of all other matters-of-fact, and hence including "our sense of the value, for its own sake, of the totality of historic fact in respect to its essential unity" (119). Whitehead's immediate example is "the subtle beauty of a flower in some isolated glade of a primeval forest." No creature with the capacity to do so has ever experienced that beauty, including the flower itself, and yet its beauty "is a grand fact in the universe" (120). Things that have no utility for us, and seemingly none for anyone else or any other thing, have intrinsic worth in their own right, each of them a grand fact in the universe. Totality so experienced, resplendent with a recognition of "the value of the details for the totality," is an experience of "holiness," of "the sacred" (120). The value of the totality is the value of its details, a value not reducible to the details but vacuous without them. To sense how precious each particular thing is, and how crucial it is to why every other particular thing is precious, is to sense how precious the totality is, how interdependent the fragile finite makings of the world. This world, the grand fact that is the totality of grand facts, is sacred because it is beyond price, holy because intrinsically precious. Qualities that are transferred in traditional religions to a supernatural Creator or Ground, Source or Guide, belong properly to the natural order, the world of particulars in communion, endlessly creating value and created by it.

At the end of *Adventures of Ideas*, Whitehead describes the power of beauty—of finite ideals—to lure us within the limitations of our ambience to seek their actualization, to make under their tutelage new matters-of-fact. But these are finite powers, functioning within a finite ambience, and in their exercise we can only succeed approximately if at all. Hence the sense of tragic beauty and the intuition of peace to which it gives rise.

The recognition of tragic beauty has been a fundamental theme of this essay, echoing Whitehead's instance that "Decay, Transition, Loss, Displacement belong to the essence of the Creative Advance" (1933: 286). Acknowledging these facts is not to suggest Macbeth's counsel of despair, however, because they are necessary features of finitude. The flower in the forest may bloom gloriously, but it might wither before it can bloom because of draught or be destroyed in a storm or eaten by a grazing herbivore. And yet it matters, as do the words in this essay, as do Romeo's words to Juliet, as does the creation of civilized societies. For they are matters-of-fact, the results of processes oriented toward the actualization of beauty, the creation of new particulars.

It is natural that we appreciate, and right that we should, the intrinsic value of each of the particular achievements of the immense congregation that comprises the totality of the world, including those that fell short of what they could have been because of their own or their ambient's limitations. These achievements, meager or cornucopic, are not only intrinsically valuable, however, but also have instrumental value as resources for our own creative efforts. These resources include especially the ideals those achievements failed in part or whole to actualize, ideals not lost but in having become integral to those efforts have been made available for its successors, including us, to actualize. In recognizing that our efforts fall short, and that they will be fleeting even should they succeed, we admit the inescapable tragedy of having ideals. However, argues Whitehead, we should also see "the tragedy as a living agent persuading the world to aim at fineness beyond the faded level of surrounding fact. Each tragedy is the disclosure of an ideal: —What might have been, and was not: What can be. The tragedy was not in vain" (286).

Loss is inevitable and all too often unredeemable, but such "gross evil" is not "tragic evil," because the tragedy is motivating, an urgent appeal to others to take up a task anew, to attempt in a new time under new conditions its completion. "The inner feeling belonging to this grasp of the service of tragedy is Peace—the purification of the emotions" (286). Totality is the community of finite things made by predecessors and making successors, sustaining and undermining one another, succeeding and failing, emerging and perishing within an open ambient of boundaries and possibilities. Peace is the sense of belonging to this community and, because tragic beauty is its essence, celebrating it. "Amid the passing of so much beauty, so much heroism, so much daring, Peace is then the intuition of permanence" (286) and "the individual whose strength of experience is founded upon this ultimate intuition, thereby is extending the influence of the source of all order" (292), which is the drive to make and preserve what matters.

Whitehead hesitates about what word he should use to name this intuition (284–85). He rejects Platonic terms like "Harmony," "Eros," "Ideas," and "Mathematical Relations" because of their "absence of 'life and motion.'" "Tenderness" and "love" emphasize the rejection of "restless egotism," but they are "too narrow" because specializations of what he wants. "Impersonality" is "too dead." So Whitehead settles for "Peace," immediately hastening to insist it is "not the negative conception of anaesthesia" but a "positive feeling which crowns the 'life and motion' of the soul." Nor is it "a hope for the future" or "an interest in present details," not a focus on importance or a focus on matter-of-fact but rather on their integration. It is a "broadening of feeling," the first effect of which is "the removal of the stress of acquisitive feeling arising from the soul's preoccupation with itself." Peace is "a surpassing of personality" through "a trust in the efficacy of Beauty," its "self-justification." It is "a grasp of infinitude, an appeal beyond boundaries" whereby "interest has been transferred to coördinations wider than personality."

Whitehead tends here to emphasize the problem of narrow self-interest so much that we might overlook that his appeal is not simply to infinitude but to totality, to the infinitude of the efficacy of Beauty. When we trust it, this teaches us that what we can do is what others could and can and will be able also to do: to create what matters, with the help of others and on their behalf. To realize that I am not alone, that my inadequacies are conditions of my humanity, and that what is true of humans is true of all things, that their value lies in what they are, and that what they are is a gift from what has preceded them, offered to them to remake as a gift for what follows after them to shape as best it can.

"At the heart of the nature of things, there are always the dream of youth and the harvest of tragedy" (296). At the heart of things. Comprising the world's essence, its very nature. Both the dream and the harvest. Always. Always ideals of what should matter, embraced and pursued. Always with results that fall short. Always lessons learned, hopes chastised. The sense of Peace is "the immediate experience of this Final Fact, with its union of Youth and Tragedy" (296). All those dreams, all those partial failures, including my dreams and my partial failures, united in one ongoing Totality stretching indefinitely in all modes and dimensions. My life finds its meaning as a matter-of-fact not only in this me here and now, not only in all the other thises and thats there have ever been and ever will be, not only in the intrinsic value of each, not only in the infinitude of those meanings. But in their union, in the great Matter-of-Fact that is the world ever-changingly what it becomes. Through this sense of the solemnity of the world, which is the intuition of Peace, "the World receives its persuasion towards such perfections as are possible for its diverse individual occasions" (296).

References

Langer, S. K. (1967) *Mind: An Essay on Human Feeling*, vol. I, Baltimore: Johns Hopkins Press.
Shakespeare, W. (2008) "The Tragedy of Macbeth," in S. Greenblatt (gen. ed.) *The Norton Shakespeare*, second edition, New York: W.W. Norton, pp. 2569–2632.
—— "The Most Excellent and Lamentable Tragedy of Romeo and Juliet," in S. Greenblatt (gen. ed.) *The Norton Shakespeare*, second edition, New York: W.W. Norton, pp. 897–972.
Whitehead, A. N. (1967 [1933]) *Adventures of Ideas*, New York: Free Press.
—— (1968 [1936]) *Modes of Thought*, New York: Free Press.

PART IV

Ecology, humans, and politics in naturalistic perspective

12

SPIRITUAL ECOLOGY AND RELIGIOUS NATURALISM

Exploring their interrelationships

Leslie E. Sponsel

Earth Day

On April 22, 1970, the first Earth Day was celebrated in the U.S. It was a national teach-in, brainchild of Gaylord Nelson, a Senator from Wisconsin. Some 20 million people assembled throughout the country in streets, parks, and auditoriums, and especially on the campuses of thousands of colleges and universities. Participants demonstrated against the degradation of nature and in favor of sustainability and a healthy environment free from pollution. A variety of many different organizations were involved that had been fighting against separate issues, such as oil spills, polluting factories and power plants, toxic waste dumps, pesticides, loss of wilderness, and wildlife species extinction. Since then celebrations have continued annually. They contributed to the establishment of the Environmental Protection Agency and the Clean Air, Clean Water, and Endangered Species federal government acts in the U.S.

Within two decades, Earth Day grew internationally to be embraced by people in 141 countries. Now more than a billion individuals participate, about one in every seven humans on the planet. Earth Day engages concerted action to change human environmental behavior to become greener and to generate appropriate improvements in government environmental policies from the local to the national levels (Christofferson 2004: 302–312).

Since 1970, a multitude of diverse secular approaches have been pursued with increasing vigor, such as environmental stewardship and management, sustainability, and recycling. These and many others are surely most worthy pursuits with numerous significant achievements (Myers and Kent 2005). Secular approaches are absolutely necessary, but most unfortunately, they have not proven sufficient to turn things around for the better as a whole. More often than not they treat specific superficial symptoms, rather than the ultimate causes of environmental problems. Furthermore, they are inherently anthropocentric with the associated arrogance including speciesism (Jensen 2016). Despite such secular efforts, environmental problems persist, some are getting worse, and new ones continue to emerge. Far more is required beyond secular approaches alone. Perhaps religion and spirituality can help significantly advance environmentalism. That is the conclusion and pursuit since the 1990s of many secular environmental organizations such as the Sierra Club, World Conservation Union (IUCN), Worldwatch Institute, and Worldwide Fund for Nature (WWF) (Ayres 1999; Gardner 2006; Speth 2008; Sponsel 2012).

Spiritual ecology

In contrast to most secular approaches, spiritual ecology cultivates environmental humility; ecocentrism; the inherent value of beings, things, and forces in nature; and even nature itself as sacred and spiritual, thereby requiring reverence as well as respect and care, attributes overlapping with deep ecology and dark green religion (cf. Taylor 2010). Nature is not viewed merely as a warehouse of resources to rapaciously exploit for profit and greed, or as a public sewer to carelessly dump into waste and pollution. From the ecocentric perspective of spiritual ecology, nature is an awesome sanctuary to cultivate spiritual development, ultimately something infinitely more valuable than any economic development (e.g., Berry 2006).

From an academic perspective, spiritual ecology focuses on the interfaces of religions and spiritualities with environments, ecologies, and environmentalisms. Each of these qualifiers is plural because spiritual ecology examines a vast, complex, diverse, and dynamic arena of interrelated phenomena (Sponsel 2012, 2014). As an umbrella category, spiritual ecology embraces other narrower approaches, such as dark green religion, deep ecology, earth spirituality, earth mysticism, ecomysticism, ecopsychology, ecospirituality, ecotheology, green religion, green spirituality, nature mysticism, nature religion, nature spirituality, religion and ecology, religion and nature, religious ecology, religious environmentalism, religious naturalism, and sacred ecology.

Many of these diverse approaches overlap and interact synergistically through reinforcing and amplifying one another. Collectively they are part of spiritual ecology as a quiet, diffuse, and nonviolent revolution that is worldwide and gaining momentum and influence (e.g., Hawkin 2007). Some refer to this as the Great Turning (Macy and Johnstone 2012: 26–27). Thus, spiritual ecology is analogous to a vast river like the Amazon with its thousands of tributaries that feed into movement downstream as the main river increases in size and power. This quiet revolution could also be revealed and documented by searching Google for the wealth of information on various approaches mentioned above, many reflecting environmental activism on the ground beyond cyberspace.

The term spiritual ecology is chosen to be as inclusive as practical. This is in contrast to the term religious ecology preferred by John Grim and Mary Evelyn Tucker (2014) as they concentrate on so-called world religions, and Bron Taylor (2010) on dark green religion and related phenomena. (A search of Amazon.com and Google would reveal other authors who use the term spiritual ecology.)

The qualifier spiritual is also used, instead of religious, to be more inclusive. While most adherents to some religion may be spiritual, many individuals are spiritual without identifying with any particular religion (Harris 2014; Saint-Laurent 2000). A survey was conducted by the Higher Education Research Institute of UCLA of 14,527 new students from 136 colleges and universities in the U.S. during 2003–10. The majority were spiritual, but not necessarily religious. A survey by the Pew Research Center on "America's Changing Religious Landscape" found that since their last project of its kind in 2007, the number of religiously unaffiliated adults rose by 19 million. Now there are around 56 million of them, a group more numerous than either Catholics or mainline Protestantism in the U.S. These two surveys demonstrate one reason why it is most important to consider spirituality as well as religion, otherwise an enormous amount of potentially relevant phenomena are ignored or neglected (cf., Gottlieb 2013; Sponsel 2012).

Humans are spiritual animals (Beauregard and O'Leary 2007; Newberg 2010). Spirituality is an elemental and often pivotal manifestation of human nature; even atheists, agnostics, non-theists, and secular humanists may be spiritual (e.g., Comte-Sponville 2008). Religiosity and spirituality resurfaced after decades of suppression in China and the U.S.S.R. Although many societies have experienced secularization, individuals still pursue spirituality including through creating alternative religions (e.g., Johnson and Ord 2012). Some kind of spirituality may develop prior to any religious knowledge because it appears to be inherent in children (Hart 2003). There are even claims that some animals may be spiritual (e.g., Goodall 2005).

Individuals and organizations may engage in spiritual ecology in one or more of three ways: intellectual (scientific and academic), emotional (spiritual and ritualistic), and practical (environmentalism including activism). These three are interrelated and often overlap in many ways and degrees.

The intellectual component of spiritual ecology grew exponentially since the 1990s. It has been advanced by numerous international, interfaith, and/or interdisciplinary conferences, publications, and other initiatives. For example, from 1996 to 1998, Tucker and Grim were the primary organizers of a series of conferences at Harvard University, each on a different world religion in relation to ecology, yielding a series of ten edited volumes with Harvard University Press. These initiatives generated in turn the Forum on Religion and Ecology (FORE) with its website and monthly newsletter, the latter now with more than 12,000 subscribers (Grim and Tucker 2014). (Today there are parallels to FORE in Australia, Canada, and Europe.) Also, Grim and Tucker were contributors to the development of the journal *Worldviews: Global Religion, Culture and Ecology*. As another example, in 2007, Taylor organized the inaugural meeting of the International Society for the Study of Religion, Nature and Culture at the University of Florida. Also he launched the *Journal for the Study of Religion, Nature and Culture* (Taylor 2005).

By now academic spiritual ecology has accumulated a history, pioneer contributors, common identity and concerns, foundational literature including the aforementioned periodicals, organizations, conferences, and websites as well as courses, graduate degree programs, and the first generations of students (Bauman et al. 2011; Gottlieb 2006a,b; Grim and Tucker 2014; Jenkins et al. 2016; Taylor 2005, 2010; Sponsel 2012, 2014). By various names, graduate studies related to spiritual ecology are available in more than two dozen institutions, such as the California Institute of Integral Studies, Drew University, University of Florida, Graduate Theological Union, Vanderbilt University, and Yale University. The American Academy of Religion has involved a religion and ecology interest group since 1991 which organizes sessions at the annual conventions.

Religions can provide unique and indispensable resources, values, motivation, venues, guidance, and rituals for transformation toward more life-sustaining and life-enhancing lifestyles, communities, and societies (Gardner 2006). In their recent textbook Grim and Tucker (2014: 86–87) describe three interpretive methods for research on religious ecology. Retrieval identifies ecologically relevant aspects of a religion, such as its ethics and rituals. Reevaluation engages aspects that are adaptive to cultivate more ecologically sensitive attitudes and practices by adherents of a religion. Reconstruction indicates how ecological principles and practices may be creatively applied to a religion in order to help deal with contemporary environmental concerns. Many purists and conservatives might object that the second and third methods could change a religion, but this neglects the fact that over millennia religions changed in response to new conditions in order to remain relevant to new generations of followers.

Turning to the emotional component of spiritual ecology, some individuals and groups perform cathartic rituals of mourning and healing for recovery from their grief over environmental degradation. For example, the Council of All Beings developed by John Seed and others focuses on enhancing empathy for other species by identifying with them through role playing (Seed et al. 2007). Such rituals may empower environmental activists, as the writings, workshops, and website of Joanna Macy demonstrate (Macy and Johnstone 2012). Rituals in spiritual ecology have been neglected by researchers, but they are an important means of confronting an era of increasing environmental anxiety, suffering, and grief.

Some traditional religious rituals are becoming more environmentally sensitive and responsible too. For instance, religious pilgrimages engage more than 200 million people annually. Recognizing this, in 2011, the Alliance of Religions and Conservation (ARC) initiated the Green Pilgrimage Network to encourage such activities to be more environmentally friendly. Accordingly, dozens of pilgrim cities, such as Assisi in Italy, are involved in long-term planning. One aim is for the pilgrim's experience to generate a greener lifestyle afterward.

Many environmentalists are ultimately spiritual ecologists, although usually this is not admitted explicitly, especially if they identify themselves as scientists (Essen 2010; Takacs 1996). Nevertheless, many environmentalists have had experiences in nature that they may recognize not merely as extraordinary, but as religious, spiritual, or mystical. Such experiences are among the contributing factors motivating them to pursue environmentalism. Recent studies have explored the religious connections of conservationists and other environmentalists, such as in the development of and/or experience in national parks and wilderness (Berry 2015; Mitchell 2016; Stoll 2015).

The activism component of spiritual ecology may be illustrated by the initiatives of three contemporary leaders. Biologist Wangari Maathai (2004, 2010) started the Green Belt Movement in Kenya in 1977. Since then more than 51 million trees have been planted in Kenya. By 1986, her project had spread to 170 countries with the planting of some 12 billion trees as a project of the United Nations Environmental Programme (UNEP). Bartholomew I, known as the Green Patriarch, leads more than 300 million Orthodox Christians. Since 1997, he has facilitated a succession of international, interdisciplinary, and interfaith symposia aboard ships sailing major water bodies of the world to publicize and remediate pollution and related environmental problems (Chryssavgis 2011). San Francisco Episcopalian Minister Sally C. Bingham (2009) launched Interfaith Power and Light. This organization promotes energy efficiency and conservation among other practical efforts in response to the challenges of global climate change, such as installation of solar panels on religious buildings. Since 2000, this project has expanded to all 50 states of the U.S. in more than 18,000 religious centers collectively engaging six million people. Each of these three individuals, like Nelson, proves that the ideas and actions of just one person can stimulate far reaching positive ramifications and consequences. As anthropologist Margaret Mead famously said, never doubt that one individual can change the world. These three individuals also illustrate some of the power, potential, and promise of spiritual ecology (cf., Ellingson 2016; Hawkin 2007; Hope and Young 2000; Sponsel 2012).

Sacred places in nature are one of the most concrete manifestations of spiritual ecology. Sacred places are specific sites, areas, and/or landscapes possessing one or more attributes that distinguish them as somehow extraordinary, usually in a religious, spiritual, or mystical sense. In them individuals may sense variously awe, mystery, power, attraction, oneness, healing, revelation, ecstasy, epiphany, and/or transformation. Whether or not an individual believes in spiritual forces and/or beings in nature, many are profoundly impacted emotionally through experiencing sacred places (Chalquist 2007; Goodenough 1998: 173; Gray 2007; Swan 1991).

Secular organizations also recognize many different sacred places in nature. For example, Sierra Club founder John Muir and others developed the system of national parks as sacred places in nature in the U.S. (Worster 2008). In turn, many other countries were inspired to establish such protected areas (Mitchell 2016). For over a decade now, ARC, the International Union for the Conservation of Nature and Natural Resources (IUCN) or World Conservation Union, and other organizations have been exploring the relevance of sacred places in nature for biodiversity conservation (e.g., Edwards and Palmer 1997; Verschuuren et al. 2010).

Such environmental actions reflect a simple but elemental, profound, and pivotal admonition by the Catholic priest and scholar Thomas Berry (2006:17): "The universe is a communion of subjects, not a collection of objects." This is the essence of spiritual ecology. If this admonition were pursued seriously in practice, then it would help challenge the rampant and rapacious capitalism, materialism, consumerism, and associated aspects of globalization that have become like a cancer infecting the planet (e.g., Meadows et al. 2004). Nature would be recognized to be spiritual as well as material, and consequently it would be approached with far more respect, reverence, and care. Probably then human environmental impact would be reduced significantly, and in many areas nature would heal itself (e.g., Weisman 2008). Spiritual ecology strives to awaken people to such matters, and to help them find their own pathway toward a greener relationship with the environment. Thereby spiritual ecology could help facilitate a transformation toward a new Ecological Age or Ecocene.

As a quiet revolution, spiritual ecology is diverse, diffuse, nonviolent, and still not generally known. It is also revolutionary in the sense that, as Berry's admonition implies, it calls for no less than a radical re-thinking, re-feeling, and re-visioning of individual lifestyles as well as communities and societies as a whole in relation to nature, a fundamental shift in consciousness and corresponding behavior (e.g., Bourne 2008; Hawkin 2007; Korten 2006). In the process, the societies and cultures of the descendants of the original indigenous peoples can serve as one source for heuristic adaptive models and inspiration. The fact that many such societies endured for centuries, some even over millennia, proves that they were sustainable ecologically, economically, and socially (IUCN 1997; Pfeiffer 2013; Sponsel 2012: 21–30).

There is some hope that this revolution may help turn things around. However, spiritual ecology can be considered radical and subversive because it challenges the power and momentum of the hegemonic and rigid status quo of many social, economic, political, religious, scientific, and academic institutions and their associated worldviews, values, attitudes, customs, and interests (Gottlieb 2006a: 215–233; Grim and Tucker 2014: 13–18; Sponsel 2012: xviii–xix; Taylor 2005: 217–220). For instance, many religious conservatives and fundamentalists are leery of spiritual ecology, fearing that it might lead to a reversion to nature worship or Paganism (cf., Harvey 2013; Higginbotham and Higginbotham 2002). These critics may give expression to ignorance, arrogance, and intolerance as well as anthropocentrism. Yet perhaps the most serious obstacle of all is the discrepancy between positive religious ideals and corresponding individual actions (e.g., Alley 2002; Wexler 2016). Such obstacles must be reduced, if not entirely resolved, if spiritual ecology is to make more substantial progress as a practical action movement beyond academia.

One means of surmounting such obstacles is to counter alienation from nature by encouraging individuals and groups to reconnect emotionally with its awesome beauty, wonders, powers, and mysteries (Bekoff 2014; Coleman 2006; Lionberger 2007). Many children in cities suffer from a nature-deficit disorder, which also needs to be remedied (Louv 2008, 2011). There is accumulating evidence from medical, psychological, and other scientific research demonstrating that nature can help restore and maintain the health of individuals, communities, and societies (Essen 2010; Selhub and Logan 2014). Ecopsychology is one field devoted to that effort, and it overlaps with spiritual ecology (Chalquist 2007).

Spiritual ecology is quite remarkable in generating collaboration in two arenas that have previously been in tension and conflict for centuries. First, it stimulates and facilitates ecumenical and interfaith engagement and collaboration, in keeping with the compelling common concern with environmental deterioration. For instance, one of the most historic events of interfaith collaboration is the "Common Declaration on Environmental Ethics" authored by Pope John Paul

II and Bartholomew I in 2002, the respective leaders of some 1.2 billion Roman Catholics and 300 million Eastern Orthodox Church members. They affirm:

> The problem is not simply economic and technical; it is moral and spiritual. A solution at the economic and technological level can be found only if we undergo, in the most radical way, an inner change of heart, which can lead to a change in lifestyle and of unsustainable patterns of consumption and production.

However, there is the question of how many followers are influenced significantly by such statements, a matter meriting research (see Chryssavgis 2011 and Lorbiecki 2014).

A second unprecedented development is that science and religion are increasingly finding common ground in the persistent and worsening environmental crises of the planet. For example, in January 1990, an international declaration was issued: "Preserving and Cherishing the Earth: An Appeal for Joint Commitment in Science and Religion." This historic appeal was signed by 32 distinguished scientists, among them astronomer Carl Sagan, botanist Peter Raven, chemist Paul J. Crutzen, climatologist Stephen H. Schneider, geologist Stephen Jay Gould, and physicist Freeman J. Dyson. One of the points that the signers assert is:

> As scientists, many of us have had profound experiences of awe and reverence before the universe. We understand that what is regarded as sacred is more likely to be treated with care and respect. Our planetary home should be so regarded. Efforts to safeguard and cherish the environment need to be infused with a vision of the sacred. At the same time, a much wider and deeper understanding of science and technology is needed. If we do not understand the problem, it is unlikely we will be able to fix it. Thus, there is a vital role for religion and science.
>
> (Sagan et al. 1990)

It is a mistake to believe that science and religion are inevitably antithetical or antipathetic, or that scientists are necessarily atheists, agnostics, or nontheists (e.g., Clayton and Schaal 2007; Clayton and Simpson 2006; Frankenberry 2008; Jaeger 2012).

The bottom line is this—if not pursued voluntarily in an informed and enlightened manner, profound transformations may well be forced on humankind by the circumstances of global ecological disasters such as climate change, and that at enormously greater expense involving human suffering as well as economic and social costs. Ultimately, the choice is between either ecocide or ecosanity. If this seems alarmist, apocalyptic, and the like, just consider the consensus in international scientific sources, such as the *Millennium Ecosystem Assessment* of the U.N. Environmental Programme (UNEP), the Intergovernmental Panel on Climate Change (IPCC), WWF's *Living Planet Report*, and Worldwatch Institute's annual *State of the World* reports. Furthermore, consider that the ecological footprint (environmental impact) of humans since WWII is actually visible in the geological record, leading to the formal recognition of a new era called the Anthropocene (Schmidtz 2016; Schwagerl 2014). Time may be running out before a threshold or tipping-point is reached, catalyzing worldwide ecocatastrophe.

It is not only foolish to frame such matters as environment versus economy; it is suicidal for the human species. If some people think that the environment is not important, then they should experiment with seeing how long they can stop breathing air. Moreover, pathological levels of air pollution in China, India, and elsewhere should be another wake-up call. Any population, economy, or society can only be as healthy as its environment.

Long-term concerns must be prioritized over short-term ones. Sustainability is far more vital than economic gain. A case in point is the island of Zanzibar off the east coast of Africa where non-traditional and unsustainable dynamite fishing in reefs has developed. Secular conservation approaches by the government and WWF over four years were unsuccessful in halting the dynamiting. Finally, WWF collaborated with the ARC, Fazlun Khalid of the Islamic Foundation for Ecology and Environmental Science, and local Muslim leaders in creating workbooks in Swahili about ecologically relevant points in the Qur'an. These were used in workshops with fishermen, and the dynamiting stopped immediately. In this case the solution was bottom-up through the local religion, instead of top down from outside secular agents (ARC 2016).

To secular initiatives spiritual ecology offers an additional approach; perhaps it may finally help turn things around much for the better. Maybe it can be an antidote to the existential dilemmas of the Anthropocene. Humans and nature are dynamic and both can be resilient (Weisman 2008). It may not yet be too late to replace maladaptive systems with adaptive ones, to find a better direction for humanity and the planet to change from the Anthropocene to the Ecocene. Religious naturalism has a vital role to play here as well as spiritual ecology in general (cf., Dalai Lama XIV 2011; De Botton 2012).

Religious naturalism

Spiritual ecology has grown and progressed to the extent that comparisons are being made among religions to identify underlying common denominators or parallels in their relevance for ecology. David Kinsley (1995: 227–232) identifies ten parallels among existing religions:

1. Many religions consider all of reality, or some of its components, to be an organic whole or a living being.
2. There is an emphasis on cultivating rapport with the local environment through developing knowledge about it and practicing reverence for it through ritual celebrations.
3. The human and nonhuman realms are directly interrelated, often recognizing some kind of kinship, and in certain cases, with animals viewed as other forms of persons or humans.
4. The appropriate relationship between humans and nature should be reciprocal; humans do not merely recognize interdependence, but also promote mutually beneficial interactions with nature.
5. Ultimately, the dichotomy between humans and their environment is nonexistent; humans are embedded in nature, integral part of the larger whole or cosmos.
6. This non-dualistic view reflects the ultimate elemental unity of all existence; nature and spirit are inseparable, there is only one reality which can be sensed and experienced.
7. This underlying unity is moral as well as physical; humans and nonhumans participate in a shared moral system; environmental issues are first and foremost ethical concerns; and nature has intrinsic as well as extrinsic values.
8. Humans should act with restraint in nature by avoiding the anthropocentric arrogance of excessive, wasteful, and destructive use of the land and other resources; and in other ways they should exercise proper behavior toward plants, animals, and other aspects of nature as sacred.
9. Harmony and balance between humans and the rest of nature must be maintained and promoted and if it is upset then restored.
10. Frequently the motivation, commitment, and intensity of environmental concerns are religious or spiritual.

[See Pederson (1998) for another list of parallels. That list is also reprinted in Grim and Tucker 2014: 12.]

These parallels resonate in various ways and degrees with the so-called world religions, Animism, Paganism, Pantheism, and religious naturalism. In the latter case, for example, Donald A. Crosby (2013: 139–140) mentions that humans are an integral part of the biotic community; have special responsibilities toward nature; and should act with reverence toward other beings and contribute to their welfare. He refers to "numinous nature" (50, 126–127) and the "sacredness of nature" (128, 130). (For related views see Goodenough 1998: 167–174 and Rue 2011: 114–116.)

Kinsley addresses Animism and several world religions, the former a belief in spiritual forces and beings in nature, the latter usually theistic, except for Buddhism in principle (Batchelor 1997). Crosby (2013: 103–104, 139) also notes that similar beliefs about nature are shared by religious naturalism and some other religions. However, there are also incompatibilities among religions, and while spiritual ecology has had other priorities, eventually these will have to be scrutinized if relevant to ecology.

Convergences in thinking about religions in relation to ecology are remarkable because they come from otherwise divergent sources. Moreover, convergences are promising for facing challenges of deleterious human environmental impacts and creating more friendly environmental relations. As religious naturalist Ursula Goodenough (1998: xv) remarks: "Without a common religious orientation, we basically don't know where to begin, nor do we know what to say or how to listen, nor are we motivated to respond." The Earth Charter Initiative (2017) is an important step in this direction too.

Another commonality between spiritual ecology and religious naturalism is the informed and critical realization that many environmental problems and crises from the local to the global levels are increasingly grave and urgent. They require a radical transformation of lifestyles and societies to avert ecocide (e.g., Rue 2011: 123–125). Jerome A. Stone (2008: 229) says it this way: "How humans can fashion a sustainable and just life for all creatures on our fragile Earth is our most pressing issue. The religious resources of naturalism provide orientation, healing, and motivation for some of us."

There are, however, significant differences between religious naturalism and other religious outlooks (e.g., Goodenough 1998: 171). Adherents of religious naturalism assert that nature alone is ultimate and sufficient for religious purpose, meaning, values, emotion, reverence, inspiration, and actions (e.g., Rue 2011: 100). Thus, most do not believe in any supernatural phenomena, transcendent and/or immanent (Crosby 2013: 2, 125, 127). Religious naturalists are usually atheists, agnostics, or non-theists, although some are theists as in variants of Pantheism (Harrison 2004; Stone 2008: 15, 227).

Some religious naturalists not only do not believe in anything supernatural, but are quick to rigorously reject the beliefs of those who do as nothing more than irrational faith and superstition. However, such confrontational atheism is extraordinarily arrogant, insensitive, disrespectful, negative, hostile, simplistic, and biased in summarily dismissing as delusional the more than 80 percent of humans who believe in some kind of supernatural phenomena, and in ignoring anything positive about religion (cf., Dawkins 2006; Haught 2008; Rue 2011: 116–122). Huston Smith, the preeminent scholar of comparative religion, asserted in the PBS series *Wisdom of Faith* in 1996: "If we take the world's enduring religions at their best, we discover the distilled wisdom of the human race."

Would the world be a better place if it became totally secularized with all religions extinct except for religious naturalism? What would be lost? Some might answer nothing, while others would point to great works of religious literature, music, art, and architecture as well as practical initiatives in caring for the suffering such as in religious charities, hospitals, orphanages, and homeless shelters (cf., Bowker 2015; Carneiro 2009; Harris 2014; Haught 2006; Smith 2000). Would anything replace religions? Some might answer that science would replace religion, government provide all social services, and nature alone would replace any belief in the supernatural (e.g., Rue 2011: 122–125). However, as mentioned previously, all of the advances in science, technology, and other secular aspects of society have failed to substantially halt environmental degradation. Furthermore, a comparison of Tibet before and since the devastating invasion and occupation by the Chinese is a prime example of the potential for secularization to generate ecocide (Sponsel 2012: 161–168). (Also, see Buckley 2014; Lafitte 2013.)

Instead of secularization and scientism, some authors are variously calling for the re-enchantment of nature (e.g., Berman 1981; Gibson 2009; McGrath 2002). On the other hand, Loyal Rue (2011: 123–128) predicts that eventually religious naturalism may prevail globally and supplant other religions because of the growing scientific power of evolutionary cosmogenesis combined with the impending worldwide environmental catastrophe and consequent disintegration of economies, societies, and polities.

Religious naturalism is grounded firmly in science, including its cosmology, evolution, and ecology (Crosby 2013: 1–18, cf., Swimme and Tucker 2011). There are religions that variously embrace science too. For instance, His Holiness the 14th Dalai Lama of Tibet repeatedly states that aspects of Buddhism not in harmony with science must be reconsidered and perhaps even rejected. On the other hand, some religious persuasions, such as extreme Christian fundamentalists, may reject much if not all science. The devil is in the details with such comparisons among the numerous diverse approaches considered by spiritual ecology. Even within religious naturalism there are many varieties and some incompatibilities (Rue 2011: 100, 122–123, 136; Stone 2008).

Spiritual ecology is more inclusive, eclectic, pluralistic, and relativistic than any other approach. It does not focus on or advocate any particular religious or spiritual tradition or practice. Ideally spiritual ecology promotes scientific and academic research about the environmental relevance of all religions and spiritualities (Sponsel 2012, 2014). In general, a researcher's own religion or lack thereof, and the validity of any religion or aspects of it, are irrelevant for neutral, objective, and empirical scientific and academic research in spiritual ecology, following anthropology's methodological principle of cultural relativism. However, the environmental consequences of religions, for better or worse, or whether adaptive or maladaptive, certainly are relevant, and any negative consequences deserve focused attention (Sponsel 2017; Wexler 2016; White 1967). Most of all, those who are religious or spiritual are encouraged to examine their own worldview, beliefs, values, attitudes, behaviors, and institutions to determine how they can relate to nature in a far more sustainable, green, just, and peaceful manner.

While in principle spiritual ecology fully embraces science, it advocates and pursues the tremendous potential of religions and spiritualities to be a positive force in relating their adherents to nature in far healthier ways, a force where secular approaches have proven insufficient. As already noted, some of the foremost leaders in spiritual ecology affirm that the environmental crisis is ultimately a spiritual and moral crisis (cf. George 2009). In contrast, ideally science is amoral and apolitical, preserving neutrality for the sake of objectivity. However, some argue that scientific facts may have moral and political implications, and that these may lead to appropriate action. Thus, Carolyn Merchant (2005: 136–137) writes:

> The main purpose of spiritual ecology is to effect a transformation of values that in turn leads to action to heal the planet. Whatever religion or form of spirituality one

practices, it is possible to find a connection to the earth and to the political work that needs to be done to change the present way of managing resources.

Conclusion

Secular approaches to environmentalism have made significant advances and are certainly crucial. However, just as certainly they have proven insufficient in fundamentally transforming lifestyles and societies to become far more environmentally friendly overall. They tend to treat particular superficial symptoms, rather than the underlying causes. The environmental crisis is a result of unrestrained urbanization, industrialization, capitalism, materialism, and consumerism. Unlimited growth on a limited base is impossible. Alienation from nature is also an important contributing factor. Moreover, ultimately the environmental crisis is a spiritual and moral crisis. Perhaps spiritual ecology can help generate the needed transformations to restore ecosanity. Through inclusivity, it embraces diverse approaches like religious naturalism. These approaches can be synergetic in positive ways. Diversity is an integral part of spiritual ecology as it is in evolution, adaptation, and adaptability. The biggest worry, however, is whether spiritual ecology as a revolutionary movement can surmount hegemonic obstacles and do so soon enough with sufficient power to avert global ecocatastrophe.

References

Note: All websites accessed on January 5, 2017.

Alley, K.D. (2002) *On the Banks of the Ganga: When Waste Water Meets a Sacred River*, Ann Arbor: University of Michigan Press.

Alliance of Religions and Conservation (ARC). (2016) Green Pilgrimage Network, Bath: Alliance of Religions and Conservation. www.arcworld.org/, www.greenpilgrimage.org/en/.

Ayres, E. (1999) *God's Last Offer: Negotiating for a Sustainable Future*, New York: Four Walls Eight Windows.

Batchelor, S. (1997) *Buddhism Without Beliefs*, New York: Riverhead Books.

Bauman, W.A., R.R. Bohannon II, and K.J. O'Brien, eds. (2011) *Grounding Religion: A Field Guide to the Study of Religion and Ecology*, New York: Routledge.

Beauregard, M., and D. O'Leary. (2007) *The Spiritual Brain: A Neuroscientist's Case for the Existence of Soul*, New York: HarperCollins Publishers.

Bekoff, M. (2014) *Rewilding Our Hearts: Building Pathways of Compassion and Coexistence*, Novato: New World Library.

Berman, M. (1981) *The Reenchantment of the World*, Ithaca: Cornell University Press.

Berry, E. (2015) *Devoted to Nature: The Religious Roots of American Environmentalism*, Berkeley: University of California Press.

Berry, T. (2006) *Evening Thoughts: Reflecting on Earth as Sacred Community*, San Francisco: Sierra Club Books (Mary Evelyn Tucker, ed.), www.thomasberry.org.

Bingham, S.G., ed. (2009) *Love God, Heal Earth: 21 Leading Religious Voices Speak Out on Our Duty to Protect the Environment*, Pittsburgh: St. Lynn's Press, www.interfaithpowerandlight.org/.

Bourne, E.J. (2008) *Global Shift: How a New Worldview Is Transforming Humanity*, Oakland: New Harbinger Publications, Inc.

Bowker, J. (2015) *Why Religions Matter*, New York: Cambridge University Press.

Buckley, M. (2014) *Meltdown in Tibet: China's Reckless Destruction of Ecosystems from the Highlands to the Deltas of Asia*, New York: Palgrave Macmillan.

Carneiro, R.L. (2009) *The Evolution of the Human Mind: From Supernaturalism to Naturalism: An Anthropological Perspective*, Clinton Corners: Eliot Werner Publications, Inc.

Chalquist, C. (2007) *Terrapsychology: Reengaging the Soul of Place*, New Orleans: Spring Journal Books.

Christofferson, B. (2004) *The Man from Clear Lake: Earth Day Founder Gaylord Nelson*, Madison: University of Wisconsin Press.

Chryssavgis, J., ed. (2011) *On Earth as in Heaven: Ecological Vision and Initiatives of Ecumenical Patriarch Bartholomew*, New York: Fordham Press, www.apostolicpilgrimage.org/the-green-patriarch.
Clayton, P., and J. Schaal, eds. (2007) *Practicing Science, Living Faith: Interviews with 12 Leading Scientists*, New York: Columbia University Press.
Clayton, P., and Z. Simpson, eds. (2006) *The Oxford Handbook of Religion and Science*, New York: Oxford University Press.
Coleman, M. (2006) *Awake in the Wild: Mindfulness in Nature as a Path of Self-Discovery*, Novato: New World Library.
Comte-Sponville, A. (2008) *The Little Book of Atheist Spirituality*, New York: Penguin Books.
Crosby, D.A. (2013) *The Thou of Nature: Religious Naturalism and Reverence for Sentient Life*, Albany: State University Press of New York.
Dali Lama XIV. (2011) *Beyond Religion: Ethics for a Whole World*, Boston: Houghton Mifflin Harcourt.
Dawkins, R. (2006) *The God Delusion*, New York: Houghton Mifflin Harcourt.
De Botton, A. (2012) *Religion for Atheists: A Non-Believer's Guide to the Uses of Religion*, New York: Vintage Books.
Earth Charter Initiative. (2017) *Earth Charter*, San Jose: University of Peace, http://earthcharter.org/.
Edwards, J., and M. Palmer. (1997) *Holy Ground: The Guide to Faith and Ecology*, Northamptonshire: Pilkington Press Ltd.
Ellingson, S. (2016) *To Care for Creation: The Emergence of the Religious Environmental Movement*, Chicago: University of Chicago Press.
Essen, C. von. (2010) *Ecomysticism: The Profound Experience of Nature as Spiritual Guide*, Rochester: Bear & Company.
Frankenberry, N.K., ed. (2008) *The Faith of Scientists in Their Own Words*, Princeton: Princeton University Press.
Gardner, G.T. (2006) *Inspiring Progress: Religion's Contribution to Sustainable Development*, New York: W.W. Norton & Company.
George, J. (2009) *The Little Green Book on Awakening*, Barrytown: Station Hill.
Gibson, J.W. (2009) *A Reenchanted World: The Quest for a New Kinship with Nature*, New York: Henry Holt and Company.
Goodall, J. (2005) "Primate Spirituality," in Bron Taylor (ed.), *Encyclopedia of Religion and Nature*, New York: Continuum 2: 1303–1306.
Goodenough, U. (1998) *The Sacred Depths of Nature*, New York: Oxford University Press.
Gottlieb, R.S., ed. (2006a) *A Greener Faith: Religious Environmentalism and Our Planet's Future*, New York: Oxford University Press.
—— (2006b) *The Oxford Handbook of Religion and Ecology*, New York: Oxford University Press.
—— (2013) *Spirituality: What It Is and Why It Matters*, New York: Oxford University Press, www.youtube.com/watch?v=BORt53UWyeY.
Gray, M. (2007) *Sacred Earth: Places of Peace and Power*, New York: Sterling Publishing Company, Inc., https://sacredsites.com/.
Grim, J.A., and M.E. Tucker. (2014) *Ecology and Religion*, Washington, D.C.: Island Press, http://fore.yale.edu/.
Harris, S. (2014) *Waking Up: A Guide to Spirituality Without Religion*, New York: Simon and Schuster.
Harrison, P. (2004) *Elements of Pantheism: Religious Reverence of Nature and the Universe*, Coral Springs: Llumina Press (Second Edition).
Hart, T. (2003) *The Secret Spiritual World of Children*, Makawao: Inner Ocean Publishing, Inc.
Harvey, G. (2013) *The Handbook of Contemporary Paganism*, Durham: Acumen Publishing Limited.
Haught, J.F. (2006) *Is Nature Enough? Meaning and Truth in the Age of Science*, New York: Cambridge University Press.
—— (2008) *God and the New Atheism: A Critical Response to Dawkins, Harris, and Hitchens*, Louisville: Westminster John Knox Press.
Hawkin, P. (2007) *Blessed Unrest: How the Largest Movement in the World Came into Being and Why No One Saw It Coming*, New York: Viking Penguin.
Higginbotham, J., and R. Higginbotham. (2002) *Paganism: An Introduction to Earth-Centered Religions*, Woodbury: Llewellyn Publications.
Hope, M., and J. Young. (2000) *Voices of Hope in the Struggle to Save the Planet*, New York: Apex Press.
International Union for the Conservation of Nature and Natural Resources (IUCN). (1997) *Indigenous Peoples and Sustainability: Cases and Actions*, Utrecht, The Netherlands: International Books.

Jaeger, J. (2012) "Bibliographic essay: Science and Religion – Renunciation or Reconciliation? A Survey of Recent Scholarly Literature," *Choice* 49(6): 1011–1020.

Jenkins, W., M.E. Tucker, and J. Grim, eds. (2016) *Routledge Handbook of Religion and Ecology*, New York: Routledge.

Jensen, D. (2016) *The Myth of Human Supremacy*, Oakland: Seven Stories Press.

Johnson, K., and D.R. Ord. (2012) *The Coming Interspiritual Age*, Vancouver: Namaste Publishing.

Kinsley, D. (1995) *Ecology and Religion: Ecological Spirituality in Cross-Cultural Perspective*, Englewood Cliffs, NJ: Prentice Hall, Inc.

Korten, D.C. (2006) *The Great Turning: From Empire to Earth Community*, Bloomfield: Kumarian Press, Inc.

Lafitte, G. (2013) *Spoiling Tibet: China and Resource Nationalism on the Roof of the World*, New York: Zed Books.

Lionberger, J. (2007) *Renewal in the Wilderness: A Spiritual Guide to Connecting with God in the Natural World*, Woodstock: Skylight Paths Publishing.

Lorbiecki, M. (2014) *Following St. Francis: John Paul II's Call for Ecological Action*, New York: Rizzoli International Publications, Inc.

Louv, R. (2008) *Last Child in the Woods: Saving Our Children from Nature-Deficit Disorder*, Chapel Hill: Algonquin Books.

—— (2011) *The Nature Principle: Reconnecting with Life in a Virtual Age*, Chapel Hill: Algonquin Books.

Maathai, W. (2004) *The Green Belt Movement: Sharing the Approach and the Experience*, New York: Lantern Books, www.greenbeltmovement.org/.

—— (2010) *Replenishing the Earth: Spiritual Values for Healing Ourselves and the World*, New York: Doubleday.

Macy, J., and C. Johnstone. (2012) *Active Hope: How to Face the Mess We're in without Going Crazy*, Novato: New World Library, www.joannamacy.net/.

McGrath, A. (2002) *The Reenchantment of Nature: The Denial of Religion and the Ecological Crisis*, New York: Doubleday.

Meadows, D., J. Randers, and D.L. Meadows. (2004) *Limits to Growth: The 30-Year Update*, White River Junction: Chelsea Green Publishing.

Merchant, C. (2005) *Radical Ecology: The Search for a Livable World*, New York: Routledge.

Mitchell, K. (2016) *Spirituality and the State: Managing Nature and Experience in America's National Parks*, New York: New York University Press.

Myers, N., and J. Kent, eds. (2005) *The New Atlas of Planet Management*, Berkeley: University of California Press.

Newberg, A. (2010) *Principles of Neurotheology*, Burlington: Ashgate.

Pedersen, K.P. (1998) "Environmental Ethics in Interreligious Perspective," in S.B. Twiss and B. Grelle (eds.), *Global Ethics: Comparative Religious Ethics and Interreligious Dialogue*, Boulder: Westview Press, pp. 253–290.

Pfeiffer, B. (2013) *Wild Earth, Wild Soul: A Manual for an Ecstatic Culture*, Winchester: Moon Books.

Rue, L. (2011) *Nature Is Enough: Religious Naturalism and the Meaning of Life*, Albany: State University of New York Press.

Sagan, Carl, et al. (1990) "Preserving and Cherishing the Earth: An Appeal for Joint Commitment in Science and Religion," Moscow: Global Forum January 1990, http://fore.yale.edu/publications/statements/preserve/.

Saint-Laurent, G.E. (2000) *Spirituality and World Religions: A Comparative Introduction*, Mountain View: Mayfield Publishing Company.

Schmidtz, O.J. (2016) *The New Ecology: Rethinking a Science of the Anthropocene*, Princeton: Princeton University Press.

Schwagerl, C. (2014) *The Anthropocene: The Human Era and How It Shapes Our Planet*, Santa Fe: Synergetic Press.

Seed, J., J. Macy, P. Fleming, and A. Naess. (2007) *Thinking Like A Mountain: Towards a Council of All Beings*, Gabriola Island: New Catalyst Books.

Selhub, E.M., and A.C. Logan. (2014) *Your Brain on Nature*, New York: Collins.

Smith, H. (2000) *Why Religion Matters: The Fate of the Human Spirit in an Age of Disbelief*, New York: HarperSanFrancisco.

Speth, G. (2008) *The Bridge at the Edge of the World: Capitalism, the Environment, and Crossing from Crisis to Sustainability*, New Haven: Yale University Press.

Sponsel, L.E. (2012) *Spiritual Ecology: A Quiet Revolution*, Santa Barbara: Praeger, http://spiritualecology.info.

—— (2014) "Feature Article: Bibliographic Essay – Spiritual Ecology: Is It the Ultimate Solution for the Environmental Crisis," *Choice* 51(8): 1339–1342, 1344–1348.

—— (2017) "Lynn White, Jr., One Catalyst in the Historical Development of Spiritual Ecology," in T. LaVasseur and A. Peterson (eds.), *Lynn White Retrospective after 50 Years*, New York: Routledge, pp. 89–102.

Stoll, M.R. (2015) *Inherit the Holy Mountain: Religion and the Rise of American Environmentalism*, New York: Oxford University Press.

Stone, J.A. (2008) *Religious Naturalism Today: The Rebirth of a Forgotten Alternative*, Albany: State University of New York Press.

Swan, J.A. (1991) *Sacred Places: How the Earth Seeks Our Friendship*, Santa Fe: Bear & Company.

Swimme, B.T., and M.E. Tucker. (2011) *Journey of the Universe*, New Haven: Yale University Press, www.journeyoftheuniverse.org/.

Takacs, D. (1996) *The Idea of Biodiversity: Philosophies of Paradise*, Baltimore: The Johns Hopkins University Press.

Taylor, B., ed. (2005) *Encyclopedia of Religion and Nature*, New York: Continuum, Volumes 1–2.

—— (2010) *Dark Green Religion: Nature, Spirituality and the Planetary Future*, Berkeley: University of California Press, www.brontaylor.com/blog/.

Verschuuren, B., R. Wild, J.A. McNeely, and G. Oviedo, eds. (2010) *Sacred Natural Sites Conserving Nature and Culture*, Washington, D.C.: Earthscan.

Weisman, A. (2008) *The World Without Us*, New York: Pacador.

Wexler, J. (2016) *When God Isn't Green: A World-Wide Journey to Places Where Religious Practice and Environmentalism Collide*, Boston: Beacon Press.

White, Lynn, Jr. (1967) "The Historic Roots of Our Ecologic Crisis," *Science* 155(3767): 1203–1207.

Worster, D. (2008) *A Passion for Nature: The Life of John Muir*, New York: Oxford University Press.

13

AFRICAN AMERICAN RELIGIOUS NATURALISM AND THE QUESTION OF THE HUMAN

Carol Wayne White

In the contemporary era, religious naturalism features a synthesis of ideas and viewpoints that depart from traditional forms of religion. These perspectives include rejecting supernaturalism in any form and following the dictums of science in understanding reality, including human life and culture. With other religious naturalists, I am convinced that any truths we discover and any meaning in life we uncover are revealed to us through the natural order (Stone 2008: 1). In this chapter, I embrace these tenets of religious naturalism within the context of African American cultural and intellectual life. I offer a naturalistic view of humans that becomes the focal point for challenging exclusionary and outdated conceptions of humanity embedded in the logic of white supremacy. I present humans as highly complex organisms, owing the lives we have to the emergence of hierarchies of natural systems. Rejecting any notion of a supernatural reality beyond nature, I rely on current developments in science to help describe humans' processes of transformative engagement with each other and with the more-than-human worlds that constitute our existence.

These theoretical convictions foreground a model of African American religious naturalism that addresses issues of justice for myriad aspects of nature. In so doing, it provides insight into a legacy of white supremacy built upon a culture–nature binary in modernist discourses, where processes of racialization have helped shape an exclusionary category of the human, designating who is properly so and who is not. In this and other historical contexts, this binary differentiation functioned to demarcate certain spheres of life as superior and others as inferior, justifying the exploitative practices of the former. Traditional humanistic discourses inflected by it have also overestimated the autonomy of human beings, positioning us outside of nature and rendering invisible our inextricable connection to other life forms and material processes. Both of these impulses—white supremacy and human exceptionalism—evoke a model of nature built on the "great chain of being" concept, and they have produced violent and harmful consequences.

This exploration of humanistic reasoning within the context of religious naturalism involves a complex yet important task. On the one hand, I seek to reconstruct the category of the human in such a way that it can address the various "isms" that have denied black (and other marginalized) subjects their rightful claims to their humanity. On the other hand, my retrieval of the human necessarily rejects the aims of traditional humanisms that position us outside of nature. Whatever conceivable notion of complex humanity I claim in this essay will be ontologically enmeshed and entangled with other forms of natural life. I first outline a trajectory of humanistic thinking in the West that provides the backdrop for this model of African American religious naturalism. Following this, I outline some of its central claims, introducing the sacred humanity concept. I then conclude with some reflections on its value in American culture and life.

Denying black humanity: white supremacy and the culture–nature binary

Addressing the National Colored Convention in 1853, Frederick Douglass offered an astute observation about the lived experiences of black people in America:

> Our white fellow-country men do not know us. They are strangers to our character, ignorant of our capacity, oblivious of our history and progress, and are misinformed as to the principles and ideas that control and guide us as a people. The great mass of American citizens estimate us as a characterless and purposeless people.
>
> (Douglass 2000: 269)

In this passage, Douglass confronts the perceived black degradation rooted in the minds of white individuals held captive by the ideology of white supremacy—what he elsewhere labels "diseased imagination" (Douglass 2016: 501). A century later, at the height of the civil rights era, James Baldwin poignantly addressed the effects of this diseased imagination with his creative use of the bastard epithet. For Baldwin, the bastard metaphor revealed the pathology inherent in many whites' refusal to embrace their familial kinship with blacks, based on the false notion of black cultural and biological inferiority: "The problem is rooted in the question of how one treats one's flesh and blood, especially one's children. The blacks are the despised and slaughtered children of the great Western house—nameless and unnamable bastards" (Baldwin 1998: 468).

Countless other African American visionaries, artists, and thinkers have expressed the conundrum of affirming life and embracing one's humanity in a world (or culture of values) where blackness—its symbolic resonance and its tactile materiality—has been the target of dehumanization processes. In recent years, the formation of the "Black Lives Matter" movement and the controversies surrounding its title have resuscitated important cultural debates on the question that stretches back to Douglass' era: When will blacks' full humanity cease being questioned and devalued in the United States?

The multivalence of dehumanizing processes and anti-life forces against black lives cannot be underestimated. Once transported onto American shores, the physical color of Africans took on symbolic significance within a cultural system of differentiation that both marked them as slaves and justified negative assessments of their humanity. With the establishment of slave laws during the colonial period, blacks were treated as objects or assets to be bought and sold, mortgaged and wagered, despised and condemned. In very few contexts were blacks regarded as human subjects with volition, feeling, and a sense of responsibility. Their slave status stripped them of many civil rights and liberties granted to all citizens of the nation (Fisher 1992; Fede 2012, 34). Along with other cultural practices, these laws were integral to an emerging white supremacist ideology that used the construct race for judging blacks' humanity against a normative model constructed by European cultures. One factor contributing to the rise of white supremacy in the U.S. was an early modern binary construct that originated in Western Europe. This construct divided human culture from nature into spheres of greater-lesser value. In *The Death of Nature*, Carolyn Merchant focuses primarily on its gender implications, asserting:

> At the root of the identification of women and animality with a lower form of human life lies the distinction between nature and culture fundamental to humanistic disciplines such as history, literature, and anthropology, which accept that distinction as an unquestioned assumption.
>
> (Merchant 1980: 132; 144)

Merchant also notes that this ideology of dualism was an integral component of Western European cultural imperialism, where the purported "civilized" races of Europe distinguished their own supposedly normative humanity against other groups they encountered in the Americas, Asia, and Africa. As an extension of the nature-culture dichotomy, racialized notions of difference led to disparaging views such as the savage Native Americans and the intellectually inferior Africans. When measured against the idealized Western bourgeois human, Africans, in particular, were found to be deficient in requisite cognitive, aesthetic, physical, and moral attributes (Eze 1996; Mills 1997; Stefancic and Delgado 2013). This epistemological framework was later sanctioned with nineteenth-century scientific studies, where notions of racial differences often presented the social inequalities between various cultural groups as reflecting the prescripts of Nature. More precisely, with the rise of scientific racism, emerging views of "black animality" appeared in influential studies. In Robert Knox's *The Races of Men: A Fragment* (Knox 1851), the slant of the brow is used to draw connections between the "Negro" and the "Oran Outan" and indicate differences between those two and the "European." In their 1854 ethnological study (*Types of Mankind*), prominent scientists Josiah C. Nott and George R. Gliddon documented their perception of objective racial hierarchies with illustrations comparing blacks to chimpanzees, gorillas, and orangutans (Nott and Gliddon 1854). Advancing a theory of polygenesis, Ernst Haeckel, a respected professor of zoology, represented the human species in a hierarchy from lowest to highest, ranking negroes among the lowest races and depicting them as savages related to apes (1876: 10). He believed people of African descent were psychologically nearer to other mammals—apes and dogs—than to the civilized Europeans, and consequently assigned a totally different value to their lives. Haeckel's evolutionary ideas were embedded within his notion of racial purity for Germans, supporting his views that the inexorable laws of evolution conferred on favored races the right to dominate others (Haeckel 1876: 332).

Some of these scientific studies contributed uniquely to the idea of whiteness as a normative category for establishing a group's humanity. While declaring the superiority of the human species in relation to other organic life, French thinker Arthur de Gobineau also advanced important distinctions within the human animal, with blacks representing the lowest form:

> I have shown the unique place in the organic world occupied by the human species, the profound physical, as well as moral differences separating it from all other kinds of living creatures. Considering it by itself, I have been able to distinguish, on physiological grounds alone, three great and clearly marked types, the black, the yellow, and the white ... the negroid variety is the lowest, and stands at the foot of the ladder. The animal character, that appears in the shape of the pelvis, is stamped on the negro from birth, and foreshadows his destiny. His intellect will always move within a very narrow circle.
>
> (de Gobineau 1915: 205–212)

With varying degrees of emphases, these perspectives proposed a gradation from civilization to barbarism, reinforcing the singular role of the Europeans as a civilizing force—a colonizing belief that is transferred to American shores (Wynters 2003).

American colonists extended the binary construction formulated in scientific racism into a unique American narrative of civilization overcoming wilderness. Encountering the deeply forested North American wilderness "the white Puritan colonists measured 'progress' and 'civilization' in terms of (among other things) 'how far a people could distance themselves from Nature'" (Spiegel 1997: 16). In diaries, personal writings, and memorials of the frontier period, colonists symbolized wilderness as an enemy to be conquered and overcome by the civilized pioneer army (Nash 2014: 27). Intermingled here are cultural, religious, and social forces that essentially made wilderness synonymous with darkness and sinister forces. As Roderick Nash contends,

> The pioneers shared the long Western tradition of imagining wild country as a moral vacuum, a cursed and chaotic wasteland. As a consequence, frontiersmen acutely sensed that they battled wild country not only for personal survival but in the name of nation, race, and God. Civilizing the New World meant enlightening darkness, ordering chaos, and changing evil into good.
>
> (Nash 2014: 24)

As part of the legacy of white supremacy in the United States, these perspectives both justified black slavery and the exploitation of the more-than-human natural worlds. As summarized by Paul Outka, this colonizing legacy consisted of whites viewing dark-skinned peoples as part of the natural world, and then proceeding to treat them with the same mixture of contempt and real exploitation that also marks American environmental history (Outka 2008: 3).

In *The Promise of Religious Naturalism*, Michael Hogue articulates an important insight into Western modernity that is intriguing, in light of the historical perspectives and accounts I have outlined:

> Just as it once was thought by many that Western modernity would eventually lead to the demise of religion, protection from and control of nature has also been a running theme of Western modernity. The paradox, in short, is that modernity has not led so much to the demise of religion but to its transformation, and that rather than insulating humanity from the risks and hazards of natural processes, modernity has in many ways led to the increasing vulnerability of nature.
>
> (Hogue 2010: 2)

Hogue's articulation of this paradox leads me to consider the potential nature of black religion's transformation after modernity. Can a transformed black religiosity effectively respond to some of the ways modernity has led to the increasing vulnerability of multifarious nature? The African American religious naturalism I propose in the next section responds to these concerns.

Toward an African American religious naturalism

In his groundbreaking volume, *Souls of Black Folk*, W. E. B. Du Bois sketched the complex unfolding of nineteenth-century African American religiosity, revealing the institutionalization of a people's hopes, fears, core values, ethical convictions, and cosmological assumptions. In this and other works, Du Bois offers a compelling view of African American religiosity as an evolving, humanistic enterprise with monumental social and communal implications (Du Bois 1989).

I share Du Bois' approach, identifying African American religiosity as the ingenuity of a people constantly striving to inhabit their humanity and eke out a meaningful existence for themselves against the backdrop of culturally coded white supremacist notions and practices. This perspective is in keeping with those aspects of traditional black religious expression (primarily Christian) in the U.S. where specific images, symbols, and rituals function to address fundamental issues of life or death to black agents intent on living fully and with dignity (West and Glaude 2003: xiii–xv).

Historically, the symbol God has functioned in African American religious culture to affirm the value of black humanity as well as the fact that all humans share in the same ontological reality as other humans (Evans 1992: 100). It has been posited as an ultimate value—indeed, an *a priori* notion—in religious expression, symbolizing the means by which particular limitations on human potentiality could be dissolved or at least addressed. However, I am not persuaded by this appeal to the God symbolism (or to any form of supernaturalism) in my reading of black religiosity. What is crucial here, I believe, is recognizing the creative energies of blacks who rejected the impoverished conceptions of their humanity used to justify slavery, Jim Crow laws, segregation, and other unjust social practices and cultural norms through the last several hundred years.

In making these claims, I am viewing African American religiosity primarily through the lens of religious functionalism, which, as Loyal Rue has suggested, places the proper focus on who actually creates and uses religion: humans (Rue 2005: 1). Religious functionalism in this context is a method of analyzing and interpreting religious experiences and expressions as natural events having natural causes. In advancing this theoretical orientation, I do not presume that religious phenomena can be completely explained; rather, I suggest that the extent of our understanding is contingent on efforts to grasp these phenomena in terms of underlying natural processes. In the Western history of ideas, this general approach is not all that new, having been advanced many times in the past. As Rue asserts, notable thinkers such as Kant, Feuerbach, Marx, Durkheim, and Freud have argued that "regardless of what religion *says* it is about, it has to *do* fundamentally with meeting the challenges to a full life" (Rue 2005: 3). With these earlier figures and contemporary naturalists like Rue, I share the conviction that religion both originates in human experience and is properly understood in natural terms, as opposed to those theories that see it emerging from some transcendent order or given by divine revelation. In this essay, however, I offer a fresh iteration of this general humanistic orientation in setting it within the context of raced living in the U.S. Given its emergence from the historical realities of slavery, black cultural and religious expression in the United States brings into relief an important point that other white humanists have failed to grasp: black slaves and their descendants have never had the luxury of asking these questions in abstraction. In surviving and making a claim about the value of life, American blacks have lived these questions into the future—both for contemporaries and for generations to come.

Through the lens of functionalism, black religious expression becomes one type of values discourse upheld by individuals and communities in their myriad struggles to humanize their existence. In short, from my perspective, the wide range of affirmations of blacks' valuable humanity reiterated from one generation to the next are more than emancipatory declarations; these affirmations provide an important shift in humanistic thinking. They offer to us an opportunity to raise important, more fundamental questions for contemporary readers: How to understand the complex human being that is being affirmed in African American liberationist discourse and that is further distorted by racialization processes? Operating on the assumption that the natural order is ultimately and finally real, I am essentially concerned about the human in its most concrete, basic terms: as a material process of nature in relationship with other forms of nature.

Sacred humanity concept

The advances of science, through both biology and physics, have served to demonstrate not only how closely linked human animals are with nature, but that we are simply one branch of a seemingly endless natural cosmos. The general view of humanity I hold, on which I build my concept of sacred humanity, presupposes this verity.[1] Donald Crosby states this insight in even more eloquent terms:

> Nature requires no explanation beyond itself. It always has existed and always will exist in some shape or form. Its constituents, principles, laws, and relations are the sole reality. This reality takes on new traits and possibilities as it evolves inexorably through time. Human beings are integral parts of nature, and they are natural beings through and through. They, like all living beings, are outcomes of biological evolution.
>
> (Crosby 2008: ix–x)

With other religious naturalists, I believe that understanding the deep history of the cosmos is profoundly important for any basic understanding of the materiality of being human, of being alive in the manner we currently find ourselves. Humans are highly complex organisms, owing the lives we have to the emergence of hierarchies of natural systems. Expressed succinctly, humans are "ultimately the manifestations of many interlocking systems—atomic, molecular, biochemical, anatomical, ecological—apart from which human existence is incomprehensible" (Rue 2005: 25). Human life is also part of an evolutionary history showing a trend toward greater complexity and consciousness. As Stephen J. Gould and other scientists have noted, there has been an increase in the genetic information in DNA and a steady increase in the ability of organisms to gather and process information about the environment and respond to it (Gould 1989; Deacon 2003; Deacon 2006).

In highlighting human animals as emergent life forms, I warn against a particular reading of this claim that concludes human beings are the triumphant summit of natural development. Rather, my position is best described by recent insights in ecological studies, aptly described by Crosby: organisms of various types, including human beings, are inextricably bound together in a web of mutual interdependence for their continual flourishing and survival as they make common if varied use of the energy of the sun (Crosby 2013: 16). Within each web, each species of animal has a niche for which it is more or less adapted, and has attributes others lack (Spiegel 1997: 22–23). This ecological bent challenges those who would use evolutionary history as the basis for deciding who is better than whom. As suggested by an earlier view of Marx and Engels, I consider human beings within the larger contexts and relations of nature, which is in constant flux and change, incessantly coming into being and passing away as every organic being simultaneously assimilates matter from without and excretes matter from within (Parsons 1977: 24). At the same time, I reject the more emphatic aspects of later Marxist discourse that looked upon the "animal" part of humanity as some sort of biological residue, a thing left over from humanity's past (Frolov 1986).

In light of these observations, the scientific epic becomes the starting point for positing an African American religious naturalism constituted by a central tenet: humans are relational processes of nature. Our inexhaustible connection or entanglement with other natural processes, or with the more-than-human, constitute the very notion of the human as such. In declaring such, I contend that our humanity is not a given, but rather an achievement. Consider that from a strictly biological perspective, humans are organisms that have slowly evolved by a process of natural selection from earlier primates. From one generation to another, the species that is alive now has gradually adapted to changing environments so that it could continue to survive. Our animality, from this perspective, is living under the influence of genes, instincts, and emotions, with the prime directive to survive and procreate.

Yet, this minimalist approach fails to consider what a few cognitive scientists, and most philosophers, humanists, and religionists tend to accentuate: our own personal experience of what it is like to be an experiencing human being. As I noted elsewhere, becoming human, or actualizing ourselves as human beings, in this sense, emerges out of an awareness and desire to be more than a conglomeration of pulsating cells. It is suggesting that our humanity is not reducible to organizational patterns or processes dominated by brain structures; nor do DNA, diet, behavior, and the environment solely structure it. In positing fundamental questions of value, meaning, and purpose to our existence, human animals become human destinies. Our coming to be human destinies is structured by a crucial question: How do we come to terms with life?

I share Ursula Goodenough's sentiment that reveling in a sense of connectedness with other living beings can be described as sacred (Goodenough 2000). On the molecular level, there is evidence to support the loftier (or religious) idea that in the very nature of life itself there is some essential joining force. This orientation toward joining with others in establishing our common humanity is what I imagine when using the phrase *sacred humanity.* Humans are, by our very constitution, relational, and our wholeness occurs within a matrix of complex interconnectedness—put another way, ways of conjoining with others that transform us. Granted, this is not your typical approach to the sacred, which admittedly is a complex word that has been used for a wide range of phenomena: places, times, persons, events, and deities. Traditionally, when people designate something as sacred, they view the thing in question as "other than ordinary." Thus, in the broadest sense of the term, the sacred has been used by scholars, especially those sympathetic to the work of Mircea Eliade, to convey the "extraordinary" (Eliade 1987).

Utilizing the tenets of religious naturalism in conjunction with values discourse, I consider humans' awareness and appreciation of our connection to "all that is," as an expression of sacrality, or what we perceive and value as ultimately important and valuable. Value in this sense refers to an organism's facility to sense whether events in its environment are more or less desirable (Dolan 2002: 1191). Minimally, this facility evokes the notion of adaptive value, which is the basic matrix of Darwinian theory (Gould 2000: 158). Within a larger ecological framework, however, this truth takes on fuller meaning, as Holmes Rolston's observations suggest: "An organism is the loci of values defended; life is otherwise unthinkable. Such organismic values are individually defended; but, ecologists insists, organisms occupy niches and are networked into biotic communities" (Rolston 2006: 911).

Humans seeking, finding, and experiencing community with others—and, in my own words, with otherness—is an essential aspect of our humanity that religious discourse tends to advance and reiterate again and again. As Ursula Goodenough writes:

> We have throughout the ages sought connection with higher
> powers in the sky or beneath the earth, or with ancestors in
> some other realm. We have also sought, and found,
> religious fellowship with one another. And now we realize
> that we are connected to all creatures. Not just in food
> chains or ecological equilibria. We share a common ancestor.
> We share genes for receptors and cell cycles and
> signal-transduction cascades. We share evolutionary

> constraints and possibilities. We are connected
> all the way down.
>
> (Goodenough 2000: 73)

Scientific theories feature the social character of cognition in animals and humans, providing various types of evidence for understanding humanity as symbol makers, creators of a world imbued with value, and as social organisms. According to Terrence Deacon, what is particularly interesting about the course of human evolution is that it has entailed the co-evolution of three emergent modalities—brain, symbolic language, and culture—with each feeding into and responding to the other two and generating particularly complex patterns and outcomes. In *The Symbolic Species*, he explores the intricate connection between the evolution of human language and our brains, or what he calls their co-evolution (Deacon 1997). The gist of Deacon's study is that language itself was part of the process that was responsible for the evolution of the brain. Language has changed the environments in which brains have evolved. We are a species that in part has been shaped by symbols, in part shaped by what we do. According to Deacon, ritual, mythology—ways of doing things that are organized conventionally, symbolically—are the hallmark of our species. Humans have transformed and even reinterpreted much of our biology through this symbolic system. So much of what we do—marriage, conflict, as in warfare, or whatever—has been transformed by this linguistic tool that has, in a sense, taken over and biased all of our interactions with the world. Expressed succinctly, our brain has evolved very differently in some regards than other species' brains and in ways that are uniquely human (Deacon 1997: 36ff; 45–46).

Based on these insights from Deacon and other scientists, we can affirm that humans seek meaning by viewing their lives in a cosmic and religious framework that is itself a human symbolic construct—the brain is part of the cosmos and a product of the cosmos. Its structures reflect the nature of the cosmos and whatever ordering and meaning-giving forces are expressed in its history (Arbib 1989; LeDoux 1996; Brothers 1997). These naturalistic views of the human indicate a complex social organism that can love, connect deeply with others, and symbolize its environment (or engage in world formation) through values and language. They also lend support to my view of human individuals as multilevel psychosomatic unities—both biological organisms and responsible selves. Here, the focus is on us humans' heightened awareness of our ability self-consciously to make decisions, act on those decisions, and take responsibility for them. This point converges with Rue's descriptive account of human beings as star-born, earth-formed creatures endowed by evolutionary processes to seek reproductive fitness under the guidance of biological, psychological, and cultural systems that have been selected for their utility in mediating adaptive behaviors. Humans maximize their chances for reproductive fitness by managing the complexity of these systems in ways that are conducive to the simultaneous achievement of personal wholeness and social coherence (Rue 2005: 75).

A consistent scientific view is that a successful life outcome consists of promoting the transmission of information conducive to maintaining an emergent dynamic logic that gives life its meaning—that is, promoting the production of emergent outcomes (called traits in biology) that collectively make their own continuation more likely. As Goodenough and Deacon assert:

> Traits common to all organisms include such non-depressing and religiously fertile capacities as end-directedness and identity maintenance; traits common to all animals include awareness and the capacity for pleasure and suffering; traits common to social beings include co-operation and meaning making; traits common to birds and

> mammals include bonding and nurturance; traits common to humans include language and its capacity to share subjective experiences, and thus to know love. Transmission of genomes is the steady background drumbeat; emergence is the music.
>
> (Deacon and Goodenough 2008: 860)

From the diseased imagination to moral imaginations

With its conception of the human as an emergent, interconnected form amid spectacular biotic diversity, the African American religious naturalism I am proposing has rich potential to work with other critical discourses seeking transformations in American culture and life. First, its new religious ideal—sacred humanity—ennobles and dignifies new images of ourselves, creatively inscripted onto the tissues, bones, and liquids of which we are constituted. In so doing, it challenges the most alienating self-other "isms" created by humans, specifically underscoring the historical insights regarding the notion of blackness as a form of otherness in a cultural formation that reifies whiteness. Further, it asserts that full liberation for all Americans requires a rejection of the dualistic, binary structures that support such problematic racialized views.

This naturalistic view also resists the lure of a generic, universal construction of "man" that has justified the devalued status of women and other subjects relegated to minority status. Enlightenment configurations of this normative human have been associated with a coherent, white, propertied, and rational subjectivity (Haraway 2004: 48). Rather than assume that gender, race, class, abled-bodiedness, and other socially derived markers provide the basis of our humanity, we should recognize them as highly complex categories constructed in contested discourses and other social practices. When these constructions are used to support racism, sexism, and other forms of cultural superiority, they become forced impositions on the wholeness of natural interrelatedness and deep genetic homology that evolution has wrought.

Moreover, this African American religious naturalism helps undermine a dominant cultural fantasy of human exceptionalism that anchors humans on one side of the Great Divide, away from all other species. This premise assumes that the human alone is not a spatial and temporal web of interspecies dependencies, and it has lent theoretical support to popular myths of the self-made individual in the U.S. African American religious naturalism rejects this fantasy, and encourages us to join with Haraway in appreciating our intricate entanglement with other material processes:

> I love the fact that human genomes can be found in only about 10 percent of all the cells that occupy the mundane space I call my body; the other 90 percent of the cells are filled with the genomes of bacteria, fungi, protists, and such, some of which play in a symphony necessary to my being alive at all, and some of which are hitching a ride and doing the rest of me, of us, no harm. I am vastly outnumbered by my tiny companions; better put, I become an adult human being in company with these tiny messmates. To be one is always to *become with* many.
>
> (Haraway 2008: 3–4)

Second, propelling citizens of our nation beyond the diseased imagination first identified by Douglass, this African American religious naturalism entertains new moral imaginations that resist the logic of white supremacy. With the sacred humanity concept as its organizing principle, African American religious naturalism directs our attention to a world evolving naturally, based on the interconnection and interaction of all of its fundamental components. In effect, it helps blur the arbitrary ontological lines that human animals have erected between themselves and other species and natural processes. We recall Hogue's assertion of religious naturalism as one viable response to the general paradox of modernity. The model of African American religious naturalism introduced here constitutes one specific case of Hogue's resolution. While it denounces black degradation and inferiority through the use of the animal other, it also goes further to analyze the speciesism that is evident in these formulations. In short, this religious worldview evokes a moral imagination that compels us to ask why is it that both racist and antiracist discourses are predicated on a repudiation of animality (Peterson 2012). One response is that the equation of blacks with co-primates that is rooted in Enlightenment racism is based on prior negative ideas about less-than-human animals (Thomas 1983: 40). In other words, species supremacy engenders the bestialization of social and political others.

Anticipating that some people will find these connections between anti-black racism and speciesism troubling, I turn to Spiegel's observation in *The Dreaded Comparison*: that the human/animal opposition makes the abjection of human others possible means that insisting on their humanity as a mode of resistance can only re-inscribe the speciesist logic that initiates their exclusion (Spiegel 1997: 30). I agree with Spiegel. Furthermore, when we continually dismiss "nonhuman" sentience, we create conditions for reducing certain human others to the status of "mere" animal life. As Spiegel notes: "Comparing the suffering of animals to that of other blacks (or any other oppressed group) is offensive only to the speciesist: one who has embraced false notions of what animals are like" (Spiegel 1997: 30, 37).

As we continue discovering and resisting the wide-ranging violence embedded in European colonialism, the sacred humanity concept in African American religious naturalism compels us to consider fully the colonization of other animals. On this matter, Jeremy Bentham offered an important insight in the eighteenth-century:

> The day has been ... in which the greater part of the species, under the denomination of slaves, have been treated by the law exactly upon the same footing as, in England for example, the ... animals are still ... The French have already discovered that the blackness of skin is no reason why a human being should be abandoned without redress to the caprice of a tormentor. It may come one day to be recognized, that the number of legs, the villosity of the skin, or the termination of the *os sacrum*, are reasons equally insufficient for abandoning a sensitive being to the same fate.... The question is not, Can they *reason*? nor, Can they *talk*? but, Can they suffer?
>
> (Bentham 1789: 283)

Bentham's emphasis on *sentience* is crucial. White supremacy and species supremacy distort the wholeness of animal sentience. One step toward decolonizing nature, toward making it less vulnerable, perhaps, is in honoring nature's sentience, which is an essential part of being alive, experiencing others, being affected by others, and experiencing wellbeing.

Finally, as we continue questioning the idea of Enlightenment progress, as well as the ethics of unrestrained development as a means of dominating nature in all its forms, the sacred humanity concept strengthens the case for addressing with concerted effort ecological degradation on various levels (Clingerman and Dixon 2011). Its theoretical claims alert us to the dangers of isolationist agendas that environmental justice advocates also resist. For example, the environmental justice movement helps make clear religious naturalism's sense of the irrefutable interconnectedness of all life when it concurrently advocates against the depletion of natural resources, challenges the policies that both create land polluted by landfills, oil refineries, and nuclear-waste repositories and force poor racial and ethnic communities to live near these sites, and fights for referendums that preserve the delicate ecosystems supporting whales and dolphins (Cole and Foster 2001). As these efforts suggest, religious naturalists and environmental justice advocates share a general maxim: harm done to any one sector of natural processes, inclusive of human organisms, is harm done to all. Inspired by this claim, a more robust environmental justice movement intentionally challenges and unmasks subtle binary differentiations that ground the most recent variations of the nature-culture continuum.

Honoring all materiality, African American religious naturalism casts aside problematic bifurcations of human materiality cast in racial and ethnic terms that often result in an "us-versus-them" mentality. With such a religious worldview, we can better identify and resist the ill-effects of white supremacy on all of us, resisting its power in determining how certain racial and ethnic bodies are treated. We can also detect and challenge the subtle processes of the racialization of nature endemic to American environmental history (Glave 2010; Finney 2014). These important ecological values are ones that social justice advocates can extend to enact important ethical, political, economic, and social changes in American life.

Note

1 Parts of this section are drawn from a fuller discussion in my book *Black Lives and Sacred Humanity: Toward an African American Religious Naturalism,* Albany, NY: Fordham Press, 2016. They are reprinted with permission from Fordham Press.

References

Arbib, M. (1989) *The Metaphorical Brain 2; Neural Networks and Beyond,* New York: John Wiley.

Baldwin, J. (1998) *Baldwin: Collected Essays*, edited by T. Morrison, New York: The Library of America.

Bentham, J. (1789; 1996) *An Introduction to The Principals of Morals and Legislation,* edited by J. H. Burns and L. A. Hart, New York: Oxford University Press.

Brothers, L. (1997) *Friday's Footprint: How Society Shapes the Human Mind,* New York: Oxford University Press.

Clingerman, F. and Dixon, M. (2011) *Placing Nature on the Borders of Religion, Philosophy, and Ethics*, Burlington, VT: Ashgate.

Cole, L. and Foster, S. (2001) *From the Ground Up: Environmental Racism and the Rise of the Environmental Justice Movement,* New York: New York University Press.

Crosby, D. (2008) *Living With Ambiguity: Religious Naturalism and the Menace of Evil,* Albany, NY: State University Press of New York.

—— (2013) *The Thou of Nature,* Albany, NY: State University Press of New York.

Deacon, T. (1997) *The Symbolic Species: The Co-Evolution of Language and the Brain*, New York: W. W. Norton.

—— (2003) "The Hierarchic Logic of Emergence: Untangling the Interdependence of Evolution and Self-Organization," in B. Weber & D. Depew (eds.), *Evolution and Learning: The Baldwin Effect Reconsidered,* Cambridge, MA: MIT Press, 273–307.

—— (2006) "Emergence: The Hole at the Wheel's Hub," in P. Clayton & P. Davies (eds.), *Re-Emergence of Emergence*, Oxford: Oxford University Press, 111–150.

Deacon, T. and Goodenough, U. (2008) "*The Sacred Emergence of Nature*," in P. Clayton (ed.), *The Oxford Handbook of Religion and Science*, New York: Oxford University Press.

Dolan, R. J. (2002) "Emotion, Cognition, and Behaviour," *Science*, 298: 1191–1194.

Douglass, F. (2000) *Frederick Douglass: Selected Speeches and Writings* (The Library of Black America series), edited by P. S. Foner and Y. Taylor, Chicago: Chicago Review Press.

—— (2016) "The Color Line" (1881), in J. Stauffer (ed.), *The Portable Frederick Douglass*, New York: Penguin Classics, 501–511.

Du Bois, W. E. B. (1903; 1989) *The Souls of Black Folk*, New York: Bantam.

Eliade, M. (1987) *The Sacred and the Profane*, Orlando, FL: Harcourt.

Evans, J. (1992) *We Have Been Believers: An African-American Systematic Theology*, Minneapolis, MN: Augsburg Fortress Press.

Eze, E. (1996) *Race and the Enlightenment*, Malden, MA: Blackwell Publishers.

Fede, A. (2012) *People without Rights: An Interpretation of the Fundamentals of the Law of Slavery in the U.S. South*, New York: Routledge.

Finney, C. (2014) *Black Faces, White Spaces: Reimagining the Relationship of African Americans to the Great Outdoors*, Chapel Hill, NC: University of North Carolina Press.

Fisher, W. III (1992) "Ideology and Imagery in the Law of Slavery: Symposium on the Law," *Chicago-Kent Law Review* 68/ 3, art. 4: 1051–1083.

Frolov, I. T. (1986) "Genes or Culture? A Marxist Perspective on Humankind," *Biology and Philosophy* 1(1): 89–107.

Glave, D. (2010) *Rooted in the Earth: Reclaiming the African American Environmental Heritage*, Chicago: Chicago Review Press.

Gobineau, A. de (1915) *An Essay on the Inequality of the Human Races*, trans. A. Collins, London: Heinemann.

Goodenough, U. (2000) *The Sacred Depths of Nature*, New York: Oxford University Press.

Gould, S. J. (1989) *Wonderful Life: The Burgess Shale and the Nature of History*, New York: W. W. Norton & Company.

—— (2000) *The Structure of Evolutionary Theory*, Cambridge, MA: Harvard University Press.

Haeckel, E. (1876) *The History of Creation: Or the Development of the Earth and Its Inhabitants by the Action of Natural Causes*, New York: Appleton.

Haraway, D. (2004) "Ecco Homo, Ain't (Ar'n't) I a Woman, and Inappropriate/d Others: The Human in a Post-Humanist Landscape," in *The Haraway Reader*, New York: Routledge, 47–61.

—— (2008) *When Species Meet*, Minneapolis, MN: University of Minnesota Press.

Hogue, M. (2010) *The Promise of Religious Naturalism*, Lanham, MD: Rowman & Littlefield.

Knox, R. (1851; 2010) *The Races of Men: A Fragment*, Whitefish, MT: Kessinger Publishing, LCC.

LeDoux, J. (1996) *The Emotional Brain: The Mysterious Underpinnings of Emotional Life*, New York: Simon and Schuster.

Manning, M. (2001) "Structural Racism and American Democracy: Historical and Theoretical Perspectives," *Souls* /1 (Winter): 6–24.

Merchant, C. (1980) *The Death of Nature: Women, Ecology and the Scientific Revolution*, New York: HarperCollins.

Mills, C. (1997) *The Racial Contract*, Ithaca, NY: Cornell University Press.

Nash, R. F. (2014) *The Wilderness and the American Mind*, 5th edition, New Haven: Yale University Press.

Nott, J. C. and Gliddon, G. R. (1854) *Types of Mankind*, Philadelphia, PA: Lippincott, Grambo & Co.

Outka, P. (2008) *Race and Nature: From Transcendence to the Harlem Renaissance*, New York: Palgrave Macmillan.

Parsons, H. L. (1977) *Marx and Engels on Ecology*, Westport, CN: Praeger Publishers.

Rolston, H. III (2006) "Environmental Ethics and Religion/Science," in P. Clayton and Z. Simpson (eds.), *The Oxford Handbook of Religion and Science*, New York: Oxford University Press.

Rue, L. (2005) *Religion Is Not About God: How Spiritual Traditions Nurture Our Biological Nature and What to Expect When They Fail*, Piscataway, NJ: Rutgers University Press.

Spiegel, M. (1997) *The Dreaded Comparison: Human and Animal Slavery*, New York: Mirror Books/IDEA.

Stefancic, J. and Delgado, R. (eds.) (2013) *Critical Race Theory: The Cutting Edge*, Philadelphia, PA: Temple University Press.

Stone, J. (2008) *Religious Naturalism Today: The Rebirth of a Forgotten Alternative*, Albany, NY: State University of New York Press.

Thomas, K. (1983) *Man and the Natural World, Changing Attitude in England 1500–1800*, London: Allen Lane.

West, C. and Glaude, E. (2003) *African American Religious Thought*, Louisville: Westminster John Knox Press.

White, C. W. (2016) *Black Lives and Sacred Humanity: Toward an African American Religious Naturalism*, New York: Fordham University Press.

Wynters, S. (2003) "Unsettling the Coloniality of Being/Power/Truth/Freedom: Towards the Human, After Man, Its Overrepresentation—An Argument," *The New Centennial Review* 3/3: 257–337.

14
A POLITICAL THEOLOGY FOR THE ANTHROPOCENE

Michael S. Hogue

Introduction

Any new constructive work on religious naturalism must engage the realities and uncertainties of life in the Anthropocene.[1] By "constructive" I refer to work that is theologically articulated, morally charged, and politically engaged. By "religious naturalism" I refer to the generative integration of naturalistic interpretations of religion and religious interpretations of nature. The "Anthropocene," which literally means "human age," signifies two ideas. First, it signals that the Earth has moved beyond the historical epoch known as the Holocene, which began roughly 12,000 years ago. This is an unsettling idea, for by providing climatic conditions necessary to agriculture and social settlement, the Holocene made it possible for human civilization to develop. The normal CO2 range during the Holocene fluctuated between 260 and 285 ppm. But we have just crossed over to 400 ppm, and given the way the carbon cycle works, no matter how much or how quickly we reduce greenhouse gas emissions, no one alive today will see CO2 levels drop below this level. The second idea signaled by the term Anthropocene is at least as disturbing. It is the claim "that human activity is largely responsible for this exit from the Holocene ... that humankind has become a global geological force in its own right" (Steffen 2011: 843). In other words, the point is not only that we have exited the Holocene, but that we booted ourselves out the door. While the Holocene provided a hospitable climatic niche conducive to the development of complex human civilization, the climatic chaos of the Anthropocene promises to make for a very different kind of experience. Life in the Anthropocene will be more uncertain, less predictable, and increasingly insecure.

In addition to signaling a geological reality the Anthropocene is also a paradoxical cultural condition; for even as it signifies the beginning of a new "human age" for the Earth, it also marks the ending of the idea of the human as separate from nature. By making the Earth *homo imago*, we have discovered ourselves as *terra bestiae*—for into whichever of Earth's systems we look, from the arctic to the atmospheric, we see a reflection of our species self-image mirrored back at us. Furthermore, although the Anthropocene names a planetary phenomenon, it must also be understood as a political problem. As a planetary phenomenon, it impacts every aspect of the Earth System. But the driving causes of the Anthropocene are historically contingent and culturally specific, emerging out of the massive escalation of carbon emissions produced by the Industrial Revolution in the eighteenth century. The Anthropocene is thus the cumulative ecological effect of roughly 250 years of a fossil-fueled political economy that originated in England and the United States.

In these ways, then, the Anthropocene is a paradox: it signifies a planetary geophysical phenomenon, a cultural condition that marks the ending of the idea of the human as separate from the rest of nature, and the apotheosis of the industrial political economy. With these general ideas in mind, my argument in this essay is motivated by, and structured around, two interrelated theses. In the section that immediately follows, I will argue that the Anthropocene paradox compels and creates the conditions for new ways of conceptualizing the intersection of theology, the political, and political theology. Then, I will use these concepts to articulate a theopolitical framework for the Anthropocene that constructively draws from the resources of religious naturalism.

Theology, the political and political theology

By signifying both the beginning of a human geological epoch for the Earth and the ending of the idea of the human difference from the rest of nature, the Anthropocene leads us to reconsider root questions of moral, religious, and political life. These include, among others, questions about value, meaning, power, and common life. As some of the deepest questions we can ask of ourselves, as individuals, communities, and as a species, how we interpret and answer these questions has tremendous implications. The ways we imagine the worth and beauty of life, our own lives as well as the lives of others, our responsibilities to human and more-than-human others, both in the present and into the future, are entangled with our understanding of the depth and scope of moral value, the limits and purposes of human power in relation to the more-than-human powers of the rest of nature, the sources and forms of human meaning, and the ideals and institutions of common life.

By revealing to us how deeply embedded and dependent our human systems are in relation to the ecological systems of Earth, the Anthropocene paradox compels us to reimagine the tasks of philosophy, theology, and moral inquiry. This entails coming to terms with the Anthropocene's dual nature as both a planetary geophysical reality and the effects of an historically particular civilizational form. For the Anthropocene is an ecospheric as well as a sociocultural condition, a problem for the whole of the present and future of human life that has been caused by a small historical subset of the species. Thus, as postcolonial historian Dipesh Chakrabarty argues, it is "impossible to understand global warming as a crisis without engaging the propositions put forward by [climate] scientists," propositions whose significance pertains to the future of the whole species (Chakrabarty 2009: 219). At the same time, however, "the story of capital, the contingent history of falling into the Anthropocene, cannot be denied by recourse to the idea of species, for the Anthropocene would not have been possible, even as a theory, without the history of industrialization" (Chakrabarty 2009: 219). It is simply untrue that the whole human species is equally responsible for the C02 and other GHG emissions that are driving the climate crisis. The most affluent and industrialized nations in the global North have not only generated far more than their share of carbon emissions, they have also capitalized the most on the economic and technological systems that drive those emissions. For instance, wealthy, developed nations emitted a cumulative total of roughly 900–950 billion metric tons of CO2 between 1800 and 2010.[2] The rest of the world, including rapidly developing countries such as Brazil, India, and China, emitted a cumulative total of only 400 billion metric tons. Although the emissions of rich nations in the global North began to escalate rapidly around 1860, the rest of the world's emissions did not begin to escalate until well after WWII. While developed nations account for only 20 percent of the world population, they are responsible for approximately

70 percent of cumulative GHG emissions.[3] As a result of their overwhelming contribution to cumulative emissions, the nations of the global North bear the burden of responsibility for the climate crisis. Furthermore, the nations that are least responsible are the most immediately and negatively impacted by climate change.[4]

Since the Anthropocene emerges out of the intersection of two histories, planetary and political, interpreting it compels us to think on multiple levels, the particular and universal. The "particular" history that needs to be engaged is the political history of capitalism and Western modernity; the "universal" history is the history of the species and the climate. The problem is that universal historical interpretations often fail to account for the diverse particular histories that run adjacent to them. Chakrabarty describes the dilemma this way:

> How do we relate to a universal history of life—to a universal thought, that is—while retaining what is of obvious value in our postcolonial suspicion of the universal? The crisis of climate change calls for thinking simultaneously on both registers to mix together the immiscible chronologies of capital and species history.
>
> (Chakrabarty 2009: 219–220)

If we seek to mix these stories, to think simultaneously on the registers of the political and the planetary, it is important to choose the right concepts. As someone working out of the American pragmatist lineage of religious naturalism, I think of concepts as tools-under-construction.[5] The meaning of concepts is thus connected to functionality. To provide the infrastructure for the constructive work in the next section, it will be helpful here to explain what I mean by, and how I am using, the concepts "theology," "the political," and "political theology."

By "theology" I refer to the construction, critique, and reconstruction of the symbols, practices, ideals, and institutions that format life-orienting religious meanings, purposes, and desires. This is a non-dogmatic mode of theology. Thus defined, theology is something one does more than something that one has. In this view, theology is not equivalent to a system of beliefs and beliefs are not reducible to cognitive propositions. When concerned with belief, this mode of theology is concerned with "belief" as the continuous process of communally forming, critically testing, and creatively transforming religious meanings, purposes, and desires. Theology is a process rather than a set of truth propositions to be possessed like property. As a process, it is not limited to critical reflection on the symbol "God." Instead, its objects are those constellations of symbols, practices ideals, and institutions that orient human life by formatting, or shaping and interconnecting, the religious meanings, purposes, and desires through which we communally negotiate the hazards and graces of life in a complex world.

By "the political" I refer to the imaginal space of reason and desire within which communities negotiate the relations of power and value that organize their common life. Thus defined, the political is not limited to traditional political objects, such as the state or party, or to formal institutions, such as legislative and judicial bodies, or to practices such as voting, assembling, and protesting. Indeed, the political is immanent within and yet not exhausted by each of these traditional political objects, institutions, and practices of politics. To refer to the political as an "imaginal" space is to suggest that it is implicitly embedded within affective, aesthetic, narrative, and ritual registers of human life in addition to its explicit presence in discourses and institutions.

By "political theology" I refer to the practice of thinking the theological and political together, both with and against one another. It is the critical and constructive engagement of the interplay between what orients us ultimately and what organizes us communally, between religious meanings and purposes and the imaginal spaces of reason and desire through which we negotiate relations of power and value in our common life. The theological and political can be and often are transposed across religious and other registers of life. Insofar as religious symbols and practices affect communal negotiations of power and value, and vice versa, there is no neat division between the religious and secular—the religious is not apolitical and the political is not areligious.

Thus construed, how can these concepts be put to work in a political theology for the Anthropocene paradox? On the one hand, the dual signification of the Anthropocene—as the beginning of a human age for the Earth and the ending of the idea of the separatism of the human species—calls for a non-anthropocentric political theology, or a political theology that engages the questions of power, value, and common life in a more-than-human world. On the other hand, the dual nature of the Anthropocene calls for a political theology that is planetary in scope and yet also politically engaged. What is called for, then, is a "bifocal" political theology. A bifocal political theology frames two different but interrelated lenses, one that focuses on the broad horizon of planetary life, and one that focuses on the particularities of local community. A bifocal theopolitics thus aims to hold in view what ultimately orients us and what immediately concerns us, the future as well as the present of life, and the breadth of planetary systems along with the depth and intensity of local life. In short, a bifocal political theology integrates an immanental theology and a horizontal politics.

Toward a political theology of religious naturalism

As I mentioned earlier, I define religious naturalism in general as a philosophy of religion that seeks to blend naturalistic interpretations of religion with religious interpretations of nature. Religious naturalism sounds like an oxymoron because these two tasks can seem to be at odds with one another. This view is based, in part, on what we might refer to as the "reductivist critique" of religious naturalism. This line of critique stems from the assumption that naturalistic accounts of religion are invariably reductive—that, for example, the attempt to explain a propensity for belief in supernatural phenomena and agents in cognitive and evolutionary terms explains religion away by eliminating the objects of religious thought and devotion. But this example itself assumes that supernatural beliefs (such as belief in miracles that contravene the most basic theories of physics, or belief in an idea of God as both discarnate and purposeful, which contradicts what we know about the embodied contingencies of consciousness) are essential ingredients of religion.

The irony of the reductivist critique is that it is itself rooted in reductive assumptions about religion. For example, as familiar as the religious idea of God may be to many of us, and as common as beliefs in other supernatural realities may be in some religious traditions, they are not ingredients in all traditions that we think of as religious, including varieties of Buddhism, Daoism, Confucianism, and even some forms of theism. In addition, the assumption that *belief* is a defining element of religion is also problematic. Being religious is surely about much more than belief—it's about the observance of rituals, the telling of stories and the transmitting of values, patterns of mourning and celebration, the communal formation of a shared identity, and much more.

Contrary to the assumptions of the reductivist critique, religious naturalists are committed to the proposition that naturalistic accounts of religion, or interpretations of religion that are inspired and constrained by the empirical, experimental, and evidentialist methods of the sciences, help to illuminate the fullness of religion. Far from "reducing" or "eliminating" religion, naturalistic approaches to religion can help to expand our understanding of its complexities and significance, as many of the other articles in this volume attest. With these ideas in mind, I interpret religious traditions as historically extended cultural systems of symbols, rituals, and beliefs that transform human groups into communities by binding them together with shared identities, meanings, and purposes. This hypothesis is useful for comparison across traditions, for interpretations of their internal diversity, and for analyses of the religious aspects of traditionally non-religious phenomena. Through story and ritual, religions establish, justify, and regulate the moral norms, behavioral conventions, and social hierarchies necessary to the development of large-scale human groups. Religions pacify and provoke; they have done great good and great harm; they bring people together and they divide them. But in the end, as the religious naturalist Loyal Rue has put it, religion is not about God (Rue 2005). Instead, religions are about us. They are historically adaptive cultural projects that transform humans who happen merely to be grouped together in space and time into communities bound together with shared identities and purposes that transcend space and time. Through mutually reinforcing networks of ritual practices, stories, values, transcendent ideals, and social institutions, religions orient individuals and communities through life's blessings and burdens.

A naturalistic philosophy of religion such as this can dissolve the dualisms between reason and faith, understanding and explanation, and nature and culture that have alienated humans from one another, from their creaturehood, and from the planet. It holds together an interpretation of the religious significance of nature and a naturalistic interpretation of religion. It affirms that nothing human, and thus nothing cultural, can be fully understood apart from its evolutionary biological context. This is not to say that evolutionary theory provides a sufficient explanation of all things human and cultural. But it is to say that evolutionary theory is a necessary aspect of understanding what humans are and what humans do. It affirms that the quest to understand more about ourselves as a biocultural species can be experienced as a religious quest, even when it is a quest driven by a naturalistic, empirical, evolutionary perspective. Rather than viewing a scientific account of religion as an impediment to religious life and experience, this philosophy of religion opens the way to a religious experience of our cosmic and planetary contexts. By demystifying religion, a naturalistic philosophy of religion can create new spaces for the religious experience of the infinite and inexhaustible mysteries of nature.

The fundamental claim in this philosophy of religion, as expressed by William David Hart, is "that *nature is all there is*," that "the whence and wherefore of all things is nature," and thus that nature "is both the things that emerge" and the processes through which they emerge (Hart 2016). To say "nature is all" is not to say nature is simple or undifferentiated. In deference to nature's complexity and ambiguity, I find it helpful to imagine nature through the medieval distinction between *natura naturans* (nature naturing, or nature creating) and *natura naturata* (nature natured, or created nature). Spinoza famously put these distinctions to work in his *Ethics* to contrast between the infinite and finite aspects of "God, or Nature" (*Deus, sive Natura*). For Spinoza, *natura naturans* refers to that aspect of God/Nature that is infinitely creative or always creating, whereas *natura naturata* refers to the finitude of singular things (i.e., particular entities or beings) and their modes (i.e., physical or mental) created by *natura naturans* (Spinoza 1994).

Other religious naturalists, including some editors and contributors to this volume, have also used these concepts in distinctive and creative ways. For example, by arguing that nature's "inexorable processes of change" are more fundamental than its "present patterns," Donald A. Crosby's philosophy of nature emphasizes the dynamic ambiguity of *natura naturans* over the relative stabilities of *natura naturata* (Crosby 2002, 2009). For Robert S. Corrington, the fundamental process within nature is the process of fissuring, "which gives rise to the most primal distinction in thought—namely, that between nature naturing and nature natured. This natural difference represents the complex 'how' of nature as it both arises *and* shapes the orders of the world" (Corrington 2016). Thus, in subtle contrast to Crosby's prioritization of *natura naturans*, what is most significant for Corrington is the alterity of *natura naturans* and *natura naturata* in their relation to one another (Corrington 1992; 1994; 1997; 2004). Inspired by Crosby's and Corrington's philosophies, I articulate the contrast between *natura naturans* and *natura naturata* into four fundamental ontological claims about nature: 1) that nature is all; 2) that nature is differentiated; 3) that nature is in process; 4) that the things and relations of nature are vulnerable and precarious.

As a religious naturalist, these claims about the natures of nature have religious significance. I interpret the significance by way of a theology of immanence. A theology of immanence signifies the idea that nature is all, that the beauty, depth, complexity, and mystery of nature is the sole context for the emergence of being and becoming, knowing and valuing, that there is no outside of nature, that nature is infinite and thus without center or periphery. Nature includes innumerable possible forms of experience and centers of value, but no center is the center of all the rest and no form encompasses all the others. Nature (i.e., *natura naturans* and *natura naturata*), rather than God, is the central religious concern within this theology of immanence, but nature itself has no center, either divine or human.

Committed to the idea that nature is all and that it is in process, and that this idea is religiously significant, an immanental theology provokes a life of contemplation, inquiry and moral practice devoted to the beauty and creativity of the forms of life and value that emerge through nature's patterns and processes. By taking *nature naturing* as the context for the discernment of meaning, value, and what orients us ultimately, and understanding human cultures as aspects of *nature natured*, an immanental theology is inquisitively open to the wisdom of diverse human cultures and forms of religiosity. This immanental theology affirms, in William David Hart's words, that the various "[c]onceptual, factual, and value distinctions that" we make as a species, "such as is/ought, good and evil, animals, humans, and God(s)" are "contingencies of our bio-cultural evolution." Thus, in an important sense, not only do "we create God and the gods," but we are ourselves "gods who shit: transcendent animals" (Hart 2016: 58).[6]

The sacred text of this immanental theology is the epic that arcs from the genesis of the Universe with the Big Bang and the swirling of the earliest cosmic elements, to the birth pangs of stars and planets and the constellation of galaxies; an epic that includes everything from the Sun's gestation of our solar system to the emergence of life to the biospheric tipping point that our own species has precipitated. An immanental theology humbly decenters the human species in relation to the infinitely broader metaphysical and aesthetic rhythms of the Universe. Rather than questing for certainty or resting in faith, it is an appreciative search for the value of things. As appreciative, this search adds value to the Universe by attending to the depth of the value that is already here. It seeks wisdom from wherever it may come, from diverse religious symbols, myths, and rituals to literature and the arts, from the intricate depths of indigenous knowledges to the mind-bending ways of the modern sciences. In an immanental theological frame, there is no "outside" of revelation—the whole of the cosmos rings with it, from the subatomic to the interstellar, from the unicellular to the civilizational.

Within this frame, nature is more than any of its names. And yet if it is all, then there is no "outside" of nature and nature has no "other." If nature has no "outside," then nature has no circumference, no form, no boundaries, and no edge. Nature is not a container that contains. It is unbounded. But if it is unbounded, if there are no boundaries or edges to nature, then nature has no center. The implications of this are profound, and profoundly unsettling, especially perhaps for common ways of thinking about ethics and moral value. For in relation to the infinite geometry of nature, all that we know and hold most dear is quite unremarkable. Our species is but one among millions of others on a single planet hitched to a middle-sized star, which is itself just one among billions and billions of others in a galaxy that is itself only one among trillions of others—in the observable universe alone.[7]

The inestimable scale of the Universe and its indeterminate acentricity problematize the persistence of human-centered moral thinking. They even bring into question fundamental aspects of moral thinking itself, insofar as the idea of "center" is instrumental to the way we think morally. Essential to common meanings of moral value is the idea that something has, or signifies, more or less of some property deemed to have value. The meaning of value has everything to do with "more and less." One can see the moral importance of this claim by imagining what life would be like if everything was of equal value. Every choice would be tragic, every decision would require the arbitrary sacrifice of one valuable thing or option for another of equal value. But we do not live this way. We order our values. We rank some more highly than others. We treat some as more important and others as less, some as intrinsic, others as instrumental. We don't all do this in the same way, using the same evaluative criteria, but these are pervasive patterns in moral thinking. We order our values and this ordering is contingent on some central value, or cluster of values, that relativizes others. In short, the ordering of values is necessary to the ordering of human moral life: getting through the kind of life that we live entails making choices at every turn to do one thing rather than another, opting for one possible future over another, aiming for certain objectives rather than others.

So, what does the acentricity of nature mean within the frame of an immanental theology? It means that a concept that has been integral to the moral ordering of human life has no objective grounding in the larger world. It is a humanly constructed concept. This does not mean it is unimportant, nor does it mean that there are no meanings or purposes in nature. Although nature itself may be unbounded by any meaning or purpose, nature is inclusive of meanings and purposes. At this point, this is merely an empirical claim rather than a moral one, or a descriptive rather than an evaluative assertion. That there are entities in nature such as humans that have purposes does not necessarily imply that those purposes are good or right. Given the epistemic constraints of naturalism, there is no possible standpoint from which we could empirically access nature as-a-whole to determine, unambiguously, whether it has an all-inclusive general purpose or not. But we can empirically describe purposes in nature wherever and whenever we observe behaviors and motivations such as desire, fear, curiosity, hunger, and intentionality. For an organism to intend some thing or another, whether that intention is mentally represented or not, means that the organism is oriented by a purpose or an interest of some kind. The satisfaction of that interest or the realization of that purpose is a value for that organism.

It can even be argued that purposes and intentions, and thus values, are present wherever there is metabolism.[8] Metabolism is a transfer of energy between organisms and their environments. Though no organism consciously controls its metabolism, its metabolic processes express themselves in behavioral intentions to secure from its environment what it needs to go on living. These intentions indicate purposes and the behavioral aim to realize purposes indicate values. While a naturalistic empirical standpoint will never allow us to know if nature as-a-whole has a central purpose, it is empirically the case that purposive behavior in nature is value-oriented.

A theology of immanence thus compels us to rethink common human assumptions about meaning and value. Among other things, it throws a wrench into the view that we should assent to meanings and commit to values only if they are absolute rather than relative. One concern driving this assumption is that the structure of moral thinking and the ordering of social systems will be undermined if meanings and values are merely circumstantial, contingent, and relative.

However, this concern is based on the fallacy that if meanings and values are relative, then they merely reflect subjective individual preferences or group prejudices. Against that assumption, this theology of immanence affirms that meanings and values are relative rather than absolute, and yet that they are also objective rather than subjective. Meanings and values are objectively grounded in nature but they are relative to diverse forms of life.

For example, insofar as the ideal of human flourishing has value to us as a species, then the value of human flourishing has objective status in the universe, even though it is species-relative. Keeping this in mind, let us say that human flourishing includes the satisfaction of basic human needs. But it also entails the exercise of the higher level cognitive, emotional, and imaginative capacities that are typical of our species. Exercising these capacities entails that we have both the freedom and the resources to do so. As a social species, we need others to help us to meet our basic needs and to secure the resources necessary to sustaining the free exercise of our various human capacities. Minimally, then, a flourishing human life is (1) a relationally enriched and enriching life in which (2) basic needs are met and (3) the resources and freedom necessary to exercising the cognitive, affective, and imaginative capacities typical of our species are secured.

These reflections support what I call a prismatic theory of meaning and value. In this theory, every consciously held meaning and value becomes a portal through which other meanings and values, whether consciously held or not, can be seen, felt, and affirmed, or not. To affirm that human flourishing is objectively good is not an anthropocentric prejudice. It is not a form of species supremacy that asserts that human values, or the good of human flourishing, should be the metric of value for all other goods. To the contrary, given all that I have said thus far, human flourishing is interdependent with the flourishing of other forms of life and the ecological systems that make and sustain the very possibility of life. The objective goodness of human flourishing is entangled with the objective value of other things, beings, and systems in the world. Human flourishing is relationally complex, objective and yet relative rather than absolute.

To understand the good of human flourishing in this way opens human religious and moral experience into a prism through which the luminous plurality of other objectively relative values in nature can be seen, felt, and affirmed. As Carol Wayne White eloquently expresses this view, "Each human birth is a glorious event, and the starting point of yet another spectacular phenomenon that helps transform the enigmatic cosmos into an even more vital, dramatic" world. As she continues, this does not mean that "the knowable universe is enlivened only through human activity," but that "humans are individual and collective destinies engaging an appreciable world" (White 2016: 40). Rather than seeing the human as the measure of all value in the world, human moral experience and imagination become prismatic vectors through which the radiance of more-than-human values can be seen. Through this prismatic theory of value, human moral evaluation appreciates—enriches, enlarges, and enlivens—the values in the more-than-human world. Paradoxically, then, it is precisely insofar as the meanings and values that orient human life are relative that they can claim to have robust objective grounding; and precisely insofar as they have robust objective grounding, the values and meanings of the rest of nature can be registered more fully and vibrantly.

So, what does this all come down to? Within the frame of this immanental theology, nature has neither a singular center nor an overarching purpose, and yet there are purposes and centers of value in nature. Although the claim that nature is all and that it is in process undermines absolutist assumptions about values, purposes, and meanings, the undermining of absolutism does not invariably lead to moral nihilism. One can be morally oriented by objectively grounded meanings, purposes, and values even if they are contingent, relative, and contextual. What is more, this contingency, relativity, and contextuality signify the fundamental precariousness of what we hold most dear—from the goodness and beauty of friendship and family to the relational complexity of flourishing life. Through the precarious significations of what we hold most dear, we come to witness and empathically feel the vulnerability of the planetary conditions of human and more-than-human life.

For these reasons, a theology of immanence holds promise as a religious and theopolitical frame for life in the Anthropocene. Religious experiences within the Anthropocene are those events, encounters, insights, those under-goings, up-endings and over-comings that throw life into suspense, stripping away the pretenses of permanency. Experiences such as these rend the veil of the ordinary; they interrupt and can sometimes transform one's life. In the words of religious naturalist Henry Nelson Wieman, religious experience widens the "bounds of awareness" and produces a "diffusive state of awareness, where habitual systems of response are resolved into an undirected, unselective aliveness for the total organism to the total event then ensuing" (Wieman 1979: 38–39). Felt at times as the "undefined awareness of the total passage of nature," religious experiences are provoked by the feeling of contact with "a far larger portion of that totality of immediate experience which constantly flows over one" (Wieman 1979: 38–39).

Religious experiences such as these contour an immanental theological frame into a form of political theology. Since we humans are embedded in nature and since, like every other thing that exists, we are unintended creatures of an exquisite latticework of natural processes that originated in mystery about 13.7 billion years ago, the answer to the question of why the Universe exists, or why there is something rather than nothing, will probably always remain out of reach of our human comprehension. But we do know that we are creatures of the stars. We do know that if the rate of acceleration of the elements out of the Big Bang varied even slightly, the Universe would not have unfolded the way that it has and continues to unfold. We know that life first emerged on our planet about 3.5 billion years ago and that since then it has been undergoing relentless transformation. We know we are latecomers to the Universe. We know that we have purposes and values, and what is more, we are consciously and empathically aware of the purposes and values of other living creatures—if our eyes and ears are open, we cannot help but witness them.

Responsibility is the moral yield of this alchemy of consciousness, empathy, and agency. It emerges through our awareness that human life is interwoven with the rest of life, that other creatural life has purpose and value, that life as such seems to be rare in the Universe, and that we as a species have far more power than any other living thing to manipulate the conditions upon which the purposes and values of life on Earth depend. By becoming a massive planetary force, geo-cultural agents, we humans have also become custodians of the values of nature. Perhaps becoming more fully human and more fully realizing our nature as *terra bestiae* has something to do with summoning the courage to face and embrace this responsibility, to welcome it as an invitation to come more fully alive by recovering our relations with more-than-human life. Though we are newborns on this far-flung planet, swirling amid the fathomless geometry of a Universe without edges, we have crossed a threshold that no other living thing on Earth has ever crossed. No living species other than our own has ever become a cause of the mass extinction of other living things.

In his evocatively titled essay, "The Outcry of Mute Things," the Jewish philosopher Hans Jonas expressed the religious significance of this realization with unmatched pathos:

> It was once religion which told us that we are all sinners, because of original sin. It is now the ecology of our planet which pronounces us all to be sinners because of the

> excessive exploits of human inventiveness. It was once religion which threatened us with a last judgment at the end of days. It is now our tortured planet which predicts the arrival of such a day without any heavenly intervention. The latest revelation—from no Mount Sinai, from no Mount of the Sermon, from no Bo (tree of Buddha)—is the outcry of mute things themselves that we must heed by curbing our powers over creation, lest we perish together on a wasteland of what was creation.
>
> (Jonas 1996: 201–202)

What will it take for us to hear this revelation, to feel and take responsibility for this loss of creatural vitality? Might the hearing of this outcry and the feeling of this loss be the event, encounter, insight, or the undergoing and upending, that will lead us to revere the contingency of what we assume to be necessary, the vulnerability of what we too often view as invulnerable, the perishability of what seems permanent? If anything might rend the veil of the ordinary meanings and values that orient our lives, and thus count as religious experience, should not this?

Thus, a political theology of religious naturalism is committed to the precariousness of *natura naturata* and the profundity of *natura naturans*. Nature is all, without center or periphery, and yet every creatural center of value is a prism of objectively relative value through which the radiance of other centers of values can be felt and honored. It provides an ideal context for the enrichment, revision, and enlargement of the life-orienting meanings, purposes, and values of religious and political life amid the uncertainties of life in the Anthropocene.

Notes

1 I articulate one such constructive option in Michael S. Hogue, *American Immanence: Democracy for an Uncertain World* (New York: Columbia University Press, 2017). Elements of this chapter are drawn from that book.

2 By "wealthy, developed nations" I refer to the Annex I nations in the United Nations Framework Convention on Climate Change (UNFCCC). See List of Annex I Parties to the Convention, United Nations Framework Convention on Climate Change, http://unfccc.int/parties_and_observers/parties/annex_i/items/2774.php (accessed December 23, 2016).

3 For data, see Carbon Dioxide Information Analysis Center (CDIAC) Data Base, United States Department of Energy, http://cdiac.ornl.gov/ (accessed December 23, 2016). See also Johannes Friedrich and Thomas Damassa, "The History of Carbon Dioxide Emissions," *World Resources Institute Blog*, May 21, 2014, www.wri.org/blog/2014/05/history-carbon-dioxide-emissions.

4 For analysis of differential impacts, see, for example, Working Group II, "Climate Change 2014: Impacts, Adaptation and Vulnerability," Assessment Report 5, Intergovernmental Panel on Climate Change (IPCC), www.ipcc.ch/report/ar5/wg2/ (accessed December 23, 2016).

5 American pragmatism is a broad tradition in American philosophy that is sometimes differentiated into classical, neopragmatic, and analytic forms. By the pragmatic lineage of religious naturalism, I refer primarily to American philosophies of religion influenced by John Dewey's pragmatic naturalism. This includes, among others, Victor Anderson, *Pragmatic Theology: Negotiating the Intersections of an American Philosophy of Religion and Public Theology* (Albany, NY: State University of New York Press, 1988); S. Morris Eames, *Pragmatic Naturalism: An Introduction* (Carbondale: Southern Illinois University Press, 1977); Philip Kitcher, *Preludes to Pragmatism: Towards a Reconstruction of Philosophy* (Oxford: Oxford University Press, 2012); and John Ryder, *The Things in Heaven and Earth: An Essay in Pragmatic Naturalism* (New York: Fordham University Press, 2013). I offer a fuller interpretation of this in *American Immanence: Democracy for an Uncertain World*.

6 Hart's reference to the human as a "god who shits" is an allusion to Ernest Becker's *The Denial of Death* (New York: The Free Press, 1997), 58.

7 A recently published study in *The Astrophysical Journal* concluded that there are 2 trillion galaxies in the observable universe, which is upwards of ten times the number of galaxies previously thought to exist.

See Davide Castelvicci, "Universe has ten times more galaxies than previously thought," *Nature: International Weekly Journal of Science*, 14 October 2016. www.nature.com/news/universe-has-ten-times-more-galaxies-than-researchers-thought-1.20809 (accessed 17 October, 2016).

8 For example, this argument was powerfully made by the German Jewish philosopher, Hans Jonas. See especially, *The Imperative of Responsibility: In Search of an Ethics for the Technological Age* (Chicago: University of Chicago Press, 1984), 79–90. Originally published as *Das Prinzip Verantwortung: Versuch einer Ethik für die technologische Zivilisation* (Frankfurt am Main: Insel Verlag, 1979). For my own analysis of this argument, see Michael S. Hogue, *The Tangled Bank: Towards an Ecotheological Ethic of Responsible Participation* (Eugene, OR: Wipf and Stock, 2008).

References

Chakrabarty, D. (2009) "The Climate of History: Four Theses," *Critical Inquiry* Vol. 35, No. 2: 197–222.

Corrington, R. S. (1992) *Nature and Spirit: An Essay in Ecstatic Naturalism*, New York: Fordham University Press.

—— (1994) *Ecstatic Naturalism: Signs of the World*, Bloomington, IN: Indiana University Press.

—— (1997) *Nature's Religion*, Lanham, MD: Rowman and Littlefield.

—— (2004) *A Semiotic Theory of Theology and Philosophy*, New York: Cambridge University Press.

—— (2016) *Deep Pantheism: Toward a New Transcendentalism*, Lanham, MD: Lexington.

Crosby, D. A. (2002) *A Religion of Nature*, Albany, New York: State University of New York Press.

—— (2009) *Living with Ambiguity: Religious Naturalism and the Menace of Evil*, Albany, NY: State University of New York Press.

Hart, W. D. (2016) "Neville's Metaphysics," *American Journal of Theology and Philosophy*, Vol. 37, No. 3: 248–262.

Hogue, M. S. (2008) *The Tangled Bank: Towards an Ecotheological Ethic of Responsible Participation*, Eugene, OR: Pickwick.

—— (2017) *American Immanence: Democracy for an Uncertain World*, New York: Columbia University Press.

Jonas, H. (1984) *The Imperative of Responsibility: In Search of an Ethics for the Technological Age*, Chicago: University of Chicago Press.

—— (1996) *Mortality and Morality: A Search for the Good after Auschwitz*, ed. Lawrence Vogel. Northwestern University Studies in Phenomenology and Existential Philosophy, Chicago: Northwestern University Press.

Rue, L. (2005) *Religion Is Not About God: How Spiritual Traditions Nurture Our Biological Nature, and What to Expect When They Fail*, New Brunswick, NJ: Rutgers University Press.

Spinoza, B. (1994) *A Spinoza Reader: The Ethics and Other Works*, ed. and trans. Edwin Curley, Princeton, NJ: Princeton University Press.

Steffen, W. et al. (2011) "The Anthropocene: Conceptual and Historical Perspectives," *Philosophical Transactions of the Royal Society A*, Vol. 369, Issue 1938.

White, C. W. (2016) *Black Lives and Sacred Humanity: Toward an African American Religious Naturalism*, New York: Fordham University Press.

Wieman, H. N. (1971) *Religious Experience and Scientific Method*, Carbondale, IL: Southern Illinois University Press.

Further reading

Davies, J. (2016) *The Birth of the Anthropocene*, Oakland, CA: University of California Press. A well-written, thorough account of the convergent geological and political histories of the Anthropocene.

Hogue, M. S. (2017) *American Immanence: Democracy for an Uncertain World*, New York: Columbia University Press. A constructive American political theology for the uncertainties of democracy in the Anthropocene.

15

PRAGMATIC NATURALISM AND PUBLIC THEOLOGY

Prospects of creative exchange

Victor Anderson

Pragmatic Theology: Negotiating the Intersections of an American Philosophy of Religion and Public Theology (1998) was the first volume published in the SUNY Series in Religion and American Life, edited by William D. Dean. Describing the aims of the series, Dean says,

> Books in this series will study the public implications of religious ideas as they operate in America's legal, educational, and political systems, in its academic philosophy and social morality; in its arts, public rituals, and styles of civility, and in its reform movements, its organized religion, and its academic theology. Necessarily these books will come from a variety of academic disciplines and styles ... *Pragmatic Theology* is an apt inaugural volume. It begins by uncovering a twentieth-century history of American pragmatic philosophy and theology and ends by erecting an American public theology.
> (Anderson 1998: ix)

S. Morris Eames describes pragmatic naturalism as the American pragmatist tradition in philosophy, which is indebted to thinkers such as Charles Sanders Peirce, William James, John Dewey, and George Herbert Mead (Eames 1977: xi–xxvii). It designates a non-reductionist version of nature to physical or material aspects, processes, and patterns. It holds that nature is all that there is but exhibits manifolds of activity, habits of behavior, and shapes of freedom, some of which are quite static, exhibiting regularity and continuity and others showing novelties and emergence. Simultaneous with nature's regularities are mutations, precarious states, development, and growth. Such is the natural world where the human organism finds itself accommodating, adapting, and adjusting as a purpose driven creature among others (Eames 1977: xi–xii, 6, 16–22). Pragmatic naturalism accounts for every determination of nature by instruments of knowledge that require no extra-scientific justification beyond empirical and physical sciences. Independent of any spiritual or theological interpretation of nature, pragmatic naturalism stands outside of religious naturalism.

Jerome A. Stone describes religious naturalism as

> the type of naturalism which affirms a set of beliefs and attitudes that there are religious aspects of this world which can be appreciated within a naturalistic framework.

> There are some events or processes in our experience that elicit responses that can be appropriately be called religious.
>
> (Stone 2008: 1)

Discussing William James, Eames argues that pragmatic naturalism may indeed meet the conditions for religious naturalism.

> If there are ideas which cannot be verified directly by scientific method, yet make some difference in human conduct, and if these ideas appear to lead to a more integrated and unified personality, then James sees no reason for not adopting them. He writes: *'If theological ideas prove to have a value for concrete life, they will be true, for pragmatism, in the sense of being good for so much. For how much more they are true will depend entirely on their relations to the other truths that also have to be acknowledged.'*
>
> (Eames 1977: 15)

Pragmatic naturalism is a version of religious naturalism when qualified by the pragmatic demands of public theology. The result is a pragmatic theology.

Public theology is the deliberate and critical use of languages and commitments of faith communities to influence substantive discourse on social and moral issues that affect our collective lives and planetary health (Anderson 2000: 51). The pragmatic demands of public theology involve understanding, transmitting, and constructing theological categories that sustain the political community in practices oriented toward the fulfillment of our spiritual and moral endeavors, responsibilities, and obligation. It also involves developing categories capable of correcting the moral conscience of the political community when public life is governed by patterns and policies that violate the radical democratic fulfillment of all and the flourishing of planetary life.

Three Christian thinkers have strongly influenced my proposal for pragmatic theology. They are Henry Nelson Wieman, James M. Gustafson, and Howard Thurman. They maintain the conviction that the languages of Christian faith, of God, grace and sin, redemption and reconciliation, can be revised in light of the pragmatic naturalist version of reality. However, while drawing on these thinkers, I do not discount the ways they differ as Christian public theologians. For instance, the confluences of Wieman and Gustafson as American empirical theologians with pragmatic naturalism appear well evident in their writings. However, Thurman's "personalism" appears a rather odd comparison to Wieman's and Gustafson's religious naturalism and revisionist theologies.

Thurman views nature in its very essence as "personal" and communicates to every entity its own personality. Moreover, "the human–divine relation is always an encounter mediated by personality, a direct subject-to-subject relation" in which "the principle of personality is universal" (Anderson 2010: 1235–1236). Personalism appears to many religious naturalists supernatural and not rigorously natural. However, in turning to Thurman, I reject the personalist metaphysics while affirming its spiritual and ethical significance for religiously appreciating nature, the world, and human worth. Therefore, turning to Wieman, Gustafson, and Thurman is dialogical and not an attempt to synthesize their thinking into a systematic statement. Still, areas of alignment exist among these Christian thinkers and the pragmatic naturalist version of nature, reality, and the world.

When discussing religion as a natural phenomenon of growth, Wieman employs the dietary metaphor:

> We have come to a time in the history of the world when the religious diet must change. This diet has always been changing throughout history, not in respect to the reality of God, but in respect to the concrete form in which these essentials are taken. At one time they may have been taken in the form of roots and berries and slugs and insects, later in the form of raw fish or bear or deer. So also in religion you can trace the many forms in which men have sought and found the reality of God. That does not mean that everything people have ever eaten truly had in it the essential nutrition.
>
> (Wieman 1996: 67–68)

This quote foregrounds several religious qualifiers to pragmatic naturalism. What we say about God comports with the reality of God in historical experience and corresponds appreciatively to human affections or senses of piety evoked by nature and manifest in worship or devotion.

Gustafson has made much of these qualifications to pragmatic naturalism. He argues that we are valuing creatures and our values are not divorced from our sensuous nature.

> Nature provides the materials and occasions for human creativity; it provides the conditions of possibility for developing our most distinctive human capacities ... The total array of human well being that religious thinkers have enumerated are evoked by our experiences of aspects of the natural world. We respond in awe and respect; nature elicits piety.
>
> (Gustafson 1981: 210)

Thurman argues that the creative integration of our powers of thinking, valuing and feeling contribute to the "total self."

> It has always seemed curious to me that man should investigate the external world, recognize its order, and make certain generalities about its behaviors which it calls laws. That he should study his own organism and discover there a kind of orderliness of inner behavior, which he seeks to correct when it acts out of character by a wide variety of ministrations, from drugs and surgery to hypnosis and faith—and yet that he should be inclined, at the same time, to regard himself as an entity apart from all the rest of creation, including his body. Man is body, but more than body, mind but more than mind; feelings, but more than feelings. Man is total; moreover he is spirit. Therefore it is not surprising that in man's spirit should be found the crucial nexus that connects him with the Creator of Life, the spirit of the living God.
>
> (Thurman 1984: 272)

This quote reveals Thurman's personalism and its stress on "personality" or "spirit" as the clue to cosmic meaning, but it also shows that, for Thurman, the human organism develops and grows into the total self, and the total self is nature's creation. What aligns Wieman, Gustafson, and Thurman as pragmatic theologians is their stress is on "transcendence," what James called the MORE in nature or Thurman's "more than" in human experience. This sense of enlargement, the "more than" in nature, is given theological or spiritual articulation and significance. As a version of religious naturalism, pragmatic theology is the intersection of an American philosophy of religion and the demands of public theology. In what follows, I take up public theology followed by pragmatic naturalism and the result, pragmatic theology as a version of religious naturalism.

Public theology

Wieman, Gustafson, and Thurman represent Christian public theologians who experienced enlivening redemptive powers in the languages of God, sin and grace, and redemption and reconciliation, which informed their thinking about nature, the world, human experience, and our spiritual and moral values and obligations toward humans and our planetary home. Public theologians inhabit a world shared by all and with shared social realities. They belong to individual publics (the church, the academy, society, and various associations) and to a generalized political community of democratic citizenship (Tracy 1981: 3–5). They regard the languages of faith as cultural in meaning, having deep moral and spiritual significance, which are not isolated from public realities and concerns. Public theologians hold ownership of languages of particular faith traditions and of languages which express the needs, interests, and goods of the generalized public. In a summary statement, Linell Cady describes the pragmatic demands of public theology.

> The task of public theology is to elicit a recognition of and commitment to the common life within which we exist. In and through the appropriation of religious symbolism, public theology seeks to nurture, deepen, and transform our common life that, while obscure and damaged, is never totally eroded. Thus public theology is not simply proposing a utopian communal vision that flies in the face of what we know about cosmic and human life. It is, rather, offering a constructive agenda that grows out of discernible features of our individual and corporate experience.
>
> (Cady 1992: 92)

Public theology is recognized by (1) its interpretative contributions to public discourse, (2) the moral agreements it shares with various public discourses, (3) the claims it makes on public life, and (4) its claims for the symbolic integration of public life with the enlargement of human and planetary flourishing. Public theology takes its departure from the symbolic universes and social practices which constitute public life in its generality and particularity. It proposes the successful transaction of public interests, needs, and goods with the tragic and ironic character of human experience. Given these realities of human experience, the test of public theology is its power to render public life spiritually meaningful, morally livable, and culturally fulfilling. The central concern of public theology is whether the languages of Christian faith (God, sin, grace, redemption, reconciliation, and the like) are sufficiently translatable for communicating genuine moral and spiritual meanings beyond the narrow boundaries of one's own fellowship and doctrinal history.

Pragmatic naturalism

Pragmatic theology represents the negotiation of pragmatic naturalism qualified by the demands of public theology. To borrow a phrase from Nelson Goodman, it is a way of "worldmaking." Pragmatic naturalism fully acknowledges our bodily immersion in natural processes, natural forces, and complexes. It rejects the reduction of life to the material/physical sphere and human activity to chemical, neurological determinants of the brain and genes. It rejects versions of reality which posit that the material and physical entities such as stone, water, air, tree, and animal all have objectively independent *ontic* statuses, as if free standing monads, independent of and prior to human thought, action, and will. Pragmatic naturalism offers an alternative version of reality. It asks that we not see the physical/material world as an unmediated background or foundation prior to and productive of thought and action. Its version of worldmaking carries within it the notion that nature is alive, breathing, feeling, and full of activity.

Goodman asks that we not slide whatever we mean by reality (either externally or mentally) into what we mean by world. He asks that we not treat any one version of reality as corresponding to our common talk about "the world" but rather as ways of worldmaking (Goodman 1978: 3–4). It is we humans who are engaged in acts of "worldmaking," and how we account for human experience and realities might best be understood as often complimenting and contrasting yet sometimes contradicting "versions." He proposes,

> Rather than speak of pictures as true or false we might better speak of theories as right or wrong; for truth of the laws of a theory is but one special feature and is often, as we have seen, overridden in importance by the cogency, and compactness and comprehensiveness, the informativeness and organizing power of the whole system. 'The truth, the whole truth and nothing but the truth' would thus be a perverse and paralyzing policy for any world maker. The whole truth would be too much; it is too vast, variable, and clogged with trivia.
>
> (Goodman 1978: 19)

Supposing, then, that there are no "undescribed, undepicted, and unperceived" worlds available for comparison, Goodman argues that having the right version is owning a preference for one version as being more right than another insofar as the version one holds is not radically incongruent with what we generally know about human experience and planetary life from physical and empirical sciences (Goodman 1978: 4).

The pragmatic naturalist version of reality envisions planetary life as constituted by manifolds of activity. Inorganic entities hold together, not by supernatural powers but by micro-nuclear energy of quantum waves, strings, and magnetic forces and as yet unknown physical forces. Organic entities are developing, growing, undergoing generation and mutation, and feeling their proximity to others (organic and non-organic). "For pragmatic naturalists, there is no being or any sphere of reality that is outside or untouched by process," says Eugene Fontinell (Fontinell 1970: 41). The pragmatic naturalist version of worldmaking "emphasizes the radical interdependence of entities as they exist concretely in the world and rejects the notion of a closed system composed of things thought to be complete-in-themselves" (Fontinell 1970: 52). Nature is processive, open, and relational. Novelty breaks through as a shape of freedom in nature. In the coming forth of being or its passing away, creativity persists in what one pragmatic theologian, Gordon D. Kaufman, calls "serendipitous creativity." Sanguine to the pragmatic theology version of religious naturalism, Kaufman's account of serendipitous creativity is quoted at length:

> Many have expressed interest in my theological reconstructions, and many have also raised serious questions about my proposals. In a number of articles in recent years I have sought to clarify some of the ambiguities and difficulties in these proposals as well as to develop them further. In the course of those activities I was led to reflect more fully on just what it might mean to think of God not as a personal being who had created the world and everything in it, and who continues to work creatively in and upon that world, but rather as neither more or less than just this creativity itself (however that is to be understood) manifest throughout the vast cosmos as we today think of it. Particularly awe-inspiring to humans, of course, has been the creativity manifest in the emergence of life on planet Earth with, on the one hand, its evolutionary expansions in countless directions and, on the other hand, the painfully slow

> evolutionary developments that at long last brought us humans—creatures capable of becoming consciously aware of this magnificent panorama of which we ourselves are part—into being, creatures who can and do stand in awe and gratitude before the serendipitous creativity that has brought all this forth and has given life ... God is nothing more or less than precisely this creativity.
>
> (Kaufman 2004: x)

Pragmatic theology

Pragmatic theology is a way of worldmaking that offers a religious perspective for resisting the ordering of our lives and those of others (human and non-human) as if they were microchips in the working of the super-computer called nature. Life and death are not without purpose, meaning, satisfaction, and equilibrium. Human agency is not epiphenomenal or an illusion generated by brain activity which the brain requires for satiating itself against the nihilistic conclusion that its own activities are pointless and non-redemptive.

The nineteenth-century theologian Friedrich Schleiermacher may not have been very far off from pragmatic theology when he insisted that the reality of God comes forth in our creaturely sense of dependence on the creator for sustaining our species in its emergence and development under nature's precarious powers and conditions (Schleiermacher 1986: 12–17). Gustafson surrenders nothing to the dark night of the soul when he stresses a vision of God as nature's power(s) bearing down on us, sustaining us, ordering human life within natural and social processes and patterns, establishing limits to human wellbeing and human endeavors without having our wellbeing and endeavors as chief ends (Gustafson 1981: 264). In Gustafson's pragmatic theology, God symbolizes nature's overwhelming powers, which bear down upon us even as they sustain, order relationships, and provide conditions of possibilities for human activity and obligations as shapes of freedom in nature. The full spiritual significance of these shapes of freedom or of transcendence is derived from our sensuous nature as religious affections, says Gustafson (Gustafson 1981: 264).

Life survives by dynamic processes and patterns and creative human capacities through which senses of religious piety are evoked. However, the senses of piety are mixed and ambiguous. The senses of awe and devotion are often accompanied by the senses of dread and danger. Senses of beauty are met with repulsion, and senses of obligation and responsibility confront senses of human inadequacy and futility. Ambiguities in senses of piety correspond to our experiencing planetary life as both determined by nature's limits and its shapes of transcendence or of freedom from the overwhelming sense of our having been fated by uncompromising determinations of nature's powers over our lives.

> Where human agency can rectify the effect of fate, can create possibilities in society and culture for those who are in despair, and can alter institutional arrangements to restrain threats to human well being and create possibilities for human flourishing, a form of redemption is occurring.
>
> (Gustafson 1981: 249)

Limits abound despite our powers of imagination and will. We are daily thwarted by the push and pull of nature, environmental conditions, our bodies, disease, and mortality. Pragmatic theology takes this fact as simply being the way things are. Moreover, this recognition affects politics and public policy making, our moral commitments to advance human wellbeing, our interventions into the ecological ordering of life, and our social endeavors to create healthy, prosperous societies where peace and justice can thrive. However, where limits abound in nature, there are also possibilities for transforming human and planetary life toward maximizing equilibrium. We move about and act in the world, adapting some conditions to cooperate toward satisfying some of our deepest aims and adjusting ourselves toward other conditions for overcoming unwarranted feelings of futility, cynicism, and fatalism. We muster resources to create safe, healthy spaces where children are provided the best chances for advancing life plans. And yet, we know well that despite our highest hopes, aims, and directives of our social institutions, not all will survive. Pragmatic theology seeks to speak meaningfully to these conditions of life on Earth. It attunes to the precarious, tragic conditions of both planetary and human life. It turns concretely to the experience of those who live life as if their backs were against a wall.

Consider Thurman's analysis of the Negro Spirituals for shapes of freedom in the experience of slaves. When analyzing Negro Spirituals, Thurman maintains the sense of ambiguity evoked by them, arguing that the spirituals speak of life and of death (Thurman 2003). The spirituals are the cultural witness of transatlantic slave trade survivors transplanted into the hostile world of chattel slavery throughout the Americas. As a pragmatic theologian, Thurman is guided by a basic question: What do people do when their backs are against the wall?

> What, then, is the fundamental significance of all these interpretations of life and death? What are these songs trying to say? They express the profound conviction that God was not done with them, that God was not done with life. The consciousness that God had not exhausted His resources did not ever leave them. This is the secret of their ascendancy over circumstances and the basis of their assurances concerning life and death. The awareness of the presence of a God who was personal, intimate, and active was the central fact of life and around it all the details of life and destiny were integrated.
>
> (Thurman 2003: 41)

In the tragic, demonic conditions of chattel slavery, the spirituals, says Thurman, speak to the ambiguities of the senses of life and death and the reality of God comes forth in the sense of blessed assurance of the "more than" power of God affecting them personally, intimately, and actively. From these shapes of freedom, the Negro spirituals present a pragmatic theology, which, Thurman says poetically, recognizes "a ceaseless search like the ebb and flow of oceans/Marks all man's days" (Thurman, 1984: 10). It is the sense of assurance that "God was not done with life." This blessed assurance was the clue to shapes of freedom in the life and destiny of a new community, a new people, a new family, and a new and enduring faith in a new humanity.

For Thurman, pragmatic theology insists that finitude meets us at every step, but so do shapes of freedom, openness, and creativity, which conspire with human agency to transform circumstances and ourselves in ways that maximize planetary equilibrium. He says,

> It is possible for the individual to move out beyond the particular context by which his life is defined and relate to other forms of life from inside their context. This means that there is a boundless realm of which all particular life is but a manifestation. This center is the living thing in human beings and animals. If a person or animal can function out of that center, then the boundaries that limit and define can be transcended.
>
> (Thurman 1986: 74)

From a boundless center, thought and action, organism and environment, and experience and culture gather mutually in interdependence and shapes of freedom. Moreover, these manifolds of freedom transcend the particularity of faith communities and enlarge the lived situations of people of faith into creative events of community. Thurman observes,

> The man who seeks community within his own spirit, who searches for it in his experiences with the literal facts of the external world, who makes this his formal intent as he seeks to bring order out of chaos of his collective life, is not going against life but will be sustained and supported by life.
>
> (Thurman 1986: 6)

Thurman points to the fact that human life is structured by choices, actions, and policies of others long passed which we did not choose for ourselves, and far too many of which have increased human misery and planetary devastation. In the language of Christian faith, these effects are called evil, even demonic. But Thurman also elicits great expectations for increasing individual and social freedoms, moral capacities, and just policies, which may be interpreted as signs of grace. Through grace, great expectations are co-present in justice denied, dreams deferred, and malicious policies oriented toward the eroding and cutting off of the democratic citizenship of all. Sin and grace attend our moral and spiritual strivings even while bringing into judgment our moral and spiritual failures of good will toward others, human and non-human. Our experiencing such gracious redemptive moments in planetary life is regarded as emergent events in nature. They must be seized, taken up, and grasped with urgency, if hope as a morally just political community and planetary equilibrium are to become concretely actualized. In such emergent events, pragmatic theology recognizes redemption and reconciliation.

Pragmatic theology greets signs of sin and grace and redemption and reconciliation with a sense of urgency, if better housing, health care, educational policies, living conditions for the poor, and better democratic participation are achievable in our present moment. The urgency is great, for as Wieman, Gustafson, and Thurman understood all too well, sin and evil crouch at the door of our moral and spiritual endeavors. Nature certainly presents limits to human agency, but these limits constitute neither sin nor evil. Sin and evil are the works and effects of our own self-directed negligence, selfishness, and overweening private interests. These human propensities cut off our capacities for recognizing transformative powers of grace, redemption, and reconciliation at work in the world because of human wickedness. "Some people," says Wieman, "would call this self-destructive propensity in man by the name of original sin." He acknowledges, "This dark side of the story of man must not be forgotten" but insists that the "self-destructive propensity rises to dominance in human life not because of some ineradicable drive called original sin, but because of the conditions under which men live" (Wieman 1982: 450–451).

Our faith communities do not escape human propensities for wickedness. They are human communities and hence natural communities, which share "many characteristics of other human communities such as nations, trade unions, and professions" (Gustafson 1961: 3). Understanding faith communities as natural communities goes a long way toward stereoscopically appreciating their capacities for enlivening downtrodden spirits, forming and reviving faith, hope, and love, while also being centers of sexual abuse, racial and ethnic distain, and parochial pride. They may be refuges for weary, hungry, homeless, sick, and seeking souls, but are not immune to being centers of exclusion of "the least of these" in need of salvation, redemption, forgiveness, and hope. As human communities, faith communities are subject to the same social and historical processes that impact other communities (Gustafson 1961: 5). They bear "the marks of a natural community, providing fulfillment of desires and needs that are common to all," providing recognition, a sense of belonging, guiding and forming identities, regulating the conduct of members, and interacting within social networks of communication, values, and concerns, mundane and ultimate (Gustafson 1961: 21).

Faith communities may form distinctive publics, but they interact with and have powerful influences on the public sphere, sometimes frustrating any reasonable account of the goods we hold in common with others, human and non-human. While providing for emotional, familiar, educational, political, and social needs of members, they also compete with other publics by staking particular claims on members' faith, loyalty, and trust. Wieman insists that conflicts between communities of faith and the public sphere need not be destructive or demonic in their effects on our common life but occasion the possibility of creativity interchange:

> The creative event, we shall find, weaves a web of meaning between individuals and groups and between the organism and its environment. Out of disruption and conflicts, which would otherwise be destructive, it creates vivifying constraints of quality if it is able to operate at all. Thus it can utilize frustration and disaster to create and weave into a web of life's meanings vivid and diversified qualities, thus adding immensely to the richness of its variety and the depth of its significant connections.
>
> (Wieman 1982: 92)

Every interchange of goods and values involves conflicts and frustrates particular goods and ends sought by individuals and groups or organisms and environments. But where mutuality exists between faith communities and the public sphere, Wieman describes such events as "creative interchange." "Creative Interchange," Wieman continues, "makes possible disagreement and diversity without hate or fear, without retaliation or estrangement, because the demands of creative interchange are contributory to the good of each" (Wieman 1982: 457). Moreover, "Creative interchange cannot save us," he says, "unless we make a religion of it. That means to give our ultimate commitment to it, accept it as the sovereign good of our lives, the source, guide and goal of our existence" (Wieman 1982: 457).

When faith communities participate with other publics and are joined by shared hopes, dreams, and concerns, mundane and ultimate, they are allies in moving our political aims toward policies that enlarge radical visions of goodwill towards the least advantaged and planetary life. They enter into processes of creative interchange with others, cooperating, through a generous pluralism of goods and values, for the sake of maximizing planetary equilibrium. Wieman writes, "Creative interchange is the only ultimate and valid basis of community because it is the one ultimate source of our humanity, also the way leading to the greatest good to be attained by human kind" (1982: 458). For Wieman, creative conflict and creative interchange are practical corollaries to the pragmatic naturalist version of finitude and transcendence in nature and contribute to the ethical aims of pragmatic theology as a way of worldmaking. As a version of religious naturalism, Wieman regards creative conflict and creative interchange as an ultimate human commitment. "This means to give our ultimate commitment to it, accept it as the sovereign good of our lives, the source, guide and goal of our existence" (Wieman 1982: 458).

Conclusion

Pragmatic theology is a version of religious naturalism. It results from pragmatic naturalism qualified by the demands of public theology. It is cognizant of the limits of human actions in a world framed by finitude and emergent shapes of freedom. It finds in the rich resources of languages of faith, categories wide enough to translate our experiences of nature in spiritually and morally transformative ways in our public life. It tests the adequacy of our talk of God, sin and grace, redemption, and reconciliation, for their powers of making sense of our bodily immersion in nature, its limits and its emergent shapes of freedom. Thus, as a version of religious naturalism, pragmatic theology acknowledges the precariousness of nature and human experience, while accenting nature's novelties, openness, and creativity toward enlarging versions and visions of human potentialities and expectation that the creative processes of nature will guide us toward maximizing human and planetary equilibrium.

References

Anderson, V. (1998) *Pragmatic Theology: Negotiating the Intersections of an American Philosophy of Religion and Public Theology*, Albany: State University of New York.

Anderson, V. (2000) "Contour of an African American Public Theology," *Journal of Theology*, Volume CIV, Summer, p. 51.

Anderson, V. (2010) "Howard Thurman" in D. Patte (ed.) *The Cambridge Dictionary of Christianity*, New York: Cambridge University Press.

Cady, L. E. (1992) *Religion, Theology and American Public Life*, Albany: State University of New York Press.

Dean, W. D. (1998) "Foreword" in V. Anderson *Pragmatic Theology: Negotiating the Intersections of an American Philosophy of Religion and Public Theology*, Albany: State University of New York.

Eames, S. M. (1977) *Pragmatic Naturalism: An Introduction*, Carbondale: Southern Illinois University Press.

Fontinell, E. (1970) *Toward a Reconstruction of Religion: A Philosophical Probe*, Garden City: Doubleday.

Goodman, N. (1978) *Ways of Worldmaking*, Indianapolis: Hackett Publishing Company.

Gustafson, J. M. (1961) *Treasure in Earthen Vessels: The Church as a Human Community*, Chicago: University of Chicago Press.

Gustafson, J. M. (1981) *Ethics from a Theocentric Perspective: Theology and Ethics*, Chicago: University of Chicago Press.

Kaufman, G. D. (2004) *In the Beginning ... Creativity*, Minneapolis: Fortress Press.

Schleiermacher, F. (1986) *The Christian Faith*, Edinburgh: T & T Clark.

Stone, J. A. (2008) *Religious Naturalism Today: The Rebirth of a Forgotten Alternative*, Albany: State University of New York.

Thurman, H. (1984) *For the Inward Journey: The Writings of Howard Thurman*, Richmond: Friends United Meeting.

Thurman, H. (1986) *The Search for Common Ground*, Cleveland: Friend United Press.

Thurman, H. (2003) "The Negro Spirituals Speak of Life and Death" in C. West and E. Glaude (eds.) *African American Religious Thought: An Anthology*, Louisville: Westminster John Knox Press.

Tracy, D. (1981) *The Analogical Imagination: Christian Theology and the Culture of Pluralism*, New York: The Crossroad Publishing Company.

Wieman, H. N. (1982) "Creative Interchange" in Broyer and Minor (eds.) *Creative Interchange*, Carbondale, IL: Southern Illinois University Press.

Wieman, H. N. (1996) "Theocentric Religion" in Peden and Stone (eds.) *The Chicago School of Theology: Pioneers in Religious Inquiry, Vol. I, The Later Chicago School, 1919–1988*, Lewiston, NY: Edwin Mellen.

Further reading

V. Anderson, *Creative Exchange: A Constructive Theology of African American Religious Experience* (Minneapolis: Fortress Press, 2008) is a pragmatic naturalist interpretation of African American religious experience. H. N. Wieman, *Man's Ultimate Commitment* (Carbondale: Southern Illinois University Press, 1958) is the comprehensive account of creative interchange in religion, social institutions, and history. J. M. Gustafson, *Ethics from a Theocentric Perspective, Volume One: Theology and Ethics* (Chicago: University of Chicago Press, 1981) is an exhaustive revisionist treatment of God from the pragmatic naturalist perspective. A. B. Pollard, III, *Mysticism and Social Change: The Witness of Howard Thurman* (New York: Peter Lang, Inc., 1992) analyzes and corrects misunderstandings of mysticism and activism in the life and practices of Howard Thurman. J. R. Shook (ed.), *Pragmatic Naturalism & Realism* (Amherst, NY: Prometheus Books, 2003) includes recent essays by philosophers worldwide on the resurgence of pragmatism and debates on realism. H. J. Recinos, *Wading Through Many Waters: Toward a Theology of Public Conversation* (Lanham, MD: Rowman and Littlefield Publishing, Inc. 2011) presents essays written by theologians representing Latino/a, African American, Asian American, Native American, and Euro-American discourse on public theology.

PART V

Religious naturalism and traditional religions

16
BUDDHISM AND RELIGIOUS NATURALISM

Jay N. Forrest

Buddhism has many ideas and practices that can be naturalized, and these should be of great interest to Religious Naturalists. Both historically and in modern times, Buddhism has shown its adaptability. But some strategies have not gone far enough. In this essay, I offer my own attempt at a Buddhist Naturalism. Buddhism, once it is naturalized, can provide Religious Naturalists a fertile field of ideas and practices to think outside a Western mindset.

Historical context

Dating the life of Siddhartha Gautama is important for understanding the historical context of his time. Traditional sources date his life from 623 to 543 BCE. But modern scholarship has placed it later. Richard Gombrich, based on his extensive research, has given 484–404 BCE as the most likely dates (Harvey 2013: 8). It is important to know about the other schools of thought at this time. In the words of Sarvepalli Radhakrishnan, "This was one of the most fertile periods of philosophy in India" and "most of the systems had their beginnings about the time of the rise of Buddhism, and developed side by side for centuries" (1957: xix).

Much is often made of the uniqueness of Buddhism, but as Radhakrishnan points out, "Hinduism, Buddhism, and Jainism, in all their branches, also accept the underlying doctrines of karma and rebirth" (1957: xxix). With a better understanding of Indian philosophy, we find that the Buddha did not create his philosophy out of thin air. He developed it within the intellectual climate of his days. As Radhakrishnan explains,

> The Buddha takes up some of the thoughts of the Upanishads and gives them a new orientation. The Buddha is not so much formulating a new scheme of metaphysics and morals as rediscovering an old norm and adapting it to the new conditions of thought and life.
>
> (1957: 272)

Jack Maguire agrees: "The Buddha's spiritual vision rose out of a predominantly Hindu culture, and his career as a teacher was geared towards refining common religious beliefs rather than overthrowing them" (2001: 26). One way to think of the relationship of Brahmanism and Buddhism is to think of Judaism and Christianity. As Dale Riepe explains, "Buddhism is to Brahmanism what Christianity is to Judaism, if we remember that Brahmanism much more than Judaism contributed to philosophical speculation and that Brahmanism much more than Judaism was receptive to the new movement" (1961: 117). This relationship is not surprising. As Riepe writes, "The Buddha, it is believed, received training along orthodox lines, probably studying the Vedas and the Upanishads" (1961: 122). Certain common teachings that were considered orthodox had the status of dogma at this time, "these included rebirth or transmigration, karma, enlightenment (*mukti*), and that life is suffering" (1961: 33). These became the building blocks of Buddhism.

But that does not mean that Siddhartha Gautama bought into the doctrines of Brahmanism wholesale. Gombrich argues "The central teachings of the Buddha came as a response to the central teachings of the old Upanishads" (2010: 31). He says that "It is a historical misconception that the Buddha took over a pre-existing doctrine of karma," rather, "The Buddha defined karma as intention; whether the intention manifested itself in physical, vocal or mental form, it was the intention alone which had a moral character: good, bad or neutral" (2010: 50–51). He also redefined the self or atman as an illusion. This redefinition also affected the doctrine of reincarnation. If there is no self, what gets reborn? The traditional answer is the stream of consciousness.

But of particular interest to Religious Naturalists is another group that predated Buddhism and existed during and after the life of Gautama. This group is known as the Carvaka or Lokayata. *The Princeton Dictionary of Buddhism* defines the Lokayata as a "Naturalist" or "Worldly" school (Buswell and Lopez 2014: 481). According to Pradeep Gokhale, this group believes that

> (a) perception is the only means to knowledge; (b) consciousness is the product of matter; (c) pleasure – rather, sensuous pleasure – is the only goal of life, which could be achieved even by immoral means; and (d) there is no God, no other world, and no life after death.
>
> (2015: 1)

However, Gokhale notes that this interpretation may be too simplistic, for "there were at least two trends in the Lokayata School or tradition, one materialist, the other skeptic" (2015: 5).

Bhupender Heera makes an interesting point: "As Lokayata is undoubtedly prior to Buddhism there is no harm in believing that Buddhism was greatly influenced by the Lokayata, that the doctrines of Buddha and Caraka almost amalgamated, and that the name Carvaka sometimes applied to Buddha" (2011: 28–29). In fact, if you think of Brahmanism as the supernaturalists and the Lokayata as the Naturalists, you could see Buddhism as a middle way between them. This explains the Buddha's rejection of an immaterial self (atman) that transmigrates from body to body. Brahmanism accepted the unchanging self or atman, the Lokayata rejected it completely, and the Buddha sought a middle way between them, that is, accepting anatman but keeping rebirth. The Buddha wanted to keep morality while jettisoning other metaphysical claims.

Before moving on, I should also mention Jainism, which was founded by Mahavira (599–527 BCE). "I suggest," writes Gombrich, "that the positive influence of Jainism on the Buddha was massive" (2009: 51). In fact, he credits the Jains "for the first ethnicized karma theory," and "not the Buddhists" (2009: 51). According to Gombrich, the doctrines the Buddha borrowed from the Jains include, "the doctrines of samsara, of the desirability of getting out of samsara, and the role that ethics played in making that escape possible" (2009: 51). It is also interesting to note the similarity of the five precepts of Buddhism with the Jain's practice of the five virtues, which, as Radhakrishnan explains, are: "(1) ahimsa (non-violence), (2) truth-speaking, (3) non-stealing, (4) chastity, and (5) non-attachment to worldly things" (1957: 251). He also points out that the Jains were the first to make non-violence "into a rule of life" (1957: 251).

After the Buddha passed away, the Sangha remained unified for a time. But as Noble Ross Reat explains,

> most accounts agree that about a hundred years after the death of the Buddha a large group of monks – known as the Mahasanghika or 'majority group' – rejected the authority of the Elders, and that as a result a major schism in Buddhism occurred.
>
> (1994: 19)

Although according to the Sthaviravada tradition, notes Reat, the first schism was "over the ten lax practices," the Sarvastivada claimed "that the cause was doctrinal" (1994: 89). From this schism there arose at least 18 different schools.

Just like Protestantism, the divisions have never stopped. Buddhism today is usually divided into three branches: the Theravada, the Mahayana, and the Vajrayana. But even this is too simplistic. Under the heading of Mahayana, there are drastically different schools. Pure Land Buddhism and Zen Buddhism are so different, it is hard to see them as part of one religion. "Buddhism," writes Rupert Gethin, "is something of an intellectual abstraction: in reality there is not one Buddhism but many Buddhisms" (1998: 2). "As the host culture gradually assimilated Buddhism," writes Maguire, "every aspect of that culture was subtly imbued with a new Buddhist character. Simultaneously, a fresh, innovative variation of the religion emerged bearing the host culture's likeness" (2001: 210).

Buddhism's encounter with modernity

Although immigrant Buddhism was already in the United States, most people see the World Parliament of Religions in 1893 as the first introduction of Buddhism to the West. It was here that Paul Carus met Soyen Shaku, who later introduced him to D.T. Suzuki, and it was partly through the writings of D. T. Suzuki that Zen Buddhism entered into the consciousness of America.

Soyen Shaku, however, did not teach traditional Buddhism. His aim was to convert Americans to Zen, and hence to accommodate Buddhism to the American Christian and scientific mind. This accommodation and modernization have not stopped. As David L. McMahan explains,

> What many Americans and Europeans often understand by the term 'Buddhism,' however, is actually a modern hybrid tradition with roots in the European Enlightenment no less than the Buddha's enlightenment, in Romanticism and transcendentalism as much as the Pali Canon, and in the clash of Asian cultures and colonial powers as much as in mindfulness and meditation.
>
> (2008: 5)

Bernard Faure agrees: "In their effort to modernize, Buddhists have sought to emphasize the compatibility of Buddhism with modern-day science, discreetly failing to comment on any areas of disagreement" (2009: 104). As he points out, many of the modern "ideas about Buddhism are not supported by tradition," but are "the result of a series of reforms in various Asian countries and of increased contact with the West. It has developed in response to colonization, the requirement to modernize, and the influence of Protestantism" (2009: 3–4).

Before we can honestly think about a naturalized version of Buddhism, we must first understand the tradition as it has come down to us. For this, we look primarily to the Buddhist Sutras. Each branch of Buddhism has its own canon. But, as Gethin explains, "The production of Mahayana sutras spans a period of some six or seven centuries" (1998: 57). Thus they are clearly not historical records of the Buddha's teaching. Reat states, "From a historical point of view it is clear that the Mahayana scriptures are highly embellished accounts of historical events, if indeed they record historical events at all" (1994: 29). The vast majority, if not all of the Mahayana scriptures, are pious fictions, for they claim to be spoken by the Buddha. What they are in reality is a living testimony to Buddhist experience and insight into that experience within the culture of the writer. They tell us what others thought about the Buddha's teachings, but they do not tell us what the Buddha actually taught. For that, we turn to the Pali Canon.

The Pali Canon is, along with the Chinese Agama, the oldest surviving record of the Buddha's teaching. Although, admits Bhikkhu Sujato and Bhikkhu Brahmali, "The texts as we have them now are not a verbatim record of the Buddha's utterances, the changes are in almost all cases details of editing and arrangement, not of doctrine or substance" (2014: 12). Furthermore, these two writers do not consider the following as Early Buddhist Texts: "Abhidhamma, Mahayana Sutras, Buddha biographies, historical chronicles, as well as the majority of the Khuddaka Nikaya and the Vinaya Pitaka" (2014: 12). But even within the accepted books, there are questions. We must remember that the Buddha's teaching was originally handed down by oral tradition. Although oral citations were much more accurate and common at this time in India, they are still not infallible. After about 400 years, the oral tradition was written down sometime in the early first century CE. But none of these early manuscripts exist today. Just like the Bible, we have copies of copies of copies, which means that the written record has been clearly corrupted to some degree. But in what follows, I will assume the general authenticity of the Digha Nikaya, Majjhima Nikaya, Samyutta Nikaya, Anguttara Nikaya, and parts of the Khuddaka Nikaya.

As Buddhism interacts with modernity, with its secular and scientific viewpoint, certain aspects of historical Buddhism will need reconsideration. We begin with the fact that gods are everywhere in the Pali canon. In early Buddhism, the gods were a part of samsara. The Buddha is really above the gods, because he alone has attained Nirvana. Gods exist, but they are stuck in the same trap as people – the trap of samsara. So you will find the Pali canon treating the gods as irrelevant. The problem of the place of the gods becomes more complicated. It is reported in Majjhima Nikaya that before becoming Siddhartha Gautama, the Buddha existed as a god in the heavenly realm known as Tusita. In the Itivuttaka the Buddha said,

> Whenever the aeon contracted I reached the plane of Streaming Radiance, and when the aeon expanded I arose in an empty Brama-mansion. And there I was a Brahma, the Great Brahma, the Unvanquished Victor, the All-seeing, the All-powerful. Thirty-six times I was Sakka, the ruler of the devas.
>
> (Ireland 1997: 165)

From this, we learn that, according to the Buddhist Sutras, the Buddha to be was god made flesh. Were these zealous monks putting words into the Buddha's mouth, or did Siddhartha Gautama actually say this?

When it was time for the future Buddha to be born, the Majjhima Nikaya says that he "passed away from the Tusita heaven and descended into his mother's womb" (Bodhi 2009: 980). At his conception, it says, "a great immeasurable light surpassing the splendour of the gods appeared in the world" (980). Then "four young deities came to guard him at the four quarters so that no human or non-human or anyone at all could harm the Bodhisatta or his mother" (981). Instantly his mother "became intrinsically virtuous" (981). As soon as he was born,

> he took seven steps facing north ... and uttered the words of the Leader of the Herd: 'I am the highest in the world; I am the best in the world; I am the foremost in the world. This is my last birth; now there is no renewal of being for me.
>
> (Bodhi 2009: 983)

Throughout, the author continually assures us, "I heard and learned this from the Blessed One's own lips." Personally, I doubt it. Rather I think this was pious propaganda to bolster trust in the Buddha's teachings. For the Religious Naturalist, all of this can be overlooked, since it probably does not affect the Buddha's main teachings.

Another difficulty is the repeated mention of psychic powers. As Vessantara (1993: 183) explains, "In the Mahayana sutras and in the Pali Canon psychic powers, including clairvoyance, clairaudience, telepathy, levitation, and memory of past lives, are frequently mentioned." The Samyutta Nikaya lists off some of the superpowers of certain monks:

> a bhikkhu wields the various kinds of spiritual power: having been one, he becomes many; having been many, he becomes one; he appears and vanishes; he goes unhindered through a wall, through a rampart, through a mountain as though through space; he dives in and out of the earth as though it were water; he walks on water without sinking as though it were earth; seated cross-legged, he travels in space like a bird; with his hands he touches and strokes the moon and sun so powerful and mighty; he exercises mastery with the body as far as the brahma world.
>
> (Bodhi 2000: 1727)

Claims of psychic ability will cause the Religious Naturalist a moment of pause. However, whether or not psychic abilities really exist, this does not require a belief in a supernatural realm.

There are two major teachings of the Buddha that will cause difficulties for Naturalists, namely, karma and rebirth. Let's first look at rebirth. Before we consider how to naturalize it, it would be best to understand what rebirth means within the Pali Canon. The Digha Nikaya says that "any who die while making the pilgrimage to these shrines with a devout heart will, at the breaking up of the body after death, be reborn in a heavenly world" (Walshe 1995: 264). Since Buddhism teaches that there is no permanent self, no soul in the Christian sense, what transfers from one life to the other? The Pali Canon is unclear, but Bhikkhu Bodhi gives the traditional answer, "The channel for the transmission of karmic influence from life to life across the sequence of rebirths is the individual stream of consciousness" (2015: 187). He further argues, "The teaching of rebirth crops up almost everywhere in the Canon, and is so closely bound to a host of other doctrines that to remove it would virtually reduce the Dhamma to tatters" (2015: 183). The problem for Religious Naturalists is that there is very little empirical evidence for rebirth. In fact, everything we know about the mind indicates that consciousness and the brain are interconnected. Change the chemicals in the brain, and consciousness is affected. Damage the brain and consciousness is diminished. Get drunk and consciousness gets drunk. There is no evidence that consciousness has any independence from the brain. As far as science is concerned, consciousness cannot exist outside of a brain.

When we turn to the subject of karma, we may seem to be on safer ground. But we must understand karma within the Pali Canon. First, it is important to understand the importance of

Karma in the Buddha's system. As Gombrich points out, karma "is fundamental to the structure of his thought" (2010: 33). But the Buddha did not define karma as the law of cause and effect. He was not interested in physics, he was interested in morality. In the Anguttara Nikaya, the Buddha said, "It is volition, bhikkhus, that I call kamma [i.e., karma]" (Bodhi 2012: 963). As Bodhi explains, "'kamma' literally means action, but technically it refers to volitional action" (2005: 146). Intentional action creates one's current life and determines one's future rebirth. This is where the Buddha departed from the Jains' understanding. They suffered for both intentional and unintentional acts, where the Buddha limited this consequence to only intentional acts.

One last troubling claim is that the Buddha was claimed to be omniscient. In *The Questions of King Milinda*, which is included in the Burmese edition of the Pali Canon, Nagasena is asked if the Buddha was omniscient. He answers,

> Yes, O king, he was. But the insight of knowledge was not always and continually (consciously) present with him. The omniscience of the Blessed One was dependent on reflection. But if he did reflect he knew whatever he wanted to know.
>
> (Davids 2007: 163)

This claim is problematic for a number of reasons. First, because it cannot be true. This is an attribute for a god, not a man. No man has or ever will be all-knowing. Second, it places the Buddha's authority beyond dispute. Such a claim makes the Buddha infallible. How can an omniscient person be wrong? The authority of the Buddha ends up being a major obstacle in secularizing and naturalizing the Dharma. Many of these beliefs, including rebirth and karma, are said to be verifiable through meditative experiences.

Strategies for responding to science and secularism

A first response to the intrusion of science on the doctrines of any religion may be to deny the science. Nobody better exemplifies this than B. Alan Wallace. He clearly draws the lines: "The metaphysical views of materialism are in fundamental conflict with the Buddhist worldview regarding the nature of the mind; if materialism is correct, then the Buddha's claims of having direct knowledge of past lives, karma, and nirvana would be invalid " (2012: 18). Since Wallace assumes that the Buddha could not have been wrong, science is wrong, because "of the materialistic assumptions saturating scientific inquiry" (2012: vii).

But Buddhism must respond to modern day science. Most have retreated to the defense that you can't scientifically disprove karma, rebirth, and the six realms. For example, the Dalai Lama has said that "if scientific analysis were conclusively to demonstrate certain claims in Buddhism to be false, then we must accept the findings of science and abandon those claims" (2005: 3). At first, this may seem reasonable, but notice where the Dalai Lama places the burden of proof. From his viewpoint, Buddhism doesn't have to produce evidence for its claims; rather science has to "conclusively" demonstrate a claim in Buddhism to be false. This is all backward. A person making a claim that asserts that something is the case, has the burden of proof. The skeptic is under no obligation to accept a mere assertion. Furthermore, the skeptic who denies the claim has no obligation to prove a negative; in many cases it is impossible to do so. Critical thinking tells us that we should reject any claim that has little or no evidence for it. Suspending judgment, which many modern Buddhist do concerning rebirth, is reserved for claims that have a balance of evidence both for and against it. So we see that by shifting the burden of proof, the Dalai Lama can continue to believe and teach reincarnation. How can you "conclusively" demonstrate that the reincarnation claim is false? As Carl Sagan states, "Claims that cannot be tested, assertions immune to disproof are veridically worthless, whatever value they may have in inspiring us or in exciting our sense of wonder" (1997: 171).

Another strategy is to be upfront about rejecting karma and rebirth. Stephen Batchelor started a revolution with his book *Buddhism Without Beliefs*. In it, he proposed, "The dharma is not something to believe but something to do" (1997: 17). He made the claim, which surely sounded like heresy to traditional Buddhists, "In accepting the idea of rebirth, the Buddha reflected the worldview of his time" (1997: 15). This led Bodhi to fire back,

> He begins as if he intended to salvage the authentic vision of the Buddha from the cultural accretions that have obscured its pristine clarity; yet, when he runs up against principles taught by the Buddha that collide with his own agenda, he does not hesitate to discard them.
>
> (2015: 85)

According to this critic, Batchelor's great crime is not accepting the doctrine of rebirth and karma "on the authority of the Buddha" (2015: 85). This will be a recurring problem for secular Buddhists.

Several years later Batchelor published his book *Confession of a Buddhist Atheist*. He calls himself a "secular, nondenominational lay Buddhist" (2011: ix). In this book he responds to the "furor" that his early book caused, "The ensuing controversy showed that Buddhists could be as fervent and irrational in their views about karma and rebirth as Christians and Muslims could be in their convictions about the existence of God" (2011: 175). He also pointed out that, "Belief in the existence of a non-physical mental agent, I realized, was a Buddhist equivalent of belief in a transcendent God" (2011: 39). He could not bring himself to accept either assertion.

Recently Batchelor has published his most systematic presentation of his views in a book entitled *After Buddhism: Rethinking the Dharma for a Secular Age*. "A secular approach to Buddhism," he writes, "is thus concerned with how the dharma can enable humans and other living beings to flourish in this biosphere, not in a hypothetical afterlife" (2015: 16). The secular Buddhism he envisions "would seek to return to the roots of the tradition and rethink and rearticulate the dharma anew" (2015: 19).

Another strategy is practiced by many psychologists, what I call the minimalist approach. In this case, psychologists have gone through the Buddhist teachings and practices and taken only those that can be easily secularized for psychological care. Here I am thinking about such therapies as Mindfulness-Based Stress Reduction, Mindfulness-Based Cognitive Therapy, Acceptance and Commitment Therapy, and Dialectical Cognitive Therapy. All you have to do is browse through the psychology section at your local bookstore to realize that mindfulness has taken psychology by storm. But minimalistic approaches are too shallow; they offer techniques for momentary suffering but offer no philosophy of life. They fail to answer the ultimate questions of life.

Buddhism and Religious Naturalism

For the Religious Naturalist, accepting Buddhism in its traditional form is impossible. We can adopt many of the meditation techniques and some of the Buddhist practices. Humanist author Rick Heller has collected and secularized 32 practices in his book *Secular Meditation*. He explains that,

> Although the group's practices are chiefly adapted from Buddhism, we exclude any beliefs and practices—such as the concept of rebirth—that conflict with mainstream

> academic science. We see the Buddha as an important cultural, literary, and historic symbol and figure but who is no more authoritative than Socrates or Benjamin Franklin.
>
> (2015: xiii)

Religious Naturalists will agree with this sentiment.

Just drawing on the *practices* of Buddhism leaves a world of philosophical ideas behind. This, I believe, would be a mistake. As Sam Harris says, "Buddhism offers a truly sophisticated, empirical approach to understanding the human mind" (2014: 20). Secular Buddhism is a step in the right direction. But the problem with secular Buddhism is that it is still Buddhism. A Buddhist, by definition, is a follower of the Buddha and his teaching. I call this "the Buddhist box." Again, the issue of appealing to the Buddha's authority enters in. I advocate moving the locus of authority to science and Naturalism. There is a great advantage in being free from the Buddhist conceptual framework. You can think outside the Buddhist box. With this freedom, one can rethink and reconstruct Buddhism along naturalistic lines. I call this *Buddhist Naturalism*. It aims to develop a fully naturalized and philosophically sophisticated reconstruction of Buddhism. This means bringing in the latest findings in all relevant sciences, including evolutionary psychology, positive psychology, neuroscience, and related disciplines. Nothing is sacred and beyond revision or rejection. The great thing about Buddhist Naturalism is that it is Naturalism, not Buddhism. That means we are not trying to change traditional Buddhism and should not offend Buddhists.

For the Religious Naturalist, explains Donald A. Crosby, nature is the focus of ultimate concern (2002: 119). Although Buddhism tends to be more environmentally friendly than some religions, primarily because of its teaching of interdependence, nature is not its ultimate concern. In Theravada Buddhism, the primary concern is one's own attainment of nirvana. Mahayana Buddhism saw the selfishness in this and expanded its concern to all sentient beings, as exemplified in the Bodhisattva vow. But Buddhist Naturalism expands this concern even further to include all of nature. Environmentalism is no longer an optional addition; it becomes a central focus. Human happiness is never chosen over the good of nature; rather a harmony with nature is aimed for. After all, nature does not need us, but we need it.

In approaching the Buddhist teachings, we no longer are looking for the authentic teachings of the Buddha, but helpful and illuminating insights from all corners. In this sense, Buddhist Naturalists have no canon, no scriptures, and no authority outside of science, reason, and human flourishing. Buddhism forms a conceptual framework that is adjusted, rethought, and reconstructed based on the best available evidence. This confronts us with the difficulties of evaluating which teachings and practices meet the naturalistic and scientific criteria, which need to be reinterpreted, and which need to be rejected. We have to be honest about the problems of decontextualizing the Buddha's teaching, the challenges of recontextualizing them, and the various difficulties of reinterpretation. None of this will be easy, and the process will never be perfected. Like all things, it will be impermanent and subject to evolution and progress.

In an effort to illustrate what this interpretive process might look like, let me give you my version of a Buddhist Naturalism. We will begin with the Four Noble Truths, which are, according to the Samyutta Nikaya,

> the noble truth of suffering ... the noble truth of the origin of suffering, it is this craving ... The noble truth of the cessation of suffering ... The noble truth of the way leading to the cessation of suffering, it is this Noble Eightfold Path.
>
> (Bodhi 2000: 1844)

I will call the Four Noble Truths *the four psychological realities*.

The first psychological reality is that unhappiness pervades life. You will notice that I translated *dukkha* as "unhappiness." This is because rendering it as stress, suffering, or pain misses the mark. Pain sounds like something physical, stress is only one form of unhappiness, and suffering sounds like it only refers to serious problems. But *dukkha* covers all forms of suffering, anguish, and distress, from the small to the large. It is a catch-all word, just like unhappiness.

The second psychological reality is that the cause of unhappiness is selfish desires. You will notice that I translated *tanha*, which literally means thirst, as "selfish desires." It is usually translated as clinging or attachment. This makes it sound like all attachment is bad, which we know from psychology is not true. Attachment is one of the most important forms of social development that occurs during infancy. It is better to think of wholesome and unwholesome attachments, rather than completely condemning all attachment. The same goes for desire; not all desire is bad. What makes a desire bad is when it hurts either me or others, including aspects of nature. "Selfish desires" catch this idea the best.

The third psychological reality is that the pervasiveness of unhappiness can be stopped by letting go of selfish desires. Although it is at least theoretically possible to be completely free from all selfish desires, most of us will not reach such completeness. But we can substantially overcome our selfish desires and reach a place where unhappiness is no longer pervasive. By saying that unhappiness is pervasive, I mean that it is present or noticeable in every part of a person's life. We notice it in levels depending on the circumstance. This unhappiness becomes acute when we lose a loved one, get divorced, lose a job, or get sick. The selfishness of our desires is what makes them painful; we want things to be different than what they are. It is the "I" and "me" that makes the desires selfish and causes our unhappiness.

The fourth psychological reality is that there is a way to overcome our unhappiness and let go of our selfish desires, and that is the eightfold transformative path. Literally, "the Eightfold Noble Path" would be the "Eightfold Ennobling Path," but since the point of the path is to transform one's character, I think it is best translated as "the eightfold transformative path."

So let's look briefly at the eightfold transformative path. The eight parts are right view, right thought, right speech, right conduct, right livelihood, right effort, right mindfulness, and right concentration. The first problem is the translation of "right." Perhaps a better translation is "wise." The eightfold transformative path, then, includes eight parts that are interconnected like a wheel. You may start by learning them in steps, but they are to be practiced as a complete interlocking system.

Wise view deals with our perspective on things, the way we look at it. You could even understand it as wise worldview. Our worldview, if it does not correspond to reality, is not wise and it is not helpful. Worldviews matter, they influence our lives and the lives of others. So the first step in the eightfold transformative path is to have a wise view of the world, based on the best scientific evidence we have.

The second step in the eightfold transformative path is wise thought. So if the first step deals with philosophy, the second step deals with psychology. "Cognitive therapy," writes David D. Burns, "is based on the simple idea that your thoughts and attitudes – and not external events – create your moods" (1999: xxxi). So, as he points out, "by learning to change your thoughts, you can change the way you feel" (1999: 4). The key here is that "every type of negative feeling results from a specific kind of negative thought" (1999: 5). Change the thought and you will change the feeling. We think ourselves unhappy because we keep "telling ourselves things that simply are not true" (1999: 7). But wise thought also includes making wise decisions. We don't always make the best decisions, and as a result we sometimes suffer or cause suffering.

The next three steps in the eightfold transformative path are wise speech, wise conduct, and wise livelihood. These deal with basic ethics. The seventh step in the eightfold transformative path is wise mindfulness. "Mindfulness," writes Jon Kabat-Zinn, "is awareness, cultivated by paying attention in a sustained and particular way: on purpose, in the present moment, and non-judgmentally" (2012: 1). This attitude, along with wise concentration, is the how of transformation.

There are other Buddhist concepts that need rethinking. One is karma. We already said that, in the words of Gombrich, "The Buddha defined karma as intention" (2010: 51). This means that karma is a psychological mechanism, not a cosmological law. From a naturalistic perspective, karma is not something that determines future rebirths, but something that has lasting psychological consequences. The best psychological translation of karma is "conditioning." The second edition of the *APA College Dictionary of Psychology* defines conditioning as, "the process by which certain kinds of experience make particular actions more or less likely" (VandenBos 2016: 87). Since the Buddha specified karma as intention, I am primarily concerned with voluntary conditioning. Everything we think, say, and do conditions future actions. We are not punished *for* our unwise deeds; we are punished *by* them. Interpreting karma as conditioning, we can see at least three ramifications. Dale S. Wright spells these ramifications out, noting that,

> (a) it [karma] shapes our character and helps determine who or what we become, (b) it helps shape others and the society in which we live, now and into the future; and (c) it encourages others to treat us in ways that correspond to our character—they will often do onto us as we have done onto them, although not always.
>
> (2016: 81)

Every choice we make opens some doors and closes others. Making unwise choices means that making wise choices later is just that much harder. Our actions have consequences, both in a guilty conscience and in molding our character, which in turn limits our future options.

Another concept that needs naturalization is nirvana. Nirvana literally means to extinguish or blow out, usually referring to a flame. As Owen Flanagan points out, "There are tame and untame conceptions of nirvana and rebirth. A tame view of nirvana would be this: nirvana involves release from unwholesome attachment and suffering" (2011: 22). Nirvana is the goal of the Buddhist system, so nirvana is happiness. But this goal can be easily misunderstood, as positive psychology discovered. It is best to translate nirvana as "inner peace," referring to the eudaimonia type of happiness. Eudaimonia is usually translated as flourishing, but within the Buddhist context, inner peace makes more sense.

Rebirth (*patisandhi*) is a little harder to naturalize, for it includes the idea of an afterlife, which Religious Naturalists generally reject. But if we consider what purpose rebirth serves within the Buddhist system, we might find a modern equivalent. In Buddhism, rebirth serves the purpose of demonstrating that the actions of our life have consequences even after we die. A modern equivalent might be our "legacy," the positive or negative influence of our lives that is passed down from one generation to the next. Every person we help or harm can start a domino effect of positive or negative repercussions. Our thoughts and actions live on in the lives that we impact. Our words and writings can influence others for good or ill. In other words, our negative words and deeds re-become, which is the literal meaning of *patisandhi*, in the hearts and minds of others. These can continue to re-become over and over again.

One of the interesting teachings of Buddhism, as it relates to science and ecology, is dependent co-arising (Pali: *pratityasamutpada*). As Joanna Macy has noted, "The systems view of causal process also reveals striking convergences with the Buddha's teaching of causality, called *paticca samuppada*, or dependent co-arising" (1991: 1). I think the best translation is "interdependence." This takes us to the heart of environmental science, which Eldon D. Enger and Bradley F. Smith define as "the study of interrelationships between humans and the natural world" (2013: 1).

This discussion is just a sampling of what a Buddhist Naturalism might look like. You will notice that Buddhism has a naturalistic bent to it. This makes it very adaptable for Religious Naturalism. Religious Naturalists tend to focus most of their attention on the Western religious and philosophical traditions. Many of the Religious Naturalists I have read tend to give only a passing mention of Buddhism. But Western thought tends to be steeped in supernatural assumptions and undercurrents, which are sometimes hard to identify and correct. Buddhism can help us to see with Eastern eyes the presuppositions and possible blind spots of Western thinking.

References

Batchelor, S. (1997) *Buddhism Without Beliefs: A Contemporary Guide to Awakening*, New York: Riverhead Books.

Batchelor, S. (2011) *Confession of a Buddhist Atheist*, New York: Spiegel & Grau Trade Paperbacks.

Batchelor, S. (2015) *After Buddhism: Rethinking the Dharma for a Secular Age*, New Haven: Yale University Press.

Bodhi, B., trans. (2000) *The Connected Discourses of the Buddha: A Translation of the Samyutta Nikaya*, Boston: Wisdom Publications.

Bodhi, B. (2005) *In the Buddha's Words: An Anthology of Discourses from the Pali Canon*, Boston: Wisdom Publications.

Bodhi, B., trans. (2009) *The Middle Length Discourses of the Buddha: A Translation of the Majjhima Nikaya*, 3rd ed. Boston: Wisdom Publications.

Bodhi, B., trans. (2012) *The Numerical Discourses of the Buddha: A Translation of the Anguttara Nikaya*, 3rd ed. Boston: Wisdom Publications.

Bodhi, B. (2015) *Investigating the Dharma: A Collection of Papers by Bhikkhu Bodhi*, Boralesgamuwa, Sri Lanka: Ajith Printers.

Burns, D. (1999) *The Feeling Good Handbook*, Revised edn., New York: A Plume Book.

Buswell, R. and D. Lopez, eds. (2014) *The Princeton Dictionary of Buddhism*, Princeton, NJ: Princeton University Press.

Crosby, D. (2002) *A Religion of Nature*, Albany, NY: State University of New York Press.

Davids, T. (2007) *The Questions of King Milinda: Part 1 & 2*, Charleston SC: Forgotten Books.

Enger, E. and B. Smith (2013) *Environmental Science: A Study of Interrelationships*, 13th edn., New York: McGraw-Hill.

Faure, B. (2009) *Unmasking Buddhism*, Malden, MA: Wiley-Blackwell.

Flanagan, O. (2011) *The Bodhisattva's Brain: Buddhism Naturalized*, Cambridge, MA: MIT Press.

Gethin, R. (1998) *The Foundations of Buddhism*, New York: Oxford University Press.

Gokhale, P. (2015) *Lokayata/Carvaka: A Philosophical Inquiry*, New Delhi: Oxford University Press.

Gombrich, R. (2009) *What the Buddha Thought*, London: Equinox.

Gombrich, R. (2010) *How Buddhism Began: The Conditioned Genesis of the Early Teaching*, New Delhi: Munshiram Manoharlal Publishers.

Harris, S. (2014) *Waking Up: A Guide to Spirituality Without Religion*, New York: Simon & Schuster.

Harvey, P. (2013) *An Introduction to Buddhism*, New York: Cambridge University Press.

Heera, B. (2011) *Uniqueness of Carvaka Philosophy in Traditional Indian Thought*, New Delhi: Decent Books.

Heller, R. (2015) *Secular Meditation: 32 Practices for Cultivating Inner Peace, Compassion, and Joy*, Novato, CA: New World Library.

Ireland, J., trans. (1997) *The Udana & The Itivuttaka*, Kandy, Sri Lanka: Buddhist Publishing House.

Kabat-Zinn, J. (2012) *Mindfulness for Beginners: Reclaiming the Present Moment – and Your Life*, Boulder, CO: Sounds True.

Lama, D. (2005) *The Universe in a Single Atom: The Convergence of Science and Spirituality*, New York: Broadway Books.

Macy, J. (1991) *Mutual Causality in Buddhism and General Systems Theory: The Dharma of Natural Systems*, 1st edn., Albany, NY: State University of New York Press.
Maguire, J. (2001) *Essential Buddhism: A Complete Guide to Beliefs and Practices*, New York: Pocket Books.
McMahan, D. L. (2008) *The Making of Buddhist Modernism*, New York: Oxford University Press.
Radhakrishnan, S. and C. Moore (1957) *A Source Book in Indian Philosophy*, Princeton, NJ: Princeton University Press.
Reat, N. (1994) *Buddhism: A History*, Fremont, CA: Jain Publishing.
Riepe, D. (1961) *The Naturalistic Tradition in Indian Thought*, Seattle: University of Washington Press.
Sagan, C. (1997) *The Demon-Haunted World: Science as a Candle in the Dark*, New York: Ballantine Books.
Sujato, B. and B. Brahmali (2014) *The Authenticity of the Early Buddhist Texts*, Kandy, Sri Lanka: Buddhist Publishing Society.
VandenBos, G., ed. (2016) *APA College Dictionary of Psychology*, 2nd edn., Washington, DC: American Psychological Association.
Vessantara (1993) *Meeting the Buddhas: A Guide to Buddhas, Bodhisattvas, and Tantric Deities*, Cambridge: Windhorse Publications.
Wallace, B. (2012) *Meditations of a Buddhist Skeptic: A Manifesto for the Mind Sciences and Contemplative Practice*, New York: Columbia University Press.
Walshe, M., trans. (1995) *The Long Discourses of the Buddha: A Translation of the Digha Nikaya*, Boston: Wisdom Publications.
Wright, D. (2016) *What Is Buddhist Enlightenment?* New York: Oxford University Press.

17

ZEN BUDDHIST PERSPECTIVES ON RELIGIOUS NATURALISM

Stephanie Kaza

The worlds of religious naturalism and Zen Buddhism are wide-ranging geographies, offering diverse journeys to wisdom and insight. The chance of a traveler from one world meeting a traveler from the other world is certainly within the realm of possibility, and they might find they share certain understandings in common. Or, depending on the history and origin of their journeys, they might be complete strangers to each other. The Zen student would not speak of a relationship with God; the religious naturalist might not be acquainted with nonduality. The degree to which science informs their understanding might be vastly different, and yet both may have experienced a profound sense of awe and appreciation for the mysterious complexity and creativity of the natural world.

In this chapter I explore the insights of Zen Buddhism relevant to religious naturalism, particularly as developed in recent Western scholarship focused on thirteenth-century Japanese Zen philosopher Eihei Dogen. Though other schools of Buddhism may also offer insight into religious naturalism, I have selected the work of Dogen because it has received considerable attention for its depth and significant attention to human-nature relations. I have not looked for broad overlaps of Buddhist thinking with the many varieties of religious naturalism, nor looked for ways to make Buddhism more compatible with religious naturalism. I note here fundamental differences in assumptions and approach that should be clearly stated, to honor the respected tradition of interreligious dialogue.

My primary texts for this study are from Dogen's main source work: *Shobogenzo, Treasury of the True Dharma Eye*, compiled in 1246 from his teaching discourses. The two fascicles I have chosen are "*Genjokoan*, Actualizing the Fundamental Point," and "*Sansui-Kyo*, Mountains and Waters Sutra." I look at key philosophical principles in these texts, paying particular attention to teachings on nonduality, mutual permeation, and the nature of existence. Eastern and Western scholars have developed a number of translations and commentaries on these texts; I work with those that make the most helpful links to a naturalistic or holistic understanding, including those by Bokusan Nishiari, Kosho Uchiyama, Masao Abe, Gudo Nishijima and Chodo Cross, and Kaz Tanahashi.

Dogen's Zen teachings

Who was Eihei Dogen and why have his teachings come to be of such interest in Western Zen? Dogen was born in 1200 in Kyoto, the Japanese imperial capital of the previous 400 years.

Japanese Buddhism had been developing for the past six centuries, having been brought to Japan from China as part of the northern school lineage. Though Dogen was entitled to a high court position due to his parents' lineage, he left home at the age of 13 to study Buddhism at Mt. Hiei and was ordained the following year. He received dharma transmission seven years later and set off for China with his teacher in 1223. He visited a number of teachers over two years, eventually settling at Rujing's monastery where he learned *shikantaza*, a form of meditation sometimes called "just sitting." After returning to Japan, he founded a small practice center in 1233, Kannondori Temple, named after Kannon, the bodhisattva of compassion. Here he wrote the *Genjokoan*, the first fascicle of his eventual life-work collection. Over the course of the next 20 years, Dogen was prolific in his teaching and writing, establishing his own monastery in the mountains at Eiheiji, now a modern pilgrimage destination for Western Zen students.

Because of his focus on pure practice, Dogen was regarded as a reformer during the Kamakura period. He called his teaching "the treasury of the true dharma eye," pointing to the true heart of awakening in an unbroken lineage back to Buddha. His spirit is reflected in a late poem he wrote to accompany a portrait painting:

> Fresh, clear spirit covers old mountain man this autumn.
> Donkey stares at the sky ceiling; glowing white moon floats.
> Nothing approaches. Nothing else included.
> Buoyant, I let myself go—filled with gruel, filled with rice.
> Lively flapping from head to tail,
> Sky above, sky beneath, cloud self, water origin.
>
> (Tanahashi 1985: 20)

After Dogen's death his religious community came to be known as the Soto school, now one of the largest orders in Japan affiliated with several major Western Zen centers, including San Francisco Zen Center. Japanese scholars brought Dogen's work to light in the twentieth century; a first English translation of the *Shobogenzo* was completed by Matsunaga Reiho in 1958. The *Genjokoan* text was used as early as the mid-1960s at San Francisco Zen Center, and in 1983 a complete four volume set of 92 fascicles was finally published, the work of Kosen Nishiyama and John Stevens with collaborators. Since then a number of scholarly commentaries and textual interpretations in English and Japanese have greatly enhanced Western understanding of Dogen's philosophy and practice approach.

Dogen's form of Zen stressed direct awakening through single-minded practice, often requiring many long hours of zazen meditation in silent retreat. For young Western students, steeped in the Vietnam war conflict, the upheaval of feminism, the lure of consumerism, the failure of idealism, Dogen's words offered an opening to something more penetratingly real. Western fascination with the exotic otherness of Japan added to the lure of Zen practice, with its obscure and highly disciplined practice forms. Japanese scholar Masao Abe introduced Dogen into Buddhist-Christian dialogues with process theologian John Cobb and others, adding to the academic foundation for Dogen studies. Dogen's haiku-like references to green mountains, plum blossoms, dragons, and fish caught the attention of early eco-Buddhists looking for deep ecology sensibilities in the world's religious traditions.

The two texts for this study are quintessential Dogen, carrying strong instructions for practice, often in inscrutable koan-like format. A footnote on one translation of *Genjokoan* offers this comment by Bokusan Nishiari, Soto priest in 1922: "This fascicle is the skin, flesh, bones, and marrow of the Founder. The fundamental teaching of the Founder's lifetime is expounded in this fascicle ... The ninety-five fascicles of the Treasury of the True Dharma Eye are the offshoots of this fascicle" (Tanahashi 1985: 244). The title, "Genjokoan," is a compilation of characters or kanji that point to something that is present or appearing yet also hidden, but in the cryptic sense of "nothing to hide, so there is nothing to reveal" (Nishiari 2011: 14). Or put another way, "A great secret is greatly apparent. What is greatly apparent is a great secret ... Because of *genjo*, things are originally complete" (Nishiari 2011: 14). Another scholar translates this idea as "the ordinary profundity of the present moment becoming the present moment" (Uchiyama 2011: 151).

Sansui-Kyo, or the Mountains and Waters Sutra, was written in 1240, during a very active period of teaching and writing for Dogen. This fascicle is of particular interest for Western students seeking Buddhist sources for environmental ethics (see Kaza and Kraft 2000). Dogen's imagery throughout the 22 verses evokes wild areas in high mountains and flowing streams, beginning with the first sentences: "Mountains and waters right now are the actualization of the ancient buddha way. Each, abiding in its phenomenal expression, realizes completeness" (Tanahashi 1985: 97). The sutra can be thought of as an eco-koan, offering an interpenetrating, co-creating view of the universe, understood through practice realization. Many of the themes introduced in the *Genjokoan* are taken up in *Sansui-Kyo*, challenging students to firmly study the way of such mountains and waters. Dogen's writing in both texts exhibits his sophisticated use of allusion, puns, and metaphor in poetic prose, meant to be absorbed rather than analyzed.

Key philosophical principles

At the core of Dogen's teaching is a consistent pointing to *nonduality* as the awakened view of reality. The opening verse of the *Genjokoan* urges students to go beyond superficial dualistic thinking on all matters.

> As all things are buddha-dharma, there is delusion and realization, practice, and birth and death, and there are buddhas and sentient beings.
> As the myriad things are without an abiding self, there is no delusion, no realization, no buddha, no sentient being, no birth and death.
> The buddha way is, basically, leaping clear of the many and the one; thus there are birth and death, delusion and realization, sentient beings and buddhas.
> Yet in attachment blossoms fall, and in aversion weeds spread.
>
> (Tanahashi 1985: 69)

This teaching of nonduality is particularly relevant in exploratory cross-philosophical dialogue, as so much of Western ethical thinking is predicated on dualistic categories. Radical nonduality challenges the limitations of relying on familiar categories of subject/object, whole/part, birth/death, truth/delusion.

The structure of this teaching is threefold. In the first statement, Dogen affirms the existence of all aspects of being, a full recognition of "what is." In the second statement, he seems to apparently contradict himself, but the key phrase here is that all beings are "without an abiding self," i.e., they have no independent existence, so in this sense they do not exist. The third statement suggests the way out of the apparent contradiction, going beyond dualistic distinctions, thus affirming the true nature of existence. Nishiari describes this as no existence separate from nonexistence, and no transcendence outside of existence and nonexistence. "Therefore being, non-being, and transcendence are each the entire world and *Genjokoan*" (Nishiari 2011: 20).

Tanahashi contrasts this three-fold structure as significantly different from the Hegelian dialectic of thesis, antithesis, and synthesis. Hegel's process involves development and progression, a narrative approach to understanding, with the outcome valued more highly than the steps leading to it. Dogen, however, gives each step of the teaching its own absolute value, with each requiring the other two for full understanding. Tanahashi summarizes this as 1) discrimination, 2) denial of discrimination, and 3) beyond discrimination and denial of it (1985: 17–18). Throughout the *Genjokoan*, Dogen emphasizes the nonduality of practice and realization, that is, no separation between living and awakening.

In verse 12, as translated by Ushiyama, Dogen suggests that the reality of life is to be seen as before the separation between being and non-being:

> We should not think that what we have attained is conceived by ourselves and known by our discriminating mind. Although complete enlightenment is immediately actualized, its intimacy is such that it does not necessarily form as a view. [In fact] viewing is not something fixed.
>
> (Uchiyama 2011: 145)

"Intimacy" (*mitsu-u*) in this sense is *before* any division into subject, the seeing, and object, the seen. The title of this fascicle is sometimes translated as "actualizing the fundamental point," i.e., becoming the true intimacy and not mistaking partial views for the full truth of reality.

This three-fold teaching is further developed in the *Sansui-Kyo* fascicle where Dogen takes mountains and rivers as the objects of contemplation. He urges students: "You should study the green mountains, using numerous worlds as your standards. You should clearly examine the green mountains' walking and your own walking" (Tanahashi 1985: 98). He consistently points out the limitations of conditioned views as "just looking through a bamboo tube at a corner of the sky" (Tanahashi 1985: 99). He explores in great detail the many aspects of mountains and waters as people tend to know them (which environmentalists can affirm) but then reminds students:

> All beings do not see mountains and waters in the same way. Some beings see water as a jeweled ornament, but they do not regard jeweled ornaments as water ... Some beings see water as wondrous blossoms, but they do not use blossoms as water. Hungry ghosts see water as raging fire or pus and blood. Dragons see water as a palace or pavilion. Some beings see water as the seven treasures or a wish-granting jewel. Some beings see water as a forest or a wall ...
>
> Thus the views of all beings are not the same. You should question this matter now ... You should pursue this beyond the limit of pursuit.
>
> (Tanahashi 1985: 102)

Throughout the sutra, Dogen seems to raise impossible contradictions, summarizing these in the concluding verse. Here he reiterates the basic three-fold teaching on nonduality, referring to Tang master Qing yuan in his statement, "An ancient buddha said, 'Mountains are mountains, waters are waters.' These words do not mean mountains are mountains; they mean mountains are mountains." For Dogen, true realization goes beyond dualistic thinking and even its opposite, to penetrate the fathomless nature of reality.

The actual experience of nonduality has been described as *mutual permeation* or mutual penetration, a sense of no separation from all the ten thousand beings and their co-existences, arising simultaneously in the present moment. One of the most popularized quotes from the *Genjokoan* instructs students to aim for exactly this interpenetration.

> To study the buddha way is to study the self. To study the self is to forget the self. To forget the self is to be actualized by myriad things. When actualized by myriad things,

> your body and mind as well as the bodies and minds of others drop away. No trace of realization remains, and this no-trace continues endlessly.
>
> (Tanahashi 1985: 70)

Uchiyama explains that the self Dogen refers to is not the Western sense of "ego" but "the reality of life prior to separation into dichotomies such as self/other or subject/object" (Uchiyama 2011: 182). To study the self is not merely to observe distracting delusional thoughts that fracture intimacy, but more correctly to "accept everything as the contents of our 'self'" (Uchiyama 2011: 182). Such awakening to the true self requires forgetting any sense of separation that reifies the self and minimizes other existences. Yet this is not an experience of "oneness" expressed in some Western philosophies, but rather a complete experience of all beings, each in their wholeness fully expressing reality in multiple and specific ways. The experience of mutual penetration becomes a profound invitation to full presence, enriched and informed by the realization of myriad things, no trace possible because no separation.

In the *Mountains and Waters Sutra*, Dogen investigates the nature of mutual penetration through engaging directly with "flowing mountains" and the "not-flowing" of water. He challenges students to experience such penetration across eons of time.

> If walking stops, buddha ancestors do not appear. If walking ends, the buddha-dharma cannot reach the present. Walking forward does not cease; walking backward does not cease. Walking forward does not obstruct walking backward. Walking backward does not obstruct walking forward. This is called the mountains' flow and the flowing mountains.
>
> (Tanahashi 1985: 98)

Dogen's students would have been able to taste this mutual penetration directly in their mountain monasteries or on their pilgrimages from temple to temple. In this sutra he exhorts them to realize the direct connection between their buddha ancestors and their own practice in present time, an unbroken lineage of mutual arising.

The *nature of existence* for Dogen is continuously arising and passing away, generation and extinction appearing in every moment in every existence. This creatively dynamic understanding of nature is marked by impermanence of all phenomena in every aspect. Nothing at all is permanent or maintains a permanent self. In verse 6 of *Genjokoan*, he says: "When you practice intimately and return to where you are, it will be clear that nothing at all has unchanging self" (Tanahashi 1985: 70). Thus impermanence is the first aspect of existence.

The second aspect is the mutually influential nature of all elements in the universe. No being lies outside the universe and all beings affect others in myriad ways throughout time and space. A popular metaphor for this aspect of existence is Indra's Net, from the Chinese Hua-Yen school of Buddhism in the tenth century. The universe is compared to an infinite strand of woven nets criss-crossing space, with an infinitely faceted jewel at each node crossing point. Each jewel reflects every other jewel in the vast net, a sea of mutual influence, changing constantly on all scales from the micro to the macro.

The third aspect of each existence is that it is completely a singular expression in that moment. In verse 7 of the *Genjokoan*, Dogen works with the elements of fire to explore this in relation to human life and death.

> Firewood becomes ash, and it does not become firewood again. Yet do not suppose that the ash is future and the firewood is past. You should understand that firewood abides

> in the phenomenal expression of firewood, which fully includes past and future and is independent of past and future. Ash fully abides in the phenomenal expression of ash, which fully includes future and past. Just as firewood does not become firewood again after it is ash, you do not return to birth after death.
>
> (Tanahashi 1985: 70–71)

Uchiyama explains the typical human error as assuming a causal before and after relationship without fully recognizing ash as ash and firewood as firewood. He shows how Dogen is speaking to a broader understanding beyond causal relationship. The take-home point is that, likewise, human life is just life. "It is neither life as a step toward becoming death, nor life that is relative and opposite death" (Uchiyama 2011: 196). Though with human eyes we may see this as some form of coming and going, even this is animated by "the coming and going of not-coming and not-going" (Uchiyama 2011: 196).

Dogen brings up yet a fourth aspect of existence as its ungraspability. Even as he teaches a nondualistic method and approach to realization, he reminds students toward the end of *Genjokoan*: "Do not suppose that what you realize becomes your knowledge and is grasped by your consciousness. Although actualized immediately, the inconceivable may not be apparent. Its appearance is beyond your knowledge" (Tanahashi 1985: 72).

Perception of realization is necessarily deluded, according to Dogen, because full realization goes beyond the limitations of human perception. Heine (2012: 60) explains how human "seeing" is drawn to what is illuminated, while the dark or unilluminated elements of existence remain concealed and therefore unknowable through typical means.

These aspects of the nature of existence are further illustrated in the *Mountains and Waters Sutra*. In the opening verse, Dogen writes:

> Mountains and waters right now are the actualization of the ancient buddha way. Each, abiding in its phenomenal expression, realizes completeness. Because mountains and waters have been active since before the Empty Eon, they are alive at this moment. Because they have been the self since before form arose they are emancipation-realization.
>
> (Tanahashi 1985: 97)

Mountains and waters themselves are a radical phenomenological expression of a multifaceted, infinitely unfolding universe of practice realization. Modern Zen teacher Daido Loori, in his *teisho* on this sutra, comments on verse 11 about water. He explains that "'Harder than diamond' expresses the unchanging Suchness of all things, or absolute reality. 'Softer than milk' refers to the "conditioned Suchness of things," e.g., the relative reality. Loori uses this seemingly impossible contrast in the qualities of water to point to the "incomprehensibility of something that is without any fixed characteristics whatsoever, without any existence, being able to give rise to conditioned existence, to the multiplicity of things" (Loori 2000: 146).

Comparing Zen Buddhist thought to religious naturalism

This preliminary exploration of key philosophical points in major fascicles of Dogen's teaching can provide a platform for examining overlaps and contrasts between religious naturalism and Zen Buddhism. To begin, I look at philosophical methods, in particular the role of scientific understanding, and experiential approaches, especially in regard to impermanence and mutual co-arising. I will further consider nature and morality in Dogen's Zen in relation to religious naturalism, as these understandings might inform human choices regarding the environment.

Philosophical methods

Both Zen and religious naturalism address and encourage depth of connection to the natural world as a source of awe, humility, and insight. For both, the natural world is seen as a fundamental crucible for human understanding, proper perspective, and guide for moral choices. For both, a monotheistic God is not a requirement for spiritual awakening. However, the two are distinctly different in considering the role of science in relation to religious context. Across a range of views within religious naturalism, modern science is taken as an important rational corrective to potential distortion in fanaticism or mysticism. Reason and clear-headed thinking are highly regarded as a true path to knowledge and, in accepting the limits to that knowledge, to humility. A number of religious naturalists draw on emerging holistic understandings from evolutionary biology, ecology, and cosmological physics, among others. Some emphasize or advocate for a shared central story that describes the human place in the cosmos, beginning with the Big Bang and continuing through the geologic stages of planet building and evolutionary phases of organic life development.

None of this was known at the time of Dogen's teaching, so he would not have had a chance to reflect on these insights from modern science. However, I believe he would likely caution students to be skeptical about accepting scientific perspectives as complete understanding. This is because Western scientific method is often based in subject/object dualities, and a foundational belief in the distinction between observer and observed. A long history of reductionist thinking has influenced Western scientific approaches to the natural world. In human/nature dualistic thinking, humans are typically perceived to be distinct from and superior to nature. Though religious naturalism may seek to correct this distortion through more holistic science, the dualistic frame is not easily dislodged.

In his essay on *Genjokoan*, Uchiyama acknowledges that, of course, "it is only natural to make a distinction between self and others. We always think, 'I am I. Others are others.' And we think the world in which we are living exists outside of ourselves" (Uchiyama 2011: 184). Yet Dogen urges students in the opening lines to give up or go beyond the view that discriminates between self and others and to study the *Buddha dharma* on this ground. Scientific method is based entirely on "the view that discriminates," thus it could never offer more than a partial view of reality. Dogen would likely consider too much reliance on scientific thinking to be an impediment to spiritual awakening.

This likelihood is further reinforced by the last sentence of the first verse, "Yet in attachment flowers fall, and in aversion weeds spread." Attachment to flowers can be interpreted as attachment to awakening, attachment to weeds as attachment to delusion. The very goal of science is to debunk delusional and distorted understanding in any scientific field, i.e., to continuously question and critique previous understandings. Attachment abounds in the ego-building of scientific careers, the publication industry, and among those seeking to use science to support their views. Dogen's caution offers the teaching on nonattachment as necessary for access to true liberatory awakening. "One needs to give up everything in order to open oneself to the ultimate truth" (Tanahashi 1985: 19). For Dogen, the path to true understanding lies in experiencing all things without any preconception.

Dogen emphasizes direct practice realization in the Mountains and Waters sutra, challenging students to see through incomplete, distorted, and conditioned views, even those of biologists, spiritualists, and storytellers.

> Even if you see mountains as grass, trees, earth, rocks, or walls, do not take this seriously or worry about it; it is not complete realization. Even if there is a moment when you view mountains as the seven treasures shining, this is not returning to the source. Even if you understand mountains as the realm where all buddhas practice, this understanding is not something to be attached to ...
> Turning an object and turning the mind is rejected by the great sage. Explaining the mind and explaining true nature is not agreeable to buddha ancestors ... There is something free from all of these understandings ... You should study this in detail.
> (Tanahashi 1985: 99)

These sutras make plain that, in Dogen's view, science is not privileged as a source of insight and may even be an impediment to true realization.

Experiential approach

Many strands of religious naturalism include a revised theistic or avowedly non-theistic or agnostic approach to awakened understanding. This arises as a critique or distinction from more traditional monotheistic approaches to spiritual realization through oneness with God and/or Jesus and other holy teachers. All of these theistic and non-theistic options would be understood as partial views in Dogen's Zen. For Dogen, the only valid path to realization is through personal direct experience "free of the conventional, discriminating frame of mind" (Tanahashi 1985: 16). As Tanahashi points out, Zen teachers use all sorts of confusing behaviors and cryptic statements to disturb students deeply enough to show them their conditioned minds. What may appear to be illogical contradictions are strategically phrased koans that serve as pedagogical tools in the non-directional curriculum of Zen teaching. Students' conditioned beliefs, whether in God or science, nature or self, offer teachable moments for the Zen teacher. Western students conditioned by dualistic thought are particularly vulnerable to mentally abstract conceptualization and idealization of spiritual goals.

The Zen experiential approach is critical for fully grasping the nature of reality as impermanent, mutually co-arising, and interdependent. The religious naturalist approach accepts the natural world as a reliable ground for religious experience, but Zen adds cautionary doubt around assumptions of permanence based on human-centered perspective. "You who study with buddhas should not be limited to human views when you are studying water" (Tanahashi 1985: 104). The religious naturalist approach typically highlights rational intellectual use of the mind, where the Zen approach leaps clear of concepts, ideas, opinions, arguments to a total body-mind experience of existence. Uchiyama speaks of this leap as "to meet everything, without exception, as part of my life is most essential to the Buddha Way" (2015: 182).

In the introduction to a related sutra, *Shinjin-Gakudo*, Learning the Truth with Body and Mind, translators Nishijima and Cross (1996: 247) explain how learning the truth in Buddhism requires both physical and mental pursuit, the two always synergistic and reflected in action. Dogen writes:

> There is the learning of truth through casting aside these kinds of mind, and there is learning of the truth through taking them up. In such instances, the truth is learned

> through thinking, and the truth is learned through not thinking ... Or there is the learning of the mind with the mind in pounding of rice and transmission of the robe ... In brief, mountains, rivers, and the Earth, and the sun, moon, and stars, are the mind ... they are not attainable or unattainable, they are beyond recognition and non-recognition, they are beyond penetrability and impenetrability, and they do not change with realization and non-realization.
>
> (Nishijima and Cross 1996: 248–249)

Underpinning Dogen's experiential approach is his insistence on the oneness of practice and attainment, *shuso-itto*—that realization does not happen without practice and that practice itself is the manifestation of realization (Abe 1992: 16–33). This is the fundamental principle in *Genjokoan* and *Sansui-kyo*, making clear that action is the expression of true understanding.

Although my understanding of religious naturalism is limited, I have not seen any parallel equation or guideline for spiritual realization. This may reflect the wide range of philosophies within religious naturalism (i.e., theistic and non-theistic) and thus a wide range of practice possibilities for realization. The Zen practice of sitting meditation, in contrast, is seen as foundational in preparing the student to break through barriers of the conditioned body/heart/mind.

Nature and morality

Dogen's collection of writings offers very little explicit instruction on how to behave in the natural world, i.e., respect life, offer kindness, do not kill, etc. However, the *Genjokoan* and *Sansui-kyo* clearly teach that humans are not separate from the universe and that mountains and waters are the human body. The most objectionable actions would be those made without this understanding, based on a mistaken sense that a tree or grasshopper's life is separate from one's own. Much of what is acceptable in Western scientific experimentation would not meet this criterion, as it is based on a hierarchy of life value with humans at the top and quite separate from life forms "below" them. Religious naturalists may not affirm this hierarchy, but also may not go as far as Dogen in understanding that every one of the ten thousand beings is "a world of buddha ancestors" (Tanahashi 1985: 107).

Dogen's Zen likewise provides few moral principles for social justice, a critique that has been made of Zen Buddhism generally. Religious naturalists have inherited the Judeo-Christian lineage with a significant body of literature supporting a religious approach to justice, following the examples of Jesus and other prophets. Such a prophetic tradition is more or less absent in Eastern religious traditions, where class analysis and racial bias are minimized in the long curve view of karmic time. Dogen is hardly a prophetic voice for modern environmental ethics. Yet his understanding of radical interdependence as central to spiritual realization has been claimed by eco-Buddhists as supporting an environmental worldview.

In a 1998 interview regarding his epic poem collection, *Mountains and Rivers Without End* (1996), Zen student Gary Snyder explains how his own insights reflect the interdependent collaboration of many selves. This includes teachers and friends of the past as well as time spent working in the engine room of a marine transport ship. He says,

> the acknowledgement that we reflect a number of selves, all of which, of course, are illusory anyway, and resolve into a non self—which is another way of speaking of the totally collaborative quality of any individual entity, namely that we are an intersection

> of influences in the present and in the past, from the present and from the past. That is, the self is a moving target.
>
> (Gonnerman 2015: 263)

Snyder's poem cycle was inspired by a Chinese scroll painting that seemed to convey the spirit of Dogen's Mountains and Waters sutra. In these poems Snyder points to the beings themselves as teachers, as wisdom guides, as sources of moral authority—Bristlecone, Coyote, Jackrabbit, Kingfisher. Many of the poems are moments of time in mountain spaces of the North Cascades, the High Sierra, or the White Mountains. Paying attention to these spirit beings is the way one learns to be ethical in place. From his Sierra foothill home in Kitkitdizze, he wrote, "The true source of compassion and ethical behavior is paradoxically none other than one's own realization of the insubstantial and ephemeral nature of everything" (Snyder 1990: 246).

While religious naturalists may look to science and spiritual experiences of awe and wonder for ethical inspiration, Dogen would find these inadequate to the full measure of human experience. He might, instead, identify the rational spaces of science and the emotional spaces of awe as potential practice fields—springboards for examining conditioned thinking. From these starting points he would urge breaking through any illusions of separation from the holographic responding universe. This would necessarily include letting go of cherished attachments to the sacred and even to the natural world as ultimate authority. For Dogen the source of humility is not scientifically verified accounts of the amazing complexity of the created, creative, and creating world. Rather, it is the absolutely stunning liberatory experience of no separate self, available in every moment and time to those who practice deeply.

Differences and overlaps between Dogen's Zen and religious naturalism

In reviewing the core principles and approaches in Dogen's writing through these two sutras, I have shown a number of areas where Zen and religious naturalism clearly do not overlap in spiritual orientation. First, the prominent reliance on science and reason in religious naturalism is absent or questioned in Zen. Second, Dogen does not rely on or give weight to a central shared creation story, as does much Western religious naturalism. If anything, Dogen's central emphasis is challenging the conditioned mind to find liberation in the present moment. Third, Dogen's Zen does not bring attention to justice concerns or principles of equity and fairness that are taken up in Western religious naturalism as an important part of spiritual engagement. This gap in the Eastern Zen heritage is being addressed by Western Buddhists in the dialogue around "socially engaged Buddhism," led by Sulak Sivaraksa, Thich Nhat Hanh, Joanna Macy, and H. H. Dalai Lama, among others.

Fourth, the two religious traditions share few religious practices in common. Zen students commit considerable time to sitting and walking meditation, bowing and chanting, silent retreat, and mindfulness practice in kitchen work and zendo caretaking. They also hear *teishos* (lectures) on key sutras, meet regularly with certified teachers, and work with religious texts such as Dogen's fascicles. In some schools, they also take up koan practice. I am not familiar enough with religious naturalism to identify parallel shared religious practices, though I imagine these may include silent retreat and other contemplative supports for cultivating a sense of awe or reverence with the natural world.

Fifth, Dogen does not use language of the "sacred" as it sets up a dualistic frame (sacred/profane), which he would see as an impediment to true realization. This leaves a big gap with religious naturalism and its emphasis on nature as sacred, and the human experience in nature as a meeting with the sacred.

From my limited exposure to the religious naturalism literature, I find several areas of intriguing possibility for dialogue around shared approaches or understandings. First, I will mention the writing around *emergentist thinking*. This sort of dynamic, co-created, co-evolving view of nature aligns well with Dogen's writing on mutual interdependence. He would see humans not only as an emergent part of nature but as an emergent process continuously unfolding, simultaneously with all other aspects of the universe. Religious naturalist thinkers in the 1920s such as Bernard Meland and Roy Wood Sellers built on the writings of William James and Alfred North Whitehead to emphasize emergence as an alternative to reductionist and dualist thinking. Meland pointed out the limits to rational thought and the need to recognize complexities beyond what human minds could conceive (Stone 2008: 132–133). In their time these writers saw the distorted implications of what had been widely accepted Cartesian reductionism and spiritual vitalism. Their critiques of the limits to these forms of thought would be shared by Dogen, and if they had known of his writing, they may have found additional inspiration for their views on emergence.

Dogen, however, might have questioned some of their proposed hierarchies of transformation, for example, from inorganic to organic to mental to social or from matter to life to mind to personality (Stone 2008: 52–56). Theories of this sort of evolutionary emergence were not of much interest to Dogen; he was focused more on human awakening and how to move students to experience the emergent nature of the entire universe. He would likely agree with Sellers in his statement that "emergence [at least in part] is brought about by the unplanned increasing organization of matter" (Stone 2008: 52). Dogen, though, would be equally interested in the increasing *dis*organization of matter and how the drive to simplification interacts with the drive to complexity. I think Dogen would have enjoyed Gordon Kaufman's expression of emergentist thinking as "trajectories of serendipitous creativity" (Stone 2008: 203–204). Kaufman was referring to the surprising, unforeseen and unexpected outcomes of dynamic events unfolding that may not always be positive.

A second area of potential dialogue could be around John Dewey's writing on *radical empiricism*. For Dewey, the starting point for all philosophy of religion is not analysis, principles, or historical perspective, but rather the actual experience that informs the practitioner. He describes the interplay between knower and known, between experience and concepts as shaping the transactional or interactive nature of religious experience. For Dewey, a key quality is "the shift in orientation from engagement with a distinct entity, ground, or process to engagement with processes within this world that have a quality that may be designated as religious, at least in a revised sense" (Stone 2008: 48). Dogen would be interested in this shift of orientation away from illusory grounds for understanding and toward a groundless and boundless experience of manifestation.

Dogen's *Mountains and Waters* sutra is rich with encouragement for radical empiricism and seems to fit well with Dewey's Western philosophical explorations. For example, he suggests that students examine closely their own walking in order to fully understand mountains walking.

> If you doubt mountains' walking, you do not know your own walking; it is not that you do not walk, but that you do not know or understand your own walking. Since you do know your own walking, you should fully know the green mountains' walking.
>
> (Tanahashi 1985: 98)

Dogen, though, may go further than Dewey intended in his admonitions to push radically beyond human mental and cultural conditioning.

> If you do not learn to be free from your superficial views, you will not be free from the body and mind of an ordinary person. Then you will not understand the land of buddha ancestors, or even the land or the palace of ordinary people.
>
> (Tanahashi 1985: 104)

A third area for future conversations with religious naturalism and Zen might be around *new animism* writings among thinkers such as Graham Harvey and Freya Mathews. Much of the older understanding of animism would be anathema to both religious naturalists and Zen Buddhists. Edward Tylor, for example, described animism in his 1871 treatise as belief in the "souls of individual creatures, capable of continued existence after the death or destruction of the body" (Tylor 1871, cited in Capper 2016: 383). In his recent article, Capper describes how the new animists define animism *relationally* rather than through dependence on a human-like soul. "Animism, for them [the new animists], means living in a community of persons, both human and nonhuman, and extending respect for the existential agency of those persons" (Capper 2016: 8–9). This is essentially a non-dualistic orientation to other-than-human life forms, something Dogen would easily support. The new animists are more interested in relational encounter and mutual respect than in transmigration of souls. This would seem to provide greater opportunity for liberatory awakening beyond dualistic views, something Capper (2016) explores in his review of three Zen thinkers as potential new animists.

Conclusion

This exploration has been undertaken in the spirit of interreligious dialogue, with an aim to shed light on religious naturalism through the lens of Zen Buddhist teacher Eihei Dogen. Because Dogen's writing has received great interest and attention among modern Western Buddhist thinkers, including those interested in environmental concerns, these teachings offer a fruitful platform for preliminary conversations with religious naturalists. Interreligious dialogue provides an opportunity to identify parallel philosophies and practices as well as distinct non-overlapping approaches.

In this particular dialogue a number of differences are profound and distinct, which could lead to unbridgeable gaps between conversation partners. However, I have also identified several areas of potential overlap that could serve as a starting point for informative dialogue. It could be that mutual interest in environmental concerns would be a motivation for sharing perspectives; this has been very productive in Buddhist-Christian interreligious dialogue (see, for example, the writings of Jay McDaniel, Rita Gross, Paul Knitter, and John Cobb).

In closing, let me offer the words of Shohaku Okumura in his study of Dogen's *Genjokoan*:

> In my understanding, this teaching of individuality and universality is the essence of the title Genjōkōan. Genjō is nothing other than kōan, and kōan is nothing other than genjō: Genjō means "reality actually and presently taking place," and kōan means "absolute truth that embraces relative truth" or "a question that true reality asks of us." So we can say that genjōkōan means "to answer the question from true reality through the practice of our everyday activity."
>
> (Okumura 2010: 21)

At the heart of this interreligious reflection has been that goal: to answer the question of true reality and find meaning in our current existence, complicated as it is, in this messy and dynamic world.

References

Abe, M. (1992) *A Study of Dogen: His Philosophy and Religion*, Albany, NY: State University of New York Press.

Capper, D. (2016) "Animism among Western Buddhists," *Contemporary Buddhism* 17/1: 30–48.

Dogen, E. (2015) *Dogen's Genjo Koan: Three Commentaries*, Berkeley, CA: Counterpoint Press.

Gonnerman, M. (ed.) (2015) *A Sense of the Whole: Reading Gary Snyder's Mountains and Rivers Without End*, Berkeley, CA: Counterpoint Press.

Heine, S. (2012) "What Is on the Other Side? Delusion and Realization in Dogen's *Genjokoan*" in Steven Heine (ed.) *Dogen: Textual and Historical Studies*, Oxford: Oxford University Press, 42–74.

Kaza, S. and K. Kraft (eds.) (2000). *Dharma Rain: Sources of Buddhist Environmentalism*, Boston, MA: Shambhala.

Loori, J. D. (2000) "River Seeing the River" in Kaza, S. and K. Kraft (eds.) *Dharma Rain: Sources of Buddhist Environmentalism*, Boston, MA: Shambhala, 141–150.

Nishiari, B. (2015) "Commentary by Bokusan Nishiari" in Dogen, E., *Dogen's Genjo Koan: Three Commentaries*, Berkeley, CA: Counterpoint Press, 11–90.

Nishijima, G. and C. Cross (trans) (1996) *Master Dogen's Shobogenzo, Book 2*, London: Windbell Publications.

Okumura, S. (2010) *Realizing* Genjokoan: *The Key to Dogen's Shobogenzo*, Somerville, MA: Wisdom Publications.

Snyder, G. (1990) *The Practice of the Wild*, San Francisco, CA: North Point Press.

Snyder, G. (1996) *Mountains and Rivers Without End*, Washington, DC: Counterpoint Press.

Stone, J. A. (2008) *Religious Naturalism Today: The Rebirth of a Forgotten Alternative*, Albany, NY: State University of New York Press.

Tanahashi, K. (ed.) (1985) *Moon in a Dewdrop: Writings of Zen Master Dogen*, San Francisco, CA: North Point Press.

Tylor, Edward B. (1871) *Primitive Culture, vol. 1*, London: John Murray.

Uchiyama, K. (2015) "Uchiyama Roshi's Teisho on Genjo Koan" in Dogen, E., *Dogen's Genjo Koan: Three Commentaries*, Berkeley, CA: Counterpoint Press, 149–223.

18

ONE SHAWNEE'S REFLECTIONS ON RELIGIOUS NATURALISM

Thomas Norton-Smith

First introductions

kiwaakomelepwa nikannaki! nitesi8o Thomas Norton-Smith. saawanwa nil:a no'ki nim'soma peleawa. That is, "Greetings to you all, my friends! I am named Thomas Norton-Smith. I am Turkey clan Shawnee." I speak for no one but myself, so any errors or misinterpretations are mine alone; I am full of mistakes. And, there will be misinterpretations, for, as I shared with Profs. Stone and Crosby, I am new to the discussion of religious naturalism. Indeed, I'm not even particularly religious; I'm just one often confused, mixed-blood Shawnee, trying to walk the right road.

I will offer reflections on religious naturalism both from an American Indian world version and from a realist Western perspective called *constructive realism*, grounded in the constructivism of Prof. Nelson Goodman (1983, 1988). While finding that there is much about religious naturalism to praise, I will argue that both constructive realism and the Native world version agree—albeit for different reasons—that religious naturalism's concept of the natural world is problematic.

I need to say something at the outset to explain this odd, apparently bifurcated little contribution. My reflections to follow will reintroduce themes from my book, *The Dance of Person and Place*—hereafter, *The Dance*—wherein I presented one possible interpretation of American Indian philosophy based upon a culturally informed realist revision of Nelson Goodman's constructivism. Thus, what will seem to be disjointed and unrelated reflections grounded within radically different philosophical traditions actually—and remarkably—have a common and interconnected history; but then, from a Native perspective, "everything is related."

With the understanding that various religious naturalists will differ in details, I will begin by discussing three tenets seemingly embraced by many. I will next rehearse my constructive realist account, which recognizes a plurality of equally privileged yet radically different world versions, and then consider how religious naturalism fares against a constructive realist critique. The section following will be devoted to important aspects of one of the world versions recognized by constructive realism—an American Indian world version—followed by its assessment of elements of religious naturalism. This will be ambitious, indeed, but with your kind attention, I believe we can do it.

Religious naturalism

As I understand it, religious naturalism embraces three tenets: (1) *Naturalism*, that all and only the natural world exists; (2) *Inherent values*, that there are values inherent in the natural world; and

(3) *Religiosity*, that the natural world is sacred and an object of reverence. Appealing to religious naturalists Donald Crosby, Loyal Rue, and Jerome Stone, I consider each of these tenets in turn.

First, religious naturalism is a *naturalism*, and as such, as Rue observes,

> [Naturalism] simply declares that whatever is natural is real and whatever is real is natural. That is, reality does not break out into two realms, the natural and the supernatural. Naturalism rejects the notion that anything at all transcends nature, except nature itself.
>
> (Rue 2011: 97)

In Crosby's words, "nothing lies behind, is the ground of, or is set over against nature. For example, there are no beings such as the *super*natural God, gods, spirits, angels, or demons envisioned in various cultures or religious traditions" (Crosby 2002: 17). Yet again, according to Stone, naturalism "involves the assertion that there seems to be no ontologically distinct and superior realm (such as God, soul or heaven) to ground, explain, or give meaning to this world" (Stone 2000: 1). That said, naturalism is "a very general metaphysical doctrine," consistent with realist or idealist metaphysical interpretations (Rue 2011: 97). Moreover, since human beings and their history and culture are clearly real, naturalism recognizes them as part of the natural world. "Nature is metaphysically ultimate, that is, there is nothing outside, beyond, or behind it. This statement includes human beings, who must be regarded as an integral part of nature" (Crosby 2002: 21). Finally, since naturalists believe that the natural world is all and only what exists, and the empirical sciences investigate the natural world—not only the "hard" sciences, but empirically based sciences of "human nature"—naturalists have confidence in the methodologies and results of the contemporary physical and social sciences. According to Rue:

> If you are a naturalist ... taking the view that the natural order is ultimately real, then you will do well to focus your inquiry on nature. We may say, then, that inquiry into nature is close to the heart of naturalism. And this means, of course, that a commitment to scientific inquiry (the study of nature) is a central concern.
>
> (Rue 2011: 101)

As Crosby observes, there is an "irreducible pluralism" in the sciences, since there is an irreducible pluralism in nature; the complex and messy human world of history and culture cannot be reduced to physics. "The sciences are unified in a common goal of understanding nature, but they remain conspicuously diverse in their ways of seeking that understanding" (Crosby 2002: 47).

The second tenet embraced by many religious naturalists is that there are values inherent in the natural world, values that are "objective, universal, and ultimate," as Rue puts it, at least in that they are a part of the natural world populated by living organisms. He argues that viability is the universal, objective, and ultimate value—the universal telos—of living things. Further, Rue holds that viability is an emergent value, emerging at the same time as life in the natural world (Rue 2011: 59). Crosby concurs, arguing that "[n]ature ... is replete with value, including religious value" (Crosby 2002: 21), and identifies a number of inherently valuable things, among them life, biological species, ecological systems, the biosphere, diversity, creativity and splendor. He also identifies disvalues in the natural world, among them death, suffering, and pain (Crosby 2002: 78–86).

Finally, religious naturalists embrace a third tenet, that the natural world is sacred—in whole or in part—and is an object worthy of religious reverence. Here I tread lightly—worried about leaving blundering, tenderfoot tracks—for the sacred is ineffable, ultimately important yet easily overlooked or misunderstood.

Defining "religious" and "spiritual" as equivalent terms, Rue regards the religious naturalist as one who "takes nature to heart," one who treats nature as vitally important, central, and something that transforms one's life. As something "taken to heart," nature becomes part of—indeed interwoven into—one's life, for nature bears heavily on fundamental matters of purpose and meaning, reordering of goals and values (Rue 2011: 110–111). Stone regards sacred things as "events, things, processes which are of overriding importance and yet are not under our control or within our power to manipulate," and because some aspects of the natural world are fundamentally important yet beyond our control, that is, *sacred*—although we may not always be aware of the sanctity of a place, thing, or event—Stone is committed to religious naturalism (Stone 2000: 4). Finally, while acknowledging the existence of intrinsic evil in nature—pain, suffering, and death—Crosby writes:

> This nature to which we intimately belong—a nature that sustains, renews, and inspires us in countless ways—can command our wholehearted religious commitment ... nature itself, when we rightly conceive of it and comprehend our role in it, can provide ample context and support for finding a purpose, value, and meaning in our lives.
>
> (Crosby 2002: 169)

In short, as I understand it, the religious naturalist sanctifies some aspect of nature—or nature itself—in the sense that nature is taken to be an object of supreme worth, perhaps even as an object of profound reverence. In words "roughly paralleling" theistic language, Rue reflects:

> Religious naturalists typically refer to nature as "sacred," in the sense of being inviolate and worthy of deep reverence. Religious naturalists are also likely to regard nature as creative and dynamic, not unlike traditional theistic talk about God. The religious naturalist will speak of nature as being, paradoxically, both intelligible and deeply mysterious, in parallel with theistic language about the revealed and hidden natures of God. For religious naturalists nature is ultimately real, the supreme giver of life, the source and destiny of all that is ... Nature is self-sufficient, sovereign, omnipotent. Theistic language about the grace of God is roughly paralleled by the religious naturalist's assertion that nature constrains and enables. Theists claim that God transcends nature, while religious naturalists—especially the emergentists among them—will say that nature transcends nature itself.
>
> (Rue 2011: 111)

Constructive realism

Constructive realism embraces three tenets: (1) *Constructivism*, that facts are fabricated and worlds constructed by the use of language and other symbol systems; (2) *Ontological pluralism*, that there are many ontologically diverse, yet equally privileged constructed worlds; and (3) *Realism*, that—contrary to nominalism about universals—*kinds* of fact fabricating and world constructing processes exist. My debt to—and sins against—Nelson Goodman should be obvious, and I thank and apologize to him here.

First, constructive realism is a *constructivism* maintaining that there are no "bare facts," that facts are created by the use of symbol systems, principally language. Consider my backyard feeder and the birds around it. There are three cardinals, and the one atop the feeder is red, so it is neither green nor blue. The feeder stands to the left of the cold frame. It is the same feeder that I filled a couple of days ago, and it's not moving. Finally, there are no persons dining at the feeder. Clearly, these are facts about my feeder, these facts make true corresponding statements about the feeder, and that my true beliefs are justified—I know my feeder is not moving.

Remarkably, because the content of my sense experiences alone underdetermines how the world *really* is, there are many odd sorts of backyard world "facts" that are consistent with my experiences. Suppose we regarded "red" as an intransitive verb like "move," and "move" as an adjective like "red." Then the true statements would be "The cardinal atop the feeder redded" and "The feeder is not move." What if, instead of distinguishing blue from green, we used the predicate "bleen"—where a thing is bleen if it is either blue or green. Then the fact would be that the cardinal atop the feeder is not bleen. "There was cardinal at the feeder" would be true, if "cardinal" were treated as a mass noun like "water." Without the relation "to the left of," "The feeder is west of the cold frame" might express the fact. Finally, in a Quinean world, wherein material objects come in day-long temporal slices, it is false that "It was the same feeder I filled a couple of days ago." By the way, astrophysicists inform that, rather than being stationary, my feeder is moving at a blazing 67,000 miles/hour. These are some odd facts indeed, for most contradict our habit of thinking and talking about the world—but not one of them is inconsistent with the content of our sense experiences![1]

Some of these odd sounding "facts" are quite at home in indigenous world versions. A Shawnee speaker might say "meci skwaawa," expressing the fact "It redded" about the cardinal, because American Indian languages like Shawnee and Choctaw lack the verb "to be," treating English adjectives like "red" as intransitive verbs (Hester 2004: 264–265; Cajete 2000: 26–27). Further, the Shawnee stem "skipaky-" applies to a thing if it is either blue or green—or better said, if the thing *bleens* (Voegelin 1939: 314). Having no notion of "to the left or right of," the Tarahumara express the fact as "The feeder is west of the cold frame," orienting the world using the cardinal directions.[2] And while "There were no persons around the feeder" is true in the Western version, in the Native world version the fact is that there were three persons around the feeder (Norton-Smith 2010: 21–22). In short, because sense experience underdetermines the way the world really is, there are many possible interpretations of the events taking place in my backyard.

So, the English speaker says "It is red" and the Shawnee speaker says "It reds"—two distinctly different "facts." However, the question, "Who's really right?" makes no sense, for what counts as a fact is determined by the way a linguistic community categorizes and organizes sense experience through the devices of a language. In general, all of the characteristics of the world—the things we understand to be objects and kinds—are relative to a particular world version, grounded in the fabrication of facts by a symbol system, principally language (Norton-Smith 2010: 19–20). The facts about my backyard scene are relative to a world constructed by my linguistic community.

Second, constructive realism embraces *ontological pluralism*, that there are many ontologically different, internally consistent, equally privileged constructed worlds—worlds whose versions satisfy certain pragmatic criteria for "rightness." It is here Goodman and I begin to part ways, for his criteria for "acceptable" versions—deductive validity and inductive rightness, utility and simplicity—are biased against non-Western world versions, including an American Indian version. That said, we agree that acceptable versions cannot be *empty*, that is, cannot employ names and predicates that fail to organize or categorize any sense experience whatsoever. "Gods" and "demons," "angels" and "Platonic forms" are labels that survive despite having no perceptual content. They label nothing, for supernatural gods, demons, and angels, and transcendental Platonic forms, are not the kinds of things that one can sense.

Finally, contrary to Goodman's nominalism with respect to kinds, constructive *realism* maintains that kinds of fact fabricating and world constructing processes exist. Again, ontological pluralism is committed to multiple ontologically diverse, yet equally privileged world versions constructed by symbol systems. This, however, is not wholly accurate, for we should regard symbol systems as linguistic *tools*; world versions are really the products of our *acts of construction with symbols*—collectings and sortings, composings and decomposings, weightings and valuings, among them—rather like the constructive acts of carpenters with hammers and saws transform the raw materials of boards and nails into a house. Now, acts of carpentry—hammering, sawing and so forth—are independent of any particular house, and many different wooden structures can be built using a single set of tools. Likewise, constructive realism maintains that the many equally privileged worlds are constructed using a common set of world constructing processes. Whether counting "two women" in an English world version, or "niiswi kweewaki" in a Shawnee world version, the same sort of collecting is required. While ontological pluralism implies that there is a multiplicity of ontologically diverse, equally privileged world versions, it does not imply that there is a multiplicity of ways of constructing worlds. Otherwise, I'd know how the Western world version was constructed—but how radically different world versions were constructed would be an inscrutable mystery.

The constructive realist critique of religious naturalism

To be blunt, from a constructive realist perspective theistic talk of supernatural beings and events—gods, miracles, and such—is either meaningless or fails in its intended reference. Assuming that the supernatural is imperceptible, "God," when referring to a *supernatural* entity, categorizes no perceptual experiences, while "miracle" refers to the unexpected experience in the natural world, but fails to refer to a supernatural influence. Thus, constructive realism wholeheartedly agrees with religious naturalism's rejection of the supernatural.

Rue writes, "Religious naturalists bear an attitude of reverence toward the universal as a whole, and toward the earth in particular—and they are disposed to expressing their reverence by affirming that nature is both sacred and mysterious" (Rue 2011: 114). Denying the supernatural, yet recognizing that one *can* experience reverence and gratitude, awe and humility with respect to the natural world is again in accord with constructive realism. "Reverence" and "gratitude" are words that categorize emotional experiences, hence are meaningful. Further, religious naturalists have naturalized theistic categories—e.g., the "spiritual," the "religious," and the "sacred"—divorcing them from supernatural meaninglessness.

Perhaps most significant, in their embrace of values in nature, Crosby and Rue have suggested a solution to a long-standing concern I've had with constructive realism, and I thank them here. The problem is that an ontological pluralism seems more at home with a moral relativism than a moral realism—but, I believe the former to be inconsistent. Moreover, among the world constructing action kinds we can clearly add "valuing" as a way to organize and categorize experience, but valuing is notoriously perspectival. However, Crosby sketches a *relational* view of values—quite consistent with constructive realism—such that values, like facts, are constructions on and of experiences. "[W]ere there no valuers," he observes, "there would be no values," and since human and nonhuman valuers are part of nature, natural values exist and are prevalent in the world (Crosby 2002: 75–77). Rue grounds his moral realism by arguing that *viability*—"the continuation and fulfillment of life"—is "a universal, objective, and ultimate value" for living creatures (Rue 2011: 59–60). We might propose, then, that human beings—as a part of nature—universally categorize as valuable those perceptual experiences that are, for them, likely to continue and fulfill their lives, so at least some "valuing" is inherent in nature and not merely perspectival.

On a less positive note, Crosby anticipates my principal constructive realist concern with religious naturalism:

> [S]ince we obviously cannot *experience* the totality of nature, encompassing everything that is and extending into the farthest reaches of space and time, as well as into the most minute, submicroscopic domains, how can we refer to that totality? If we cannot refer to it, how can our conception of it be meaningful?
>
> (Crosby 2002: 18)

The problem, expressed as a constructive realist, is that meaningful words organize and categorize sense experiences—facts are fabricated through the devices of a language. "God" and similar words purporting to refer to supernatural entities, categorize no experiences, hence are meaningless. But at first blush, the word "nature"—understood by religious naturalists as referring to "everything that is and extending into the farthest reaches of space and time, as well as into the most minute, submicroscopic domains"—seems to fare better, but only a bit. Clearly, the sense experiences I have while catfishing in Couwachenink—on old Lenape ground—are a part of the extension and meaning of "nature." However, isn't the vast majority of the natural world to which religious naturalists purport to refer both temporally and spatially beyond my experience?

Crosby's correct reply is that talk of the whole of nature may be abstract, but that doesn't imply that theoretical term "nature" is vacuous. (Crosby 2002: 18). Indeed, the general term "cardinal"—which categorizes and organizes my backyard experiences—is, as well, a nonempty abstraction. "We obviously do not experience nature as a whole; all of our experiences of it are partial and fragmentary at best, but the abstractions of broad-ranging conceptual schemes can help fill in the gaps" (Crosby 2002: 26). The "broad-ranging conceptual scheme"—the *model of nature* Crosby develops—"is intended to order and elucidate important aspects of our experience of nature" (Crosby 2002: 26). His "conceptual and imaginative" model of nature intends to capture its plurality and diversity, its regularity and unrelenting change, and the "inexorable, disruptive workings of chance and novelty" (Crosby 2002: 34). Indeed, for Crosby the term "nature"

> Suggests a dynamic, restless energy of growth, nurture, productivity, and change. It points to nature as the fruitful womb of all that is, has been, or ever will be. This etymology even hints at the wonderful power of nature to produce and sustain myriad forms of life, here on earth, in all likelihood in other regions of space, and probably in other epochs. These aspects of the concept of nature are for me the ones that are the most definitive, awesome, and compelling.
>
> (Crosby 2002: 42)

While not vacuous, I believe that Crosby's conceptual and imaginative model of nature suffers a problem similar to its supernatural counterparts, as Rue's earlier parallel theistic language about nature reveals. I'm reminded of bygone days when, as a mathematics student, I puzzled over the cardinality of the natural numbers. The numbers are, of course, a countably infinite set—remarkably, the same cardinality as the rational numbers, despite the fact that the former are a proper subset of the latter—which made me wonder what sort of concept of the "infinite" one can have. I certainly have some notion of the infinite—but nothing like the concept of the cardinality of the natural numbers—an actual or completed countable infinity. At best, "the infinite" for me denotes the more pedestrian notion that "every number has a successor," that is, "I could keep counting"—a concept of a potential infinity. Now, don't get me wrong; I was taught to *say* lots of things about the actual infinite. But that does not imply that I have a clear conception of it—imaginative or otherwise.

Likewise, it seems to me that Crosby and Rue's conception of nature as "the universe as a whole," "the totality of nature," or even its power to produce and sustain life, "here on earth, in all likelihood in other regions of space, and probably in other epochs" are as inconceivable as the actual infinity—although religious naturalists *say* lots about it. That does not imply, of course, that the concept of "nature" is vacuous. I am suggesting, however, that "nature" cannot refer to what religious naturalists believe it refers; perhaps the *inconceivability* of nature is the source of mystery and awe. This, by the way, is not a crushing objection to religious naturalism; indeed, from a constructive realist perspective, it fares quite well.[3]

An American Indian world version

I argued in *The Dance* that an American Indian world is numbered among the ontologically diverse, internally consistent, yet equally privileged actual worlds. As one surveys various Native traditions, four common themes appear—which constructive realism sees as world constructing principles: (1) *Relatedness*, that everything in the natural world is interrelated; (2) *An expansive conception of persons*, that both human and nonhuman beings can be persons; (3) *The semantic potency of performance*, that performance with symbols is the principal vehicle of meaning; and (4) *Circularity*, that time and space are ordered in cycles and circles. I also cautioned that because *incommensurable* worlds are constructed by radically different languages, it will be difficult to describe—let alone wholly comprehend—the Native version, for the description will impose Western categories on the American Indian world. But, we'll do our best.

I'll begin by sharing an old Menominee story, "The Indian and the Frogs" to introduce elements of a Native world. At first blush, it seems merely to be a story about one Indian's encounter with "nature"—but, it teaches far more:

> Once an Indian had a revelation from the head of all frogs and toads. In the early spring, when all the frogs and toads thaw out they sing and shout more noisily than at any other time of the year. This Indian made it a practice to listen to the frogs every spring when they first began, as he admired their songs, and wanted to learn something from them. He would stand near the puddles, marshes, and lakes to hear them better, and once when night came he lay right down to hear them.
>
> In the morning, when he woke up, the frogs spoke to him saying: "We are not all happy, but in very deep sadness. You seem to like our crying but this is our reason for weeping. In early spring, when we first thaw out and revive we wail for our dead, for lots of us don't wake up from our winter sleep. Now you will cry in your turn as we did."
>
> Sure enough, the next spring the Indian's wife and children all died, and the Indian died likewise, to pay for his curiosity to hear the multitude of frogs.
>
> (Skinner and Satterlee 1996: 82)

Perhaps the most striking thing for one unfamiliar with indigenous stories is that the frogs speak. But why shouldn't they converse with the Indian as other persons do? The *equality* of humans and other creatures in the natural world has been long misinterpreted as a lowering of human beings to the status of other beings—a mark of "savagery" according to J. W. Powell

(1877). However, Indians strongly affirm and raise other creatures to the status of *persons*. In the Native world, a *person* is an animate being standing in a nexus of relationships with, relationships sustained by respect for other persons—clearly a circular definition, but Indians don't mind. And, of course, the Indian in the story is in a relationship with the frogs and toads, and so is in "nature." There is no artificial distinction between the human and non-human worlds; this is no mere encounter with the natural world seen as another realm.

At this point, however, I need to say a word about *animation*, a category so important in the Native world that it is marked grammatically in Algonquin languages, much like gender is in many Western languages. For Indians an animate being is one experienced as having and discharging *power*—including, importantly, the power to transform. Thus, among the usual candidates, Natives regard some rather unusual entities as animate—plants and animals, sacred places and physical forces, the sun and moon, even Mother Earth, herself—and, of course, tobacco and flint.

The next striking thing is the Indian's punishment for his "curiosity"—something that baffles the Western ethicist, for he neither harmed nor intended to harm the frogs. However, in a world where everything is related, the Indian's behavior made him *brother* to the frogs, and so he and his family shared their fate. But, what "behavior" merited such harsh punishment? The dynamic nexus of relationships between persons of various sorts and powers in the Native world is kept in equilibrium only by the respectful exchange of gifts—an instance of meaningful performances with symbols. The Indian, however, sought to acquire the frogs' knowledge without their permission, *stealing* what should only be *given* with the obligatory gift of gratitude in return. Because of his disrespect for other persons, the Indian and his family shared the frogs' fate. The story, therefore, does not tell of the Indian's mere encounter with nature; instead we see that his thoughts and actions—like those of all human and non-human persons alike—create and recreate the dynamic, unfixed, and unfinished "nature" to which he—and we—are inextricably bound.

The Menominee story teaches yet another lesson, that Indians live in a moral universe wherein everything thought, said, or done—no matter how "innocent"—has a moral dimension. Imagine the Western moral human world of motive and action extended to and multiplied by a plethora of other non-human persons with all of the needs and desires, volitions and beliefs that human persons have. But that moral universe is a "universe" of limited scope; my "universe" encompasses only the human and non-human persons to whom I am related: my family and friends, my ancestors and descendants, my tribe and its sacred places, and the green, water, winged, and four-legged people who nourish me—the catfish gifted to me by Couwachenink—and, of course, the Grandmother. Walking the right road is being mindful of and fulfilling my obligations to my human and non-human relatives—including the Menominee story that has taught me so much, and I thank it here.

A Native critique of religious naturalism

I begin by reiterating my earlier caveat about incommensurable worlds constructed by radically different languages: Examples of radical differences in Western and indigenous ontology abound. These traditions categorize and recognize the value of the animate and inanimate differently. Witness that flint, pipe, and tobacco are animate in American Indian traditions and animation is important enough to be marked grammatically in Algonquin languages. Persons are categorized differently. The sun, moon, Earth, cardinal directions, and other critters are persons for Indians. Attributes are conceived differently. Without the verb "to be" in many indigenous languages, Western properties such as "red" denoted by predicates are understood as activities denoted by intransitive verbs—"redding." What the Western ontologist understands as disjointed and unconnected material events, the Indian ontologist knows are intimately related and interconnected moral events. The Western ontologist observes two natural events to abstract the similarities in form while the Indian ontologist focuses on their individual uniqueness. And if differences in ontology weren't impediment enough, there are radical differences in corresponding notions of truth and verification, utility and knowledge. Even the logics in each tradition are different. Comparing the worlds of Native religion and religious naturalism, therefore, will be especially difficult and fraught with dangers. Perhaps only a fool—or the trickster Coyote—would try!

Native traditions tend not to draw sharp distinctions between literature and philosophy, science and religion (Burkhart 2004: 22). Nor are there sharp distinctions between fact and value, subject and object, the spiritual and material, or this-worldly and otherworldly (Deloria 2012: 201). Unlike the Western tradition, Indians' collection of information about the natural world is comprehensive, arising from immediate experience. Moreover, the Western tradition recognizes only objective and replicable experiences as verifying; yet no experience—even the private and unique—is discounted in Indian traditions (Deloria 1999: 66). Thus, the Native experience of "spirit"—whether in visions, dreams, or in sacred places—is *genuine* (Deloria 1999: 355–356).

In this light, it might seem remarkable that I believe that Indians would agree with religious naturalists that there is no supernatural world, that only the natural world exists. Speaking of the Ojibwa, Hallowell opined that "[i]t is unfortunate that the natural-supernatural dichotomy has been so persistently invoked by many anthropologists in describing the outlook of peoples in cultures other than our own" (Hallowell 1960: 28). Yet, indigenous people are not shy about talk of "manitoo," "mana," or "wakan"—usually translated as "spirit." However, I argued in *The Dance* that the air of the supernatural is imposed on the Native world version because of a mistake in translation: All of the Western supernatural connotations wedded to "spirit" are imposed on their indigenous counterparts. I suggested that "life force" is a better translation (Norton-Smith 2010: 86–89). Deloria suggests "power" or "energy" (Deloria 2012: 203). Neither uses "supernatural" as a translation.

For Indians, the natural world of experience is the only world that exists, hence, there is no supernatural world—but in the natural world we most certainly experience the power of sacred places, places with the power to move us in profound ways. For Natives these places are animate—they have power and discharge that power to affect us. So interpreted, there are "spirits" in the natural world. Nor, agreeing with religious naturalism, is the human world distinct from the natural world—another false distinction by Native lights. Western science, philosophy, and religions—save religious naturalism—view humans as not only superior to but different in kind from other critters in the natural world. Indeed, through ceremony human beings help shape and create the unfixed and unfinished world—also through the unceremonious burning of fossil fuels.

With regard to values in nature, Rue argues that certain values—certain "meanings"—are emergent and inherent. On his "strong emergentist" view, the natural world is now filled with *tele*, purposive activities: "The astonishing claim here is that a pointless process created the conditions for the pointfulness of biological functions. A universe with no telos, a universe without an agenda, just *inadvertently* made possible the spontaneous emergence of purposeful activity (Rue 2011: 54). Crosby argues that, while "[n]ature as such lacks purpose, sentience, or consciousness ... it is not devoid of value but is replete with value, including religious value" (Crosby 2002: 21). Nature is "replete with value," because values are relational and contextual, that is, you cannot have values without valuers—both human and non-human—and these valuers value *life* and the conditions that sustain life (Crosby 2002: 74–85). An assumption uniting both Rue and Crosby is that Western natural science teaches that over time life evolved from a lifeless and purposeless material world of causal processes, so values emerged hand-in-hand with the emergence of life.

Deloria frames the Native response: "It cannot be argued that the universe is moral or has a moral purpose without simultaneously maintaining that the universe is alive. The old Indians had no problem with this concept because they experienced life in everything" (Deloria 1999: 49). Deloria motivates three observations. First, if you are living in a moral universe—populated by all sorts of human and non-human persons—wherein there is no absolute distinction between fact and value, then it goes without saying that values are inherent in nature, as are facts. But not so much, I think, the rather pedestrian biological values of life or viability. *Respect*, without which the Native world of equilibrium would dissolve into chaos, must be the fundamental moral attitude and an inherent value. Second, if the mark of animation is having power or energy and discharging that power—as experienced by the "old Indians"—then the animate extends beyond the narrow biological Western scientific category: There was power and energy, animation and purpose before the emergence of biological life. Finally, when time and space are organized cyclically and circularly, it is not clear to what extent it makes sense to say that life emerged "over time"—a linear, not cyclical ordering of time.

Having mentioned "Western natural science," I would be remiss should I not say a word about the way in which religious naturalists view the methodologies and findings of science and how it differs from the Native view. While correctly arguing that the natural sciences alone cannot give a comprehensive account of nature, Prof. Crosby acknowledges the importance of those methodologies and findings: "Past and continuing contributions of the natural sciences to our understanding of nature are thus of immense, indispensable importance and ought always to be taken into account in philosophical reflections about nature" (Crosby 2002: 45). Rue concurs, because religious naturalism is a *naturalism*, naturalists believe that the natural order is ultimately real, and science studies the natural order, so religious naturalists are committed to scientific inquiry into the nature of nature (Rue 2011: 101).

Now, Deloria and other Indians are famously hostile to Western science—far more than I am, by the way. However, I will take this opportunity to observe that the methods, findings, and *purpose* of Western science are radically different from Native "science." As I wrote in *The Dance*, the Western tradition constructs an inert, material, mechanical, law-like natural world. The role of science is to explain, predict, and ultimately conquer, manipulate, and exploit that natural resource. Yet, in a dynamic, animate, interrelated moral universe, wherein our actions and choices create and recreate the world, the role of Native "science" is to show how one ought to behave—what is necessary to walk the right road. The frogs again illustrate: Using a Western scientific methodology and purpose, the Indian observed the frogs hoping to learn something from them—rather like contemporary bio-colonialists who appropriate and exploit indigenous knowledge of medicinal plants (Whitt 2009). Forgetting that knowledge is given as a gift—not something to be disrespectfully stolen—the Indian learned an indigenous "scientific" finding.

Finally, there is no doubt that Indians agree with religious naturalism's religiosity with respect to nature. It is commonplace in the literature that certain sites in the natural world are regarded as sacred by indigenous people. Indeed, writers like Deloria distinguish Abrahamic religions like Christianity from "tribal religions," arguing that the former are grounded by sacred events in time while that latter are based upon sacred places in space (Deloria 1994). Moreover, Deloria correctly observes that many Western religions consider the natural world to be corrupt and corrupting—to be an unredeemable realm to which we are condemned until death (Deloria 1994: 80–81). To its credit, religious naturalism does not follow the Western religious practice of vilifying the natural world; indeed, reiterating Rue's earlier assessment: "Religious naturalists typically refer to nature as 'sacred,' in the sense of being inviolate and worthy of deep reverence" (Rue 2011: 111). But despite this virtue by Native lights, religious naturalism shares a different feature with other Western religions that distinguishes it from its Native counterpart. The followers of Abrahamic religions, reverencing sacred events in the distant past—for example, the Christian resurrection—are reverencing a mere *conceptualization* of the sacred event, a conceptualization shaped by and abstracted from its original context—its time and place. Likewise, the "nature" reverenced by religious naturalists is a conceptualization, an abstract concept—as Crosby earlier shared—rather like the abstract notion of *humanity*. But, just as I cannot love an abstract conception of *humanity*, I cannot reverence an abstract conception of *nature*. "[Indigenous] peoples always have a *concrete* reference—the natural world—and the adherents of the world religions continually deal with abstract and ideal situations on and intellectual plane" (Deloria 2012: 211).

That said—and knowing well the dangers in comparing incommensurable world versions—it seems that religious naturalism is closer in "spirit" to Native religions than other Western supernatural, temporal, doctrinal, dogmatic, and event-grounded religious traditions. "The great bond of experience with nature," observes Deloria,

> incorporates the emotional and intuitive dimensions of our lives much better than do the precise creeds, doctrines, and dogmas of the great world religions and as such provides us with continuing meaning as long as we treat our apprehension of the great mystery with respect.
>
> (2012: 207)

A closing reflection

As a dawning religious tradition within a Western world version, religious naturalism fares well in relation to a constructive realist critique; and, to the extent that it can be compared to a radically different indigenous world version, it again fares well. Constructive realism and the Native view it recognizes both agree that there is no supernatural world; that the human world is a part of nature; and that values are inherent in the natural world. However, religious naturalism's over-conceptualized and abstracted notion of "nature" is suspect on a constructive realist analysis, and goes well beyond the concrete, immediate experience of nature grounding Indian religious traditions. A sacred nature, a nature reverenced, must be—literally and figuratively—close to home. pesalo no'ki tanakia![4]

Notes

1 I understand "sense" and "perceptual experiences" to include visual and auditory experiences and the like, but also Humean sorts of "inward impressions." Specifically, in response to a concern raised by Prof. Crosby, I maintain that it makes perfect sense to speak of experiences characterized as "religious." Indeed, Prof. Rue develops a naturalized account of the "religious" or "spiritual person" (Rue 2011: 110–112).

2 At the American Philosophical Association Pacific Division Meetings in 2014, Lera Boroditsky shared that a "bleen-like" predicate is so common in indigenous languages that anthropologists have adopted the predicate "grue" to refer to it. She also shared these findings about the Tarahumara.

3 My friend Jerome Stone clarified his view to me in the following way: "I would say that the totality of nature is conceivable, abstract but conceivable, but not experiencable ... The religious dimension of my own naturalism alternates between conceiving of *nature-as-a-whole* and appreciating *that particular* bald eagle swooping over the Wisconsin River. Hence there is, for me, as I use language to describe my reactions, an alternation between relating to nature-as-a-whole and specific natural events." Indeed, the whole of nature is beyond our experience; but unlike Stone's view, it is also beyond my conception. This, perhaps, says more about *my* limited ability to conceive the abstract and unlimited than about the

conceivability of the nature-as-a-whole. After all, my mathematics professors had little trouble discussing Cantor's universe of infinities.

4 My research for this chapter was supported by Kent State University at Stark; one could not be a part of a more splendid institution.

References

Burkhart, B. (2004) "What Coyote and Thales Can Teach Us: An Outline of American Indian Epistemology," in A. Waters (ed.) *American Indian Thought*, Oxford: Blackwell Publishing, pp. 15–26.

Cajete, G. (2000) *Native Science*, Santa Fe, NM: Clear Light Publishers.

Crosby, D. (2002) *A Religion of Nature*, Albany NY: State University of New York Press.

Deloria, V. (1988) *Ways of Worldmaking*, Indianapolis: Hackett.

—— (1994) *God is Red: A Native View of Religion*, Golden, CO: Fulcrum.

—— (1999) "If You Think About It, You Will See That It Is True," in B. Deloria, K. Foehner, and S. Scinta (eds.) *Spirit and Reason*, Golden, CO: Fulcrum, pp. 40–60.

—— (1999) "Relativity, Relatedness, and Reality," in B. Deloria, K. Foehner, and S. Scinta (eds.) *Spirit and Reason*, Golden, CO: Fulcrum, pp. 32–39.

—— (2012) *The Metaphysics of Modern Existence*, Golden, CO: Fulcrum.

Goodman, N. (1983) *Fact, Fiction, and Forecast*, Cambridge, MA: Harvard University Press.

—— (1988) *Ways of Worldmaking*, Indianapolis, IN: Hackett.

Hallowell, A. (1960) "Ojibwa Ontology, Behavior, and World View," in S. Diamond (ed.) *Culture in History: Essays in Honor of Paul Radin*, New York: Columbia University Press, pp. 19–52.

Hester, T. (2004) "On Philosophical Discourse: Some Intercultural Musings," in A. Waters (ed.) *American Indian Thought*, Oxford: Blackwell Publishing, pp. 263–267.

Norton-Smith, T. (2010) *The Dance of Person and Place*, Albany, NY: State University of New York Press.

Powell, J. W. (1877) *Outlines of the Philosophy of the North American Indians,* New York: Douglas Taylor, Book, Job and Law Printer.

Rue, R. (2011) *Nature Is Enough*, Albany NY: State University of New York Press.

Skinner, A., and J. Satterlee (1996) "The Man Who Loved the Frog Songs," *The Telling of the World,* New York: Stewart, Tabori & Chang, p. 82.

Stone, J. (2000) "What Is Religious Naturalism?" *Journal of Liberal Religion*, vol. 2, no.1. Available from: Meadville Lombard Theological School Online [19 October 2016].

Voegelin, C. F. (1939) "Shawnee Stems and the Jacob P. Dunn Miami Dictionary," *Indiana Historical Society Prehistory Research Series*, vol.1, no.8, pp. 289–341.

Whitt, L. (2009) *Science, Colonialism, and Indigenous Peoples*, Cambridge: Cambridge University Press.

19

DAO AND WATER

Rethinking Daoism as naturalism

Jea Sophia Oh

In *Dao De Jing*, Chapter 1, Lao Tsu teaches, "the *Dao* that can be told is not the eternal *Dao*" (Feng and English 1985: 9). Perhaps this sounds vague and mysterious. Nature is the manifestation of the *Dao* while the condition of its manifestation is its inner stillness (*xing*), the power within. It is the Way (*Dao*) of nature. It is not strange to say that nature comes out of nature. The *Dao* is everywhere and flows through everything. Therefore, the *Dao* is everything. It is not awkward to use naturalistic terms to describe *Dao* as the ultimate mystery that is manifested in nature natured (*natura naturata*) via nature naturing (*natura naturans*).

In *Dao De Jing* Chapter 8, Lao Tsu teaches, "The highest goodness resembles water. Water greatly benefits myriad things without contention. It stays in places that people dislike. Therefore it is similar to the *Dao*" (1985: 23). Reading *Dao De Jing* Chapter 8, this essay rethinks Daoism as religious naturalism. I dive into the water of *Dao* and find how *Dao* and water resemble each other to say that nature is *Dao* inasmuch as *Dao* is nature. Then, I will ask an analytic question: "If *Dao* is like water, what is *Dao* of water?" The water metaphor in *Dao De Jing* signifies nature naturing as a gender neutral word, whereas water has been interpreted by many feminist scholars as a feminine metaphor and identified as the Primal Mother. Water in the *Dao De Jing* is non-binary and neutral and means neither female nor male. In contrast with the feminist essentialist notion of Daoism, I will suggest "fecundity" as the inner power of nature in myriad things, including human males and more-than-human nature, which cannot be reduced to femininity alone. Finally, the *Dao* of nature will be presented as the basis for an ethic of nature that humans should follow in their relations with other humans and with nature as a whole.

Dao and nature

The Dao that can be spoken is not the eternal Dao.
The name that can be named is not the eternal name.
The nameless is the origin of Heaven and Earth.
The named is the mother of myriad things.
Thus, constantly free of desire.
One observes its wonders.
Constantly filled with desire.

One observes its manifestations.
These two emerge together but differ in name.
The unity is said to be the mystery.
Mystery of mysteries, the door to all wonders.

(1985: 9)

Dao is ineffable according to *Dao De Jing* Chapter 1: "The *Dao* that can be spoken is not the eternal *Dao*" (1985:9). In other words, *Dao* can be understood negatively as what never appears as such nor ever can be exactly apprehended and defined.

Likewise, Nicholas Cusa in his final work, *De apice theriae* (1464), presents *posse ipsum* (possibility itself) as the most appropriate name for God: "that without which nothing whatsoever can be, or live, or understand" (Cusa and Bond 1997: 58). *Posse ipsum* is Nicholas Cusa's favorite name of God, which the ecotheologian Catherine Keller applies to creation as *chaosmos* of *posse ipsum* in her recent book, *Cloud of the Impossible*:

> For the nuance of beginning, for the fly in the abyss, for the mindful, wrenching and collective materializations of the im/possible now: is there time? Is there space? Only in an unknown that does not terrify us, the practice of an apophatic entanglement.
>
> (Keller 2015: 49)

The ineffability of *Dao* and such Western apophatic discourse are comparable in their radical inability to describe what the ultimate reality is.

The ineffability of *Dao* lies at the base of nature. Nature for Daoism is the mysterious sanctuary that humans should not ruin because of its sacredness: "The universe is sacred. You cannot improve it. If you try to change it, you will ruin it. If you try to hold it, you will lose it" (1985: 49). We read about the aesthetic quality of nature in Daoism throughout the chapters of *Dao De Jing*. The beauty of nature in *Dao De Jing* is simultaneously manifested and hidden. These two faces of *Dao* are mysteriously interwoven through nature's folding, unfolding, and enfolding just as the moon waxes and wanes but still remains one and the same, just as the mountains wear the colors of four seasons and in this way express the beauty of nature. The beauty of nature is celebrated throughout the 81 chapters of the *Dao De Jing*.

Dao of water

Supreme good is like water.
Water greatly benefits the myriad things, without conflict.
It flows through places that multitudes detest.
Thereby it is close to the Way.
A good dwelling is on the ground.
A good mind is deep.
A good gift is kind.
A good word is sincere.
A good ruler is just.
A good worker is able.
A good deed is timely.
Where there is no conflict, there is no fault.

(1985: 23)

In Dao De Jing Chapter 8, Lao Tsu *states, "The highest good is like water.* Supreme good is like water. Water greatly benefits the myriad things, without conflict. It flows through places that multitudes detest. Thereby it is close to the Way" (1985: 23). In this passage water is equated with the *Dao* and proclaimed the highest good. Water is described as "the deep valley spirit that never dies. It is woman, the primal mother, the gateway of the sky and the earth" in *Dao De Jing* Chapter 6 (1985: 17). It is the power of nature, therefore, *yin*. Nevertheless, *yin* is not just femininity but "fecundity." Beyond the dichotomy of masculinity and femininity, everything is endlessly creating myriad things in the process of becoming. In this sense every entity, both at the micro and macro levels, is depicted as "a maternal body: birthing, dying, and renewing itself" (Oh 2011: 50). This deep hidden power exists in nature and works not only between women and men but also in and through every living and non-living aspect of nature.

The numerous becomings of life as multiplicities of multiplicity are written in *Dao De Jing* as myriad things (*wan wu*, 萬物). *Dao de Jing* Chapter 51 teaches that myriad things arise from *Dao* and are manifested in myriad things. *Dao* is metaphorically depicted as water that is the mother of ten thousand things in Chapter 25 (1985: 57). The Daoist water is like the chaotic *tehom* in the Book of Genesis in the Hebrew Bible that Keller describes as an oceanic and female-associated *chaos*:

> The poems of the *Dao De Jing* were composing of the primal water a nothingness that is a cornucopia that never runs dry. It is the deep source of everything – it is nothing, and yet in everything. The *Dao* invokes a flowing, formless infinity that nurtures all things without lording it over anything.
>
> (Keller 2003: 14)

Keller's interpretation might be applied to *Dao De Jing* Chapter 25:

> Something mysteriously formed. Born before heaven and earth. In the silence and the void, standing alone and unchanging, Ever present and in motion. Perhaps it is the mother of ten thousand things. I do not know its name. Call it *Dao*. Being great, it flows. It flows far away. Having gone far, it returns.... Humans follow the earth. Earth follows heaven. Heaven follows the *Dao*. *Dao* follows what is natural.
>
> (1985: 57)

Dao of nature

The term nature, *zìrán* (自然) in Chinese is a combination of zì (自, self) and *rán* (然, as it does). Thus, the idea of nature is "becoming-so" or "of-itself," the natural state. The closest translation of *zìrán* is "self-so-ing." This term is used to refer to "spontaneity" as well as "spontaneous act as against artificiality" (Oh 2015: 108). Thus, nature has both intentionality and spontaneity. Life is everywhere in everything, as we see with weeds growing in between cement blocks of sidewalks that barely contain soil. Life extends and grows in quantity and evolves and seeks to be better, that is, to be qualitatively abundant and happy.

Most of us have seen dandelion seeds flying in the air at one time or another. When you blow them into the air, the unfolding of dandelion seeds does not seem to move only vertically but also horizontally like parachutes move through air. This can be called a "rhizomatic movement" in the language of Gilles Deleuze. Dandelion seeds blow rhizomatically without a program or a plan and create multiplicities. Deleuze and Felix Guattari assert that multiplicities are rhizomatic and expose arborescent pseudo-multiplicities for what they are (Deleuze and Guattari 2007: 6).[1] Deleuze and Guattari present the "rhizome" as a nomadic and fluid movement that rejects a pivotal center. With this figure of the "rhizome," they envision structure-free-movements that spread towards multiple exteriors and become unpredictably permutated by coming into contact with whatever lies at their exteriors. The rhizome is an achronological system where non-categorizable singularities and multiplicities traverse the fixed boundaries without being arranged and schematized by any central order (2007: 7). The dandelion seed is a good analogy for Deleuzian rhizomatic movements, given the fact that Deleuze and Guattari observe that there are neither points nor positions but only lines in a rhizome such as those found in the structure of "tree or root" (2007: 7).

A dandelion flower is born of a seed among thousands of seeds and creates numerous seeds that again blow away to the air through the continual process of convergence and dissemination. A singularity creates a multiplicity that is composed of configurations of fuzzy, flexible, and vibrating lines with indeterminable trajectories. In other words, a multiplicity is necessary for regeneration. Because nature is differentially interrelated rather than unified in any absolute sense, it continually produces itself through new combinations with heterogeneous elements. Yet, there are particular finite elements and interrelations of elements produced in the continuous movements of becoming (Hayden 2008: 26).

> *The Dao begot one.*
> *One begot two.*
> *Two begot three.*
> *And three begot the ten thousand things.*
>
> (*Dao De Jing*, Chapter 42, 1985: 89)

The endless fecundity of nature in Daoism is similar to Robert Corrington's concept of nature. According to him, in the aspect of nature natured we can speak of nature as the sum of all complexes. In the other aspect of nature naturing, nature is the active source of all of the complexes. He states, "The innumerable complexes manifest as nature natured are themselves located and ordered by the sheer power of nature in its naturing" (Corrington 1991: 349). Corrington recognizes nature natured as equivalent to the orders of creation. Corrington's nature natured has a certain autonomy from the creative impulses that sustain its orders. "Nature in its naturing can be understood as the continuing acts of creation by and through which the world is sustained against the recurrent threats of nonbeing." While "self-so-ing" is the positive sense of nature (自然, *zìrán*), the negative sense of *zìrán* might be "self-othering" which can be expressed in Lao Tsu's concept of non-doing (無爲, *wu wei*). The positive side of this negative aspect of non-doing is yielding, helping others, and encouraging others on the ways to their own "self-so-ing." Therefore, it is "self-othering." In this aspect, nature as "self-so-ing (自然, *zìrán*)" and as *wu-wei* (無爲) are closely related and deeply interconnected.

Dao De Jing Chapter 40 teaches, "Returning (反, *fan*) is the motion of the *Dao*. Yielding is the way of the *Dao*. The ten thousand things are born of beings. Being is born of not being" (1985: 85). The *Dao* of nature is non-doing rather than doing. It suggests the active passivity of negation of violence against life's spontaneity. Therefore, it is negation of negation, double negation, because doing non-doing is the negation of violence that is negation of nature's Way (*Dao*). Non-doing means not doing intentionally but doing spontaneously in a natural way. Change and spontaneity reflect the nature of life. When one follows its own nature, one does not need forceful interaction to intervene and support its action. Non-doing (無爲, *wuwei*), therefore, refers to "doing non-doing" (爲無爲, *wei-wuwei*), which proceeds freely and spontaneously from one's own nature.

The negative practice of doing non-doing can be comparable to Corrington's selving when he points out that "nature struggles to give birth to selves" (1991: 353). Nature's self-transcending power is the theme of Corrington's ecstatic naturalism. He writes, "Selving is a gift of that dimension of God that drives toward the future and the manifestation of greater actuality and consciousness" (1991: 354). For Corrington, selving is god-ing, for it manifests *Dao* in myriad things. Corrington envisions an eschatology of "autonomy as *theonomy*" that is practical rather than merely abstract. The divine lure is most fully manifest in the eschatological core of personal and social transformation. Corrington hopes for a movement of God toward the community of selves in which all heteronomy (alien law) is transformed into a true *theonomy*:

> God is still finite in this dimension of God's nature.... God can provide a lure within which the finite self may find strength to overcome its previous limitations, but God cannot force the self toward a transformed and deepened autonomy.... The divine lure is persuasive rather than coercive.
>
> (1991: 356)

Thus, *theonomy* is indeed closely similar to *Dao* of nature.

Understanding Daoism as religious naturalism, we should recognize that religious naturalism is older than its twentieth-century forms and is anticipated by Daoism much earlier on. Daoism was formed in 550 BCE, and *Dao De Jing* was written sometime in the third or fourth century BCE, while Corrington's *Ecstatic Naturalism: Signs of the World* was published in 1994, and Donald Crosby's *Living with Ambiguity* was published in 2008. Even though religious naturalism has its historical roots in Spinoza (1632–77), Daoism already existed in his lifetime. It would also be wrong to claim that only contemporary religious naturalists have brought to light the naturalistic aspects of Daoism since Daoism is the Way of nature.

In September 2016, the Obama administration temporarily blocked the Dakota Access Pipeline (DAPL). Native Americans and environmental activists protesting the DAPL supported by the US Army Corps have argued that the 1,172-mile pipeline would damage sacred lands and could contaminate the tribe's water source.

> The Standing Rock Sioux Tribe won a major victory on Sunday (December 4, 2016) in its battle to block an oil pipeline being built near its reservation when the Department of the Army announced that it would not allow the pipeline to be drilled under a dammed section of the Missouri River.
>
> (Healy and Fandos, *The New York Times*, Dec. 2016)

However, Donald J. Trump has supported finishing the 1,170-mile pipeline, which crosses four states and is almost complete. The DAPL threatens our climate, water, and land.

Dao De Jing Chapter 8 teaches that the *Dao* is like water. We humans are a part of nature on earth, the largest part of which is water. Nature's sufferings are undeniable, and among them are many caused by human interventions. Thus, confronting the predictable water crises in the Dakota Access Pipeline, the Standing Rock Protest by indigenous people and environmental activists involves a lot more than protecting water. Water gives life to everything without any effort. Therefore, "Water is the *Dao*." Ruining water means going against the *Dao*.

At this juncture, we need to ask, what is the *Dao* of nature? *What Would Nature Do*? We need to shift the pipelines to the Way of nature, that is, *Dao*. *Dao De Jing* Chapter 48 offers a profound lesson for confronting our ecological crisis:

> *In the pursuit of learning, every day something is acquired.*
> *In the pursuit of Dao, every day something is dropped.*
> *Less and less is done.*
> *Until non-action is achieved.*
> *When nothing is done, nothing is left undone.*
> *The world is ruled by letting things take their course.*
> *It cannot be ruled by interfering.*
>
> (1985: 100)

Note

1 According to Deleuze and Guattari, puppet strings as a rhizome or multiplicity are tied not to the supposed will of a puppeteer but to a multiplicity of nerve fibers, which from another puppet in other dimensions are connected to the first. G. Deleuze and F. Guattari (2007) *A Thousand Plateaus: Capitalism and Schizophrenia*, Minneapolis: University of Minnesota Press.

References

Corrington, R. S. (1991) "Divine Nature," in A. Marsoobian, K. Wallace, and R. S. Corrington (eds.) *Nature's Perspectives: Prospects for Ordinal Metaphysics*, Albany: State University of New York Press.

Crosby, D. A. (2013) *The Thou of Nature: Religious Naturalism and Reverence for Sentient Life*, Albany: State University of New York Press.

Cusa, N. and H. L. Bond (1997) *Nicholas of Cusa: Selected Spiritual Writings*, New Jersey: Paulist Press.

Deleuze, G. and F. Guattari (2007) *A Thousand Plateaus: Capitalism and Schizophrenia*, Minneapolis: University of Minnesota Press.

Hayden, P. (2008) "Gilles Deleuze and Naturalism: A Convergence with Ecological Theory and Politics," in B. Herzogenrath (ed.) *An [Un]Likely Alliance: Thinking Environment[s] with Deleuze and Guttari*, Newcastle: Cambridge Scholars Publishing.

Healy, J. and N. Fandos (2016) "Protesters Gain Victory in Fight Over Dakota Access Oil Pipeline," *The New York Times*, December. 4.

Keller, C. (2003) *Face of the Deep: A Theology of Becoming*, London: Routledge.

—— (2015) *Cloud of the Impossible: Negative Theology and Planetary Entanglement*, New York: Columbia University Press.

Lao Tsu (1985) *Tao Te Ching*, Gia-Fu Feng and J. English (eds.), New York: Vintage Books.

Oh, J. S. (2011) *A Postcolonial Theology of Life: Planetarity East and West*, Upland: Sopher Press.

—— (2015) "Nature's Spontaneity and Intentionality," in L. Niemoczynski and N. T. Nguyen (eds.) *A Philosophy of Sacred Nature: Prospects for Ecstatic Naturalism*, Lanham, MD: Lexington Books.

Further reading

Girardot, N. J., J. Miller, and L. Xiaogan (eds.) *Daoism and Ecology: Ways Within a Cosmic Landscape*, Cambridge, MA: Harvard University Press.

20
A CHRISTIAN RELIGIOUS NATURALISM

Karl E. Peters

For more than two decades the contemporary movement of religious naturalism has been growing. Developed by such authors as Ursula Goodenough (2000), Charley Hardwick (1996), Jerome A. Stone (1992, 2008), Don Crosby (2002, 2008, 2015), Karl E. Peters (2002), Chet Raymo (2008), Loyal Rue (2012), Carol Wayne White (2016), and other members of the Religious Naturalist Association (2017), today's religious naturalism is on solid intellectual footing. One of the things that remain to be done is to explore how established world religions might be understood and practiced naturalistically. This Handbook offers work relating naturalism to Buddhism, Confucianism, Daoism, Shawnee religion, and Judaism. However, many assume that Christianity is not very amenable to being expressed in a naturalistic framework. From my experience with liberal or progressive Christians, I have concluded that many would be open to a naturalistic understanding of a Christian life-way if only they had a coherent presentation of ideas that were both naturalistic and offered helpful understandings and guidance as what it means to follow Jesus. This essay is an attempt to provide that understanding.

There are a few theologians who might be considered Christian naturalists. Jerome Stone gives a fine account of major figures that include both the idea of theism and a focus on the significance of Jesus (Stone 2011). Three of these are Henry Nelson Wieman (1964), Charley Hardwick (1996), and Gordon Kaufman (2004, 2006). I am indebted to their thinking, and some of their ideas will be reflected in this essay.

However, rather than discussing the work of others, this essay will construct the beginnings of a possible systematic theology. In what follows I will try to show that Christian thought and practice can be developed with the following naturalistic assumptions: first, reality is the space-time world in which we live; second, knowing is through empirical methods—scientific, everyday sense experience, and historical critical research; and third, the contemporary findings of science are relevant to how we might think and live. With these assumptions we will examine how three foci of Christianity can be understood. First, God and the world can be conceived pragmatically as two ways of viewing the same reality, as creative interactions and as what is created. Second, humans and salvation can be understood in terms of dynamic systems that are mal-functioning becoming well-functioning. Third, Jesus can be portrayed as a non-violent revolutionary, an exemplar for social and individual transformation. Each focus will be introduced with a sketch of changing understandings in Western history about the world and God, human beings and salvation, and Jesus. The recognition of historically changing understandings provides a context for exploring a Christian naturalism.

Naturalistic theism—changing understandings of God and the world

Naturalistic theism is part of a major shift in Western thought—from thinking of everything as individual substances to thinking of everything as events or processes. This is a shift into dynamic relational thinking, namely, that everything is constituted by relationships and in turn becomes part of constituting others. While there are stable relational structures or systems that maintain their integrity for a long time, there also are relational or system dynamics that allow for the creation of new stable systems.

In *A God that Could Be Real* Nancy Abrams presents the historically changing relationship between cosmology and theology in the Western world (Abrams 2016: 7–17). The biblical cosmology of a flat earth, with waters under and above the earth and a dome of heaven beyond the waters above, fits well with the notion of a personal God—a patriarchal father-king who resides in the highest heaven, rules over the world, influences the course of events, sometimes with the aid of messengers (angels) and, finally, a divine-human son. Similarly, the Ptolemaic cosmology of the Middle Ages, with the earth as the center of the universe surrounded by spheres of planets and stars with possible movement up and down through them, allows for a personal divine ruler to direct the course of events. However, for Isaac Newton, the Enlightenment scientific genius of universal gravitational relationships in the sun-centered planetary system of Copernicus and Galileo, humans lived on an average planet circling the sun, which was an average star. No place is different from any other in the Newtonian cosmos; space goes on, perhaps forever; and there is no physical location for God.

It was not until the twentieth century—not even 100 years ago—that scientists were able to determine that the universe had to be thought of as "infinitely" more vast and wondrous than had ever been conceived in Western thought. Edwin Hubble used the Doppler Effect to reason that the entire universe was expanding, leading to the notion of an originating point of singularity metaphorically labeled the "Big Bang." Calculating backwards from the present, scientists have been able to create a scientifically grounded story of the "epic of creation." Around 13.8 billion years ago a point of singularity rapidly inflated; then it began to expand and cool, allowing the creation of hydrogen, helium, and a small amount of lithium. From the "Big Bang" the universe grew to at least 100 billion galaxies each with an average of at least 100 billion stars. Countless stars were born, and in nuclear fusion they created more helium, and carbon, neon, oxygen, silicon, and other elements up to iron. Then, depending on the mass of the star, when it uses up all its fuel, a star can collapse to a dwarf star, or explode as a nova or supernova. In these massive explosions all the remaining elements are created and ejected into space. In space, free floating hydrogen combines with oxygen to form water, and free-floating silicon combines with oxygen to form sand. Out of this "stardust" of elements and molecules more generations of stars were formed, until 4.6 billion years ago our solar system was born. Elements and simple molecules from space created our earth. On earth, organic life emerged from non-life, followed by sentient life, intelligent life, and about 200,000 years ago our own species, *Homo sapiens*. As our species evolved, so did tool making, agriculture, social structures, cultures, religions, and most recently the sciences, which are providing the ideas we now are considering—including this sketch of the creation of the cosmos based on twentieth-century science.

Then, beginning in the 1980s and '90s this science-based construction of the universe expanded in a way that makes the above picture almost unbelievable (Abrams and Primack 2011: 39–66). First, the 100 billion galaxies, each with an average of 100 billion stars, making up visible matter is only about 5 percent of the entire universe. The discovery of dark matter, evidenced by measuring gravitational attractions more prevalent than can be accounted for by the visible universe, is another 25 percent. The remaining 70 percent is thought to be dark energy, perhaps a property of space itself, which is accelerating the expansion of the universe ever more rapidly. In the distant future an observer from our own solar system would only see our own galaxy and its satellites along with the Andromeda galaxy. In fact, an observer from any galaxy would see only their own and nearby galaxies. According to astrophysicist Joel Primack, from any observational viewpoint the universe would only be as "large" as a few galaxies. The increasing rate of universe expansion would make it impossible for light and other forms of radiation to travel the ever-increasing distance from any particular galaxy to make its observation possible. Primack says that we live in a special time when we can scientifically understand the vast, amazing universe we inhabit (Primack, personal communication).

What does this twentieth and twenty-first century cosmology do for theology? Simply put, it creates a huge "housing problem for God." A God beyond an ever more rapidly expanding universe is no longer conceivable. Neither is God within the universe—if God is conceived of as some kind of being, force, energy, or spiritual reality that exists alongside the physical world. The current scientific story of creation, and its physicalism makes it impossible to locate God as a distinct reality within or separate from the world as known by today's science.

Yet, in a dynamic relational understanding of things, it may be possible to think of God and World as two ways of looking at the same thing. Wieman (1964) distinguishes between creative good or creativity and that which is created, created good. In line with this, we can say that creative good signifies interactions among subatomic particles, atoms, molecules, and organisms that over time give rise to the emergence of new molecules and life forms or created good. Creative good also signifies cultural interactions among people and existing social structures and patterns that lead to the emergence of new social institutions, ways of living, values, religious ideas and practices, and scientific understandings. These in turn are created goods.

However, creativity and its products are not separate. They are the same thing viewed in two different ways. As soon as stable forms (atomic, chemical, organic, social, and conceptual) are created, these interact with other created forms, and the interactions give rise to more new created forms. Created good becomes part of ongoing creative good. In a dynamic relational view of things, World in all its relatively stable manifestations becomes God the creative source of all things. And God is the World in its creative interacting.

If God and World are really the same reality, why bother making the distinction at all? The distinction makes a practical (pragmatic) difference in how we live. When we consider the World, created good, we have the tendency to want to preserve what has been created. When we consider God, creative good, we are open to and even welcome the possibilities of as yet unknown, new, emerging good. For example, we live in a frightening era of climate change, largely caused by human activity. If we focus on what has already been created and hope to conserve the rich and diverse world in which we live, we will be seriously troubled by the destruction of habitats, the increasing rate of species extinction, and the increasing social turmoil as people migrate away from rising sea levels and from draught spreading across land masses around the world. We will try to preserve cities as they are inundated by flooding, develop more irrigation systems for crops in drought stricken areas, and generally try to maintain our lives the way they have been. Losing created good is indeed significant—a global tragedy. On the other hand, if we focus on creative good, we are reminded of how previous planetary changes with mass extinctions led to the proliferation and flourishing of new species. Sixty-six million years ago, when the dinosaurs went extinct as the result of an asteroid colliding with the earth, a small, rat-like mammal, the tree shrew, began to flourish and evolve, eventually leading to our own species. Of course, creativity does not "promise" the same good as before. Human living may be completely transformed. Something else may become dominant on the planet. But new good will emerge. Creative interactions will continue to work, bringing about things we cannot foresee. It may even be a new form of ecological civilization in which humans and other species live in a more loving, just, and peaceful world. We really cannot know. Our hope lies not in maintaining the status quo but being open to new, yet unforeseen, good.

If God and World are two ways of looking at the same reality, then we can say that as the world evolves, so does God. As created good becomes part of further creativity, that creativity exhibits features related to the world at a particular time. An implication of this is that, before *Homo sapiens* evolved, the creativity of the world—God—was non-personal. Further, in the continuing physical-chemical-biological aspects of the current world—on planet earth—it remains non-personal (Kaufman 2006: 52; Peters 2013: 584). Only in humans does the full range and power of personality formerly associated with a personal God emerge. This means that before human persons, God is not intentional and does not know consequences of actions. Thus, God is not responsible for evil—evil itself being something constructed by and relevant only to humans. This eliminates the problem of theodicy.

However, once creativity has produced humans as created good that can intentionally and knowingly participate in creativity, then it is possible to assign value to what creativity has done in the history of the universe and what it is doing now. Concepts of good and evil make sense when creativity evolves to its human form. So do concepts of salvation and of what humans must be saved from.

Changing conceptions of humans and the world

If one asks how earlier thinkers in Western thought have approached the idea of what a human being is, two things are apparent. First is the idea of substance. All things including humans are substances, which means they are not composed of internal relations but are one kind of thing that is externally related to others. The second is that the model of a human being is the adult human, especially the adult male. There seems to have been little sense of developmental change from child to adult. In Western art before the Enlightenment, children were usually depicted as miniature adults, if they were depicted at all. There was no concept of childhood (Aries 1962; Clarke 2004). Furthermore, in the two creation stories of Genesis, humans are created as adults.

There are, however, some significant examples of thinking about human individuals as systems in ancient philosophy and Christian thought: Plato's threefold division of the person into reason, will, and appetites (Plato 2016); Aristotle's idea of virtues and vices in relation to a golden mean between the extremes (Aristotle 2012); Paul's self-understanding of being at war with himself, not doing the good he seeks but instead doing the evil he does not want to do (Romans 7:14–15, *HarperCollins Study Bible* 2003); Augustine's internal struggle between two wills inside himself that were in conflict, "one carnal, one spiritual" (Augustine 1998, Book 8); and Paul's image of the body of Christ with many parts each supporting the whole or at odds with each other in a Christian community (1 Corinthians 12:12–31). Of course, these are adult human systems.

Beginning in the nineteenth century the understanding of humans became much more complex. First, the rise of evolutionary thinking led to the idea that every living thing, including a human being, is an evolved, dynamic system. Second, twentieth-century psychology has come to see humans as internally complex—from the thinking of Freud and Jung to a contemporary view that humans consist of interacting feelings, emotions, behaviors, and thoughts, analogous to how members of a family interact. An individual human is an internal family with subsystems or sub-personalities. All have a positive role, especially when led by a calm, creative, compassionate core center of experience that Richard Schwartz calls "Self" (Schwartz 2001). Further, family systems theory, developed by Murray Bowen, suggests that individuals cannot be understood in isolation from one another. Families are systems of interconnected and interdependent individuals, none of whom can be understood in isolation from the system (Bowen 1978). Third, late twentieth-century technological developments in understanding the human genome, body chemistry, and the human brain have led to the idea that a human individual is an evolved complex, dynamic system of cells, organs, chemicals, and nerve fibers that carry information about interacting with the wider social and natural environment. Each human is a complex system that develops from fetus, to infant, to teenager, to young adult, to middle-aged adult, and through old age.

In the context of cosmic evolution we can say that each of us is a complex created good, constituted of energy from the big bang, atoms and molecules created in exploding stars, self-replicating molecules found in all of life on earth, DNA lineages derived from our living ancestors, brains that have evolved from reptiles and mammals, and beliefs and practices that are legacies from the cultures in which we live. Also in an evolutionary context, each of us is a participant within the wider evolved creativity of the universe that is taking place on our planet. In Darwinian terms there are new variations occurring in all parts of the human system, from chromosomes and genes, cells throughout the body, patterns of activity in our complex brains (Calvin 1990, 2014). As we grow through our life cycles from fetus to old age, interactions with others (and within ourselves) shape and reshape our behavior and the kind of person we are.

Melvin Konner has developed a multi-causal framework for understanding the complexity of human behavior. He suggests that there are nine kinds of causes arranged in three types. First, there are remote or evolutionary causes: (1) the phylogenetic constraints because the organism is of a particular type; (2) ecological/demographic causes resulting from an organism being adapted to a particular environment; and (3) resulting from the first two causes, the individual's genome that falls within a certain spectrum of variation for its species. Second, there are intermediate or developmental causes: (4) embryonic/maturational processes guided by the genome throughout life; (5) environmental effects in critical or sensitive periods of development; (6) and ongoing environmental effects, such as stress, trauma, and various kinds of social reinforcement that operate throughout life. Finally, there are more immediate causes: (7) longer-term physiological causes, such as hormones, that are outcomes of gene expression in response to environmental contingencies and that operate for minutes or days; (8) short-term physiological effects such as neural circuits and transmitters that operate from milliseconds to minutes and are the immediate internal causes of behavior; and (9) the immediate external causes that are elicitors or releasers that precipitate the behavior (Konner 2010: 28–9; see also Peters 2008: 681, 686–97, where, following Konner 2002: 175, 234, I apply these to murderous rage).

Space does not permit the detailed unpacking of Konner's multi-causal process. However, two things illustrate its significance. They are effects taking place during sensitive periods of human maturation. First, because the evolved pelvic structure of the human female results in a narrow birth canal, a human being just born does not have a fully developed brain. This contributes to a period of child dependency on parents and other adults. Studies show that if a parent or family, including the infant, is undergoing considerable, chronic stress, and if there is a lack of adult nurturing during the first three years of life, a child's brain does not develop properly.

> Science tells us that young children who experience significantly limited caregiver responsiveness may sustain a range of adverse physical and mental consequences that actually produce more widespread developmental impairments than overt physical abuse. These can include cognitive delays, stunting of physical growth, impairments in executive function and self-regulation skills, and disruptions of the body's stress response.
>
> (Center for the Developing Child, 2017)

This leads to further health and behavioral problems as the child grows older. Second, there is a critical time of development during the "teen-age" years, from 13 through mid-twenties. The last part of the brain to mature is the frontal cortex, which is the brain's executive subsystem that enables teens and young adults to foresee consequences of their actions and rationally control them. The not yet complete development of the frontal cortex shows up in erratic and risky behavior that is sometimes harmful to the person and to others. Excess party drinking, experimenting with drugs, risky relationships, and reckless driving—all partly result from the brain immaturity of the teen–young adult brain as it interacts with others in social systems (Jensen and Nutt 2015).

Naturalistic salvation

In this context we can ask, what does Christianity do? An answer is that it offers "salvation." What can this mean from a naturalistic perspective? Salvation cannot be from hell to heaven or even from some state of suffering in this world to peace and love in "the next." Christian naturalism affirms that salvation must be here and now in this world.

One root of the idea of salvation is the Latin word *salvus*. According to the *Latin Dictionary* (2017), *salvus* means well, unharmed, safe, alive, and sound. It means to heal or make whole. In this sense, salvation is "well-being." However, the idea of well-being is too static. In a dynamic relational naturalistic understanding, we can think of salvation as well-functioning. Ideally, human beings would be systems that are functioning well, within other well-functioning systems—families, organizations, societies, ecosystems, and planet earth. This is an extension of an Aristotelean understanding of "happiness." Happiness is not mere pleasure. Rather, "happiness seems rather to be a satisfaction that arises from functioning well, fulfilling our capacities." If we ask what the well-functioning of human beings is, we can say:

> we are *bodies* in need of good nourishment; we are centers of *sensation*, of appetites and passions, in need of good experience; and we are *minds* in need of good thinking. When all of these dimensions are functioning well, then we are functioning well as humans.
>
> (Ferré 2001: 31)

Christianity complements this understanding with its wisdom about ideal ends and life-guiding principles and values. Ideal ends can be symbolized by metaphors such as the peaceable kingdom where "the wolf lives with the lamb" (Isaiah 11:6) and a "new heaven and earth" (Revelation 21:1) where God is present among humans. Guiding principles and values are expressed in the Ten Commandments (Exodus 20:1–17), Micah's "what does the Lord require of you but to do justice and to love kindness and to walk humbly with your God?" (Micah 6:8), and "love your neighbor as yourself" (Mark 12:31).

These ideals and values do not always square with the ways humans live. How can it happen that we often are not functioning well? What leads to not functioning well? Generally, it is when some part of a system takes over other parts creating what Walter Wink calls a "domination system" (Wink 2003). For an individual person, what takes over may be a part of the body that is diseased. This could be the result of invasion of other organisms evoking the complex immune system, a "mis-take" in the DNA replication of body cells that is not corrected and leads to runaway growth of particular types of cells in cancer, or a disease such as sickle cell anemia that is part of the human's evolutionary legacy and that may be adaptive in some circumstances but not others (Nesse and Williams 2012). For a family, environmental circumstances of stressful poverty and isolation from outside support may contribute to child neglect and abuse, especially in single-parent homes. For business systems, a financial crisis may lead the Board of Directors to hire what appears to be a strong leader. This leader promises to take care of everything but turns out to be a psychopath—with no empathy but getting joy out of manipulating others, and seeking his or her own aggrandizement at the expense of the business and its employees (Babiak and Hare 2009). For social-political systems the leadership may become authoritarian and dominate the system at the expense of many of its people. Imperial Rome at the time of Jesus, we shall see, was such a domination system, which under Caesar tightly controlled the far-flung areas of the Roman Empire, including Judea (Aslan 2014).

When systems are not functioning, whether a human body, a family, a local organization, a city, a country, or a world civilization, we can say it is divided, conflicted, and malfunctioning. In Plato's *Republic* the system is no longer "just." We can say the same thing in our contemporary world: racial, sexist, and ageist are all terms that point to systems not functioning well. Because of discrimination, some do not have a voice and consequently cannot contribute their unique talents and share the benefits of the system. In Christian terms we might say the system is "fallen," "missing the mark," or in a state of "sin." Systems that are extremely out of balance and not functioning well are the antithesis of what William James describes as "conversion":

> To be converted, to be regenerated, to receive grace, to experience religion, to gain assurance, are so many phrases which denote the process, gradual or sudden, by which a self hitherto divided, and consciously wrong inferior and unhappy, becomes unified and consciously right superior and happy, in consequence of its firmer hold upon religious realities. This at least is what conversion signifies in general terms, whether or not we believe that a direct divine operation is needed to bring such a moral change about.
>
> (James 1997: 160)

Salvation from a naturalistic perspective is the transformation of a system of any kind and complexity to function well again in a new form of living. The conditions that lead to malfunctioning may still be present, but through creative transformation a system may come into a new kind of wholeness. For example, a person suffering a terminal illness such as cancer might undergo a transformation so that he or she can still function well. Even though physically impaired, he or she may engage in new meaningful social relationships in a caring community. As my first wife was dying of cancer, new ways of sharing and growing in love between ourselves and with others emerged (Peters 2002: 113–118).

To become relatively well-functioning, however, one must be open to the future. In Hardwick's terms it is living by grace rather than in a state of sin. In Wieman's terms it is not trying to preserve created good that can no longer be preserved but living in creativity that opens up new possibilities for good even when a system can no longer be maintained as it was. From a Christian perspective it is engaging not the man Jesus—who like all humans was a created good—but rather being open to the "Christ event"—the creativity working in the new interrelationships that Jesus catalyzed as he interacted with his followers.

Jesus as a way of salvation in naturalistic Christianity

Through the centuries of Western thought there have been many ways of symbolizing the meaning of Jesus. Historian Jaroslav Pelikan presents many metaphors of Jesus that show his significance for the Christian community in the changing, wider cultural context (1985). Some of these metaphors are biblically based: "Rabbi" in relation to Judaism at the time of Jesus; "Light to the Gentiles" as Christianity spreads beyond the boundaries of Palestine; "King of Kings" in the context of the Roman Empire; the "Son of Man" who revealed both the promise of human life and the power of evil in fifth century Christian psychology and anthropology; "Christ Crucified" of the Middle Ages when the suffering of Jesus on the cross became the primary image of salvation in Christianity (see also Parker and Brock 2008: 223ff.); and the "Prince of Peace" in the resurgence of pacifism among sixteenth-century Anabaptists. Other metaphors reflect philosophical developments in the culture and the response of Christian theologians: the Assisi to transform the way of Christian living and the institutional Church. Still others are rooted in the culture itself: the "Universal Man" of the Renaissance, the "Mirror of Eternal Truth" in the Protestant Reformation, the "Teacher of Common Sense Morals" during the Enlightenment, the "Poet of the Spirit" in the nineteenth-century Romantic Movement, and the "Liberator" in the social gospel and human rights movements of the last two centuries. One might add Dietrich Bonhoeffer's "the Man for Others" in the context of Nazi Germany (Bonhoeffer 1972: 382; see also Beck 2010). Note how these metaphors have become more naturalistic in the last four centuries since the science-based Enlightenment.

Another way of seeing a changing understanding of Jesus is suggested by Gordon Kaufman, who developed the idea of historical trajectories. There are countless historical trajectories emerging as a result of creative interactions—cosmic, biological, and human-cultural. Out of particular trajectories, new trajectories emerge. Examples are the emergence of our solar system from an earlier generation supernova and the creation of new species through Darwinian variation and selection. Likewise, human history contains many cultural trajectories. Each can give rise to further trajectories, such as Christianity emerging from Judaism in the context of the Greco-Roman world. Within Christianity, Kaufman writes of two different historical trajectories stemming from Jesus. One trajectory leads to a supernatural understanding of Jesus. "Jesus trajectory$_1$" begins with his baptism by John the Baptist, his teachings of the coming Kingdom of God, and his crucifixion. Next there is among his followers the emergence of the belief that God has raised him from the dead and that he is the Son of God who is bringing God's Kingdom. The third conceptual step in this trajectory is emergence of the conviction that Jesus ascended to heaven and that in his life on earth he was the incarnation of God. Finally, Jesus trajectory$_1$ culminates in the Church's development of the doctrine of the Trinity, in which the Son (Jesus Christ) has equal divine status with God Father and the Holy Spirit (Kaufman 2006: 11). This trajectory assumes a dualism of the natural and supernatural, so that the person Jesus becomes divinized and understood to be fully God and fully human. As fully divine and fully human, "the God-Man," he is the mediator between human beings and God.

In the context of this dualistic concept of mediator, Jesus' death was understood as the sacrifice for the sins of the world, the substitutionary theory of atonement. Beginning with the Jewish notion of animal and plant sacrifice as a way of sharing and maintaining a relationship with God, the crucifixion of Jesus became understood as a way of restoring the relationship between sinful human beings and the Creator. In the eleventh century this view was developed by Anselm of Canterbury into a formal argument for the "substitutionary theory of atonement." God was the Divine King to whom humans owed *complete* obedience. Humans inevitably disobeyed God. Because they owed everything to God in the first place, they could never compensate for their falling away into a state of sin. Only a "God-Man" could satisfy God's demand for justice and restore the relationship: as a human being Jesus was the one to make the sacrifice; as God, he was capable offering a level of obedience that compensated for all the sins of the world (Platcher 1990: 142–144). Even though few today know about Anselm's argument for "Why God Became Man," the idea of substitutionary atonement, Jesus "dying for my sins," is the most common understanding of salvation for Christians today. It also is the most common understanding for those who reject the supernaturalism of Christianity.

The second Christian trajectory, "Jesus trajectory$_2$" leads to a naturalistic-humanistic understanding of Jesus, represented by Kaufman himself. This trajectory becomes clearer after the rise of historical, biblical scholarship in the nineteenth, twentieth, and twenty-first centuries. This scholarship enables a reconstruction of Jesus and his significance for today that is in keeping with the idea of God viewed as the interactions in the world that are creative of new forms of cosmic, biological, and social existence. In Jesus this non-personal creativity takes shape in the form of a personal, normative moral standard. Kaufman writes:

> The reconstruction of Jesus' ministry, leading up to his death as a dangerous rabble-rouser in Roman Palestine, is plausible historically and presents us with a Jesus in many respects still quite attractive: his forthright challenge to the conventional religion of his time; his forceful preaching punctuated with striking parables; his beautiful vision of the coming kingdom of God in which the sick are healed and the poor are cared for, and the outcast and despised are welcomed to the dinner table; his radical emphasis on love as the overarching posture within which humans should live their lives – love of God, love of neighbor, indeed love of enemies; his unwavering conviction that he must not respond violently against those who were forcing upon him the bitter death of crucifixion; his profound hope that God was bringing in a New Age.
>
> (Kaufman 2006: 21)

Kaufman's depiction of trajectory$_2$ illustrates a second theory of atonement that is part of the history of Christian thought—Peter Abelard's twelfth century "moral exemplar" theory (Platcher 1990: 144–145). In this theory Jesus' life and death is an expression of God's love—an inspiring example of how a Christian should live in everyday life. Consistent with this theory is a view of Jesus as a non-violent revolutionary. Those who are non-violent revolutionaries (as was Mohandas Gandhi and Martin Luther King) neither passively accept unjust systems or engage in violence to overthrow them. Instead they lead a non-violent protest as Jesus did against the Roman Empire. According to New Testament scholar Marcus Borg, Jesus was a "nonviolent revolutionary," challenging the "domination system" of his day: the Roman Empire that had been accommodated by some Jewish leaders (Borg 2006). This idea is illustrated by Jesus proclaiming that the Kingdom of God was occurring among the people he was with, eating with and healing social outcasts, and telling stories that encouraged people to look at themselves and society in new ways—for example, the parables of the Good Samaritan and the Prodigal Son. It also is illustrated by his developing what New Testament scholar Walter Wink calls "Jesus's third way"—a path between passively submitting or violently responding to evil (Wink 2003). This is the path of nonviolent resistance in the face of unjust systems of domination.

Jesus was an exemplar of this third way during his final week in Jerusalem. The Palm Sunday procession, the ejection of the money changers out of the Temple, the debates with Jewish leaders—all are actions of a nonviolent revolutionary protesting the domination system of his day on behalf of the poor and oppressed. The opening act of this week is the Palm Sunday procession. This was not the only procession at that time. To guard against things getting out of hand among the Jews, the Roman governor Pilate

> rode into the city from ... the west, at the head of ... imperial cavalry and foot soldiers arriving to reinforce the garrison on the Temple Mount. They did so each year at Passover, coming to Jerusalem from Maritima, the city on the Mediterranean coast from which the Roman governor administered Judea and Samaria.

Jesus came into Jerusalem in another procession from the east. The biblical texts tell us that this was not accidental. It was a procession that Jesus planned. According to Borg,

> his decision to enter the city as he did was what we could call a planned political demonstration, a counter demonstration. The juxtaposition of these two processions embodies the central conflict of Jesus's last week: the kingdom of God or the kingdom of imperial domination ... two visions of life on earth.
>
> (Borg 2006: 232)

This type of living offers a way of overcoming oppression in all areas of life, in all kinds of dynamic systems: family, churches, communities, and nations, so that the well-functioning of all can be restored. It is even a way of dealing with conflicted parts or subsystems within an individual. Instead of simply acquiescing to out-of-control inner desires or angrily rejecting them, one can calmly, compassionately, and courageously connect with them and enable them to be creatively transformed so that the entire person functions well again. With the "moral exemplar" theory of atonement, one is inspired by Jesus' example (and by that of others like him) to live a "Christ-like" life.

This process involves a transformation into a new way of living. Central to Christianity is the transformation known as the resurrection of Jesus. How can the resurrection be understood naturalistically? Wieman distinguishes between the man Jesus and the creative interactions Jesus had with his followers. It was not the man Jesus who was divine; the man Jesus was human, a created good. However, in the relationships Jesus catalyzed with and among his followers, an open, undiscriminating, creative love was present. When Jesus was crucified by the Roman imperial domination system that was crucifying all kinds of people it saw as a political threat, this creative love was liberated. What rose from the dead was not the man Jesus but the creativity of the "Christ event" in the midst of his followers as a new community of love began to be formed. Creative interaction became present in history in a new way. Jesus's interchange with his disciples so transformed them that they became capable of having such interchange with one another (Wieman 1964: 39–40.)

Christian naturalism shows a way in which we can reshape our thinking about God and the world, and about human beings and salvation in a this-worldly context of science. Christian naturalism also shows us a way to live—following Jesus as an exemplar and living creatively in the "Christ event" of unconditional love and justice in all our social, economic, and environmental relationships.

References

Abrams, N. E. (2016) *A God That Could Be Real: Spirituality, Science, and the Future of Our Planet*, Boston: Beacon Press.

Abrams, N. E. and J. R. Primack (2011) *The New Universe and the Human Future*, New Haven, CT: Yale University Press.

Aries, P. (1962) *Centuries of Childhood: A Social History of Family Life*, R. Baldick, trans., New York: Alfred A. Knopf.

Aristotle (2012) *Aristotle's Nicomachean Ethics*, R. C. Bartlett and S. D. Collins, trans., Chicago: University of Chicago Press.

Aslan, R. (2014) *Zealot: The Life and Times of Jesus of Nazareth*, New York: Random House.

Augustine (1998) *Confessions*, H. Chadwick, trans., New York: Oxford University Press.

Babiak, P. and R. D. Hare (2009) *Snakes in Suits: When Psychopaths Go to Work*, New York: HarperCollins.

Beck, R. (2010) "Letters from Cell 92: Part 6, 'The Man for Others,'" *Experimental Theology*. Dec. 19. http://experimentaltheology.blogspot.com/2010/12/letters.

Bonhoeffer, D. (1972) *Letters and Papers from Prison*, Eberhard Bethge ed., New York: Macmillan.

Borg, M. (2006) *Jesus: Uncovering the Life, Teachings, and Relevance of a Religious Revolutionary*, New York: HarperOne.

Bowen, M. (1978) *Family Therapy in Clinical Practice*, Northvale, NH: Jason Aronson Inc.

Calvin, W. (1990) *The Cerebral Symphony: Seashore Reflections on the Structure of Consciousness*, New York: Bantam Books.

Calvin, W. (2014) *How Brains Think: Evolving Intelligence Then and Now (Science Masters)*, New York: Basic Books.

Center for the Developing Child (2017) http://developingchild.harvard.edu/science/deep-dives/neglect/, accessed 2/15/17.

Clarke, J. (2004) "Histories of Childhood," in D. Wyse ed. *Childhood Studies: An Introduction*, Malden, MA: Blackwell, 3–12.

Crosby, D. A. (2002) *A Religion of Nature*, Albany, NY: State University of New York Press.

Crosby, D. A. (2008) *Living with Ambiguity: Religious Naturalism and the Menace of Evil*, Albany, NY: State University of New York Press.

Crosby, D. A. (2015) *Nature as Sacred Ground: A Metaphysics for Religious Naturalism*, Albany, NY: State University of New York Press.

Ferré, F. (2001) *Living and Value*, Albany, NY: State University of New York Press.

Goodenough, U. (2000) *Sacred Depths of Nature*, New York: Oxford University Press.

Hardwick, C. (1996) *Events of Grace: Naturalism, Existentialism, and Theology*, Cambridge: Cambridge University Press.

HarperCollins Study Bible (1993) New Revised Standard Version, New York: HarperCollins.

James, W. (1997) *Varieties of Religious Experience*, New York: Simon & Schuster, Touchstone.

Jensen F. E. and A. E. Nutt (2015) *The Teenage Brain: A Neuroscientist's Survival Guide to Raising Adolescents and Young Adults*, New York: Harper.

Kaufman, G. (2004) *In the Beginning—Creativity*, Minneapolis, MN: Augsburg Fortress Publishers.

Kaufman, G. (2006) *Jesus and Creativity*, Minneapolis, MN: Fortress Press.

Konnor, M. (2002) *The Tangled Wing: Biological Constraints on the Human Spirit*, New York: Henry Holt.

Konnor, M. (2010) *The Evolution of Childhood: Relationships, Emotion, Mind*, Cambridge, MA: The Belknap Press of Harvard University Press.

Latin Dictionary (2017) www.latin-dictionary.org/latin/english/meaning/salvus, accessed 2/15/17.

Nesse, R. M. and G. C. Williams (2012) *Why We Get Sick: the New Science of Darwinian Medicine*, New York: Vintage Books.

Parker, R. A. and R. N. Brock (2009) *Saving Paradise: How Christianity Traded Love of This World for Crucifixion and Empire*, Boston, MA: Beacon Press.

Pelikan, J. (1985) *Jesus Through the Centuries: His Place in the History of Culture*, New Haven, CT: Yale University Press.

Peters, K. E. (2002) *Dancing with the Sacred: Evolution, Ecology, and God*, Harrisburg, PA: Trinity Press International.

Peters, K. E. (2006) "*Spiritual Transformation and Healing in Light of an Evolutionary Theology*," in *Spiritual Transformation and Healing: Anthropological, Religious, Medical and Biological Perspectives*, Joan Koss-Chioino and Philip Hefner, eds. Lanham, MD: Altamira Press.

Peters, K E. (2008) "Understanding and Responding to Human Evil: A Multicausal Approach," *Zygon: Journal of Religion and Science*, 43/3 (September): 681–704.
Peters, K. E. (2013) "A Christian Naturalism: Developing the Thinking of Gordon Kaufman," *Zygon: Journal of Religion and Science*, 48/3 (September): 578–591.
Platcher, W. C. (1990) *A History of Christian Theology: An Introduction*, Philadelphia, PA: Westminster Press.
Plato (2016) *The Republic of Plato*, A. Bloom, trans., New York: Basic Books.
Raymo, C. (2008) *When God Is Gone, Everything Is Holy: The Making of a Religious Naturalist*, Notre Dame, IN: Ave Maria Press, Sorin Books.
Religious Naturalist Association (2017) http://religious-naturalist-association.org, accessed 2/13/17.
Rue, L. (2012) *Nature is Enough: Religious Naturalism and the Meaning of Life*, Albany, NY: State University of New York Press.
Schwartz, R. C. (2001) *Introduction to the Internal Family Systems Model*, Oak Park, IL: Trailheads Publications.
Stone, J. A. (1992) *The Minimalist Vision of Transcendence: A Naturalist Philosophy of Religion*, Albany, NY: State University of New York Press.
Stone, J. A. (2008) *Religious Naturalism Today: The Rebirth of a Forgotten Alternative*, Albany, NY: State University of New York Press.
Stone, J. A. (2011) "Is a 'Christian Naturalism' Possible? Exploring the Boundaries of a Tradition," *American Journal of Theology & Philosophy*, 32/3 (September 2011): 205–220.
White, C. W. (2016) *Black Lives and Sacred Humanity: Toward an African American Religious Naturalism*, New York: Fordham University Press.
Wieman, H. N. (1964) *The Source of Human Good*, Carbondale, IL: Southern University Press.
Wink, W. (2003) *Jesus and Non-Violence: A Third Way*, Minneapolis, MN: Fortress Press.

21
RELIGIOUS NATURALISM
Hindu perspectives

Varadaraja V. Raman

Introduction

Religions are organized frameworks that have emerged in most human communities and cultures. Theism, which is part of many religions, affirms a personified transcendent principle to account for the emergence and existence of the world of experience. That supreme principle is referred to as *God* in English and in equivalent terms in other languages. Theism endows God with every positive human capacity and quality imaginable, such as power, knowledge, love, and mercy. These are all attributes of God, each a million-fold to their asymptotic perfection. Not many human beings have claimed to have seen God in a personal way, and fewer still have reported they have interacted or conversed with Him. Yet the idea of God is wide-spread, and firm belief in His existence has been a powerful force in humanity's cultural history.

It is difficult to trace the origins of the God concept to any precise way. But it is possible to trace the periods and places relating to the origins of some of the major religions. What we do know is that most religions emerged from the sayings of charismatic elders and shamans, prophets and sages. Equally importantly, religions have endured because multitudes have continued to have implicit faith in the idea that the truths preached by the founders were revealed to them from a mysterious higher source. More exactly, the initiators of religions are reported to have had visions of God or of His messengers, and came to know about the hereafter from sources that are not within reach of the average person.

The scriptures of religions explore and expound on the nature of God and His many qualities. He is lauded and worshiped by groups and communities who gather together periodically for this purpose. In this context many texts emerged: sacred writings, canonical prayers, moving psalms, and music for singing: all of which serve in the canonical services of religions. These are generally conducted in sanctified centers whose appearances range from modest to magnificent. Though not always explicitly viewed as such, the aesthetic aspects of religion have contributed immensely to religious experience.

All of this must be borne in mind in founding any new religious movement, because it has proven to be effective in enticing members into the fold and in propagating the message of the religion.

Initially, nearly all religions were located within regional boundaries circumscribed by language and ethnic commonalty: Norse religion in Scandinavian countries, Greek religion in the

Hellenistic world, Isis worship in Egypt, Judaism in Judea, Confucianism in China, Shinto in Japan, Hinduism, Buddhism, Jainism, and Sikhism in India, Islam in the Arabian peninsula, and so on. In due course geographical confinements melted way. Today multitudes belonging to various religions have spread worldwide. Their competition for adherents in new locales give rise to conflicts and confrontations.

In the modern age, a new religion need not be anchored to a specific people or nation. It can become global right from the start. For example, even if religious naturalism (RN) does not have millions of members right away, it can already have members from all over the world. Another major advantage of RN is that most major traditional religions have a naturalistic component. What this means is that members of all faiths can subscribe to important tenets of RN even if they do not move away fully from the religion to which they are currently affiliated.

The localized and spread-out aspects of God

All through the ages people have raised questions about the basic assumptions of religions. Assuming God's existence, a question in earth-bound minds relates to where that God is located. Religions often give a double answer to this question: God is omnipresent, i.e., everywhere or God is in Heaven, a far removed location beyond the blue sky. This answer is not unlike the physicist's view of the electron that it is at one point in its space-time, localized particle aspect, but it is also everywhere in its spread-out wave aspect. Thus, both the microcosmic electron that can be seen through the eye and instruments of physics and the macrocosmic God recognized in the vision of the religious person through the eye of faith have this dual presence: one local (spatially confined) and the other unbounded (spatially expansive).

Using the analogy with quantum entities, one way of looking at RN is to focus on the Divine not as an anthropomorphic individual (the particle aspect of the electron) but as universally spread out (wave aspect). The second aspect is the manifestation in the natural world: not just in the terrestrial land and water, plants and trees, birds and beasts, but also beyond earth and air, in stars and planets, galaxies and matter, palpable and dark. So the awakened religious naturalist does not need church or synagogue, temple or mosque to feel God's presence, but recognizes Divinity here and there, everywhere. This idea is expressed in a verse in the *Bhagavad Gita* (6.30) where the voice of the Divine says:

> Who sees me everywhere,
> And sees everything in me:
> To such a one I am never lost,
> Nor is he lost to me.
>
> (Radhakrishnan 2010: 204)

Here the Gita does not say that God is everywhere, but refers to those who see the Divine everywhere. The first is a claim about the existence of God. The second is an affirmation that when one sees a grander principle manifest in every perceived entity, that is what seeing God is all about.

Seeing God everywhere does not mean that one sees Krishna playing on his flute under every tree or standing near every cow. Rather it means that one admires, appreciates, and recognizes all Creation as a manifestation of the Divine. Whether it is a rosebud or a rotten tomato, a mountain or a microbe, a stone or a star: one feels every aspect of nature as reflecting a touch of the Divine. This perspective certainly resonates with that of RN.

The Bhagavad Gita also says that God is never lost to such a person. If one experiences directly or conceptually that everything one sees is made up of atoms and molecules, then those entities will never be lost to that person. However, if one is aware of them only in textbooks and classrooms, then it would be like people seeing God only in places of worship or scriptures. Finally, the person who sees God in all of nature is never lost to God, because that person is always a conscious part of the natural world.

Experiential dimension and divine representation

There is more to religions than dogmas and doctrine. Aside from their assertions about the nature of God, religious frameworks have rich experiential aspects. In this respect religions are not like the sciences: collections of knowledge and information about the world. They are rather like art and poetry, music and dance: not just things to see and know about but, more importantly, to feel and enjoy. They endow the participant with a sense of fulfillment that comes from contemplation of the cosmic presence or entity that is personified, pictured, or abstractly conceived as the God of the tradition. Prayer and worship modes are bridges by which the individual attempts to connect with the Cosmic Whole and merge with that totality. This merger arises from the recognition that there is more to existence than eating, drinking, sleeping, and procreating. Every religion is an urge to move beyond physical existence. The coloring of the mind of the devout is the spiritual experience that one seeks in places consecrated to the Divine.

Some religions speak of God in abstract terms, and some use iconic representations of God. Others, like Buddhism and Jainism, say nothing at all about a personal God. But the description of God in all theistic religions is invariably anthropic: He is loving and kind, just and merciful, powerful and knowing, and so on. In the Abrahamic religions God is essentially masculine. In others, like Greek and Roman, ancient Egyptian and Hindu, God is also symbolized as a feminine principle.

The concept of an all-powerful God probably arose from a need to find a cause for the wonder of rain and thunder, the splendor of fruits and flowers, the majesty of mountains and oceans, and the grandeur of the universe with stars and galaxies, unending space, and much more: in short, as an enormously versatile Author of nature and natural phenomena. All religions look upon the natural world as the Creation of that omnipotent God. Religions also believe that one of God's goals in bringing about this wondrous Creation is to make it enjoyable for human beings. That is why the human being is reckoned as the pinnacle of Creation. That is also why religions worship God: a mark of gratitude for the experience of life. Without reckoning a source, it is difficult to say "Thank you!"

Identification of God with nature

There is nothing morally objectionable in affirming a personified God as the ultimate cause of everything. To the extent that such an affirmation does not hurt fellow humans and is not forcibly imposed, it may be even appreciated as one regards works of art and poetry. Nevertheless, over the centuries many have questioned the notion of a God of any kind. However, instead of rejecting the God(s) whom billions of people continue to worship, one may extend the notion of God.

There are a number of reasons why thoughtful people may wish to do this. First, the idea of a personal God is an unverifiable hypothesis at best in the scientific meaning of verification. Though theologians have offered proofs for God, most of these have been questioned by non-members of the group. These proofs have strengthened a belief that is already there, and seldom convinced unbelievers. When a theologian writes about God, only the God of one's own religion is in his mind. Then again, logical consistency in a system of thought is no guarantee that it reflects an aspect of reality. Not one of the logically consistent theorems of plane geometry that are impeccably demonstrated is valid in the Non-Euclidean space in which we live.

Next, belief in a traditionally prescribed God often leads to inner contradictions. The most glaring of these is reconciliation of an omnipotent and all-merciful God with the suffering and pain humanity experiences daily. The existence of evil in a God-created world, of disease and despair in the realm of the all-merciful, is not unlike seeing filth and squalor in the living quarters of a billionaire. The problem of theodicy has not been solved in a rationally acceptable way in any traditional religious framework. One religion attributes it to the original sin, another to the law of karma, a third to God's arbitrary decision to make the world such as it is; none of these is totally satisfactory. It is extremely difficult to reconcile the idea of a personal God with that fact that the capacity for cruelty and unreason seems to be an intrinsic part of what is alleged to be the finest life-form created by God in His image in the Biblical tradition, or an aspect of whom is ever-present in every human being, per the Upanishadic vision. One would expect more rational and less hurtful specimens from the omnipotent creator than the ones we see.

Then again, unlike the blue sky or the red planet, the personified portraits of God are invariably culture-dependent. Though individuals—usually the founders of religions—have claimed to have seen God or heard His messengers, the God has generally been the One of the particular tradition to whom the mystic or the prophet has been exposed. God appears invariably in an indoctrinated garb. Thus, the Jewish prophets had visions of YHWH, rather than one of Shiva; St. Francis of Assisi had visions of Jesus, rather than of Amaterasu-ò-Mi-Kami, Bernadette of Lourdes saw Virgin Mary and not Lakshmi, Meera Bai's vision was of Lord Krishna and not of Norse God Odin, the Prophet of Islam encountered the archangel Jabril of the Abrahamic tradition rather than Narada, the Hindu messenger-sage, and so on. Celestials invariably appear with the attributes presented in specific religious frameworks.

Finally, the God idea in various religions has led to some values that are no longer acceptable. These include the notion of a chosen people, of upper and lower castes, of God appearing as the son of a virgin, of superior (believing) and inferior (unbelieving) people. Most of all, every religion claims to be the only path to salvation. The emphatic proclamation of there being only one God is theologically known as monotheism. It should more correctly be called *moutheism*: there is only *my* God: YHWH, Vishnu, Shiva, Jesus, Allah, and so forth. It is a fact of history that moutheism has provoked countless sectarian persecutions, denominational intolerance, iconoclastic vandalism, and mindless wars, some of which have resulted in cultural genocide. Even with centuries of modern science and enlightenment values, ancient and medieval belligerence towards members of other religions has not been altogether eliminated.

It is important to note that none of this is proof against the existence of an unfathomable God, belief in which is likely to be always there in human culture. It is safe to have it, as long as God is declared unfathomable by all religions and deserving of only local cultural images and names for personal and group fulfillment, and not for global imposition. More importantly, in the interest of harmony and mutual respect, adherents to all religions would do well to acknowledge the difficulties in accepting the time-honored notions of regional and culturally constrained gods.

That vast numbers of people all over the world are looking for some religious affiliation or other cannot be denied. The currently available faith systems have many positive aspects and are deeply fulfilling to many. However, they strike many sensitive and intelligent people as conceptually inadequate. More seriously, their practice is wrought with a good many glaring moral derelictions. In this context, humanity's goal should be not so much to reject time-honored religions and a revered God, nor to demolish beautiful traditions that have enriched human culture, but to explore ways in which religious fulfillment can be achieved in more enlightened, meaningful, and non-hurtful ways.

Theoretical and practical side of religions

There is a huge chasm between the analytic ivory-tower thinkers who dissect and analyze every idea of God and look for coherence and consistency in the conclusions derived from it and down-to-earth religious practitioners who kneel and pray, meditate and sing hymns and psalms, and silently appeal to celestial powers in times of dire want or danger. Indeed the relevance of traditional religions and gods is seen in times of collective jubilation and of individual sorrow; on days of feasts and festivals, and in times of helplessness and bereavement.

Skeptical thinkers and anti-religious writers point out that traditional religions have instigated people to irrational worldviews, unconscionable behaviors, and anachronistic value-systems. This may all be very true. However, it is no less important to remember that traditional religions have also enabled human beings to rise to levels and modes of experience that are enriching and uplifting. They have prompted millions to caring and compassion, charity and service. It must also be emphasized that many elements of the positive potential of religions have arisen from religious *attitudes* rather than religious *beliefs*, which are invariably linked to the name and nature of the God one takes to be the only one there is. The point not always realized by religious practitioners is that the positive values and behaviors arising from religions can also be brought into play without all the doctrines about God listed in sacred texts.

Religious naturalism

The question then arises: Is it possible to arouse and foster the positive religious values and experiences without subscribing to traditional gods? In the view of many, the answer to this is *yes*. Furthermore, considering the many disastrous side-effects of traditional religious frameworks, it is worthwhile propagating a worldview that incorporates the best religious attitudes without the unhappy and often unavoidable consequences inherent in the narrow visions of traditional religions.

One theoretical aspect of this approach is the following: One adopts the view that nature is not the creation, but the manifestation of an abstract Cosmos. The abstract Cosmos is like the theoretical pre-Big Bang Void which erupted into space and time, matter and energy. We do not have the faintest idea of what that pre-Big Bang non-entity is or was, except in our imagination as a concept of total nothingness: the *shúnyata* in Buddhist terminology. It may be regarded as the equivalent of the Chaos of ancient Greek thought, nirgunabrahman of Hindu cosmology, or as the God of traditional religions: after all, it is omnipotent, powerful beyond imagination in its stupendous manifestation, and omnipresent in that it pervades the length, breadth, and depth of the universe. In other words, one may simply substitute the God of religions with nature: her inexorable laws and the many manifestations and transformations that result from those laws. In cultural historical terms, ancient thinkers symbolized the abstraction of the indescribable cosmic entity as the Divine. This was certainly what happened in the (Hindu) Vedic framework.

After recognizing nature as the all-encompassing totality of all that exists, both perceived and as yet undetected, we transfer our feelings of respect and reverence for the abstract God of traditional religions to nature, which we see all around and on which we depend for our very existence. This is one perspective on the framework of religious naturalism.

This approach is not different from delighting in, honoring, and celebrating great works of art for all that they offer to the human experience. We can be emotionally jolted by a play and moved deeply by a poem without inquiring into the private life of the poet or the playwright. We can be deeply touched by a symphony or a raga, a sculpture or a painting without knowing about the composer or the artist. It is no less meaningful to pay homage to the work as to be arguing about the biography of the author. A book of maxims is far more relevant and valuable than the date and deeds of whoever put it together. Moreover, given that there is no inkling from observed data that there even was a sentient anthropoid Being who consciously brought about all the order and chaos, events, and mishaps—from terrestrial hardships and harmony to stellar supernovas and galactic recessions—this approach is more reasonable and less fraught with potential for sectarian hurt.

Thus, the essence of RN lies in the transference of focus from an abstract anthropoid Creator to the palpable splendor of nature, which is at once simple and complex, minute and mammoth, stable and fleeting. Whether the perceived Creation implies an unperceived Creator may or may not be an interesting question, but it is often irrelevant to leading a decent, meaningful, and useful life. Indeed, though in theory the question may seem to be of only metaphysical interest, whether or not there can be religious peace and harmony among the upholders of religious beliefs depends to a large extent on how one answers it.

The Hindu framework

As in all religions, the Hindu world also has many layers, dimensions, and responses to the world of experience. The Hindu framework has a strong spiritual/mystical undercurrent that seeks oneness with the Cosmic Whole. It envisions Divinity in multiple forms and with multiple names, directing the world at large. Like all religions it has its magical dimension too. Again, as elsewhere, the Hindu world has many sacred writings that have been given varying interpretations by different sages and scholars over the centuries. Divergences in interpretations lead to sectarian divisions, but they also keep cultures alive and vibrant.

What interests us here is that there are several passages in sacred Indic writings, as well as in the writings of the poets and commentators of the tradition that bear the stamp of the religious naturalistic approach. This suggests that over the ages many thinkers perceived the world as religious naturalists. But since they were/are formed by worldviews and thought currents that were/are predominantly mystical, magical, and spiritual, even the deepest feelings of religious naturalism were often presented through a lens colored by supernatural imageries.

One may detect this tendency in other ancient visions as well. The choice for the modern mind is between abandoning altogether traditional views of humanity and the world and embracing the more recent (in human history) scientific-materialistic worldview; or seeing in utterances of earlier religious schemes ideas and insights, explicit or subtle, that resonate with the naturalistic framework. The latter approach not only enables the traditional practitioner to better appreciate the world through the eyes of RN, but also encourages the religious naturalist to regard traditional religions with the respect they deserve. It would seem that in the changing world in which we live, this is a far more appropriate and realistic way to go.

This is especially possible with many Vedic hymns, the ultimate sources of Hindu religious formulations. The sage-poets who composed the Vedas lived in the midst of nature, surrounded by lakes and rivers, mountains and meadows, sunshine and dawn, twilight and moonlight, the sky and the stars. They experienced the rising and setting of the sun, rain-showers and thunder, whistling winds and stormy weather. They witnessed all this from the core of their being even as they meditated on the unfathomable mystery whence all this originates. To them, if there was anything in the world worthy of reverence and worship, it was nature in all its richness, majesty, and splendor.

Nature can be seen and felt, provoking awe and admiration, but it cannot be extolled in the abstract. One cannot pay homage to the sun if it is seen merely as a red and burning stone far away, a ball of fire, or even as a nuclear furnace. Such descriptions may be appropriate for static and indifferent neuter nouns, but they fall short of poetry, and poetry is what makes the human spirit transform a catalogue of facts into lived experience. As a poet who grew up in the Lake District, William Wordsworth was touched by nature as Vedic rishis were. One cannot write a love letter to a daffodil, but if we personify it, as Robert Herrick did, we can speak to fair daffodils and tell them that we weep to see them fade away so soon. That is precisely what Vedic sage-poets did in order to convey their exuberance at the splendor of nature. Poetry demands that raw and apparently mindless elements be personified so they may be addressed in one way or another. List the forces and principles of nature as inert and inanimate entities, and you can produce an impressive anthology of things in nature. Display the locations and configurations of stars and planets and you will get a map of the sky. But you need to feel the celestial glitters as entities that can hear you in order to sing "Twinkle, twinkle little star!"

So it was with the awakened rishis of yore. Their visions made them see every aspect of nature as a living entity. By shifting to this mode, nature is adorned with the charming and colorful robe of poetry. This capacity breathes life into naturalism and transforms it into religious naturalism. Non-poetic naturalism may be satisfying as long as we are in the data-collection and explanatory enterprises. But poetic sensitivity brings us to a different level where nature throbs with communing beauty and becomes fulfilling for the aesthetic experiencer.

So the poets were inspired to invoke the earth as a son would speak with reverence to his mother in a simple phrase: *máta bhúmih putroham pritivyah*: "Earth the Mother, I am the son of that earth." There is a promise to preserve and protect the snow-clad mountains and the dark-hued soil. The phrase reads like an allegiance that ecologists would make today. The point to note is that naturalism here is not an affirmation that there is nothing beyond nature, but more importantly the sanctification of nature, the attitude to nature that is touched by reverence, which makes it religious naturalism.

In the Brahmanas, often known for their ritualism, there are speculative ideas on nature. However, as one eminent scholar on the subject put it, "the ritualistic tendency actually swallowed up what little philosophy the later parts of the Vedic hymns were trying to express, but there are unmistakable marks that this tendency existed and worked" (Dasgupta 1922: 292). Here we may recall Kaegi's description of Agni (Fire), which "lies concealed in the softer wood, as in a chamber, until, called forth by the rubbing in the early morning hour, he suddenly springs forth in gleaming brightness" (Kaegi 1886: 85).

Invocations to river and water

So it is that Vedic poets addressed the earth as mother and described themselves as her progeny: *Mátá bhúmi putro aham prithivyah.* Another hymn extols the river Sarasvati as mother: *Ambitame, naditame, devitame, Sarasvati.* This is why the confluence of streams into a single mighty river reminds the poet of calves rushing to their maternal milk cow. Here is how a personified Ganga (the venerated river in India) is described in the Mahabharata:

> [T]he daughter of the snowy mountains ... came down; her whirlpools were raging, and she was teeming with fishes and sharks.... She directed her course towards the sea, separated herself into three streams, and her water was bestrewn with piles of froth, which looked like so many rows of white ganders. And crooked and tortuous in the movement of her body, at places; and at others, stumbling as it were, covered with foam as with a robe, she went forward like a drunken woman. Elsewhere by virtue of the roar of her waters, she uttered loud sounds.
>
> (Ganguli 1981: II, 235)

Or again, we read in the Atharva Veda: "May the water that flows down from the snow clad mountains bring you happiness. May the water that flows in the rivers bring you happiness. May the swift flowing streams bring you happiness. May the water of the monsoon bring you happiness" (Griffith 1899, Kanda 19).

> Here is a long invocation to water in the Rig Veda:
> Waters! You are health giving,
> Give us energy, so that
> We may look on great delight.
> Give us a portion of the sap, the most
> Beneficent you have,
> Like mothers longing with love.
> So gladly do we go with you
> To the home for which you make us live.
>
> (Griffith 1896: Bk 10, CXXIX-3)

Invocation to dawn

Dawn is described in Vedic poetry as another goddess who is invoked in about 20 hymns. In the words of one scholar,

> The personification is but slight, the physical phenomenon always being present to the mind of the poet. Decked in gay attire like a dancer, clothed in light, she appears in the east and unveils her charms. Rising resplendent as from a bath she comes with light, driving away the darkness and removing the black robe of night. She is young, being born again and again, though ancient. Shining with a uniform hue, she wastes away the life of mortals. She illumines the ends of the sky when she awakes; she opens the gates of heaven; her radiant beams appear like herds of cattle. She drives away evil dreams, evil spirits, and the hated darkness. She discloses the treasures concealed by darkness, and distributes them bountifully. She awakens every living being to motion. When Ushas shines forth, the birds, fly up from their nests and men seek nourishment. Day by day appearing at the appointed place, she never infringes the ordinance of nature and of the gods. She renders good service to the gods by awakening all worshippers and causing the sacrificial fires to be kindled. She brings the gods to drink the Soma

draught. She is borne on a shining car, drawn by ruddy steeds or kine, which probably represent the red rays of morning.

(Griffith 1896: Bk 6, LXV)

Invocation to herbs

Of all the wonders in nature, perhaps the most important for humans are plants and trees, grains and vegetables. Every religious naturalist would acknowledge their presence on the planet, for it is they that sustain all life forms here below by serving as the channels through which solar energy feeds into bodies that reproduce and propagate. We find in Vedic poetry eloquent praise of herbs. The sage-poet refers to herbs as having sprouted even before the gods: herbs that sprang up in time of old, three ages earlier than the gods. Then he goes on to wonder how even the notion of God can come in the human mind without herbs. So the *Hymn to Herbs* addresses herbs as mothers, for they are ultimately the progenitors of all life. The hymn refers to the healing power of plants and refers to the countless ways in which they arise. It appeals to the medicinal powers of herbs. One who has stored herbs is like a king in a crowd, it says. He is a physician, because with the herbs he can cure diseases. So he is called a fiend-layer and chaser of disease (Griffith 1896: Bk 10, XCVII).

What is to be noted in all these references is that an important aspect of nature is made the equivalent of God, and this is an important dimension of RN: the shift to nature what traditional religions described as God. Herbs are life-giving, powerful, and medicinal. They sustain life here on earth. So they are declared to be sacred and worthy of reverence, and to be deserving of address and worship in poetic meters and metaphors.

Mini-nature

There is more to nature than trees and rivers, the sky and the stars. Deserts and prairies are grand arenas of nature. But the religious naturalist is stirred as much by the little islets of nature as by its holistic sweeps. Consider for example this short reflection in *Periyapuranam* by an ancient (eleventh to twelfth century CE) Tamil poet musing on the harmony in nature:

In the shimmering light of the ashen waters,
In that region of sugar-canes, full,
Within the banks of the juice-filled field
Constrained by ramps is the plowing bull.
To bear its tiny progeny
A crab from the mud moves and climbs
Over a furrow by plowshares torn.
A lovely lotus exudes honey
For the little ones that are born.

(Chekkizhar 2010)

This poem is about a scene in a sugar-cane field where furrows are made by plowshares drawn by an ox. In that marshy land the poet sees a mother crab slowly finding her way to a safe spot where she gives birth to her little ones. When this happens, a lotus plant gives out some nectar on which the newborn creatures feed. Only the keen eye of one who has great sensitivity for all creatures great and small could make this careful observation, and only a gifted poet could express it in rhythmic verse that unfortunately loses its original charm in translation. Aside from the beautiful description of an insignificant episode in nature, the poet is trying to say here that all nature is in harmony and that in the village he is describing, everyone cared for one another.

What makes this utterly worldly verse remarkable is that it occurs in a sacred work in the Tamil world. Known as *Periya Purânam*, this is a compendium of the lives of 63 Tamil saints of the Saiva tradition. The point to note is that in these lines we see RN not as a philosophy but in action.

The philosophical aspect of nature: prakriti

The Sanskrit term for nature is *prakriti*. The associated concept is quite complex. It has a simple meaning as well as sophisticated metaphysical connotations. The simple meaning is that which is the source of everything that comes into being. Thus in Hindu thought the notion of nature includes both what is observed or perceived and that which is the source of all that is perceived; in other words, individual entities as well as a conceptual totality from which everything emerged.

Prakriti is always dynamic. It is always in transformation. This is similar to the current understanding of physics by which the universe is an arena of incessant energy transformations. Every phenomenon involves changes in energy forms: potential to kinetic, heat to light, light to motion, and so on. In a sense prakriti is manifested energy. But the key idea is that nature per se is an empty notion when regarded as an entity that has no observable existence. The idea may be interesting to speculate upon, like a universe before the Big Bang or galaxies that have receded away beyond recognition for good, but is of no practical interest.

Prakriti becomes meaningful, relevant, and useful only when it is mapped on to human consciousness. This is not a metaphysical mumbo-jumbo but a simple incontrovertible fact, as true as the fact that the furniture in the bedroom is, for all practical purposes, irrelevant and non-existent for the person who is sound asleep on the bed. In other words, meaningfully associated with prakriti is an experiencing principle. This experiencing principle is human consciousness, which grasps prakriti. This is not to say there is no prakriti without consciousness, but rather to remind us that for prakriti to become meaningful or relevant it needs to be perceived. Or else, it would be like a play on the stage in an empty theater, or an encyclopedia without a reader.

Two levels of mapping prakriti

Consider some aspect of prakriti, say electromagnetic waves. When these impinge on the human retina they are first registered in the neuronal realm in a mode that is different from the oscillating electric and magnetic fields. If the waves are within the visible spectrum the mapping results in the sensation of sight. Some of the waves are transformed into the colors of the visible spectrum. We may call this first order or purely physical mapping. This occurs in the brains of many animals.

Now there is another level at which the mapping is transformed in highly evolved brains like that of the human. This is the level of consciousness. On this plane, a second-order mapping occurs. Here the sensation of light and colors leads to aesthetic experiences and joy. It is important to distinguish this from the purely physical sensations, usually described as *qualia*. How consciousness arises and what is its source are interesting questions that have occupied philosophers, metaphysicians, neuroscientists, and psychologists for a long time. But irrespective of the theories to explain consciousness, there remains the empirical fact that consciousness induces a level of experience that transcends mere qualia. Seeing the sunset is not the same as exclaiming at its beauty; smelling the rose is not the same as delighting in its fragrance.

Naturalism linked to atheism

It is no secret that naturalism is often associated with atheism. Though this is not the general case in the Hindu world, there are certain schools of naturalistic philosophy that explicitly deny the existence of God. The most famous of these is associated with the name of Cārvāka, a personage about whom very little is known but whose writings are referred to by various historically-known philosophers. There is a historical fiction on his life, entitled *The Curse of Charvaka*. His followers came to be known as Cārvākas. According to them, only the physical world existed, and there was nothing transcendent. This world was made up of the four basic classical elements: Earth, Water, Fire, and Air. They rejected the notion of ākāśa (Greek: aether) as a fifth element because it was said to be intangible like soul and ethereal beings.

Cārvākas maintained that there was no such thing as a vital principle. Like Julien Offray de La Mettrie of the eighteenth century who wrote his famous *L'Homme Machine* (Thomson 2011), they believed that the basic elements that give rise to various material substances in the world also generate the phenomena of life and consciousness. In other words, everything we see and become aware of, including awareness itself, is a product of the elements combined in various proportions.

An associated thesis of Cārvākas was that every substance formed from the elements has its characteristic property, which they referred to as its *svabbhāva*. This is similar to the quantum mechanical view that every quantum system has its own characteristic properties: its Eigen Functions as they are called. Even the elements have their svabháva: thus the svabháva of water is cold and that of fire is hot.

An unusual feature of Cārvāka naturalism is that it rules out causality. Contrary to received wisdom by which every occurrence has a cause, according to this system, acausal events can occur. At first blush this might seem unscientific. But we do know that even in the physical world, in phenomena such as radioactive decay and the generation of virtual particles, statistical and random factors play a more fundamental role than strict causality. Moreover, in the arena of thought-generation and human events, we simply do not have a strictly cause-effect structure. Thus, the Cārvākas may be called irreligious naturalists, for they were naturalists while being explicitly anti-religious.

Concluding thoughts

Over the ages many thinkers in all cultures have rejected the idea of a Supernature, let alone a supernatural being reigning from up there. In spite of their efforts to eradicate such beliefs, and notwithstanding the many unhealthy consequences of fervent beliefs when carried to an extreme, religions continue to persist with undiminished vigor among billions of people. This is a significant and intriguing phenomenon in human culture. One explanation for this fact could be that religions attend to some very important needs in being fully human. Some of these needs have little to do with logical consistency or impact on others.

The goal of RN should be to offer to humanity an opportunity for religious experience that is pan-human in content, transcending cultural and local affiliations, and yet respectful of some of the basic religious needs, the most important of which is the admission that we are finite beings in a world that spans infinity. RN does well in its respect and reverence for the phenomenal world of which we are all apparently particular parts. But, in reality, we are woven together in one seamless web.

In the tug between indoctrinated faith and reasoned worldview, success for the skeptics is not assured if RN decries and marginalizes every aspect of traditional religions. Religious naturalism can play an important role if, instead of rejecting religious views altogether, it tries to see in the sacred writings of the tradition's hints for regarding nature as the principle that deserves to be respected and revered. Unlike invisible, intangible, and abstract gods, nature is very visible and palpable. It is as grand as anything one can imagine, as full of splendor and wonder as the gods of religions. It stretches from the minute microcosm to the magnificent macrocosm. In every scripture of humankind one can find references to nature. When these references are interpreted in the framework of science and humanity, they can be of enormous appeal to most people. However, this approach needs to be supplemented by commonly agreed upon rites and rituals that are meaningful, humanistic, and periodic

In this spirit that I conclude with a prayer for Peace in the Yajur Veda (12) that is chanted every day in the Hindu world:

> Aum dyauḥ śānti-rantarikṣaṁ
> śāntiḥ-pṛthivī śānti-rāpaḥ śānti-roṣadhayaḥ
> śāntiḥ-vanaspatayaḥ śānti-rviśvedevāḥ
> śānti-rbrahma śāntiḥ
> sarvaṁ śāntiḥ śāntireva śāntiḥ
> sā mā śānti-redhi
> Aum śāntiḥ, śāntiḥ, śāntiḥ
>
> (Griffith 1899, 36:17)

> May there be
> Peace in the whole sky and in ethereal space;
> Peace on earth, Peace in the waters and in the herbs;
> Peace in trees and in creepers.
> Peace in the whole universe.
> May peace be in the Supreme Being.
>
> (My translation)

References

Chekkizhar (2010) In V. V. Raman, trans. *Periya Puranam*. www.siddha.com.my/forum/ubbthreads.php?ubb=showflat&Number=120).

Dasgupta, S. (1922) *A History of Indian Philosophy*, Vol I, 292, Cambridge: Cambridge University Press.

Ganguli, K. M. (ed.) (1981) *The Mahabharata of Krishna-Dwaipayana Vyasa*, New Delhi: Munshiram Manoharlal Publishers.

Griffith, R. T. H. (1896) *The Hymns of the Rig Veda, Translated with a Popular Commentary*, Benares: H. J. Lazarus.

Griffith, R. T. H. (1899) *The Texts of the White Yajur Veda*, Benares: H. J. Lazarus.

Kaegi, A. (1886) *The Rig Veda: The Oldest Literature of the Indians*, Boston: Ginn and Co.

Radhakrishnan, S. (ed.) (2010) *The Bhagavadgita*, India: Harper Collins Publishers of India.

Thompson, A. (2011) *Machine Man and Other Writings*, Cambridge: Cambridge University Press.

22
NATURALIZING RELIGION[1]

Loyal Rue

An ancient and provocative Chinese expression of well-wishing suggests that one is fortunate to live in interesting times. What makes this expression provocative is that it leaves us on our own to work out what counts as "interesting." Susan Sontag once remarked that the only persons whom she found interesting were those who were engaged in a process of self-transformation. Historians would readily agree with Sontag, although their embrace of the principle might take the collective form: the most "interesting times" in history are those involving dramatic cultural transformations. William James, too, had things to say about the interest factor: the most interesting and important thing about a person is his or her worldview.

Our own period in history is nothing if not interesting, and for my money the most interesting thing about it is the momentous transformation we are now experiencing—both individually and collectively—in the ways we picture the world and ourselves. To say that a period in history is interesting is not, however, to say it is comforting. Far from it. Momentous transformations of worldviews can be challenging, unsettling, tumultuous, even terrifying. But they can also be promising, enlightening, revitalizing and liberating.

The essays assembled in this volume call our attention to an unfolding transformation in the domain of religion. The term *religious naturalism* will strike many as an oxymoron because in common parlance *religious* normally entails the metaphysics of supernaturalism, while *naturalism* often connotes the anti-religious perspective of atheism. But religious naturalists will insist that genuinely committed naturalists can honestly admit to having religious sensibilities. We should expect to see a fair amount of diversity among naturalists in the ways these religious sensibilities find expression. Some individuals come to a religious naturalist view as refugees from traditional religious orientations. Others come as seekers who are drawn by the promise of finding new layers of meaning in their experience of the natural world. Our back-stories are diverse, and it is inevitable that our struggles to explore and express new meanings will reflect past experiences, sometimes in surprising ways. How could the efforts of these refugees and seekers to find a more satisfying worldview fail to be interesting?

I fancy myself as both a refugee and a seeker. As a refugee, I took flight from the world of Lutheranism. I was raised in the Midwest by third-generation Norwegian immigrants. My parents both came from farming families but left the farm to take advantage of educational opportunities. My father became a Lutheran pastor, and my mother was trained in a teacher's college and, later, in a Bible institute. Our family was a marinade of Lutheran piety. My father died in his early forties, leaving my mother to raise four young children—I was five at the time—with meager means. Fierce determination, hard work, good luck and unwavering faith enabled my mother to beat the odds against keeping the family intact and well provisioned. The Lutheran Church was her most valued resource, and without its support she would never have managed to carry on from day to day. The religious life was not a weekend outing for our family: it was daily, hourly and deadly serious.

As I reflect on the religious context of my youth I am struck with the same mix of feelings you might have when revisiting your childhood home or your first schoolroom: a deep sense of gratitude tempered by the realization that these things have nothing more to offer. My life was in many ways shaped by Norwegian-American Lutheran pietism, for which I will always be grateful, but I am now as alien to its way of thinking as a refugee can possibly be. In some ways, however, the distance from my religious past may not be as great as it first appears. That is to say, the *substance* of my beliefs and attitudes are light years away from those of my religious heritage, but the *forms* of my seeking will not be unfamiliar to an old-fashioned Lutheran.

As a seeker, I have been in pursuit of all the usual high-minded goals: truth, wisdom, fulfillment, serenity, self-understanding, justice and a meaningful life. Unable to find satisfaction in these pursuits within the tradition of conventional theism, I began to wonder how things would look from a non-theistic, naturalist, perspective. Seeking is less a rigorous discipline than it is a posture of honesty and openness to worlds of experience—one's own and, as far as possible, the experience of others. It is a vague enterprise with no reliable strategies or rules to follow. Seekers stumble blindly ahead with no clear sense of direction. It is only in retrospect that their narratives achieve coherence. My own story of transformation has been one of rethinking, or *naturalizing*, what I take to be the central themes (forms) of conventional theism. In the following pages I will attempt to show how these themes functioned in my previous life as a practicing supernaturalist Christian, and then explain how I now think about them as a religious naturalist.

God and Creation

In my youth I acquired the traditional concept of a personal deity. God the Father was the all-powerful, all-knowing, benevolent, eternal, and invisible creator and sustainer of the world. He (!) was the final explanation for every fact and the final adjudication for every value. God created and loved the world; he was everywhere and knew everything, even my deepest secrets. As a child I was never afraid of God, for how could one fear a Father that kept one safe from all harm? My siblings and I were assured by our mother that God was especially protective toward children who had lost their earthly father, a thought that gave me comfort and confidence. If ever I needed to feel special I would consider that I was on God's list of favorites—and probably quite high on the list, too, in view of the fact that my father had been a man of the cloth. No matter how confusing things might get, I would be safe, my family would be safe, and the entire creation would be safe. All contingencies were covered. I could fit God into the picture to explain anything at all, and I could ask God to arrange things as they might suit my desires. It was exactly like having a magic lamp with an obliging genie in it.

The problem, of course, was that it was pure fantasy, and anyone bothering to give it a few moments of serious thought would know this. But thinking seriously about God can be distressing. For one thing, it can make you feel guilty if you have been taught that having faith in God is the highest of human virtues, and that harboring doubts about God makes you a bad person. And beyond this, having doubts about God creates cognitive dissonance of the most intense sort. Part of you wants the good old God to remain firmly in place, on call whenever needed.

But another part knows that the good old God is a logical monstrosity, believed in for the sake of comfort and convenience.

I find it impossible to pinpoint the time when I first had doubts about God. Most children, I suspect, have their first doubts about God round about the time they begin to have doubts about Santa Claus. This is reasonable, since God and Santa Claus share many similar magical attributes. The critical difference is that doubts about Santa Claus are allowed to run full-course, whereas doubts about God are carefully explained away. While I was growing up much care was given to settling doubts about God, but almost nothing was offered for the sake of saving Santa. Whenever concerns about God arose I would simply ask my mother and she would provide satisfying answers. Inevitably there were times when her answers were incomplete, or when they simply failed to convince me. Much later I came to realize that in answering my God questions my mother had been following the ancient tradition of *apologetic theology*, that is, the practice of explaining how the concept of God is perfectly consistent with whatever it was that created doubts about it.

Eventually the moment arrived when I quit bringing doubts to my mother and took them straight to books about philosophy, theology and psychology. I found academic theology books especially helpful because here I found religious people who rejected the concept of a personal God without apology, replacing it with carefully crafted abstract ways to think about the sacred. Academic theology, as it happened, was a useful resource for relieving me from bouts of cognitive dissonance. Ultimately, however, I concluded that even academic theology was not enough to save God.

I reasoned that a person in my position was left with few options. The first was the apologetic option of loading up ad hoc explanations to explain away my doubts. I entertained this option for a considerable period of time, until it became too burdensome. What could possibly make perpetual ad hoc-ery satisfying in the long run? Eventually apologetic theology begins to look as tortured and tiresome as efforts to save Santa Claus.

Another option would be to abandon the conventional concept of a personal God in favor of something more subtle and abstract—perhaps a concept that emphasizes mystery, or a metaphysical principle that places God safely beyond the scope of inquiry. For example, Rudolf Otto described God as *mysterium tremendum et fascinans*, and Paul Tillich defined God as *the ground of being*. These strategies for saving the concept of God have the appearance of being highly effective, but at the end of the day they disappoint. Here's the problem: if you have a concrete conception of God (e.g., all-powerful, all-knowing, benevolent creator), then it lacks coherence and plausibility, but if you replace it with a high abstraction (e.g., *mysterium tremendum* or *ground of being*), then the concept becomes flaccid and unable to perform the essential work of explanation and justification.

I have finally settled on a third option, which is to relinquish the concept of God altogether. This includes rejecting the strategy of some of my naturalist friends who simply redefine the word *God* as "the creativity of the cosmos."[2] I have no strong objection to this approach, except to say that it is reprehensibly misleading. If the word *God* adds nothing to one's understanding or appreciation of the creativity of the cosmos, then what would be the point of using it?

If this were a theological tribunal I would have to confess that I have become an atheist, although I feel a twinge of discomfort with this label. It's not because I nurture a residue of old-fashioned theism, but rather because there is so much opprobrium attached to the term. In common parlance atheists are grouped together with terrorists, murderers and kidnappers—and this despite the fact that these offenses correlate far more closely with theism than with atheism. I much prefer the term *non-theist* to atheist, but I will gladly confess to atheism if you admit to being an atheist as well. The point here is that everybody is an atheist from someone's point of view. The early Christians were accused of atheism, and Islam accuses all non-Muslims of being heretics and infidels. So there.

Being an atheist does not, however, preclude the possibility of being genuinely religious. Not taking a metaphysical fiction to heart does not mean that I take nothing to heart, or that I deny the sacred. As a religious naturalist, I take nature to heart and affirm the mystery and sanctity of creation. In my own reflections on these matters I have gathered a sense that the theistic distinction between creator and creation introduces an artificial and unattractive gap between the natural and the sacred, with the effect (for me, at least) of diminishing the value of creation. Theism suggests that the creation counts for nothing apart from God—without God matter is mere grunge. As a naturalist, and as a material being, I hope to be excused for taking offense at this outrageous idea.

It remains to say something about the place of worship in connection with the theme of God and Creation. All religious naturalists are united by a reverence for nature, but they are not united by any distinctive patterns for expressing their reverence. That is, there are no universal practices that might describe the worship of all religious naturalists. In this respect religious naturalism is rather like Hinduism. Hindu traditions honor human diversity in the religious life by encouraging individuals to seek paths of piety that suit their own dispositions and temperaments. Intellectuals who are drawn to the contemplative life of study and reflection will practice the disciplines of *jñana yoga*; practical take-charge types who cannot sit still or withdraw from social activity will find the path of *karma yoga* more agreeable; while passionate, sensitive and emotionally expressive individuals will find *bhakti yoga* most satisfying. Religious naturalism honors human diversity in the same way. Thus, some religious naturalists might be found barefoot in a forest clearing, holding hands in a circle and singing praises to unspoiled wilderness. But others may be found alone in urban libraries or laboratories, deepening their understanding and appreciation for some particular feature of the natural world. Yet others will be found expressing their reverence for nature by sitting in endless subcommittee meetings, hammering out resolutions and legislative proposals. Nature, lest we forget, is everything and everywhere, spoiled and unspoiled by human activity. What matters most for the religious life of religious naturalists is imagination and experimentation.

If I were a Hindu I would probably follow the path of contemplation. What I seek, primarily, is the stimulation of wonder and insight, which for me calls for a lot of solitary reflection. I live on the banks of a lovely river, offering me a sort of sanctuary where I occasionally find opportunities for what might be called, very loosely, worshipful meditation. There is nothing resembling ritual discipline in these reflective events: I merely sit and gaze, usually at the flickering flames of a campfire or the rippling patterns of the water. Once I become relaxed my mind spontaneously wanders into the realms of the very large and the very small. I consider the size and the age of the universe, the billions of galaxies drifting aimlessly apart, and the billions of stars in each galaxy. I consider the history of our own star, circled by careless obedient planets, and then the Earth, with its abundant materials continuously recycling themselves. How many Earth-like planets are there in the universe? How unlikely is it to find places like this, capable of generating and supporting life? How many events, what sorts of events, had to take place before the emergence of life became possible? How close can I get to comprehending the fact that trillions of lifeless molecules—each one a universe in itself—are generating the life I now feel? What would the complete history of any particular atom look like? Eventually I am brought back to the middle realm, to my own body and the campfire. Here is an event: a relation between a living system and a pile of logs, both generating heat, both burning up. How did such an event come to be? What if it hadn't? How will it change things?

I have discovered that these reflective occasions always have a renewing, refreshing effect on me. I have discovered that they invariably alter my attitudes and my perspective on the events of the day. And on very rare occasions, when the realms of the very large and the very small are brought together in a fleeting thought, I have discovered that it is possible to feel embraced by nature.

Sin and grace

Sin and grace, sin and grace; sin and grace. Such was the persistent litany of my childhood. At the core of Norwegian Lutheran piety is the unromantic doctrine that humanity is essentially corrupt. Human beings are sinful creatures by their origins: conceived in sin, born into sin and destined to live sinful lives. To grow up Lutheran is to inherit a heavy burden of guilt. I hasten to add that the original sin of humankind is only half the story. The other half is that God in his mercy has bestowed his grace on the faithful, thereby dropping the charges. Human beings are simultaneously guilty and acquitted (*simul justus et peccator*).

After dinner each evening my mother would lead the family in devotions. This amounted to reading a passage from the Bible, followed by an illustrative story, and finally punctuated with a prayer, the length of which was a precise measure of my mother's energy level. The prayers always ended on the same note: *Lord, we are sinful creatures and deserve eternal punishment; we beseech thee to be gracious unto us*. Sin and grace were drilled into us each day. To her credit, my mother had the good sense to emphasize the grace as much as the sin, and she was careful to remind us that we should not take our guilt personally—it was inherited, originally, from Adam and Eve.

Somehow I managed to offload the doctrine of original sin at an early age. This may have been due to conflicting signals coming from church and school. The church was sending the message that all humans are declared guilty even before birth, whereas in school we learned that the United States does not prosecute innocent people, and that all persons are considered innocent until proven guilty. In America Adam and Eve would have to bear the burden of sin on their own. The whole idea of hereditary guilt appeared counter-intuitive to me. To inherit the shape of one's nose or the color of one's hair was sensible enough, but how could someone inherit guilt? I wasn't buying any of it.

Much later in life I came to appreciate that the doctrine of original sin, ensconced as it was in bogus metaphysical terms, expressed an important insight about human nature. To say that human beings are sinful by nature is simply an archaic way of saying that, despite our most heroic efforts, we cannot avoid breaking rules. The reason for this is not that we are corrupt by nature, but rather that our behavior is governed by a plurality of motivational systems that inevitably conflict with one another. We come into the world with physiological drive systems urging us to satisfy hunger and thirst, to keep warm and dry, and so on. Taking nourishment and staying warm are "goods" scripted into our nature. But imagine that you are desperately hungry and the only way to get food is to dash out into the freezing rain to fetch it. In such circumstances it is impossible to achieve both goods simultaneously, which means that one has to be sacrificed for the sake of the other. But this is merely the beginning of complications. In addition to our physiological drive systems we inherit a suite of emotional systems, which also find themselves competing with one another. Suppose that two friends both come to you asking for your last morsel of food. You happen to feel grateful to person A for a recent favor, but person B's hunger is noticeably more acute, leaving you torn between gratitude and sympathy, unable to satisfy both emotional mandates. One good must give way to the other. And if you happen to be hungry yourself then your physiological drive system enters into the competition along with the emotional systems. To be human means to be faced with difficult choices like these on a daily basis, making it virtually impossible to achieve good and avoid evil in every case. We are, by our origins, destined to sin.

Matters become even more complicated as our cultural context indoctrinates us with a set of learned moral values. Now the business of good and evil involves more than the gut reactions of drive systems and emotional systems. The acquisition of a learned morality adds yet another layer, one that requires deliberate judgments over moral principles. With so many competing variables—physiological, emotional, rational—factored into motivation, it becomes a certainty that every human being will be guilty of transgressions on a regular basis.

Once the concept of original sin is naturalized it becomes clear that it represents a sound insight about human nature. But now we must consider the other half of the story: the bit about God's grace. Can the doctrine of grace be naturalized? Yes, of course it can. The Christian doctrine of sin and grace says that we are all judged guilty by God's Law, and consequently we all deserve to be damned. But God in his mercy gives us another chance to lead a righteous life. Grace is construed here as an undeserved gift. There is nothing uniquely Christian about the doctrine of grace—it amounts to an archaic and culturally specific variation on a principle of restitution that is characteristic of every stable social order. It is obvious that every society must establish and enforce a moral order. Failure here would result in anarchy. But the other half of the story is that every society must provide a moral reset button for offenders. Consider for a moment how things would look in a society that fails to offer restitution to moral offenders. Assuming that everyone is inevitably guilty of wrong-doing (as I have argued), it would appear also that virtually everyone would sooner or later become a scornful victim of wrong-doers. Can you think of anyone who has never done wrong or been wronged? Having no provisions for moral restitution would be a recipe for a dysfunctional social order, where everyone is simultaneously burdened with guilt and seething with resentment. Where sin is universal grace is necessary. Obviously, moral offenses vary in gravity, and there will be extreme cases where moral restitution appears counterproductive. But these are exceptions to the general dynamics of blame and forgiveness that characterize every functional social system.

I have completely rejected the doctrines of sin and grace as a function of God's judgment and mercy, but naturalized versions of these doctrines occupy a central place in my understanding of religious naturalism. In practical terms, I would like to think that naturalizing sin and grace might help to make people more understanding and forgiving. When you realize that everyone is wired up in ways that make them vulnerable to moral transgressions on a regular basis, then you may find yourself a bit more inclined to lighten up and cut your neighbor some slack. If we can manage to be more forgiving, then there's a good chance that those in our acquaintance might reciprocate.

There is one further aspect of the doctrine of grace that impresses me as appropriate to religious naturalism. If grace is understood as unwarranted benefit, then it is certainly the case that nature is full of grace. I have done nothing to deserve the air I breathe or the soils and waters that produce the food I eat. I have done nothing to deserve the energy I exploit from the sun, or the ongoing services rendered by the Earth's many decomposers. These, and so much more, come to my benefit as the free gifts of nature, and I would take a low view of myself if ever I failed to be grateful for them, or if I failed to act in ways befitting my gratitude. Thanks be to Nature!

Death and salvation

My siblings and I were forced to come to terms with death at a very early age. The "terms" were that our father had been "called home to glory" by God, and that we would join him in heaven if we remained faithful. It would be the most joyous reunion imaginable. My mother, who remained a widow for the next 22 years, kept the death recent and the reunion imminent with daily references to both events. Death and salvation were not occasional subjects in our home. Nor were they ever separated. In my young mind they were synonymous. I cannot recall ever having the slightest fear of death. What's to fear about the most joyous reunion ever?

Although I never feared death I had lots of questions about it, which my mother took care to answer, always with an air of confidence and anticipation. At some obscured point in time I started to entertain questions about heaven that I sensed would not be welcomed, so I kept them to myself. In particular, there were two insoluble problems that kept returning to me. The first might be called "the dilemma of charitable homicide" and went as follows. Two important principles captured my attention in the course of my religious education. One was that good Christians should be prepared to make sacrifices for the welfare of others, and the other was that eternal life in heaven would be infinitely better than life on Earth. Occasionally, when I combined these principles, I came up with the conclusion that I should murder my brother. Sending my brother to heaven would be the best imaginable thing I could do for him. I would of course sacrifice my own salvation in the process, but the Christian life was all about self-sacrifice anyway. The logic of this argument impressed me as unassailable, yet I resisted the urge to act on it because the conclusion was intensely counter-intuitive.

The other problem concerned the heavenly reunion with my father. We were taught that bodies went to the grave and souls went to heaven. The great reunion, then, would be a reunion of souls. At some point I started to have concerns about how my father and I would recognize each other in heaven. There were also problems about communication—if we didn't have bodies, then we wouldn't have eyes, ears or voices. Even beyond that, there was this problem: if our souls communicated somehow directly, then how could my father distinguish between genuine messages from me and mere hallucinations of such messages?

The concept of heaven was a consistent source of curiosity and confusion throughout my youth. I had always assumed that everyone wanted to live, and if that is true—which seemed obvious to me—then it must follow that everyone would want to live forever in heaven. But at some point I began to question the assumption that eternal life in an obscure supernatural realm was something worth longing for. In order to answer such a question we would need a lot of specific information about the afterlife. Who would opt for a long vacation in a remote place without first gathering information about it? But specific information about the afterlife is something we rarely get, and the little we do get sounds ridiculous (streets of gold? Eighty thousand servants and 72 virgins?). I can recall being told that heaven is desirable because it will be a place where all our desires are fulfilled. This seems to be what many people have vaguely in mind when the subject of heaven comes up. But try to imagine a place where every person always gets exactly what he or she wants. It would be sheer hell! Besides, how long would it take before we got bored with such a hedonic paradise? The point here is that nobody has the faintest idea what they're talking about when they get onto the topic of an afterlife. And this being the case, no one should be obligated to take such talk seriously.

In the course of time and inquiry, my supernaturalist beliefs about death and salvation were displaced entirely by naturalist beliefs. Interestingly, I have remained fearless about death. I am perfectly willing to attribute this fortunate fact to my early religious education, but I am also aware that it has no bearing on the truth of my early beliefs. My present beliefs about death and salvation are typical of naturalistic materialism. Immortal souls, I believe, do not exist, and when the body dies the person (the self) will be annihilated. Just as we did not exist 1,000 years before our birth, so we shall not exist beyond our death.

I have come to view death as a gift, implicit in the process of life itself. It is quite natural to think about death as a problem; indeed, in most cultural traditions death is *the* problem of human existence. But from the perspective of evolutionary wisdom death is not a problem at all; it is a *solution*, and a fairly recent one at that. Prior to the emergence of sexual reproduction there was nothing inevitable about death. It happened all the time, of course, but it did not need to. Many species of organisms need never to die. Single-cell organisms reproduce by simply dividing into halves, each half becoming a distinct individual capable of further subdivision. Death is not a part of the picture, as both halves go on living, enjoying virtual immortality. Death has no sting to an amoeba.

It is only among sexually reproducing multi-cellular organisms that death enters the picture as a certainty. The reason has to do with the divergence of cells into the germ line and various soma lines. The germ line (reproductive cells) produces informed seed for the next generation, while the many soma lines diverge for the production of various body parts. The body negotiates a living in the environment, thus enabling the germ line to do its own job of bringing forth more seed-bearing organisms. The strategy is simple and elegant: the soma lines are instruments of the germ line. Having performed its duty to the germ line, the body becomes redundant and eventually dies, making room for the next generation. But the germ line continues immortally onward in subsequent generations. The death of the body is an essential part of the strategy.

In the wisdom of this scheme it becomes difficult to view death in a negative sense. The inevitability of my death is now held to be a necessary condition for the life I now have—a mere entrance fee to be paid on the way out. If there were no death there would be no soma lines, and without soma lines there would be no possibility of an embodied person—no memories, no loves, no joys, no wonder or wisdom, no longing or learning. These are among the many splendors of the body, and for these we must die. *We must die because we get to live.* To the extent that I cherish my life, therefore, I have reason to be profoundly grateful for my death. When I have occasion to mourn the death of others I will try to absorb the loss in what I have gained from them. I will try to understand my grief as a measure of my gratitude. And when I have occasion to consider the fact of my own death I will attempt to think large. I will try to see that a soma-centered story of the self is a small and impoverished view, and that the life within me was first quickened among the primordial organisms appearing on Earth nearly four billion years ago. I will affirm that all lives, no less my own, are instruments of life itself; and by these measures I will submerge the absurdity of death in gratitude for the wonder and wisdom of life.

In my upbringing the idea of death and salvation were so closely linked that they became synonymous. Salvation was something that happened—or didn't—when a person died. Salvation was also understood to be a discrete, atomic event—it happened one soul at a time. Religious naturalism finds no place for either of these ideas. If the concept of salvation is to have any meaning for religious naturalists it must be something for the living, and it must include social systems as well as individual persons. A naturalized concept of salvation is molecular, not atomic.

Religious naturalism of the sort I envision will understand salvation in terms of safeguarding the life-giving potential of the biosphere. If there are synonyms for salvation they will be sustainability and viability. For humans in particular, salvation will be a matter of achieving personal wholeness and social coherence simultaneously, which is what I meant by the molecular characteristic. We save individuals by enhancing the solidarity of their communities, and we save communities by nurturing and enabling their individual members. To succeed in achieving these twin goals is all the glory we shall ever need.

Faith, hope and love

There is no supernaturalist monopoly on faith, hope and love. Naturalists embody these virtues as well. For naturalists, however, faith cannot mean believing without evidence, hope cannot mean the prospect of overcoming death, and love cannot mean that we are unconditionally loved.

Imagine yourself a teacher with a pair of very persistent students to whom you have assigned a math problem. The first student comes with a wrong answer, and you say, "wrong answer, try again." But the student stubbornly continues to insist that the answer is correct, despite your judgment. The second student also comes with a wrong answer, and you say, "wrong answer, try again." So the student does try again, and again, and yet again, despite previous unsuccessful efforts. Here we are presented with two forms of faith: one continues to believe despite the evidence, and the other continues to inquire despite failures.

The religious tradition of my upbringing, sad to say, made a high virtue of the first form of faith. Martin Luther insisted that "scripture alone" was enough to certify the truth of Christian faith. In other words, the answer was itself the evidence. Luther also hated philosophical inquiry, calling reason "the devil's greatest whore." My mother followed Luther in this teaching, but my siblings and I were quick to notice that it was only in matters of religion that she refused to consider evidence. In every other domain of life she remained a strict evidential rationalist.

Religious naturalism has no place for faith as belief despite evidence. However, it considers the other form of faith—persistence in pursuit of truth—to be a mark of good character. I have come to regard this second form of faith as much more consistent with the deepest teachings of most religious traditions. At their best, all religions are life-affirming. Basically, they all admonish us to persist in living, to "keep on keepin' on," to pick ourselves up, dust ourselves off and start all over again. I see this as the only plausible reading of the Christian story of crucifixion and resurrection: it's not over yet; there's more to come; defeat is not final; it's always darkest before dawn; you can do it. These familiar expressions take us to the heart of faith, to the credo of continuation, the essential algorithm of viability. Faith, for religious naturalism, echoes the slogan for a recent Harvard class reunion: "In spite of everything ... Yes!"

Hope is what happens when one takes to heart the best possible future outcomes. It is important to distinguish sharply between hoping and wishing. Hope is always contingent on real possibilities, whereas wishing is not. You may wish that John McCain had won the election of 2008, but it makes no sense to hope for it. Hope is contingent on what is possible, which means that genuine hope will always be well informed. To long for the continuation of life on Earth beyond the death of our sun is to nurture a false hope. Likewise, you are hoping falsely if you look forward to having thoughts and experiences after the material of your brain has decomposed.

I was raised on a packet of false hopes—that I would live forever, that I would be reunited with my father, that I was special in the eyes of God, and I would be less than honest if I denied that losing these hopes was a painful and confusing process. But suffering and loss are often transformative: they have a way of clearing the air and opening one to the adventure of new insights and the exhilaration of a fresh perspective. Losing faith brings the promise of finding it.

Religious naturalists encourage genuine, informed hope. Like most people, they will hope to leave behind a world that is more just, more beautiful, more viable and better understood than the one they found. They hope that scientific inquiry will continue to advance, for in the process of learning they see heightened possibilities for achieving these goals.

Many religious naturalists in my acquaintance see prospects for hope embedded within the concept of emergence. Emergence theory offers a plausible alternative to the radical extremes of supernatural agency, on one side, and hard-core reductionism on the other. Models of emergence demonstrate that unpredictable and irreducible novelty can occur at different levels in the organization of matter. Changes in material relations may impose unanticipated constraints on systems, enabling new material constructs and new properties of matter to emerge. On the principle that laws of nature amount to formulations of the regularity observed in material properties, we may say with justification that genuinely new laws of nature emerge as new properties of matter emerge. If some naturalists appear to be overly optimistic about the long-term future possibilities for life it may be due to a belief that, in time, some impossible events become improbable events, some improbable events become probable events, and some probable events become certainties. Religious naturalists who are informed by emergence theory know full well that they cannot hope beyond the possibilities, but they can, and do, hope for emergent possibilities. And such hope—even vague, speculative hope—is what sustains faith.

I had the good fortune of growing up with a clear sense that I was loved unconditionally. God loved me and my mother loved me, and I was confident that they would go on loving me no matter what offenses I might commit—and I committed my share! But my mother has passed away, and God has vanished from my imagination, with the result that I no longer feel the sense of being loved unconditionally. But I do not feel diminished in any way by this loss. I know that while my mother lived my siblings and I occupied the center of her universe of meaning, and I know that this is no longer the case. I miss my mother's presence, but somehow I don't miss being the cherished center of anyone's universe.

I strongly agree that the experience of being loved is critical for one's sense of self-esteem, but I am not convinced that feeling loved unconditionally is essential for personal wholeness, and in some ways it might even be counterproductive. To my mind, the capacity to love unconditionally is far more important for personal wholeness than being loved unconditionally. To love as God is presumed to love is more important than being loved by God. To love unconditionally is to love without regard for oneself, to invest oneself in goals and outcomes that have no bearing on one's own interests. Those who live for themselves alone have nothing much to live for. To genuinely love another person is to take their interests to heart in ways that involve self-sacrifice. To genuinely love peace and justice is to take them to heart in ways that relativize the self. My kind of religious naturalism—like all religious traditions at their best—seeks ways to inspire this self-transcending love.

Notes

1 This essay was revised from the final chapter of my book entitled *Nature Is Enough*, Albany, NY: State University of New York Press, 2011.

2 See Stuart Kaufmann, *Reinventing the Sacred*, New York: Basic Books, 2008, and Karl Peters, *Dancing With the Sacred*, Harrisburg, PA: Trinity Press International, 2002.

23
A JEWISH PERSPECTIVE ON RELIGIOUS NATURALISM

Dan Solomon

Introduction

I grew up as a science buff and self-proclaimed "Jewish atheist." Still, I attended religious school, studied Hebrew, and prepared for my Bar Mitzvah. I had doubts, but was blessed with a liberal, philosophical rabbi. I told him that I didn't believe in God, though I felt a kind of religious awe for the beauty of the natural world and the laws of science. The rabbi replied, "Oh, you believe in Spinoza's God, which is the same as Nature; so did Einstein." At 12 years old, without having a name for it, I had discovered that I was a religious naturalist. Not finding the language of scripture and traditional worship relevant to my naturalistic worldview, I drifted away from congregational Judaism. I continued to identify as Jewish, celebrating Passover and Hanukkah at home, and studying Jewish literature as part of my broader philosophical education.

Religious naturalism is rich in intellectual resources, but it is lacking in models for personal development, within a lived tradition. That is why I began with the story of my own religious development, arising out of contemporary American Jewish experience. My aim is to find a way toward religious naturalism growing out of that experience, rather than any essentialist "authentic" Jewish voice. I hope that non-Jewish religious naturalists will also find it relevant, since I believe it parallels the condition of a growing number of people from every religious background.

In *Religious Naturalism Today* (Stone 2008), Jerome Stone reviews a variety of approaches to religious naturalism, including Jewish ones ranging from the austere metaphysical naturalism of Spinoza to Mordecai Kaplan's view of Judaism as a civilization. Other Jewish approaches relating to religious naturalism have included Sherwin Wine's Humanistic Judaism; and philosophers Henry Levinson and Hilary Putnam, who emphasized the experiential value of Jewish practice over history, theology, and metaphysics.

I will critique these approaches as to how they develop naturalistic religion from Jewish resources. These resources include the Biblical story of the Jewish People; the transcendent God; festivals rooted in pre-biblical agricultural traditions; ethics emphasizing humanistic values of justice and mercy; and a critical approach to scripture. I will relate this to a view of religious naturalism drawing on Loyal Rue, where a successful religion is based on a narrative core, supported by five ancillary strategies: intellectual, experiential, ritual, aesthetic, and institutional (Rue 2006: 126–128).

Religious naturalism

I begin with a straightforward definition of naturalism, as

> a set of beliefs and attitudes that focuses on this world ... it involves the negative assertion that there seems to be no ontologically distinct and superior realm (such as God, soul, or heaven) to ground, explain, or give meaning to this world.
>
> (Stone 2008: 1)

Another view I find useful is that of Jack Cohen, a follower of Mordecai Kaplan, where naturalism is the "disposition to believe that any phenomenon can be explained by appeal to general laws confirmable either by observation or by inference from observation"; and "nature will be conceived of as the totality of reality—its substance, functioning, and principles of operation—including man and his spiritual qualities" (Cohen 1958: 21).

I view religion very broadly, as the framework we use to create meaning and purpose in our lives. Religion relates our beliefs about reality, our attitudes towards reality, and our choices of how to live in reality. Religion doesn't have to tell us what reality is, but it does have to connect with whatever we consider reality to be. In my naturalistic view, everything is part of the natural world, including human beings and human culture. So, my religion is grounded in my understanding and experience of nature.

Given the above, my definition of religious naturalism follows that of Stone (2008: 1), as "a type of naturalism which affirms ... that there are religious aspects of this world which can be appreciated within a naturalistic framework." In application, my approach is like that of Ursula Goodenough (1998), in making nature my locus of concern nature, rather than humanity or a particular human tradition. Finally, I follow Stone (2008: 7–8) in finding a subtle but definite difference of attitude between many religious humanists and my understanding of religious naturalism. This will be discussed further in the section on Humanistic Judaism.

Jews, Judaism, and naturalism

It wasn't until the end of the eighteenth century, when European Jews began to be freed from medieval ghettoization, that a distinction between Jewish peoplehood and Jewish religion was recognized. This is particularly evident in the Reform Movement, which emphasized the idea of Judaism as a religion, and deemphasized ethnicity, in hopes of enabling full participation in the national loyalties that were developing (Meyer 1988). The mostly secular early Zionists also maintained this separation, in aiming to develop a new Jewish culture in Palestine. While Mordecai Kaplan famously questioned this distinction, as will be discussed, it is still important for many American Jews. A recent Pew survey (2013), for example, found that 22 percent of self-identified Jews say they are not Jewish by religion. Even among "Jews by Religion," 55 percent say that ancestry and culture matter more to being Jewish than religion does.

The Pew survey unfortunately doesn't include information about supernatural beliefs, as opposed to belief in "God," which is undefined. The fact that a much larger percentage of Jews, 23 percent, than in the general population indicate no belief in God supports the impression that Jews tend to be more naturalistic than non-Jews. Interestingly, only 47 percent of Jews of "no religion" say they don't believe in God; I wonder if these include many of the famous "spiritual but not religious" folks. On the other hand, 16 percent of those who are religious say they don't believe in God, so there is some overlap in these categories. This suggests that a significant number of American Jews might be identified as religious naturalists.

Previous approaches to Naturalistic Judaism

Stone (2008) has discussed a number of Jewish thinkers who might be characterized as religious naturalists. Here, I will just highlight those ideas that I find most useful, and focus on the Jewish aspects, rather than ideas *vis-à-vis* naturalism. I refer the reader to Stone for the latter. Note that I'm grouping loosely under denominations or movements, not individuals as Stone does.

Secular and unaffiliated Jews

The first and foremost of historic naturalistic Jews was Baruch Spinoza, whose "God, that is, Nature" has defined pantheism. Another was the philosopher Samuel Alexander (though he did attend a Reform synagogue). Many of the science writers who fed my childhood passion for science, such as Isaac Asimov and Carl Sagan, were Jewish, and followed Einstein in taking an interest in what I would call a religious attitude towards nature. A number of secular Jewish movements, or movements founded by Jews, such as Socialist and Zionist groups, and Ethical Culture, have emphasized the ethical tradition within Judaism. Just as liberal Jewish religious movements provide resources for religious naturalism, without themselves fitting the definition, so have these secular movements been influential in maintaining Jewish connections among many who personally might indeed fit my more robust definition of religious naturalism.

Reform Judaism

The Reform movement, from its beginnings in Germany, has included leaders and thinkers who eschewed supernaturalism. For example, Israel Jacobson, sometimes called the founder of the Reform movement, to maintain decorum, banned "superstitious" customs such as breaking a glass at weddings (Meyer 1988). On the other hand, the strong emphasis on "ethical monotheism" in Reform leads it to a more humanistic than naturalistic attitude. At the extreme this can be seen in what, in the twenty-first century, seems like an archaic denigration of nature; as with Samuel Hirsch, who contrasted Jewish active (ethical) religiosity with "pagan" subordination to nature (Meyer 1988).

Reconstructionism

Nothing less than a naturalistic reconstruction of Judaism was the project of Mordecai Kaplan, possibly the most influential naturalist voice in American Jewry. In *Judaism as a Civilization* (Kaplan 1967), he detailed his plan for an American Judaism rooted in Jewish culture, but re-imagined naturalistically. His key idea was that Judaism and Jewish culture (which he called civilization) were inextricable, or even identical. He criticized Reform Judaism for being more a religious philosophy than a living religion. He emphasized the need for cultural distinctness, to "hold one's interest" in Judaism, against alternatives (Kaplan 1967: 178). This led to his proposal for a "reconstruction" of Jewish civilization, maintaining Jewish distinctness by adhering as much as possible to traditional practice, while reinterpreting traditional concepts in keeping with modern ideas.

Kaplan advocated replacing the traditional allegorical method for interpreting traditional concepts with a functional method. This pragmatic approach involves identifying traditional values and practices "which are spiritually significant for our day," and finding equivalents relevant to our contemporary civilization (Kaplan 1967: 385–389). Obviously, this leaves a lot of room for picking and choosing, and in practice Kaplan's own sense of "spiritually significant" is what dictated the course of Reconstructionism. These most significant, i.e., sacred, elements of Jewish tradition are the sancta, including God, Torah, festivals, and ritual observance. But these must be reinterpreted within the context of lived experience, and continue to develop as culture evolves. Again, the reinterpretation leaves a lot of room for variation, especially if the "lived experience" of different Jews is different. This is why maintaining a Jewish "civilization" of shared experience was critical to Kaplan's project. This is also why change must be gradual, within the organic life of the Jewish people. For example, in keeping with modern universalism, he removed references to "the Chosen People" from the liturgy. However, instead of removing references to God, he provided various naturalistic reinterpretations of the "God-idea."

Kaplan felt that the continuation of Jewish civilization requires a God-idea, whose significance derives from the conduct it leads to, not its specific theological expression. Further, this God-idea must in some sense be personal, since "the modern thinker tends to base his conception of God upon the cosmic implications of human personality" (Kaplan 1967: 397). Kaplan's expositions of his own God-idea, or rather God-ideas, are creative and varied. First, there is God as seen in "the element of creativity, which is not accounted for in the so-called laws of nature" (Kaplan 1967: 316). Then we have "the reaction by which man's will-to-live overcomes the fears and miseries that only a being of his mental capacity can know" (Kaplan 1967: 330). Finally, "there is something in the nature of life which expresses itself in human personality, which evokes ideals, which sends men on the quest of personal and social salvation." Stone (2008: 111–119) provides a more complete exposition of Kaplan's God-ideas.

Given Kaplan's admission that the specific expression of a God-idea is not important, it is questionable how seriously his own statements should be taken. In fact, Kaplan is quoted as saying in 1915 (early in his career) that "Judaism, to be significant to modern man or woman, can no longer afford to speak in the language of theology" (Cohen 1958: xviii). Similarly, Levinson (2001) points to Kaplan's "predicate theology" as converting the noun "God" to the predicate (adjective) "divine." Thus, instead of "God is just, merciful, and forgiving," one would say "justice, mercy, and forgiveness are divine." It seems that Kaplan didn't always heed his own warning as to the "extent the mental habit of hypostasis (the tendency to treat qualities, attributes, relationships as though they had a separate existence) has been responsible for the contradictions and ambiguities that have discredited the conception of God and driven many to atheism" (Kaplan 1962: 21). Levinson (2001: 6) even suggested that "as a religious naturalist, Kaplan would have been better off stopping short of giving any theological explanations for anything."

In *The Meaning of God in Modern Jewish Religion*, Kaplan (1962) presents a reconstructed version of another Jewish sanctum, the cycle of festivals. Following the functional method, he insists that "they must have meanings which will guide and inspire us. It is, therefore, necessary to identify the Sabbath and Festivals with specific ideas and ideals which play an important part in the psychology of the modern man" (Kaplan 1962: 39). He then, in a series of chapters with titles like "God as the Power that makes for Social Regeneration," embarks on a project of reinterpreting the traditional Jewish God in ways that might be invoked in particular festivals.

The most glaring problem with Kaplan's project of reconstructing American Judaism as a civilization is that, in the century since he started developing his ideas, it has not happened. Thriving, organic Jewish cultures exist in Israel and in Orthodox enclaves such as those in and around New York City; but liberal American Judaism has moved on, into the melting pot. The largest Jewish denomination is still Reform, with 35 percent, while Reconstructionism has less than 1 percent of American Jews (Pew 2013). ("No Denomination" is second, with 30 percent.) Even more to the point, the Reconstructionist congregations are not noticeably distinct from other groups; they are often viewed as Conservative liturgy with Reform beliefs. Besides the assimilationist *Zeitgeist*, reasons for this failure—or perhaps incomplete success—of Kaplan's project can be found in his ideas themselves.

Kaplan's elements of group separateness are not enough to mark a civilization or even a culture. They might be seen instead as identity markers that are, in practice, mixed and matched with a variety of other elements in a given individual's life. Even members of the same Jewish congregation will not share the same set of these; just as they will not share the same God-idea. Then there is the need to take the multiple identities of American Jews into account. My children and I are proudly Jewish, though my mother was of Orthodox Christian heritage, and my children and stepchildren are of Vietnamese and Japanese, as well as Jewish, heritage. American Jews are at the forefront of liberal American pluralism, where a "mix-and-match" approach to culture prevails. This is reflected, for example, in "Nones" who have a Christmas tree, and the growing number of Christians who shop for churches where they find a comfortable environment, rather than adhering to an ancestral tradition.

Another problem, from a naturalistic point of view, is the use of traditional language in Reconstructionist services. Most of them do not hold traditional God-ideas, but most of them also do not have, and may not be interested in, specific interpretations. This seems to be in keeping with Kaplan's claim that the idea matters less than the practice, but the result is vagueness, not naturalism. Stone has raised additional questions about Kaplan's naturalism, for example in calling the God-process "trans-natural" (Stone 2008: 118). Levinson suggests that "Kaplan made claims about what he called 'transhistorical, transcultural, and eternal' entities that were values ... hardwired into the ways of the world," and "identified 'truth,' 'justice,' and 'love' as ways of being that operated independently in the universe." "This cosmic and super—or extra-human side of Kaplan's religious naturalism had few if any influential followers among American Jewish thinkers" (Levinson 2001: 6).

Two American Jewish pragmatists

Hilary Putnam is an interesting case of a major American-Jewish philosopher, a naturalist influenced by pragmatism, who began Jewish practice late in life. In *Jewish Philosophy as a Guide to Life* (Putnam 2008), he discusses three twentieth-century Jewish philosophers, Franz Rosenzweig, Martin Buber, and Emmanuel Levinas, who were not naturalists, but whose ideas he found of use in his own quest. As a "nonreductive naturalist," he held that higher levels of reality, like the moral one, had to be understood in their own terms. With John Dewey, he saw values and ideals, and even God, as having a kind of reality. While subjective, as deriving from humans, "which values and ideals enable us to grow and flourish is not a mere matter of 'subjective opinion'; it is something one can be wrong or right about" (Putnam 2008: 101). So, he did "not see reality as morally indifferent; reality, as Dewey saw, *makes demands* on us" (Putnam 2008: 6). The philosophers Putnam discusses emphasized this aspect of Jewish ethics, and Buber's approach to this will be discussed further later in this paper.

Another American Jewish philosopher in the pragmatic tradition was Henry Levinson. His "Festive Jewish Naturalism," as touched on in a couple of articles mainly dealing with other topics, seems to be a promising contribution to Jewish religious naturalism (Levinson 2001, 2006).

In these articles, as discussed by Stone (2008: 179–182), a variety of Jewish and other ideas are offered in the spirit of Festive Jewish Naturalism, which is engagingly humorous, but substantive. I find his exposition useful and a kindred project to my own collection of resources for Jewish religious naturalism. Unfortunately, no further, more systematic writings by Levinson on this subject seem to be available, and he is no longer living.

Humanistic Judaism

As Reconstructionism is associated with Mordecai Kaplan, so Humanistic Judaism is molded in the image of its founder, Sherwin Wine. As the name implies, Humanistic Judaism is based on (1) strict Humanism, eschewing God-language; and (2) Judaism, in Kaplan's sense of a "civilization" or culture. For Humanistic Judaism, a Jew is simply someone who identifies with Jewish culture.

The most distinctive aspect of Humanistic Judaism is the insistence on saying what one believes, and believing what one says. This means discarding or radically revising much of the traditional liturgy, especially to remove all reference to deity. Contrasting his approach with Reconstructionism, Wine asked, "Why bother to change one little item in the service when the whole concept of a worship experience where people talk to God for three hours is inconsistent with an impersonal deity? How can any reasonable person talk to creative energy?" (Wine 1985: 68).

In keeping with this strict nontheism, Wine and his followers have modified the Hebrew words of the traditional blessings. For example, the central "Sh'ma Yisrael, Adonai Eloheinu, Adonai Ehad" (Hear, O Israel, the Lord our God, the Lord is One) becomes "Sh'ma Yisrael, Ehad Ameinu, Adam Ehad" (Hear, O Israel, our People are one, Humanity is One). Wine relegated the Torah from its traditional central place in the holy ark at the front of the sanctuary, for readings during worship services, to the library, for study as literature.

In Humanistic Judaism, as in fact for many secular Jews, traditional festivals and life-cycle events continue to be celebrated, but without theistic language. Many of the Jewish holidays lend themselves easily to humanistic practice, with Shabbat being a day of rest, and Sukkot a week to celebrate nature and the harvest, eating meals in a "booth" in the backyard. Others have obvious humanistic themes, such as Passover's story of freedom, and Hanukkah's of fighting for one's beliefs (even while recognizing the irony that the Maccabees were themselves intolerant zealots).

The most important festivals, the "High Holidays" of Rosh Hashanah and Yom Kippur, take a bit more humanistic work, but there are rich themes of renewal and repentance available in these celebrations. So, following the example of Kaplan's functional method, but much more thoroughly, Humanistic Judaism identified "spiritually significant" values in traditional practices, and found "equivalents relevant to our contemporary civilization."

From the point of view of religious naturalism, Humanistic Judaism shares the shortcomings of humanism in general, as discussed by Stone (2008: 71). In particular, there is a strong emphasis on human self-reliance, sometimes joined with a sense that humans are alone in a hostile world. Just as one example, Sherwin Wine's *Staying Sane in a Crazy World* includes the following, as part of a credo:

> I believe that we live in a crazy world ... that the strength to cope with a crazy world comes from within ourselves, from the undiscovered power we have to look reality in the face and to go on living. I believe that the best faith is faith in oneself, and that the sign of this faith is that we allow our reasoning mind to discipline our action.
>
> (Wine 1995: Preamble)

Another characteristic of Humanistic Judaism is ambivalence as to whether it is religious or secular. Some Humanistic Jewish rabbis and congregations say they are religious humanists, and others are staunchly secular. It is not clear whether this is an advantage, in providing flexibility, or a disadvantage, as courting confusion. Going along with the ambivalence over secular vs. religious, Humanistic Judaism also shows a reluctance to use words such as "sacred," "holy," and "spirituality"; in contrast to many writers on religious naturalism who continue to find them useful. The congregational model that Wine and others instituted is clearly following in the religious path laid by predecessors, culminating in Kaplan. The sticking point for many is the notion that religion requires God-belief. In any case, these ambivalences seem to disqualify Humanistic Judaism as a candidate for a true religious naturalism. However, I feel that Humanistic Judaism is closer to what religious naturalistic practice would have to be like than the other Jewish groups discussed here. There is no reason that a Humanistic Jewish community could not evolve into a "humanistic religious naturalism," as presented by William Murry (2006). A healthy infusion of full-blooded naturalism, in Goodenough's sense, with the accompanying humility in the face of nature, might be all that it takes.

Resources for Jewish religious naturalism

One can be a Jewish religious naturalist within any of the movements mentioned in this chapter, or following the ideas of the individual thinkers. However, I believe that none of them are suited to maintaining a robust religious naturalism within a Jewish framework. Instead, I think the latter requires:

1. Starting from a science-based, naturalistic viewpoint, which I believe is the only way our globalized civilization will survive this twenty-first century. The alternative, of beginning with Jewish heritage, has been seen to result in any original naturalistic intent being fragile, as traditional language reinforces traditional non-naturalistic responses.
2. Following Goodenough in "exploring the religious potential of the scientific understanding of nature," rather than taking "a venture in theological reconstruction" (Stone 2008: 161).
3. Making use of Jewish resources suitable for naturalistic appropriation. Judaism is to be viewed as a heritage to use and adapt, rather than as a tradition to maintain or reject. In Stone's words, this should be "a dialog between the tradition (as faithfully reconstructed as possible, albeit from our perspective) and our own viewpoint, requiring the autonomy and integrity of our own viewpoint and the challenge of tradition" (Stone 2008: 171). So, Kaplan's "reconstruction" would just be the first step, in providing a partner for the conversation to begin.
4. Framing human relation to the ultimate in the traditional Jewish polarity of the transcendent vs. immanent aspects of God; but naturalized by drawing on the contrasting philosophies of Buber and Spinoza.

A robust religious naturalism, with the focus on nature, includes humanity as an integral part of nature. Similarly, Jewish heritage is taken as a natural part of human heritage. Though our tribal nature leads us to dwell in our particularistic traditions, naturalism indicates a universalist view. Jewish heritage is seen as one among many strands of human heritage, with an increasing number of Americans recognizing themselves as woven from multiple strands. Even those who imagine that they are "simply" WASP, Vietnamese, or whatever, might be surprised to see a genetic and cultural analysis of where their ancestors really came from!

So, I look for resources in the heritage of Jewish civilization, which a Jewish naturalist can use to live religiously in our contemporary global civilization. Levinson (2001: 6) quotes Kaplan as saying, "sancta ... provide Jewish religious naturalists with what Daniel Dennett calls memes, or vehicles of cultural memory with which to keep spinning our cultural webs." But memes do not have to exist in culture-sized packages; they can indeed be spun into webs, or woven into tapestries or braids. In the following sections, I will review key elements of the Jewish heritage, as I see them as fitting into a robust religious naturalism.

God

First, we should deal with the elephant in the room. As in the parable of the six blind philosophers, arguing over their different senses of what the elephant is, any theology that claims to have defined, or described, God, seems to be missing the big picture. Ignoring altogether the conventionally theistic views of God, one can find a wide variety of plausibly naturalistic God-ideas: "Love," "the Power for Good," "Creativity," and others mentioned by Stone (2008), for example. None of these seems to fit all of the usages in the Hebrew bible, nor to be compelling enough to stop one short, to feel that "here is God." Rather, there is an alternative take on our parable: six blind women get together, discuss their differences, and conclude that there is a being greater than their limited awareness, whose parts manifest the features recorded separately by each of them. This metaphor is an apt one for the long history of "God-ideas," from personal genie to abstract "ground of all Being," which have been applied to the ultimate reality, our source and sustenance. A big-picture solution would have to satisfy Dowd's (2008: 119) criterion, that "any understanding of 'God' that does not at least mean 'Ultimate Reality' or 'the Wholeness of Reality' (measurable and nonmeasurable) is, I suggest, a trivialized, inadequate notion of the divine." So, I start from Spinoza's God or Nature, the one Substance (i.e., Wholeness of Reality) in which all things inhere.

Though the familiar, personal God of so many Biblical stories, and of the traditional liturgy, is far from naturalistic, there are some Jewish ideas concerning God that can be adapted for Jewish religious naturalism. For those eschewing God-language, the tradition that the Hebrew, personal name of God is not to be spoken, is especially relevant. Many of the standard metaphors are not suitable for clear naturalistic use, but some are, and others can be developed, as exemplified by Dowd. Related to this move is the notion that God is beyond understanding, as in the famous "I am that I am" passage in Exodus (3:14); or in Job, where God must be obeyed, though seemingly beyond human notions of Good and Evil.

Job also highlights another important issue, that of faith. In Jewish tradition, faith is not equated with belief; it is trust; and, at least for the more thoughtful, it is not even naive trust that "everything will work out fine." It's more like Job's attitude, that "though he slay me, yet will I trust in him." At the time Job was written, there was no belief in an afterlife, so this meant essentially that "whatever happens, happens, and I will accept it." This is something that a naturalist can relate to. Even though life, nature, reality will eventually kill me, I still trust that reality will move on. I have a feeling that things make sense, and that there is an order, from fundamental physics to human society, that allows us to understand the world, and strive for better lives and a better future.

There is indeed a pervasive sense of personality in the Biblical God (or God-ideas), and this fact must be addressed. Borrowing Kaplan's functional method to look for the key elements "which are spiritually significant for our day," Martin Buber's (1970) sense of God as residing in the "I-Thou" of personal relationship seems most promising. Buber insisted that God was not to be spoken *of*, but rather spoken *to*. This seems radically non-naturalistic, and in fact Buber cannot be classified as a naturalist. But he also held that an "I-Thou" relationship can be had with all manner of non-human beings, including cats, trees, and a cliff of mica. Thus, the I-Thou relationship does not require an equal, fully human-like personality on both sides, and does not require belief in God. As Buber wrote:

> Whoever knows the world as something to be utilized knows God the same way. His prayers are a way of unburdening himself—and fall into the ears of the void. HE is godless—not the 'atheist' who from the night and longing of his garret window addresses the nameless.
>
> (Buber 1970: 156)

Buber emphasized that the I-Thou relationship involves an ethical demand, for us to be present to the other, whomever or whatever that might be.

Combining the approaches of Spinoza and Buber, we can derive a conception similar to that of Bernard Meland's "alternation of two approaches to reality" (Stone 2008: 88). This gives us a polarity of the transcendent, impersonal "God, that is, Nature" of Spinoza; and Buber's intensely personal, immanent God of relationship. Again, we need to find a balance, rather than gravitating compulsively to either of these poles. I believe that a religious naturalism rooted in a scientific understanding of nature, combined with an aesthetic sense of our existential rootedness in nature, can provide the balance beam between these two poles.

Torah and the narrative tradition

Torah for Jews represents more than its basic definition as the first five books of the Hebrew bible. It represents law, teaching, commandment, including the supplementary "oral law" that tradition holds was handed down from time of Moses, finally to be written down in the Talmud. It truly is at the center of Rue's mythic core of traditional Judaism, and is proclaimed in familiar liturgy as "a tree of life for those who hold fast to it" (Proverbs 3:18). An emphasis on the word, and on study, for the Jewish people in general, and not just priests and scholars, has been integral to Jewish culture at least since Ezra read the Torah to the people at the dedication of the second Temple and had translators available to ensure that they understood it.

For a naturalist, much of this tradition will have to be left behind, radically reworked, or simply treated as edifying fiction. But, this very reworking can be seen to be part of another key Jewish tradition, that of Midrash, or interpretation. The literary history of the Jews can be viewed as one long and growing commentary on the core set of narratives in the Torah, beginning with Haftorah—the rest of the books of the Hebrew bible. There is no need for a fixed, authoritative naturalistic interpretation, either. The authorities themselves, the rabbis, developed from the start a tradition of sometimes rancorous debate. They included both sides of many arguments in the Talmud to demonstrate that the meaning of Torah requires interpretation, and the result is a human product, not a voice from God. So, Jewish religious naturalism simply joins this continuing discussion, from a naturalistic viewpoint.

For Jewish religious naturalists, the true history of the Jewish people is also an important resource. Sherwin Wine's last book *A Provocative People: A Secular History of the Jews* (Wine 2012), published posthumously, was just such a history. Finally, as with all religious naturalists, Jewish religious naturalists can make use of "Everybody's Story" (Rue 2000), and the perspective this provides on the real place of Jewish history within it.

Israel

The third leg of the traditional central triad of Judaism, together with God and Torah, is Israel, the Jewish people. Even though Kaplan's attempt to create an American-Jewish "civilization" was impractical, the idea of Jewish peoplehood has been essential for Jewish continuity among religious and secular Jews alike. In fact, this is why there is such a large group of people who continue to identify as Jews, while claiming no religious affiliation to Judaism. A useful aspect of this fact for Jewish religious naturalism is that it provides plenty of non-supernatural Jewish resources, even if it is just the "lox and bagel" Judaism of serving traditional foods. Naturally, these aspects would have to be integrated with the broader narrative and ritual practices, and not just stand in isolation.

The importance of the Jewish community in Jewish religious life highlights another important feature of Judaism that can be put to naturalistic use. Rather than seeking individual "salvation," the "world to come" is traditionally viewed as one for Jews as a people, and, ideally at least, for all people. While there are great variations, such as the influence of Christian ideas, the communal tradition seems most useful for a Jewish religious naturalism.

Prayer and ritual

Prayer would seem to have little place in naturalistic practice. One option, though, is to follow Goodenough and Woodruff in replacing traditional theistic prayers with "Mindful Reverence" (Stone 2008: 162). This lends itself nicely to the Jewish tradition of reciting blessings for a myriad of daily activities; for example hand washing, tree planting, and appreciating natural objects. Of course, the traditional wording praises God for providing whatever it is that is being appreciated. A naturalist could simply take this as an opportunity for mindfulness and gratitude, without saying a blessing at all. Alternatively, God could be replaced with Nature or a natural process, in the kind of "revolutionary step" advocated by Cohen (1958: 157), in finding "new prayers or readings ... that can serve to tap deep-rooted feelings."

Nonverbal rituals are also important in Judaism. In fact, while those holding most strictly to Jewish traditions are typically called "Orthodox," their own preferred term is "Observant," meaning that they observe all of the Torah's commandments, including a whole set of rituals. Observance has the advantage of allowing degrees, so that the middle-of-the-road Jews called "Conservative" can be described in terms of strictness of observance. The Liberal Jewish movements are defined as those in which observance is optional, though possibly of varying degrees of symbolic importance. Nonverbal rituals seem to be especially appropriate for Jewish religious naturalism, as these are more easily reinterpreted naturalistically than blessings, prayers, and scriptural readings. Some kind of reinterpretation does seem necessary, though. "While ... abstract ideas are not likely to make a profound impression on the minds of most people unless accompanied by some activity, meaningless activities dissociated from any significant ideas are no less ineffectual" (Kaplan 1962: 38).

Festivals and life-cycle events

As mentioned previously, Humanistic and secular Jews have adapted Jewish festivals and life-cycle commemorations to a humanistic, naturalistic world-view. A good example of such adaptation is provided in *God Optional Judaism* (Seid 2001), whose central sections are "Reclaiming Our Jewish Holidays" and "Reclaiming Life-Cycle Observances." Though the author, "Secular Rabbi" Judith Seid, is not directly associated with Humanistic Judaism, she received her rabbinic training at the International Institute for Secular Humanistic Judaism, founded by Sherwin Wine.

Among the Jewish festivals are Passover, Shavuot, and Sukkot, which originated in ancient harvest festivals, and were later adapted by the rabbinic tradition to a more complicated theology following the "chosen people" narrative in Torah. Practices deriving from pre-literate agricultural traditions, such as eating ritual foods, are readily adapted to naturalistic, nontheistic practice. Music of all kinds is an almost ubiquitous feature of Jewish celebrations. This is also quite adaptable, requiring at most adjustments in wording.

As to life-cycle observances, I can provide a few personal examples. My wife and I were married in a humanistic Jewish ceremony, with a traditional chuppah (wedding canopy), wine, and the breaking of a glass. The language was entirely nontheistic, but followed the traditional patterns and invoked the traditional values. Again, at a casual Saturday evening chapel service for an IRAS (Institute on Religion in an Age of Science) conference, with an audience from a variety of religious backgrounds, I celebrated Havdalah, the traditional ceremony marking the end of the Sabbath. This includes wonderfully evocative rituals, including singing, passing a container of fragrant spices, lighting a braided, multi-colored candle, and finally, dousing the candle in a glass of wine. All of these were related to the occasion of the beginning of a conference week of study, play, and socializing; as well as to the Jewish significance.

Ethics

At the core of all forms of Jewish culture, no matter how secular, is the ethical demand of the prophetic voice. From a naturalistic point of view, it is helpful to have a model like that of Putnam, referenced previously, which follows Buber and others in recognizing ethical demands as being placed on us by reality, i.e., nature. With this interpretation, the traditional Jewish symbol "hineni," here I am, as the response given by Abraham, Jacob, Moses, and other Biblical characters to God's demands, can be adapted for religious naturalism.

Though the details of the ancient moral code can no longer be taken as authoritative, the key moral attributes of mercy and justice continue to guide secular and Humanistic Jews. This dichotomy is echoed by a number of contemporary writers on religious naturalism or ethics. Rue (2000: 117), for example, views religion as promoting both individual integrity and social cohesion.

A distinctive Jewish ethical resource is the concept of "Tikkun Olam," repairing or improving the World. Though the ethical call is to *me* personally, the project is one for all of us together, and relates to the wellbeing of the entire world. The Kabbalists developed this idea long after the Hebrew and Christian canons were set, but it is rooted in the strong social sense in Judaism. It has been embraced by liberal Jews and applied to the call for social action. In this sense, rather than the original mystical Kabbalistic sense, it can be of value for Jewish religious naturalists.

Conclusion

After meeting five theoretical objections to "a religion of nature," Crosby (2002: 155–157) raises a final, practical objection; that religious naturalism has no practicing communities or institutional structures. He suggests building on existing nature oriented religions, such as Taoism, Shinto, and Native American ones.

That this communal and institutional aspect is missing in religious naturalism seems related to the problem of connecting the big picture with the personal one. For religious naturalism, the sense of our lives must be found in view of the natural reality to which we belong. The narratives that inform each individual's identity must tell the whole story, beginning with "everybody's story," evolving through the people(s) from which that particular person descends, and culminating in the life narrative of the individual. Each chapter of this story has been told in various ways. For religious naturalism, it seems that the big and the small pictures are given: "everybody's story" as informed by science and one's own experience. But the connective between these two, cultural heritage, is not so easily addressed in a general exposition of religious naturalism.

The exposition of Jewish religious naturalism in this paper has aimed at providing this needed connective. The polarity of ideas of God or Reality in Buber *versus* Spinoza can be taken as a symbol of the need to integrate the various narratives. Following Rue's account of religion (Rue 2006), we have seen that the mythic core is provided by a rich narrative tradition, including a tradition of interpretation that Jewish religious naturalists can take advantage of. Rue's ancillary strategies can also be seen as being available from Jewish resources, as well as from the general religious naturalistic resources discussed elsewhere in this Handbook. The intellectual dimension is especially strong in Jewish naturalists including Kaplan, Cohen, and Wine. One example of an experiential strategy for Jewish religious naturalism is converting occasions for traditional blessings into exercises in mindfulness. Jewish heritage is rich in ritual strategies, and ways of adapting these for religious naturalism have been presented. A few aesthetic strategies, such as the multi-sensory Havdalah ceremony, have also been discussed. Music has just been briefly mentioned, precisely because it is so easily adapted. Institutional strategies for Jewish religious naturalism might follow the example of other pioneering liberal Jewish branches. For example, an existing Humanistic Jewish community might evolve into a true Jewish religious naturalism, in the sense advocated here.

References

American Humanist Association. (2017) *What Is Humanism?* Retrieved 9 January 2017, from https://americanhumanist.org/what-is-humanism/

Buber, M. (1970) *I and Thou*. Trans. W. Kaufmann. New York: Scribner's Sons.

Cohen, J. J. (1958) *The Case for Religious Naturalism: A Philosophy for the Modern Jew*. New York: The Reconstructionist Press.

Crosby, D. A. (2002) *A Religion of Nature*. Albany, NY: State University of New York Press.

Dowd, M. (2008) *Thank God for Evolution: How the Marriage of Science and Religion Will Transform Your Life and Our World*. New York: Viking.

Goodenough, U. (1998) *The Sacred Depths of Nature*. New York: Oxford University Press.

Kaplan, M. M. (1962) *The Meaning of God in Modern Jewish Religion*. New York: Reconstructionist Press.

—— (1967) *Judaism As A Civilization: Toward a Reconstruction of American-Jewish Life*. New York: Schocken Books.

Levinson, H. S. (2001) "Festive Naturalism and *The Legends of the Jews*," *Harvard Divinity School Bulletin* 30 (Summer/Fall 2001).

—— (2006) "Festive Religious Naturalism and Richard Bernstein's Work on Freud and Arendt," in S. G. Davaney and W. G. Frisina, eds. *The Pragmatic Century: Conversations with Richard J. Bernstein*, Albany, NY: State University of New York Press, pp. 115–130.

Meyer, M. A. (1988) *Response to Modernity: A History of the Reform Movement in Judaism*. New York: Oxford University Press.

Murry, W. R. (2006) *Reason and Reverence: Religious Humanism for the 21st Century*. Boston, MA: Skinner House Books.

Pew Research Center. (2013) *A Portrait of Jewish Americans*. Retrieved 6 January 2017, from www.pewforum.org/2013/10/01/jewish-american-beliefs-attitudes-culture-survey/

Putnam, H. (2008) *Jewish Philosophy as a Guide to Life: Rosenzweig, Buber, Levinas, Wittgenstein*. Bloomington, IN: Indiana University Press.

Rue, L. D. (2000) *Everybody's Story: Wising Up to the Epic of Evolution*. Albany, NY: State University of New York Press.
—— (2005) *Religion Is Not About God: How Spiritual Traditions Nurture our Biological Nature and What to Expect When They Fail*. New Brunswick, NJ: Rutgers University Press.
Seid, J. (2001) *God Optional Judaism: Alternatives for Cultural Jews Who Love Their History, Heritage, and Community*. New York: Citadel Press.
Stone, J. A. (2008) *Religious Naturalism Today*. Albany, NY: State University of New York Press.
Wine, S. T. (1985) *Judaism Beyond God*. Farmington Hills, MI: Society for Humanistic Judaism.
—— (1995) *Staying Sane in a Crazy World: A Guide to Rational Living*. Birmingham, MI: The Center for New Thinking.
—— (2012) *A Provocative People: A Secular History of the Jews*. Farmington Hills, MI: International Institute for Secular Humanistic Judaism.

24
CONFUCIANISM AS A FORM OF RELIGIOUS NATURALISM

Mary Evelyn Tucker

Introduction

The art of Confucian religious naturalism might be described as discovering one's cosmological being amid daily affairs. For the Confucian the ordinary is the locus of the extraordinary; the secular is the sacred; the transcendent is in the immanent. What distinguishes Confucianism is an all-encompassing cosmological context that grounds its world-affirming orientation for humanity. This is not a tradition seeking liberation outside the world, but one that affirms the spirituality of becoming more fully human within the world. The way of immanence is the Confucian way.[1]

The means of self-transformation is through cultivation of oneself in relation to others and to the natural world. This cultivation is seen in connection with a tradition of scholarly reflection embedded in a commitment to the value of culture and its myriad expressions. It aims to promote flourishing social relations, effective educational systems, sustainable agricultural patterns, and humane political governance within the context of the dynamic, life-giving processes of the universe.

One may hasten to add that, while subject to debate, aspects of transcendence are not entirely absent in this tradition, for example, in the idea of Heaven in classical Confucianism or the Supreme Ultimate in later Neo-Confucianism.[2] However, the emphasis of Confucian spirituality is on cultivating one's Heavenly-endowed nature in relation to other humans and to the universe itself. There is no impulse to escape *samsara*, the cycles of suffering as in Hinduism or Buddhism or to seek other-worldly salvation as in Judaism, Christianity, or Islam. Rather, the microcosm of the self and the macrocosm of the universe are implicitly and explicitly seen as aspects of a unified but ever-changing reality.

The seamless web of immanence and transcendence in this tradition thus creates a unique form of spiritual praxis among the world's religions. There is no ontological split between the supernatural and the natural orders. Indeed, this may be identified as one of the distinctive contributions of Confucianism, both historically and in its modern revived forms.

How to describe this form of religious naturalism is part of the challenge to give the reader an overview of the remarkable array of Confucian thought from the classical period to the contemporary period. We hope that by examining these distinctive forms of Confucianism the very notion of religious naturalism will be broadened and enriched as a result. It is important to realize that the broad aspirations of Confucianism were not always realized. Indeed, like all pre-modern religious traditions, hierarchical and autocratic aspects often dominated.

What is Confucian religious naturalism?

Among the world's religious traditions Confucianism has the distinction of being the tradition that is least understood as having religious or spiritual aspects. Part of the complexity of the problem regarding the religious nature of Confucianism lies in sorting out a series of interlocking questions. Foremost among them is how one defines Confucianism—as a political system, as ethical teachings, as social norms, as a humanistic philosophy, or as a religious worldview.[3] We acknowledge all of these features as being part of Confucianism. However, we aim here to explore Confucianism not necessarily as a "religion" per se, but as a religious naturalism with distinctive spiritual dimensions.

We are refraining from using the term "religion" to describe Confucianism because "religion" tends to be associated with formal institutional structures and most often with characteristics of Western religions such as theism, personal salvation, and natural/supernatural dichotomies.[4] The term "religion" may thus obscure rather than clarify the distinctive religious and spiritual dimensions of Confucian naturalism.[5] Therefore, instead of claiming Confucianism as a religion (which is problematic in itself for many people), we are suggesting that Confucianism is a religious naturalism with a cosmological orientation.[6] This cosmological orientation is realized in the connection of the microcosm of the self to the macrocosm of the universe through spiritual practices of communitarian ethics, self-transformation, and ritual relatedness.

Religious naturalism in the Confucian context is that which gives humans a comprehensive and defining orientation to ultimate concerns. Spirituality is that which provides expression for the deep yearnings of the human for relatedness to these ultimate concerns. While a religious worldview may be assumed as part of a given set of cultural ideals and practices into which one is born, spirituality may be seen as the vehicle of attainment of these ideals. Confucian religious naturalism is distinguished by its cosmological context in which humans complete the triad of Heaven and Earth. Confucian spirituality requires discipline and practice along with spontaneity and creativity. Confucian spirituality establishes different ethical responsibilities for specific human relations, deepens subjectivity in its methods of self-cultivation, and celebrates communion of cosmic and human forces in its ritual connections. It aims to situate human creativity amid concentric circles of interdependent creativity from the person to the larger universe.

One way to appreciate the distinctiveness of Confucian religious naturalism and its spiritual expressions is to observe broad characteristics of religions with a common geographical place of origin. In this spirit it is significant to note that the flowering of the world's religions, which occurred in the sixth century B.C.E., was labeled by Karl Jaspers as the Axial Age (Jaspers 1953). This period can be characterized as having three major centers of origin: those in West Asia—Judaism, Christianity, Islam; those in South Asia—Hinduism, Jainism, Buddhism; and those in East Asia—Confucianism and Daoism. The first can be described as prophetic and historically-based religions; the second can be seen as mystical religions and religions of liberation; the third can be understood as religious naturalisms of cosmic and social harmony (Berthrong 1994). It is precisely the interaction of the cosmic and social that underlies the spiritual dynamics of Confucian naturalism.

The appeal of Confucian religious naturalism

As David Keightley observes:

> the strength and endurance of the Confucian tradition, ostensibly secular though its manifestations frequently were, cannot be fully explained, or its true nature understood, unless we take into account the religious commitment which assisted at that tradition's birth and which continued to sustain it.
>
> (Keightly 1978)

Clearly, Confucian thought had an appeal to individuals and groups in East Asia for centuries beyond its political or ideological uses. Individual scholars and teachers engaged in the study and practice of Confucianism for intellectual inspiration, personal edification, spiritual growth, and ritual expression. We can see this in the spread of Confucianism to Korea, Japan, and Vietnam. This was especially evident in Japan where there was no civil service examination system to advance personal careers. In the Tokugawa period (1603–1868), for example, many Japanese scholars and teachers studied Confucianism for its inherent value and assisted its spread in the society by establishing schools (Rubinger 1982).

Confucianism is more than the conventional stereotype of a model for creating social order and political stability sometimes used for oppressive or autocratic ends. While Confucianism aimed to establish stable and harmonious societies, it also encouraged personal and public reform, along with the reexamination of moral principles and spiritual practices appropriate to different contexts (de Bary 1953). This is evident in Confucian moral and political theory, from the early classical concept of the rectification of names in the *Analects* to Mencius' qualified notion of the right to revolution. It is likewise seen in the later Neo-Confucian practice of delivering remonstrating lectures to the Emperor and, when necessary, withdrawing one's services from an unresponsive or corrupt government.

On a personal level, the whole process of self-cultivation in Confucian spiritual practice was aimed at achieving authenticity and sincerity through conscientious study, critical self-examination, continual effort, and a willingness to change oneself (de Bary 1991). "Learning for oneself," not simply absorbing ideas uncritically or trying to impress others, was considered essential to this process (de Bary 1991). Thus, authenticity could only be realized by constant transformation so as to bring oneself into consonance with the creative and generative powers of Heaven and Earth (Tu 1989). These teachings sought to inculcate a process in tune with the dynamic, cosmological workings of nature. It thus affirmed change as a positive force in the natural order and in human affairs. This process of harmonizing with changes in the universe can be identified as a major wellspring of Confucian spirituality expressed in various forms of self-cultivation. This is at the heart of Confucian religious naturalism.

The focus on the positive aspects of change can be seen in each period of Confucianism as well as in its spread to other geographical contexts. Change in self, society, and cosmos was affirmed and celebrated from the early formative period, which produced the *Classic of Changes* (*Yi Ching*). Later Han Confucianism emphasized the vitality of correspondences between the human and the various elements in nature.[7] Eleventh and twelfth century Sung Neo-Confucianism stressed the creativity of Heaven and Earth. Confucian spirituality in all its diverse expressions was seen in East Asia as a powerful means of personal transformation. Furthermore, it was a potential instrument of establishing social harmony and political order through communitarian ethics and ritual practices. It emphasized moral transformation that rippled outward across concentric circles rather than the external imposition of legalistic and bureaucratic restraints. It was precisely this point that differentiated the Confucian aspirations and ideals from those of the Legalists, such as Han Fei Zi, who felt that humans could be restrained by law and changed by punishment.[8] Confucianism is a tradition that has endured for more than two and half millennia in varied historical, geographical, and cultural contexts. It is still undergoing transformation and revitalization in its contemporary forms.[9]

Overview of the historical development of Confucianism

The Confucian tradition has assumed distinctive expressions in China, Korea, Japan, Vietnam, Hong Kong, Taiwan, and Singapore. Viewing Confucianism as a singular tradition is problematic because of its geographic spread, its historical development, and its varied forms, ranging from Imperial State Confucianism to local and familial Confucianism. Nonetheless, this overview will try to make some distinctions in the various kinds of Confucianism in order to highlight aspects of its religious naturalism.[10]

While originating in the first millennium B.C.E. in China, the tradition includes the transmission and transformation of Confucianism that took place in different East Asian cultural and geographical contexts. In examining the reasons for its spread and its appeal, it is important to highlight the ways in which it interacted with native traditions in China and across East Asia. For example, Confucianism responded to and mingled with Daoism and Buddhism in China, with shamanism in Korea, and with Shinto in Japan.[11] The borrowing and creative interaction among the various religious traditions in East Asia needs to be underscored. Indeed, the so-called unity and syncretism of the three traditions of Confucianism, Daoism, and Buddhism in China should be noted. This was especially pronounced in the Ming (1368–1644) and Qing (1644–1911) periods (Berling 1980).

While recognizing this dynamic cross fertilization of religious traditions in East Asia, we can also identify historically four major periods of Confucian thought and practice. The first stage in China is that of classical Confucianism, which ranges from approximately the sixth to the second century B.C.E. This is the period of the flourishing of the early Confucian thinkers, namely Confucius and Mencius. The second period is that of Han Confucianism when the classical tradition was shaped into a political orthodoxy under the Han Empire (202 B.C.E–220 C.E.) and began to spread to other parts of East Asia. This period saw the development of the theory of correspondences of the microcosm of the human world with the macrocosm of the natural world. The third major period is the Neo-Confucian era from the eleventh to the early twentieth century. This includes the comprehensive synthesis of Zhu Xi in the eleventh century and the distinctive contributions of Wang Yangming in the fifteenth and sixteenth centuries. The influence of Confucianism as an educational and philosophical system was felt throughout many parts of East Asia during this period. The last phase is that of New Confucianism in the twentieth century, which represents a revival of the tradition under the influence of scholars who came to Taiwan and Hong Kong after Mao's ascendancy in 1949.[12] Four decades later, in October 1989, the International Confucian Society held two major conferences in Beijing and in Confucius' birthplace, Qufu, to explore the future of the Confucian Way. These conferences were intended to mark the 2,540th anniversary of Confucius' birth and they both signified the interest of Confucian practitioners in looking toward the future.

Categories for the study of Confucianism

It may be helpful to distinguish various kinds of Confucianism so as to reframe the questions surrounding the emergence and manipulation of political ideologies and separate them from the spiritual dimensions of Confucian religious naturalism. At the same time, we can acknowledge the ambiguous nature of many religions or philosophies in their frequent appropriation for manipulative or distorted ends.

Let us identify some broad descriptive categories of Confucianism that are distinctive from, yet overlap with, Confucian religious naturalism:

1. *Political Confucianism* refers to State or Imperial Confucianism, especially in its Chinese form, and involves such institutions as the civil service examination system and the larger government bureaucracy from the local level to the various ranks of court ministers. In Korea, Confucian bureaucratic government was adapted in the Koryo dynasty and in 958 C.E. the civil service examination system was adopted as a means of selecting officials. Confucianism was further established as official orthodoxy under the Yi dynasty in 1392 C.E. and civil service examinations were inaugurated. In Japan there were no civil service examinations but Confucian ideas were used in the Nara government and in Prince Shotoku's "Constitution" of 719 C.E., as well as in legitimizing the Tokugawa Shogunate and later in the Meiji government's "Imperial Rescript on Education."
2. *Social Confucianism* alludes to what one might call family-based or human relations oriented Confucianism. This involves the complex interactions of individuals with others both within and outside of the family. It has been described by Tu Weiming as a cultural DNA, or "habits of the heart" passed on from one generation to the next through the family. These interactions both reflect and create the intricate patterns of obligations and responsibilities that permeate East Asian society. In Japan, for example, these patterns are expressed in concepts such as *on* and *giri* (mutual obligations and debts requiring repayment).
3. *Educational Confucianism* encompasses public and private learning in schools, in families, and by individual scholars and teachers. It refers, although not exclusively, to the curriculum of study of the Four Books—the *Great Learning*, the *Mean*, the *Analects*, and the *Mencius*—selected as the canon by Zhu Xi. This was used as the basis of the civil service examination system in China and Korea. Educational Confucianism incorporates the adaptation of that curriculum to other educational institutions and venues in East Asia. In Japan and Korea, for example, it includes the various schools set up both privately and by national and provincial governments, especially in Yi dynasty Korea and Tokugawa Japan. In addition, it refers to some of the moral training that continued to be part of the educational system in Korea and Japan in the twentieth century (de Bary and Chaffee 1989). Educational Confucianism can be said to go beyond schools, institutions, and curriculum to include at its heart the notion of learning as a means of self-cultivation, an approach that is emphasized in the *Analects* and *Mencius* (de Bary 1991).
4. *Economic Confucianism* describes business company forms of Confucianism in the modern period and merchant-related Confucianism in the pre-modern period, especially in Qing China, Yi Korea, and Tokugawa Japan (for example, Najita 1987). It includes the idea of familialism and loyalty as critical principles for the transmission of family-based Confucian values into organizational structures within the business community. This seems to be particularly widespread in East Asia, especially in the last 50 years (Tu 1996; see also Dore, Tu, and Kim in Kreiner 1996). It also includes the transmission across the society of values often associated with Confucianism such as frugality, loyalty, and industriousness.

Worldview and ethos: organic cosmology and communitarian ethics

Confucian religious naturalism, while by no means singular or uniform, is one that can be described as having an organismic cosmology (Needham 1956: 291–293) characterized as a "continuity of being" (Tu 1985) within an "immanental cosmos" (Ames and Hall 1987: 12–17). There is no clear separation, as in the Western religions, between a transcendent, other-worldly order and an immanent, this-worldly orientation. As the *Mean* (*Zhongyong*) states: "The Way of Heaven and Earth can be described in one sentence: They are without any doubleness and so they produce things in an unfathomable way. The Way of Heaven and Earth is extensive, deep, high, brilliant, infinite, and lasting" (Chan 1963: 109).

Without an ontological gap between this world and another world there emerges an appreciation for the seamless interaction of humans with the universe. The Confucian cosmological worldview is one that embraces a fluid and dynamic continuity of being. In terms of ethos or ethics this involves working out the deep interconnections of Heaven, Earth, and humans. This profound symbolic expression of the triadic intercommunion of an immanental cosmos is invoked repeatedly in both the Confucian and Neo-Confucian texts cited across East Asia.[13] As Tu Weiming notes, this cosmology is neither theocentric nor anthropocentric, but rather anthropocosmic (Tu 1989: 102–107). In this sense the emphasis is not exclusively on the divine, nor on humans, as is the prevailing model in the West. Rather, the comprehensive interaction of Heaven, Earth, and humans is what is underscored by the term *anthropocosmic*. Thus the worldview of an organic cosmology creates a context for the intricate communitarian model of social ethics that distinguishes East Asian societies.

The mutual attraction of things for one another in both the human and natural worlds gives rise to an embedded ethical system of reciprocal relationships. The instinctive qualities of the human heart toward commiseration and empathy is what is nurtured and expressed in human relations and ritual practices (Mencius 2 A:6 in de Bary and Bloom 1999). The human is not an isolated individual in need of redemption by a personal God, but is deeply embedded in a network of life-giving and life-sustaining relationships and rituals. Within this organic universe the human is viewed as a microcosm of the macrocosm where one's actions affect the larger whole, like ripples in a pond as expressed in the *Great Learning*. Thus, there is a relational resonance of personal and cosmic communion animated by authenticity (*cheng*) as illustrated in the *Mean* (de Bary and Bloom 1999: Chapters 22, 25, 26). The individual is intrinsically linked via rituals to various communities, beginning with the natural bonding of the family and stretching out to include the social-political order and embracing the symbolic community of Heaven and Earth (Berry 2003). Humans achieve their fullest identity as members of the great triad with Heaven and Earth. Within this triad Heaven is a guiding moral presence, Earth is a vital moral force, and humans are co-creators of a humane and moral social-political order.

Cosmology and cultivation: creativity of heaven and transformation of humans

Confucian religious naturalism embraces a vast cosmological order that is distinguished by the creativity of Heaven as a life-giving force that is ceaselessly self-generating.[14] Similar to Whiteheadian process thought, the Confucian universe is seen as an unfolding, creative process, not as a static, inert mechanistic system controlled by an absent or remote deity (Berthrong 1994). As a protecting, sustaining, and transforming force, Heaven helps to bring all humans to their natural fulfillment as cosmological. This is because humans are imprinted with a Heavenly-endowed nature, which enables them to transform themselves through self-cultivation (de Bary and Bloom 1999: Chapter 1).

The ethos, then, of this creative cosmology is one that encourages education, learning, and self-transformation. The optimistic view of humans as receiving a Heavenly nature results in a Confucian educational and family ethos, which ideally creates a value system for nurturing innate human goodness and the creative transformation of individual potential. This ethos is one that encourages a filial sense of repayment to Heaven for the gift of life and for a Heavenly-bestowed nature. The way to repay these gifts is through ongoing moral cultivation for the betterment of self and society. The symbol or model that joins this aspect of the worldview and ethos together is the noble person (*junzi*), or the sage (*sheng*), who "hears" the will of Heaven and is able to embody it naturally in the ongoing process of learning and self-cultivation. The sage is thus the highest embodiment of the spiritual aspirations of the Confucian tradition (Taylor 1990).

Vitalism of the Earth and co-creativity of humans: cosmological correspondences and human ritual

The creativity of Heaven in the Confucian cosmological worldview is paralleled by the vitalism of the natural world. From the early text of the *Classic of Changes* (*Yijing*), through the Neo-Confucian reappropriation of this classic, the sense of the vitality of the natural world infuses many of the Confucian writings (Smith, Bol, Adler, and Wyatt 1990). This vitality is understood as part of the seasonal cycles of nature, rather than as the developmental, evolving universe discussed by contemporary process philosophers and theologians. It is expressed in an elaborate series of correspondences (seasonal, directional, elemental) and rituals, which in Han Confucianism were seen as patterns suggestive of the careful regulation needed in the social and political realms (Henderson 1984; Rosemont 1984; Queen 1996; Eno 1990). This cosmological view of the integral cycles of nature reinforces an ethos of cooperating with those processes through establishing a harmonious society and government with appropriate ritual structures. The rituals reflect the patterned structures of the natural world and bind humans to one another, to the ancestral world, and to the cosmos at large.

The vital material force (*qi*) of the universe is that which joins humans and nature, unifying their worldview and ethos and giving humans the potential to become co-creators with the universe (Tu 1989: 70, 78, 98, 102, 106). As Mencius notes, it is *qi* that unites rightness (ethos) and the Way (worldview), filling the whole space between Heaven and Earth (Mencius 2A:2 in de Bary and Bloom 1999). The moral imperative of Confucianism, then, is to make appropriate ethical and ritual choices linked to the creative powers of the Way and thus contribute to the betterment of social and political order.

Confucian religious naturalism, then, affirms change, as is manifest in the creativity of Heaven and in the vitality of Earth. In particular, the varied and dynamic patterns of cosmological change are celebrated as part of a life-giving universe. Rituals and music are designed to harmonize with these cosmic changes and to assist the process of personal transformation. Rituals help to join the worldview of cosmic change with the ethos of human changes in society, thus harmonizing the natural and human orders. Rituals and music are a means of creating grace, beauty, and accord. Thus, the natural cosmological structures of the Earth provide a counterpoint for an ethos of social patterns expressed in ritual behavior and music. Harmonizing with the universe in a cosmological sense is balanced by an ethos of reciprocal resonance in human relations and expressed in the patterned behavior of rituals.

Conclusion

Confucian religious naturalism encompasses a dynamic cosmological orientation that is interwoven with spiritual expressions in the form of communitarian ethics of the society, self-cultivation of the person, and ritual expressions integrating self, society, and cosmos. This tapestry of spiritual integration, which has had a long and rich history in China and in other countries of East Asia, deserves further study. We trust such studies will also point the way toward future forms of Confucian religious naturalism in new and creative expressions.

Notes

1 In this essay, except where noted, I am using the terms *Confucian* and *Confucianism* to refer to the tradition in a broad sense without necessarily distinguishing between the early classical Confucian expressions and the later Neo-Confucian forms in China, Korea, and Japan.

2 As Tu Weiming puts it, "Despite the difficulty of conceptualizing transcendence as radical otherness, the Confucian commitment to ultimate self-transformation necessarily involves a transcendent dimension" (Tu 1985: 137). This is not "radical transcendence but immanence with a transcendent dimension" (Tu 1989: 121). See also similar arguments made earlier by Liu Shu-hsien (Liu 1972: 45–52). Roger Ames and David Hall have argued that the Confucian tradition, especially in its classical forms does not focus on transcendence. See their books *Thinking Through Confucius* and *Thinking from the Han: Self, Truth and Transcendence in Chinese and Western Culture*. See also Ames (1984).

3 Liu Shu-shien and others have observed that Confucianism as a cultural ideal embodying certain spiritual values and aspirations should be differentiated from Confucianism as embedded in social and political ideologies and institutions. See his chapter in Tu (1996). See also Liu (1998).

4 While one could utilize certain Western definitions of religion to illustrate that Confucianism is a religion, these definitions may limit the understanding of the nature of Confucian spirituality. For example, we can draw on both Paul Tillich's and Frederick Streng's definitions of religion. For Tillich, religion focuses on ultimate concern, while Streng suggests that religion is a means of ultimate transformation. See Tillich (1957, 1952) and Streng (1985). Both of these broad definitions are applicable to Confucianism. Ultimate concern in Confucianism is evident when a person is responding to the will of Heaven that is discovered in one's Heavenly—endowed nature and manifest in temporal affairs. Ultimate transformation in Confucianism involves modes of self-cultivation, which are intellectual, spiritual, and moral. The goal here is to become more fully human, namely, more deeply empathetic and more comprehensively compassionate. Ultimate transformation leads one toward sagehood. Still, this attainment is within the phenomenal world, not apart from it, and for the benefit of the larger society, not for one's salvation alone. This distinguishes Confucian religiosity from Western forms of religion. See also Smith (1963) for a discussion of the nature of religion. Articulating the Confucian worldview (both philosophically and religiously) apart from Western categories has been the concern of Roger Ames and David Hall, who suggest that Confucianism is at once non-theistic and profoundly religious. See their commentary on the *Mean* in *Focusing the Familiar.*

5 Tu Weiming has observed: "The problem of whether Neo-Confucianism is a religion should not be confused with the more significant question: what does it mean to be religious in the Neo-Confucian community? The solution to the former often depends on the particular interpretive position we choose to take on what constitutes the paradigmatic example of a religion, which may have little to do with our knowledge about Neo-Confucianism as a spiritual tradition" (Tu 1985: 132).

6 I am indebted to the work of Wm. Theodore de Bary and Tu Weiming in this area. For one of the first comprehensive discussions of the religious dimensions of Confucianism see de Bary's introduction to *The Unfolding of Neo-Confucianism*. Similarly, see Tu Weiming's *Centrality and Commonality: An Essay on Confucian Religiousness*. In addition, the work of P.J. Ivanhoe and Roger Ames and David Hall has been significant. See P.J. Ivanhoe, *Confucian Moral Self Cultivation* and Roger Ames and David Hall's numerous books including their most recent, *Focusing the Familiar.*

7 It is important to note that this ordering of cosmos and society can have both life-enhancing and life-constraining dimensions. When used as political ideology in the Han period the record becomes more mixed.

8 The Confucians were, however, caught in matters of pragmatic politics of governance that often required not only an appeal to personal moral transformation and ritual practice as a means of restraint, but also recognized that law and punishment had their function, although as a secondary measure.

9 Many of the writings of Western Confucian scholars are being translated into Chinese as part of the renewed interest in Confucianism in China. These include works by Wm. Theodore de Bary, Tu

Weiming, Roger Ames and David Hall, Robert Neville, John Berthrong, and two volumes on *Confucian Spirituality* edited by Tu Weiming and Mary Evelyn Tucker.

10 Clearly the tension of acknowledging the historical particularity of Confucianism along with identifying certain overarching religious elements in the tradition is present here.

11 For example, in Japan Confucianism linked itself to Shinto during the seventeenth century, was separated from it by the nativists of the eighteenth century, and was rejoined to Shinto again in the late nineteenth century. Japanese Confucianism as a worldview and as a form of spiritual cultivation is still part of many of the New Religions in Japan and deserves further study. See Helen Hardacre's discussion of "The World View of the New Religions" in *Kurozumikyo and the New Religions of Japan* (Chapter 1).

12 Tu Weiming speaks of the New Confucians as the "Third Epoch of Confucian Humanism" after the classical and Neo-Confucian periods. See *Confucianism: The Dynamics of Tradition*, edited by Irene Eber, pp. 3–21. John Berthrong has outlined six periods of Confucianism that separate out the Han, Tang, and later Qing Evidential Learning. See *All Under Heaven*, pp. 77–83 and 191–192.

13 These include, among others, the *Book of Changes* (Third Appendix), the *Book of Ritual* (7th Chapter), the *Mean* (Chapter 22), Dong Zhongshu, *Luxuriant Gems of the Spring and Autumn Annals* (Chapter 44), the *Diagram of the Great Ultimate* of Zhou Dunyi, the *Western Inscription* of Zhang Zai, and the *Commentary on the Great Learning* by Wang Yangming. See these texts in Wm. Theodore de Bary and Irene Bloom, eds., *The Sources of Chinese Tradition*.

14 *Book of Changes* Appendix HI 2:1/8. See also the chapter on "Creative Principle" in Hellmut Wilhelm, *Heaven, Earth and Man in the Book of Changes*. The Neo-Confucians frequently refer to the productive and reproductive forces of the universe (Ch. *sheng sheng*, Jp. *sei sei*).

References

Ames, R. (1984) "Religiousness in Classical Confucianism: A Comparative Analysis," in *Asian Culture Quarterly* 12/ 2: 7–23.

—— and Hall, D. (1987) *Thinking Through Confucius*, Albany, NY: State University of New York.

—— and Hall, D. (1998) *Thinking from the Han: Self, Truth and Transcendence in Chinese and Western Culture*, Albany, NY: State University of New York Press.

—— and Hall, D. (2001) *Focusing the Familiar*, Albany, NY: State University of New York Press.

Berling, J. (1980) *The Syncretic Religion of Lin Chao-en*, New York: Columbia University Press.

Berry, T. (2003) "Affectivity in Classical Confucian Tradition," in *Confucian Spirituality*, Tu, W. and Tucker, M.E. (ed.), New York: The Crossroad Publishing Company.

Berthrong, J. (1994) *All Under Heaven: Transforming Paradigms in Confucian-Christian Dialogue*, Albany, NY: State University of New York Press.

Chan, W. trans. (1963) *A Source Book in Chinese Philosophy*, Princeton, NJ: Princeton University Press.

de Bary, W. (1953) "A Reappraisal of Neo-Confucianism," in A. Wright (ed.), *Studies in Chinese Thought. The American Anthropological Association* 55/5, Part 2, Memoir No. 75 (December).

—— (1975) *The Unfolding of Neo-Confucianism*, New York: Columbia University Press.

—— (1991) *Learning for One's Self*, New York: Columbia University Press.

—— and Bloom, I. eds. (1999) *The Sources of Chinese Tradition* (2 vols.), New York: Columbia University Press.

—— and Chaffee, J. (1989) *Neo-Confucian Education: The Formative Stage*, Berkeley, CA: University of California Press.

Eber, I., ed. (1986) *Confucianism: The Dynamics of Tradition*, New York: Macmillan.

Eno, R. (1990) *The Confucian Creation of Heaven*, Albany, NY: State University of New York Press.

Hardacre, H. (1986) "The World View of the New Religions," in *Kurozumikyo and the New Religions of Japan*, Princeton, NJ: Princeton University Press.

Henderson, J. (1984) *The Development and Decline of Chinese Cosmology*, New York: Columbia University Press.

Ivanhoe, P.J. (2000) *Confucian Moral Self Cultivation*, Indianapolis, IN: Hackett.

Jaspers, K. (1953) *Origin and Goal of History*, London: Routledge and Kegan Paul.

Keightley, D. (1978) "The Religious Commitment: Shang Theology and the Genesis of Chinese Political Culture," in *History of Religions* 17: 211–224.

Kreiner, J. ed. (1996) *The Impact of Traditional Thought on Present Day Japan*, Munich: Iudicium-Verl.

Liu, S. (1972) "The Confucian Approach to the Problem of Transcendence and Immanence," in *Philosophy East and West* 22/1: 45–52.

—— (1998) *Understanding Confucian Philosophy: Classical and Sung-Ming*, Westport, CT: Praeger.

Najita, T. (1987) *Visions of Virtue in Tokugawa Japan: The Kaitokudo Merchant Academy of Osaka*, Chicago: University of Chicago Press.

Needham, J. (1956) *Civilization in China* (Vol. 2), Cambridge: Cambridge University Press.

Queen, S. (1996) *From Chronicle to Canon: The Hermeneutics of the Spring and Autumn According to Tung Chung-shu*, New York: Cambridge University Press.

Rosemont, H. ed. (1984) *Explorations in Early Chinese Cosmology*, Chico, CA: Scholars Press.

Rubinger, R. (1982) *Private Academies of Tokugawa Japan*, Princeton, NJ: Princeton University Press.

Smith, K., Bol, P., Adler, J., and Wyatt, D. (1990) *Sung Dynasty Uses of the I Ching*, Princeton: Princeton University Press.

Smith, W. (1963) *The Meaning and End of Religion*, New York: Macmillan.

Streng, F. (1985) *Understanding Religious Life*, Belmont, CA: Wadsworth Publishing Co.

Taylor, R. (1990) *The Religious Dimensions of Confucianism*, Albany, NY: State University of New York Press.

Tillich, P. (1952) *The Courage to Be*, New York: Harper and Row.

—— (1957) *The Dynamics of Faith*, New York: Harper and Row.

Tu, W. (1985) *Confucian Thought: Selfhood as Creative Transformation*, Albany, NY: State University of New York Press.

—— (1989) *Centrality and Commonality*, Albany, NY: State University of New York Press.

—— (1996) *Confucian Traditions in East Asian Modernity: Moral Education and Economic Culture in Japan and the Four Mini Dragons*, Cambridge, MA: Harvard University Press.

—— and Tucker, M.E. (2003) *Confucian Spirituality* (2 vols.), New York: The Crossroad Publishing Company.

Tucker, M.E. (1989) *Moral and Spiritual Cultivation in Japanese Neo-Confucianism: The Life and Thought of Kaibara Ekken*, Albany, NY: State University of New York Press.

—— (trans) (2007) *The Philosophy of Qi: The Record of Great Doubts* (Kaibara Ekken), New York: Columbia University Press.

—— and Berthrong, J. (1998) *Confucianism and Ecology: The Interrelation of Heaven, Earth, and Humans*, Cambridge, MA: Harvard Center for the Study of World Religions.

Wilhelm, H. (1977) *Heaven, Earth and Man in the Book of Changes*, Seattle, WA: University of Washington Press.

PART VI

Putting religious naturalism into practice

25

RELIGIOUS NATURALISM AND THE SPIRIT OF QUERY

Taking adult religious education in a new direction

Pamela C. Crosby

Rigorous and conscientious inquiry is a key element in flourishing societies and organizations (Wildman 2010: 172). As Aristotle argued, we cannot always rely on experts to make decisions for us in every perplexing situation so we must learn to judge for ourselves what is the right thing to do. Such mature judgment, developed by experiences as self-directed decision-makers, is crucial in the development of good moral character (Aristotle 1109b 30–1115a 5).

If educating citizens to be thoughtful inquirers is necessary for a thriving society, then it follows that religious movements, institutions, and organizations—which continue to play significant roles in the lives of many persons today—should foster communities of inquiry. While faith institutions and organizations offer different types of worship programs, activities, and service and social opportunities, the ideal place for religious groups to engage in inquiry is within their religious education programs. Although religious education leaders of faith institutions in the USA consider adult religious education as a vital factor for the growth and development of their communities (Rowland 2007), it is not clear to what extent they value rigorous inquiry as a primary factor in spiritual and moral development.

Because Religious Naturalism (RN) is a burgeoning movement and unique in its religious outlook in fundamental ways compared to most Western traditions, now is the time to lay the groundwork for conceiving and generating adult religious educational models that draw from the successes of other educational programs and to take into account RN's specific characteristics. In this chapter, I construct a framework for RN adult education as a particular kind of inquiry to be used in the context of a class setting. This type of inquiry, based on the philosopher Justus Buchler's works, is a species of the genus *query*, a systematic approach to judgment and investigation (e.g., Buchler 1985:7). In addition to the philosophical and practical distinctiveness of inquiry as query, I provide contemporary research findings that support and advance Buchler's theory of query. While many types of religious practice including rituals, devotionals, and mediations are essential in a comprehensive adult education model, the focus of this chapter is on inquiry as practiced in a class or group discussion.

Religion of Nature

As in all religious movements, RN takes many forms and perspectives. Religion of Nature, as professed by Donald A. Crosby, is one type of RN perspective. For Crosby, RN is the "recognition that to be is to be natural and the conviction that nature in all of its forms and manifestations is a proper focus of religious commitment." It does not include "anything other than nature, meaning that, for religious naturalism, there is no such thing as supposed *super*natural being, beings, regions, revelations, origins, purposes, destinies" (Crosby 2014: 5).

According to Religion of Nature, the universe has always existed in some way, most likely in previous multiple forms, and thus, there is no need for a creator to explain its existence. Neither does Religion of Nature subscribe to a form of pantheism or panentheism; there is no divine existence required or acknowledged as a ground of nature. In addition, nature itself—as a whole—is without purpose. Although there is nothing real outside of nature, such as a deity or spirit, and there is no overall purpose of nature, nature itself can be a religious ultimate, an object of intensely satisfying and enriching devotion and faith. The basis of such devotion concerns nature's life-giving, healing, and sustaining power and its beauty and wonder.

Although such a nontheist outlook is in contrast to traditional Western and Middle Eastern religions, nontheist religions have inspired spiritual devotion and practice throughout the ages and throughout the world for millions of persons. Such religions include Advaita Vedanta Hinduism, Theravada Buddhism, and Daoism. As these religions have incorporated rituals and practices, so too can Religion of Nature. Acts of religious devotion include innumerable expressions of the profound recognition that life is precious. As human beings, we can reflect with reason and appreciation the wonders of nature with its almost infinite diversity of magnificent creatures with whom we share our home. As a result, religious practice might involve meditations and rituals of gratefulness for nature as our nurturing home and celebrations of its mysterious abundance and dynamism. While such gratefulness and celebration are appropriate, nature is not a personal being and is not suitable as a presence to be worshiped.

Closely related to rituals that express appreciation and celebration of nature, is another important practice of Religion of Nature: the articulation and demonstration of a genuine commitment to advancing nature's well-being including the flourishing of its creatures and their ecological habitats. While nature is not purposeful in itself as a whole, diverse purposes are present within nature because life in its various forms strives to create, communicate, solve problems, and act in distinctive ways. Religious purpose and meaning for individuals who sincerely practice Religion of Nature are focused on their responsibilities to nature's various needs in the present rather than on yearnings for immortality and heavenly reward. A different way of viewing immortality for Religion of Nature adherents is the influence of the good they do for others in the span of their lives that replicates and multiplies after they die (Crosby 2014: 21).

As followers of RN should be committed to working toward ecological flourishing for all of its creatures, so should they also be committed to expressing openness toward other religious outlooks. As Crosby emphasizes: "No one religious faith however compelling or conclusive it may seem is adequate to capture the full range of the mysteries of life or the meanings, prospects, and demands of human existence" (Crosby 2014: 24). The responsibility to care for the world and its creatures involves caring for others who differ from us in countless respects, including their secular worldviews or other kinds of religious faith.

Yet, Crosby admits that we must not neglect the power of nature to cause harm in numerous ways to numerous beings. While nature's laws are generally beneficial to sentient beings and their ecological homes, they can have harmful effects as well. Furthermore, human freedom allows persons to make evil decisions and perform evil actions. Human beings wreak devastating suffering and loss on one another and the world. Thus, human evil is a reality that must be confronted.

In light of the fact that there is no deity persons can petition for help, responsibility to care for the world and its inhabitants lies with human beings—such responsibility cannot be placed otherwise, argues Crosby. To have considerable enthusiasm, inspiration, or compassion is not enough to make substantial improvements. Religious naturalists must have the astute reasoning, resourcefulness, imagination, and moral insight that can effect substantial beneficial changes in the life of the planet and its inhabitants. As we shall see, the means to acquire such qualities (according to Justus Buchler) is by learning to engage in query.

Introduction to query

A central concept of Buchler's theory of query is *judgment*. Buchler explains that a person must inescapably take various positions throughout her life, and when she does, she makes judgments. Judgments create the history that defines who she is and the net result of all the commitments she has made thus far. Making a judgment for Buchler is much broader than a mental process. It includes saying (assertive judging), making (exhibitive judging), or doing (active judging) something; it is an "appraisal" or a "pronouncement" (Buchler 1985:13).

As each one of these modes of judgment describes a particular function that distinguishes it from the other modes, so is each one characterized by how it is evaluated. An *assertion* (saying something) is evaluated in its most fundamental form on the basis of its truth or falsity. Assertion is not merely a form of discourse. Nodding can be a mode of judging assertively, for example. *Exhibitive judgment* is judging by ordering elements, materials, and/or features into a structure. A structure or contrivance is evaluated on the basis of its being good or bad, and includes the selection of elements and their relations to each other and to the arrangement. Judging *actively* is behaving in some manner. An act is assessed on the basis of it rightness in conduct—particularly in terms of moral rightness (Buchler 1985: 27–28). A *product*, another technical term for Buchler, is the outcome of a process of judging and is often used by Buchler interchangeably with judgment (Buchler: 1985: 3–10).

Query is the "ramifying" of any mode of judgment "systematically" (Buchler 1985: 58). A judgment is ramified when it combines with other judgments to create a more complex whole or process. Although many conditions are involved in the query process, I highlight the following conditions.

First, query involves methodic *reflexive* and *social communication*. Reflexive communication actualizes what Buchler calls the *reflexive community*, which is the individual in dialog with herself. It is "self-directed" communication with the self, being both "interrogator and advocate" in ongoing critical selection, evaluation, and validation that add to the many dimensions of her distinctive world (Buchler 1985: 7).

When engaging in query, the person experiences "insight," "inspiration," and "intuition," which are "characteristic events" that make up the "interrogative temper," a condition more like "hunger" than just "puzzlement" (Buchler 1985: 75). The interrogative temper is not a phase in the process of query, but an outcome, as the person seeks to discern the most suitable possibilities and to "concretize" them as she "hovers between alternative and choice, between selection and evaluation" (Buchler 1985: 77, 70).

A diligent participant in a class discussion in a community of query exemplifies this striving. In "What Is a Discussion?" Buchler sheds light on what participants in this type of discussion might look like. Students search for answers but also question answers, question the process, and challenge alleged support of claims. Included in the discussion process is thoughtful listening with time for reflection while tracing paths; manipulating and assimilating the product; and discerning obstacles, struggles, and one's own emotional responses. Observing how such navigation takes place, together with the elements that comprise it, is what Buchler calls *ideational awareness*, an important aim of the classroom discussion. In short, it is awareness of how ideas and theories are born, formulated, rejected, affirmed, and modified; it is the witnessing of their potentials and limits (Buchler 1954: 8).

Social communication, the second kind of communication, is communication with others. The products and process of social communication extend the self because they influence the self and are influenced by it (Buchler 1985: 7; 1951: 53). Query cannot be something that one possesses as one's own, and the more accessible to others the products of query become, the greater the chance for their complexity and achievement. The process of query thrives in collaborative conditions of a social community with its diverse perspectives.

As part of the social community in a discussion process, fellow inquirers develop and establish collaboratively a product.[1] Unlike a lecture, students can experience how a product is created, augmented, modified, or rejected and how this process affects persons in a group setting (Buchler 1954: 8, 13). While Buchler does not advocate imposing strict rules or structure when facilitating discussions, he emphasizes that a successful discussion depends upon a skilled, knowledgeable, perceptive, and creative teacher as a critical member of the social community of inquiry, so that she can stimulate query and its flourishing. While teachers should be adept at handling problems that often arise in a group discussion, such as when participants become unruly or dominate the conversation, the teacher is much more than a facilitator. She contributes to the discussion in valuable ways including modeling the discussion process as an exemplary inquirer/querer.

Communication in reflexive or social form is part of the process of *creating meaning*, another condition of query. Meaning occurs when a person asserts, makes, or does something that relates to something significant in her life, when she judges assertively, exhibitively, or actively in relation to some other meaningful thing. Meaning is dependent upon *articulation*. Articulation is not strictly linguistic; it is not just a part of the assertive judgmental process but is a ramification of any mode of judgment, thus a changing of the product in a substantial way. Any mode of judgment can articulate another mode of judgment to create meaning such as *making* something related to an *action*; *doing* something related to an *assertion*; or *saying* something related to an array or *art* display. For example, we can present a historical event by means of an artwork. Articulation, when it involves purposeful and systematic ramifications, happens at each moment of query, where judgment is intensified or new possibilities are created and examined (Buchler 1985: 84–88).

Unlike listening to a lecture, when concepts are often introduced too hastily for many listeners, participating in a discussion can provide the proper conditions for the articulation of ideas. Also in a lecture, information is often presented as a finished product to listeners who may have little or no commitment to engagement. Although lectures can be efficient ways to transmit information, they can create tempting situations for students to be inattentive. This lack of focus and lack of time for reflection adversely affect the amount of knowledge listeners can articulate (Buchler 1954: 8–13).

While creating and finding meaning are essential for authentic engagement in a discussion, meaning by itself is not sufficient as an indicator of appropriate growth and maturity. Meaningful engagement should connect to desired outcomes of the particular course of study as well as to those of the mission, purpose, and broader aims of the institution that relate to student learning and growth. Buchler emphasizes that a demanding reading program is indispensable for a meaningful discussion when assignments align with appropriate outcomes. They are a means for authors with relevant expertise to enter the community of query to be acknowledged and challenged.

The *increase of complexity*, a third condition of query highlighted here, is closely connected to articulation and meaning, and its multi-pathways are analogous to the growth of tree branches as they sprout and create new offshoots. The branches change in number, character, and direction, pressing against other trees and vegetation—adding their growth and crowding out other growth, adjusting to their surroundings in new ways and compelling the world to adjust to them. During processes of creativity, destruction, novelty, and continuity, each element of the process affects the whole and vice versa. The process is not simply an accumulation of elements.

In a discussion taking place in a community of query, there is an increase in the complexity of method, involvement, content, and results. Its complexity calls for new and more difficult vocabulary, effective methods, and challenging claims and theories that bring about multifaceted growth. Discussion as a process of query is not a bull session, a gossip fest, or a summary of current events. While there is no predetermination of where the path of query proceeds, its method is intentional and its purpose is to gain new insights by means of ongoing query.

The fourth condition of query highlighted here is the use of *more than one mode of judgment*. Although one mode of judgment usually plays a dominant role in a query process, rarely if ever is only one mode of judgment involved. Because query is a holistic process that occurs in the midst of life itself, the modes of judgment are not neatly separated.

Class discussion is primarily a form of assertive judging, and Buchler views inquiry as a systematic form of assertion because it involves affirming and rejecting claims. Yet, students are also involved in devising and arranging ideas into novel items (exhibitive mode), such as methods, viewpoints, arguments, theories, and hypotheses. Discussion products can also include different types of viewpoints, elaborations of definitions, and an increase in awareness of various aspects. Thus, products are brought about continually throughout any discussion, and they often are not assertions. In short, says Buchler, products in a class discussion can be outcomes of an "envisioning or of an exhibiting than of an affirming" (Buchler 1954: 14).

A final condition discussed here is *relative fulfillment*. Query persists beyond any product and can do so in two ways: (1) a product of query often arouses the need for further query in terms of novel relations to the current product, and/or (2) a product of query often stimulates a need or desire for a new product (Buchler 1985: 62). Thus, the spirit of query goes on even though at times it may be suspended after a period of selection and evaluation. The significance, consequences, or meaning of a product of query are never final or exhaustive. The value of a product of query should be evaluated on how it stimulates further action and further query.

During a discussion in a community of inquiry as a form of query, each product is part of a larger product, and each product depends on other products that were produced before it. As methods, ideas, and needs change, they influence future products. Buchler cautions that teachers often confuse a product with a conclusion or resolution. Some think that nothing has been accomplished unless resolution is final. Yet, Buchler argues that teachers and students should not aim at some fixed resolution. A resolution should be only a relative conclusion—being reached after much struggle and growth and in accordance with the conditions of query, becoming a springboard to new products of query.

Community of inquiry

How does contemporary research on learning in higher education and adult education support Buchler's theory of class discussion by way of inquiry as a means of query? Here are some insights from leading scholars who research the theory and practice of inquiry as educational process.

In the 1980s, Matthew Lipman, Buchler's former student at Columbia and a scholar of Dewey and Peirce, popularized the term *community of inquiry* in educational research. For Lipman, communities of inquiry can foster critical thinking skills in students of all ages. They involve a quest for meaning, inclusiveness, "shared cognitions," and thinking for oneself (Lipman 2003: 94). The *community of inquiry framework* (CoI), developed in 2000 and drawn from the works of Peirce, Dewey, Lipman, and others, is often used for designing interactive learning environments based on collaborative inquiry in the higher education setting. The goal of CoI is for students to create meaning and validate understanding while solving problems collaboratively in a community of trust, common purpose, and interdependence. Because CoI closely resembles Buchler's community of inquiry as query model, CoI's research findings are helpful in supporting Buchler's educational theory and constitute a link between his work and adult religious education research and practice. CoI presents a methodological approach that can assist adult educators in designing, implementing, and evaluating a community of inquiry as a form of query.

CoI's methodological approach to thinking and learning has three interdependent elements: *cognitive presence, social presence*, and *teaching presence* (Garrison 2016: 58–63). *Cognitive presence* (which compares with Buchler's reflexive communication and ideational awareness) involves students dealing with a *triggering event* that poses some problem and causes a state of discontent (Garrison, Anderson, and Archer 2000: 98). Next, there is investigation in response to this discontent (compares with Buchler's interrogative temper), involving a search for relevant information and possible options in the context of reflection and clarification, followed by integration of the relevant findings into a coherent concept, belief, or hypothesis (compares with Buchler's meaning and articulation). Finally, resolution is achieved, but the resolution is often a means for further inquiry (compares with query's relative fulfillment).

The second element of CoI is *social presence*, the capacity of students to identify with their group and the group's purpose in an openly participatory climate where no one perspective dominates (Garrison 2016: 60–61). CoI integrates the reflective process (compares with query's reflexive communication) with collaborative and interactive questioning and problem-solving (compares with Buchler's social communication). In other words, what is personally meaningful is shared and evaluated in a community of discourse and reflection.

CoI's concept of *teaching presence*, the third element, is much broader than Buchler's concept of teacher (Garrison 2016: 61–62). For CoI, teaching presence indicates an equally shared responsibility of monitoring and directing the learning of the group with the encouragement and guidance of the community. Such shared responsibility is the goal of the formally assigned instructor. Because teaching presence involves shared responsibility, students have dual roles of teacher and learner that involve metacognitive awareness that includes "self and co-regulation" activities and "recursively moving between the private and shared worlds" (Garrison 2016: 65).

Students monitor their own personal multiple paths of thought, and such metacognitive awareness compares with Buchler's ideational awareness. Like query, the learning process becomes more complex as more perspectives are involved. Students monitor ways their ideas interact and react with those of others. During the monitoring process, they apply corrective approaches to feedback from others and formulate new insights.

More research is needed to show how CoI resembles query in incorporating what Buchler describes as other modes of judgment: exhibitive (e.g., art and music) and active (e.g., character education, service learning, sports, administration), which would further the complexity and effectiveness of inquiry that exemplifies the assertive mode of judgment.

Cooperative inquiry

Co-operative inquiry (CI) is a systematic, democratic approach to inquiry that provides a group setting for adults to generate knowledge, which is applied to achieve practical goals (Reason 1999: 207; Kasl and Yorks 2010: 315). Participants examine their individual experience and knowledge and that of their group as a whole. CI involves adult participants in small groups who are both co-researchers and co-subjects. As co-researchers, they generate ideas, plan, administer, and assess a project, and as co-subjects, they participate in its activity.

Initiators establish a group by determining who has interest in taking part in particular activities and/or learning more about particular topics or concerns. Often recruitment is by some type of invitation letter. A co-operative group is ideally about 12 persons. Groups decide the number of meetings and meeting times. Some groups might meet for only a weekend, while others might meet for much longer. Group participants decide what types of roles are needed and who assumes these roles, keeping in mind that roles evolve. Because some topics can help raise deep concerns that often introduce emotional reactions, the group plans ways to deal with conflict and anguish as they arise.

An important feature of CI is the *four phases of action and reflection* (Reason 1999: 209–210). *Phase one* involves a small group organizing to address an issue that members all think is important (compares with Buchler's reflexive and social communication). They explore relevant aspects of the issue and formulate a set of questions or propositions they want to investigate. They examine beliefs, data, and theories critically, understanding that complete objectivity is impossible.

In *phase two*, participants commence action (compares with Buchler's active mode of judgment) and begin to record the consequences of their actions and outcomes. By *phase three*, the participants are fully immersed in the activity's experience. Their attitudes and actions can be various. They may expand on their experience in more profound ways and their understanding may become more complex, more perplexing, and more authentic (compares with Buchler's increase in complexity). However, they may reject their original proposal and seek alternate actions and understanding. It is possible also that any number of participants will lose their identification or interest with the group or activity for some reason.

Instead of just focusing only on the outcomes of the group and becoming too attached to goals, participants focus also on what they are actually learning themselves (compares with Buchler's ideational awareness). In light of learning and improving as ongoing goals, the group prepares for disorder, contingency, and risk so that spontaneity and creativity can be managed positively.

In *phase four*, the participants return to their group to consider their posed questions. They may alter their questions, reject them, or pose new ones and propose a new action/reflection cycle that may include changes in variables in any phase based on feedback from the earlier cycle and may include six to ten cycles (compares with Buchler's relative fulfillment).

Another important feature of CI is its *four different types of knowing* (Reason 1999: 211). First, *experiential knowing* is a face-to-face engagement with a person, process, place, or thing. It involves a kind of appreciation of, recognition of, and holistic experience with another (compares with Buchler's social communication). *Presentational knowing* (compares with Buchler's exhibitive mode of judgment) comes out of experiential knowing and provides the first form of expression through fine, performing, or literary art. *Propositional knowing* (compares with Buchler's assertive mode of judgment) involves knowing by means of ideas and theories, expressed in propositional statements. *Practical knowing* (compares with Buchler's active mode of judgment) is knowing expressed in skillful useful action.

According to CI, learning is more genuine, profound, and sound if it is consistent with these four ways of knowing, supported by means of experience, expressed through art and stories, comprehended by virtue of ideas and theories, and embodied and expressed by action. Participants learn how the four forms of knowing overlap and differ and how these are interdependent upon the learning without one way of knowing being dominant (in contrast to query, when one mode of judgment is always dominant in the learning process). At different stages of the process, more action or reflection is needed, but too much reflection hinders action, and too little reflection leads to problems ranging from disorder, poor results, and wasted efforts.

During the repetition of the four-phased cycle incorporating four ways of knowing, participants apply research in *convergent* ways, that is, they examine in more detail the same issues. Participants also apply research in *divergent* ways, that is, they explore other issues in consecutive cycles (Reason 1999: 212). Often both ways (comparing with Buchler's ramification and articulation of judgment) are used in the advancement of inquiry. The downside to CI is that participants can continue in cycles without any creative or critical effort. On the other hand, persons can be tempted to ignore the cyclical format and stick with a traditional linear process that is simpler, more familiar, and more predictable.

Drawing from other research on adult education are the following two important insights. One important aspect of adult learning is that adults bring to the educational setting a rich array of resources and skills, including problem-solving, which they have drawn from life experiences (Knowles, Holton, and Swanson 2015: 90). Another critical insight is the importance of a conducive climate for adult religious education, which includes emotional support of the group and its members along with physical comfort in the group setting (McKenzie and Harton 2002: 205–208).

Online inquiry and learning

Because online discussion opportunities are likely to be the most popular venue for RN adult religious inquiry since the number of religious naturalists is small and dispersed, it is useful to look at what researchers of online inquiry have found that works most effectively. The findings of CoI, which were discussed earlier that support an inquiry framework, are especially useful because of CoI's emphasis on online interaction as a valuable learning environment.

Online learning is communication by means of networked devices to achieve learning outcomes (Garrison 2016: 43). It provides a means to involve participants in the learning process without the limitations of time and space. As technology has advanced, more opportunities for two-way learning have emerged. Two-way learning has immense advantages over one-way learning that characterized distance education in its early days.

Online learning can accommodate asynchronous and synchronous types of communication, including text, oral, graphic, pictorial, and video. Researchers are finding that there are many advantages for particular types of learning in an asynchronous learning environment. For example, written communication in an online discussion board allows students to have time to summon, organize, synthesize, and generate ideas and theories before responding to others. In addition to decreasing demands on memory, it allows for greater opportunity for resourcefulness. Synchronous environments in which all assigned students are expected to participate in real time include chat rooms, teleconferences, videoconferences, telephone conferences, and webcasts. Therefore, it is possible for individuals to participate as inquirers asynchronously by means of written interaction and synchronously by means of oral interaction.

Blended learning integrates face-to-face communication (that is, communication without any mediating technology) with various online approaches and technologies. Such a hybrid format allows for spontaneous oral dialogue and the building of community support while providing needed time for reflection, analysis, and creativity that an asynchronous written format can support (Garrison and Kanuka 2004: 97). Another advantage is an administrative one in that the number of scheduled face-to-face meetings can be reduced and collaborative work can be done online in variously sized groups. One example might be planning face-to-face interaction at the beginning of a class in order to build trust and cohesion and set the groundwork for later deeper learning through asynchronous, written, critical discussion.

A further example might involve students listening to an online lecture beforehand and then coming to class to participate in face-to-face discussion, followed by further reflection and dialogue through an online discussion board. Every learning context has unique characteristics and requires ongoing monitoring and adjustments in order for members and leaders to approach the ideal blended environment.

Moral implications of inquiry as a practice of query

A religious naturalist model for adult education can provide unique approaches to inquiry relating to the sacred relationship of humans with the natural world, including all of its creatures, and concerning efforts that seek ecological sustainability and social justice. On the one hand, an RN approach to adult religious education should involve an inquiry model that is action-based in order to embolden and mobilize its faith community to address environmental issues while contributing to the spiritual development of its members. On the other hand, any advocacy group is likely to be much more effective if it is driven by rigorous inquiry so that it can channel its advocacy in more methodological, purposeful, and insightful ways.

Another important factor should be considered when proposing a theory of inquiry as query for RN adult education. While religious practice and belief are not reducible to moral actions, a RN adult education framework should clearly articulate the moral implications of inquiry as query.

For Buchler, there is no religious dogma on which to found ethical behavior because there is no privileged center from which any individual views the world. There is no all-seeing, all-knowing being to whom persons can appeal to justify religious and ethical codes of conduct. However, the ability to judge proper moral conduct is necessary in order for individuals to be moral beings. Being committed to an end or a means to an end, as opposed to other ends or means, is often a matter of moral choice. While systematic moral action is described as query in active mode, for Buchler, the process of query in any mode can have moral implications

Whenever there are conflicting perspectives about proper action, the morally responsible person learns to evaluate perspectives purposefully, including her own, without regard to ego. The spirit of query calls for inclusion of relevant judgments while never viewing any process involving moral judgments as final. Query's moral grounding is in its purposive and methodical ramification of judgments that are evaluated on the basis of their validation by individuals and community in light of resourceful, reasoned, and democratic challenges and advocacy (Buchler 1951: 161, 167).

In light of query's moral implications, Buchler disagrees with those who accuse naturalistic ethics as rendering all moral decisions the same level and who say no one decision can have any more justification than the other without a deity to offer it moral grounding. Furthermore, he criticizes religious dogma that rejects the fundamental elements of validation in query. Adherents to orthodox religions, he says, seek to isolate themselves from the outside world and dwell on ways to uphold what they view as fixed principles at the price of interrogative investigation (Buchler 1951: 161).

Religious Naturalist Community of Query

In this section I propose a framework I call Religious Naturalist Community of Query (RNCQ) as one approach to inquiry for RN. I present various ways to characterize it. In most instances, there is considerable overlap of content in the sections.

I. Conditions for learning

A. Interrogative reflexive and social communication

- Seeking answers to questions and resolutions to problems
- Applying cognitive tools such as reasons and rules
- Applying cognitive skills such as comparing and contrasting
- Addressing concerns
- Responding to a sense of wonder
- Listening and reflecting
- Determining appropriate resources
- Dialoging with self
- Tracing a multipath of query
- Revising, rejecting, recreating, re-proposing, assessing
- Collaborating in creating a product

B. Purposive articulation for meaning

- Connecting to life activities and experiences and prior learning
- Restating, expanding, and ramifying concepts and tasks
- Elaborating, amplifying, and augmenting
- Framing action with overall purpose
- Relating specific needs to personal needs
- Identifying relevant data and support of claims
- Interpreting content and actions
- Manipulating, assimilating, and adjusting

C. Increase in complexity

- Envisioning potentials
- Determining most appropriate possibilities
- Augmenting possibilities
- Assuming both researcher and subject roles
- Identifying connections and disconnections
- Creating new connections and eliminating other ones
- Devising and arranging elements into new more complex structures
- Planning for and taking new risks
- Moving in multiple trajectories

D. Integration of three modes of judgment

- Affirming and disconfirming claims, theories, and arguments (assertive)
- Assuming and defending positions (assertive)
- Explaining different viewpoints (exhibitive)
- Expanding models and definitions (exhibitive)
- Designing new arrangements (exhibitive)
- Performing actions (active)
- Leading initiatives (active)
- Participating in mental and physical activities (all three modes)

E. Relative fulfillment

- Moving toward initial conclusions and resolutions
- Transforming ends as means
- Producing products that are parts of larger products or shaping future products

II. Three essential elements

A. Rigorous study and/or research preparation

- Responding to new and challenging content and tasks
- Acquiring new skills
- Learning and applying new vocabulary, concepts, and models
- Reading about new theories, opinions, and historical contexts
- Taking notes and comparing notes
- Framing new learning in context with past experience
- Identifying themes, claims, arguments, and supporting details
- Identifying examples and counter-examples

B. Shared facilitation: sharing responsibilities such as

- Identifying group setting and parameters
- Identifying outcomes
- Designing, questioning, and guiding the process and methods
- Presenting arguments and theories
- Directing the learning
- Monitoring and adjusting
- Explaining and clarifying
- Planning for conflict
- Identifying and responding to struggles
- Reacting to criticism

C. Conducive climate for learning

- Planning for ways to handle conflict
- Mutually supporting others' contributions
- Providing physical and emotional comfort and needs

- Engaging in freedom of expression
- Providing constructive criticism
- Presenting and responding respectfully and honestly to others

III. Five phase cycle

A. Tentative ground rules

- Selecting format (face-to-face, online, or hybrid)
- Deciding the length of the course or study
- Choosing a venue or venues
- Selecting a number of participants
- Outlining management of discussion procedure, including dealing with conflict

B. Choice of topic, question, problem, or goal

- Exploring relevant aspects of issues
- Presenting possible beliefs and theories for consideration
- Formulating sets of questions or propositions for investigation

C. Concept mapping and task analysis

- Mapping paths, concepts, categories, and connections
- Identifying general approaches
- Identifying specific actions
- Correcting approaches from feedback (if in a succeeding cycle)
- Accessing claims critically and avoiding logical fallacies
- Devising and refuting arguments
- Selecting resources

D. Projects or activities: testing ideas and theories (or searching for answers) in new contexts by

- Forging partnerships with agencies and/or religious institutions
- Helping to launch local initiatives
- Writing articles and opinion pieces
- Creating websites
- Volunteering in schools
- Organizing protests
- Sponsoring events
- Meeting with government officials

E. Assessment

- Measuring effectiveness
- Recording consequences and outcomes
- Evaluating criteria for assessment
- Adjusting on the basis of findings and feedback

- Selecting what to leave behind and what to use for further query
- Formulating new proposals
- Considering alternate approaches, resources, and assessments
- Considering new questions

IV. Suggested curricula resources

- Philosophical, religious, literary, historical, ethical, and scientific texts
- Spiritual texts (various religious and secular sources)
- Articles and books on environmental justice, activism, policy, science
- Policy position statements and statement proposals
- Environmental laws and their implications
- Social justice proposals and policies

V. Suggested communication techniques for learning

- Brainstorming, reflecting, and collaborating by means of text, oral, graphic, pictorial, and video communication
- Defending and assessing actions and arguments in chat rooms and discussion boards
- Reaching out to the community by means of teleconferences, videoconferences, telephone conferences, and webcasts
- Viewing and presenting live and recorded talks via online streaming, online video, and face-to-face
- Building trust and understanding with fellow inquirers and community members with face-to-face interaction in seminars and workshops

VI. Sample questions for discussion

- What sorts of practices and rituals can we build on and create anew that will enrich our appreciation and celebration of nature?
- How can we create various purposes within nature—including those that relate to problem-solving, creating, communicating, acting, and being happy—that matter to us, to the earth, and to our fellow creatures?
- How can we articulate and demonstrate a genuine commitment to promoting the well-being of nature, such as nurturing the welfare of its creatures and helping their ecological habitats to thrive?
- How can we devote ourselves not only to fulfilling current needs but future needs of our earth and its beings as a whole rather than being focused on immortality for ourselves?
- How do we investigate the implications of human freedom, human evil, and natural evil in a world where there is no deity to whom persons can petition for help?
- How can we address acts of moral evil (e.g., overuse of resources, fracking, destruction of native people's lands) that create acts of natural evil (e.g., climate disruption, earthquakes, increase in intensity of hurricanes)?
- How can we determine ways to demonstrate toleration while welcoming other religious outlooks in order to enrich our own religious faith and build community with the perspectives of others about the mysteries of life, its meanings and demands?

Suggested activities

- Engaging in discourse with other faith groups and institutions on similarities and differences among various faith beliefs and practices relating to environmental justice in conjunction with a shared community project of recycling and waste reduction
- Exploring global and local environmental issues relating to fair trade and supporting one relevant local issue with political demonstrations, literary and/or art projects, or promotional materials
- Researching local options for energy efficiency and conservation and presenting at local schools or young people's groups a short play dramatizing possible problems and inviting suggestions for possible solutions and assessing these suggestions
- Welcoming representatives of other faiths as co-researchers of historical and geographical impacts of drought or flooding and creating together shared rituals and ceremonies relating to water as metaphor
- Investigating causes of human and nonhuman suffering and visiting places that display art objects that express this theme
- Discussing religious and philosophical implications of nature as a religious ultimate and assuming roles as advocates of and challengers to this belief in a debate
- Examining rituals and their histories and devising rituals appropriate for RN
- Identifying the miraculous or wondrous in nature, researching examples of such wonders, and expressing appreciation through music and poetry
- Examining philosophical and religious issues relating to environmental justice and advocacy and inviting others to engage in community dialogue
- Exploring the concepts of moral and natural evil and assembling in a rural area or city park to meditate in the midst of non-human surroundings on human interpretations of good and evil
- Researching the roles of supernatural beliefs in religious history and the implications these roles have for religious outlooks and social practice
- Identifying logical fallacies and published examples of these fallacies and planning community workshops and online seminars on effective argumentation relating to environmental issues

Conclusion

RNCQ is only one approach that might provide a starting point for religious naturalists to plan and implement adult religious education programs and initiatives. Scholars and leaders are encouraged to apply, evaluate, and build upon this proposed framework in order to enrich and extend engagement in inquiry as a form of query. If Religious Naturalists value rigorous inquiry, then future development that contributes to the augmentation, feedback correctives, and enrichment of research and practice in RN adult education is a critical factor in the success of RN as an emerging movement.

Note

1 In this journal article "What Is a Discussion," whose audience is primarily educators, Buchler uses only the term *product*, and not *judgment* (Buchler 1954: 9–14).

References

Aristotle (1985) *Nicomachean Ethics*, trans. T. Irwin, Indianapolis, IN: Hackett Publishing Company.
Buchler, J. (1985) *Nature and Judgment*, Lanham, MD: United Press of America.
—— (1951) *Toward a General Theory of Human Judgment*, New York: Columbia University Press.
—— (1954) "What Is a Discussion?" *Journal of General Education* (October) 8/1: 7–17.
Crosby, D. (2014) "Religious Naturalism and Its Place in the Family of Religions," *The Fourth R* (January–February) 27/1, 5–8; 20–21; 24.
Garrison, D. R. (2016) *Thinking Collaboratively: Learning in a Community of Inquiry*, New York: Routledge.
Garrison, D. R. and H. Kanuka (2004) "Blended Learning: Uncovering Its Transformative Potential in Higher Education," *Higher Education* 7: 95–105.
Garrison, D. R., T. Anderson and W. Archer (2000) "Text-Based Environment: Computer Conferencing in Higher Education," *The Internet and Higher Education* 2/2–3: 87–105.
Kasl, E. and L. Yorks (2010) "Whose Inquiry Is This Anyway? Money, Power, Reports, and Collaborative Inquiry," *Adult Education Quarterly* 60/4: 315–338.
Knowles, M. S., E. F. Holton III, and R. Swanson (2015) *The Adult Learner: The Definitive Classic in Adult Education and Human Resource Development*, 8th edn, New York: Routledge.
Lipman, M. (2003) *Thinking in Education*, 2nd edn, Cambridge: Cambridge University Press.
McKenzie, L. and R. M. Harton (2002) *The Religious Education of Adults*, Macon, GA: Smyth and Helwys.
Reason, P. (1999) "Integrating Action and Reflection Through Co-operative Inquiry," *Management Learning* 30/2: 207–226.
Rowland, M. L. (2007) "Faith and Adult Learning," *Adult Learning* (Winter/Spring) 18/1–2: 4–5.
Wildman, W. J. (2010) *Religious Philosophy as Multidisciplinary Comparative Inquiry: Envisioning a Future for the Philosophy of Religion*, Albany, NY: State University of New York Press.

26

BRINGING RELIGIOUS NATURALISTS TOGETHER ONLINE

Ursula Goodenough, Michael Cavanaugh, and Todd Macalister

The Religious Naturalist Association

What do the following have in common?

> Farmer, architect, chef, speech pathologist, homemaker, scientist, magician, mental-health counselor, writer, puppeteer, banker, minister, filmmaker, student, veterinarian, social worker, scholar, ski instructor, composer, warehouse worker, accountant, teacher, sculptor, bus driver, painter, lawyer, publisher, truck driver, salesman, physician, carpenter, librarian, software engineer, midwife, administrator, housecleaner, editor, and performance musician.

The answer: This is a partial list of the > 400 persons who have become members of the online Religious Naturalist Association (RNA),[1] a not-for-profit 501(c)(3) entity brought together by Michael Cavanaugh in 2014. Its stated goals are 1) to create a worldwide "home" for those of us who self-identify as religious naturalists and 2) to encourage the development and spread the awareness of a religious naturalist orientation.

The authors of this chapter were elected by the RNA Board to serve as its first president (UG), secretary (MC), and communications director (TM). Membership is free. We maintain a website that provides information and resources, including essays and descriptions of creative projects, with links, that examine aspects of a religious naturalist orientation. We also host a members-only Facebook site,[2] an online open discussion forum via Google Groups, and a discussion group of past and present clergy who are based in traditional denominations.

We have been thrilled by this list, it being our first indication that an RN orientation can extend well beyond the academy, beloved as the academy is to us all. Even better, these religious naturalists hail from 47 (49) states (and territories) in the US and from 24 countries.

This chapter serves as a report from the field in 2017. In a decade or two, it will be clear whether the online approach takes off or stalls out, and whether we have found viable and meaningful ways to supplement our online experience with face-to-face experiences with one another. Meanwhile, we are having a wonderful time engaging in the experiment.

Matters of terminology: the R-word

RNA is not the first online 501(c)(3) association to promote a like-minded orientation. In 1996, Paul Harrison launched what is variously called Scientific Pantheism or the World Pantheism Movement (WPM),[3] where its history is recorded[4] and its two very active Facebook sites[5] are open to all. More recently, Daniel Strain launched the Spiritual Naturalist Society[6] with a Facebook site.[6] A compendium of other similar online offerings is found at the RNA website.[7]

The most notable distinction between RNA and these other groups is that RNA uses the adjective religious. WPM explicitly emphasizes that it is "spiritual but not religious," a trope that has earned its own Wikipedia entry.[8] Taking this to its perhaps inevitable endpoint, the commercial website Patheos,[9] which self-describes as "hosting the conversation on faith," includes a "Nonreligious channel" [9] devoted to "the largest-growing segment of the population—those who are neither spiritual nor religious," which the Pew foundation estimates to comprise 21 percent of the US population.[10]

Hence, adopting the adjective *religious* would appear to be a swimming-upstream decision, and those of us promoting an RN orientation encounter the "I would never use the R-word" response on a regular basis. When probed, many of these responses prove to arise from a conflation of "religious" and "religion" and an antipathy towards a particular traditional religion that has expanded to include all religions.

We see two ways to think about this. One, which RNA has adopted,[11] is to suggest that "religious" be considered the large "catch-all" term. In this view, being religious entails the adoption of a core narrative (in this case, a naturalistic worldview based on findings from modern science) and its elicitation of three kinds of responses: 1) the interpretive (the philosophical/existential meanings of the worldview); 2) the spiritual (e.g., awe, gratitude, humility, and reverence); and 3) the moral/ethical (e.g., responsibility, fairness, cooperation, community), where a key RN focus is Ecomorality and Social Justice. Under this rubric, being spiritual becomes a facet of, rather than a substitution for, being religious.

The other response is to say that it does not matter, that persons who take the natural world seriously in launching their meaning systems are all members of the same community, regardless of what they call themselves. To the extent that "religious" has positive connotations for some, the community is to that extent expanded.

We are fully on-board with such a plurality of descriptors. Naming issues have diced and sabotaged all too many communal and religious trajectories. What is important is that all members of these communities take Nature to heart (Rue 2011: 109).

Matters of terminology: -ism vs. -ist

We intentionally elected to call our organization the Religious Natural*ist* rather than the Religious Natural*ism* Association in recognition of the present-day antipathy, which we share, to the creedal/doctrinaire connotations of -isms. The term *Naturalism* is particularly vulnerable on this axis since it has been adopted in the past to describe various philosophical and literary mindsets or "schools"[12] that put forward various truth claims.

A religious naturalist, we suggest, seeks to synthesize his/her interpretive, spiritual, and moral responses to the natural world into a coherent whole, a synthesis that functions as his/her version of religious naturalism, where the vocabulary, metaphors, and meanings that emerge from that search are not expected to conform to some external received credo. Rue (2006: 364) lifts up this feature of RN as well:

> It may be enough that particular versions of religious naturalism are prepared to install their own conceptual integrations of reality and value. Perhaps the only critical point for religious naturalism is that evolutionary cosmology becomes integrated with eco-centric morality by means of some conceptual device. Whatever metaphors do the trick are fine, so long as they don't compromise the principles of naturalism... . Religious naturalists must remain open to a range of options.

Such an approach is not unlike that of the Unitarian-Universalists (UU),[13] with an important distinction: religious naturalists elect to adopt the story of Nature as their core narrative, whereas the UU tradition is centered on—indeed takes pride in—its absence of a core narrative.

The option of becoming a subgroup of a pre-existing institution

There are three general options for creating a community of religious naturalists: 1) Become a traditional religion with clergy-equivalents, brick-and-mortar churches, seminaries, and so on. This could conceivably come to pass in the distant future, but it is obviously not an option given our current numbers and resources. 2) Become a subgroup—a sect equivalent—operating within an existing traditional institution. 3) Form online associations.

There have been several experiments with Option 2. A group of religious naturalists who regularly attended the Star Island Conferences of the Institution on Religion in an Age of Science (IRAS)[14] during the 1990s successfully petitioned the IRAS leadership to create a Religious Naturalism Interest Group (IRASRN) with an attendant email tie-in. RN subgroups have at various times also been formed within UU congregations. Such initiatives have the important merit of bringing actual people together. However, there prove to be many challenges inherent in sustaining a subgroup within a larger entity pursuing its own agenda, and none of the RN-subgroup initiatives that we know of have had a long half-life. Moreover, as expanded below, such groupings do not include those religious naturalists without proximity to and/or interest in becoming a member of the larger institution.

Why RN online?

An obvious disadvantage of an online organization is that it does not bring actual people into the same building or around the same campfire, a feature nurtured by religious institutions for many millennia. That said, there are of course countless opportunities available in one's community to join like-minded groups—in sports, in environmental and social activism, in book clubs, in music-making, in the work sphere, and yes, in traditional churches—so that to ground one's religious life in a virtual community in no way cuts someone off from experiencing actual people. The advantages of an online RN organization are presented as follows.

The diversity and global reach parameter

We have already lifted up the wide range of professions, avocations, and geographies that RNA has attracted. A similar spectrum doubtless characterizes our sister organizations.

Here are two testimonials that speak to this parameter: "As someone who is disabled, on-line e-mail communication is all I am capable of these days. Group discussion would be more fulfilling, but as I live in a remote rural part of Wales, communal gathering is also scarcely practicable for me." "For me sitting out here at the end of the world (South Africa), this group is a lifeline." The writers of these testimonials are isolated by geography and mobility, but it is worth noting that many persons have become comfortable self-describing as introverts: the merriment of a church social event can energize some and drain others. Online community can be a lifeline for introverts as well.

The time and expense parameter

Church membership has traditionally entailed a significant commitment of time (attending services, participation in governance) and expense (annual giving, offerings). Surveys of those who have "left the church" usually include responses to the effect that a person did not have enough time for it, or did not choose to spend money that way.

At such time that RN picks up enough steam to attempt to influence global affairs, time commitments and financial resources may enter the mix, but until then, a religious naturalist is free to explore her/his orientation, for free, in numerous contexts, including walking in the woods, reading a book, going to an art or science museum, engaging in meditative practice, attending a concert or a workshop, playing with children or friends, or participating in online conversation.

The online discussion parameter

A subset of RNA members has signed up for RNAnet, our Google Groups discussion forum. Of these, the majority "lurk," and a minority post—a ratio that obtains in other such forums familiar to us. We find these conversations to be deeply stimulating and productive. As posters, we treasure the clarity of mind that comes from putting vague intimations into coherent sentences, and we treasure the insights obtained as we read how others put things together, even if we are content to be lurkers. Happily, too, none of us is "trapped" into listening to a viewpoint that fails to interest or resonate; we can simply click to the next posting, an option not available when one is literally trapped in a dull sermon or in an adult-ed gathering in a church basement.

It is also important to lift up the affinity that one can develop with online "strangers." True, each of us is presenting only selected facets of who we are when we are online, but that is usually the case in face-to-face interactions with others as well, exceptions being close family and long-time friends. Indeed, one often feels most comfortable posting expressions of doubt and confusion in a quasi-anonymous but trusted setting.

And personality abounds! On the equivalent IRASnet that has been in place for several decades, most of us previously knew one another in-person from Star Island conferences, but a European woman joined us online for several years before attending a conference, and meeting her in the flesh, while wonderful, was in many ways an affirmation of who we already knew her to be.

The website resources parameter

We pay a great deal of attention to our website, developed and overseen by our excellent webmaster Terry Findlay, considering it both an introduction to the RN orientation for first-timers and a resource for both members and non-members. The founders and leaders of IRAS, our parent organization, were primarily academic scholars or clergy, and this tradition is reflected in our Board of Advisors[15] and in our efforts to provide informational resources. Our book pages, for example, list numerous offerings categorized as Nature, Human Nature Mind and Culture, Philosophy and Religion, Ecomorality, and Varieties of Religious Naturalism.

All naturalist orientations face the same headwinds: our central story—our Core Mythos in Rue's framing—is poorly taught in most schools, and is often either misunderstood or actively undermined or "disbelieved." For example, Citizens for Objective Public Education recently sued the Kansas State Board of Education for its adoption of the Next Generation Science Standards (COPE et al. 2016). According to the plaintiffs, the Standards "establish[ed] and endorse[d] a non-theistic religious worldview in violation of the Constitution... . Do theistic parents and children have standing to complain if the goal of the state is to cause their children to embrace a 'nontheistic religious worldview that is materialistic/atheistic'?"

Countering this perennial pressure is a growing movement to teach what has come to be called *Big History*[16]: human history in the context of biological and cosmological evolution. Websites promoting this trajectory include the Deep Time Journey Network,[17] Journey of the Universe,[18] The Great Story,[19] and the Big History Project.[20] Support for such initiatives is of course vital to RN, it being impossible to generate religious responses to a story that one does not know.

Joining online organizations

An unexpected difficulty that we have encountered is a reluctance to join online groups. It turns out that some persons have provided their names and e-addresses to sites that have gone on to publicize their membership, or send unwelcomed emails, or provide their lists to other groups. Once someone has had or heard about such experiences, hesitation can often trump affinity. Compounding these specific "cybersecurity issues" is a chronic reluctance to affiliate in general, documented in Robert Putnam's book *Bowling Alone* (2000), where concerns include discovering that other members are, for whatever reasons, "different from what I expected."

We know of no way to counter these issues, but console ourselves with the hope that an encounter with our website, even if it does not result in signing up for membership, serves as an introduction to an RN orientation that may bear fruit down the road.

RN evangelism

The other ongoing challenge we face is the matter of publicity. After an initial flood of pre-committed religious naturalists, we continue to add members incrementally, largely by word-of-mouth or via search engines that direct persons with the correct key words to our site. What else, if anything, should we be doing? A friend quipped that all it would take would be for some super-celebrity to announce having joined RNA on a Facebook page or in a tweet, and we'd be swamped.

In religious contexts, "evangelism" is often used instead of "publicity." Evangelism is usually associated with the spreading of the Christian gospel, but it can be more generally defined as "zealous advocacy of a cause." In our times, evangelism is widely and effectively practiced via online communication. Indeed, traditional religions supplement their own complements of clergy and houses of worship with extensive online activity.[21]

Returning to the goals of RNA, the second goal reads: "to encourage the development and spread the awareness of a religious naturalist orientation." An evangelical statement indeed!

Loyal Rue offers a framework for RN evangelism at the close of his important book, *Religion Is Not About God* (2006: 341–368). He first evokes various plausible (and dire) Doomsday scenarios resultant from "humans having failed to acknowledge and embrace their true status as natural beings." He then suggests that those emerging from the chaos of said holocaust will seek a religious naturalist orientation that provides an integrated vision of cosmology and eco-centric morality. "Religious naturalism treats the integrity of natural systems as an absolute value, implied by the principle that any vision of the good life presupposes life, and that life presupposes the integrity of natural systems." He goes on: "Religious naturalism is already in the air, but it is not yet a robust mythic tradition ... and it is unlikely to become a dominant influence until the events of history render alternate mythic visions irrelevant and unpalatable."

While it may prove to be the case that the global adoption of an RN orientation will not happen without the kick-start of global catastrophe, the optimists in our midst envision an alternate sequence, wherein global adoption of this staves off global catastrophe. In this scenario, RN evangelism becomes a moral responsibility, even if we are only now struggling to figure out how best to deploy it.

References

COPE et al. v. Kansas State Board of Education et al. (U.S Supreme Court denied cert. November 14, 2016).

Putnam, R.D. (2000) *Bowling Alone*, New York: Simon and Schuster.

Rue, L.D. (2006) *Religion Is Not about God: How Spiritual Traditions Nurture Our Biological Nature and What to Expect When They Fail*, New Brunswick, NJ: Rutgers University Press.

Rue, L.D. (2011) *Nature Is Enough: Religious Naturalism and the Meaning of Life*, Albany, NY: State University of New York Press.

Web links

1. Religious Naturalist Association (RNA) http://religious-naturalist-association.org
2. RNA Facebook site www.facebook.com/groups/323349644530093/
3. Scientific Pantheism or the World Pantheism Movement (WPM) www.pantheism.net/
4. History of WPM www.pantheism.net/about/history
5. WPM Facebook sites www.facebook.com/groups/2230619808/, www.facebook.com/Pantheism
6. Spiritual Naturalist Society http://spiritualnaturalistsociety.org; Facebook site www.facebook.com/groups/spiritualnaturalistfriends/
7. Compendium of similar online offerings http://religious-naturalist-association.org/varieties-of-religious-naturalism-websites/
8. Spiritual but not religious https://en.wikipedia.org/wiki/Spiritual_but_not_religious
9. Patheos website www.patheos.com; "Nonreligious channel" www.patheos.com/Nonreligious
10. Pew Foundation. "A Closer Look at America's Rapidly Growing Religious 'Nones'" www.pewresearch.org/fact-tank/2015/05/13/a-closer-look-at-americas-rapidly-growing-religious-nones/.
11. "Who Is a Religious Naturalist?" http://religious-naturalist-association.org/what-is-religious-naturalism/
12. Naturalism (philosophy) https://en.wikipedia.org/wiki/Naturalism_(philosophy)
13. Unitarian-Universalist (UU) Association www.uua.org
14. IRAS www.iras.org
15. Board of Advisors http://religious-naturalist-association.org/officers-and-advisors/
16. Big History https://en.wikipedia.org/wiki/Big_History

17. Deep Time Journey Network https://deeptimejourney.org
18. Journey of the Universe www.journeyoftheuniverse.org
19. The Great Story http://thegreatstory.org/home.html
20. Big History Project https://school.bighistoryproject.com/bhplive
21. Religion and the Internet https://en.wikipedia.org/wiki/Religion_and_the_Internet

27
WHITHER RELIGIOUS NATURALISM?

Walter B. Gulick

What might be the future of religious naturalism? Might it take on some communal incarnation of its own, or is its destiny best understood as a theoretical perspective adhered to by dissociated individuals? Might it spice up existing religious traditions and thus have long-term influence, or is it an airy intellectualism that will soon fade away?

Religious naturalism: distinctive aspects

Reliable answers to these questions depend on the conception of religious naturalism employed. The essays in this *Handbook* demonstrate that religious naturalist insights come packaged in a variety of forms. Some scholars restrict their understanding of what is religious to shared beliefs and actions that appeal to supernatural agency (see, for instance Whitehouse 2004: 2). In this view, religious naturalism, rejecting supernaturalism, would be an oxymoron. However, all varieties of religious naturalism support such functions as promoting social solidarity, offering wisdom about how to live well, providing psychological support during the transitions and crises of life, promoting social justice and general welfare, and so on. While these functions can be addressed on a purely secular basis, the addition of a religious orientation to naturalism introduces a more comprehensive and committed approach to these vital concerns.

Yet for many, the rejection of supernaturalism is the most dramatic characteristic of religious naturalism—an essentially negative stance. Is there a positive, emotionally evocative commonality underlying the various expressions of religious naturalism that might attract new adherents yet be affirmed by a strong majority of those who consider themselves religious naturalists? Here is my candidate definition that attempts to incorporate those attributes:

> Religious naturalism is a richly endowed naturalistic worldview that seeks to develop and sustain existentially and morally positive relations with self, others, and the world. Scientific understanding of the structures and processes of empirical reality provide the framework within which its thought and practices unfold. But there are also normative dimensions of human experience that transcend scientific determination and are crucial for human flourishing. If shorn of dogmatism, authoritarianism, and any form of supernaturalism, the rich apprehensions and practices of some religious traditions offer powerful communal resources in support of excellence in living. Religious

> naturalism appreciates how religious sensibility can nurture and validate such life-affirming emotions as gratitude, wonder, reverence, and compassion—all of which are healthy expressions of meaning in existence. In allegiance to the highest moral ideals, it also affirms actions that further justice, support non-violence, build caring solidarity, and honor wisdom wherever it is found.

If this delineation accurately reflects the views of most religious naturalists, it would appear that it contains attractive, existentially meaningful sentiments that can fund some sort of future embodiment. But what sort?

A logical first step toward answering that question would be to describe what makes religious naturalism so distinctive that it merits special attention.[1] First, it takes the verifiable understanding of reality opened up by science as its ground. The Epic of Evolution as set forth by Thomas Berry, Loyal Rue, Ursula Goodenough, and many others provides the basic narrative account of how things have emerged in the cosmos. In this story, revelation is displaced from its traditional supernatural locus and transferred to a combination of scientific discovery, cultural wisdom, and spiritual insight. Second, it unequivocally rejects any form of supernaturalism. But that does not mean it is committed to single level ontological reductionism or some variety of scientism. The notions of emergence and pluralism have a crucial place within religious naturalism. Third, the evidence-based curiosity that is a hallmark of science is also a defining characteristic of religious naturalism. When evidence about best policies is missing or ambiguous, religious naturalism treads lightly and is tentative. Dogmatic stasis has no place in religious naturalism. Attention to reliable evidence extends beyond grand theory to practical issues of daily living.

Some questions

What sort of impact has religious naturalism had to date? Well, there are members of various religious denominations and traditions that accept some form of religious naturalism. Acceptance of themes from religious naturalism seems common within Unitarian-Universalism and religious humanism. Still, when one surveys the broad scope of religious thought and practice, religious naturalism plays a tiny role. Maybe this is because to date it has evolved as a fairly abstract intellectual program. Rarely has a novel theoretical vision of religious ideas become influential outside a major religious tradition. Charismatic leaders such as the Buddha, Jesus, Mohammed, and Martin Luther have created new religious traditions, but they initially acted as reformers within a historical and social context of religious belief and practice. Naturalistic views have ordinarily been taken as contrary to and outside any religious tradition. Convincing others that there is a place for religious sensibility within a naturalist worldview is thus a formidable challenge.

So when contemplating the future of religious naturalism, several issues need to be considered, which I will list in the form of questions. 1) Are the winsome aspects of religious naturalism noted in the earlier definition sufficient to foster a drive among its adherents to seek an ongoing institutional home? 2) Even more basically, to what extent do religious naturalists think institutional or communal form enhances or advances their ideas and goals? Is there an evangelistic thrust inherent in religious naturalism? 3) Are there already thinkers and intellectual movements that share enough of the worldview of religious naturalism that it would be advantageous to engage them as potential partners or allies for the good of the cause? 4) How might the discernments of religious naturalism be developed and packaged so that it would have a good chance of becoming a flourishing association with its own identity? 5) If taking on its own organizational identity proves not to be a live option, are there viable structures whereby the ideals of religious naturalism might be creatively combined with living religious traditions?

It is too early in the development of religious naturalism to give any reliable answer to the first two questions, important though they are. So in the balance of this essay I will first describe some thinkers and movements that have views mostly compatible with religious naturalism (question 3). Then I will describe the most sustained effort to date to locate religious symbols, rituals, and moral practices inherent in religious naturalism that have the power to generate a self-contained tradition (question 4). Finally, I will describe a structure within which key ideas of religious naturalism may be melded with existing religious traditions (question 5). Such a fusion of religious practices with a naturalistic worldview may offer the most likely way in which religious naturalism can increase its influence.

Relationship to possible partners or allies

So far I have assumed that religious naturalism ideally needs either to become accepted as a legitimate theological option by some current and significant religious institution(s) or assume its own institutional embodiment. The essays in Part VI of this *Handbook* tend to sustain the view that novel religious visions become vital when grafted onto existing religious traditions with vibrant spiritual practices, ritualized enactments, and communal celebrations. What earlier and present day thinkers outside religious naturalism are influential and inspiring, yet have enough insights compatible with it to be candidates for alliance? What religious movements might be recruited as allies to help secure a substantial role for religious naturalism? In this section I will limit my comments to philosophical and theological views that have emerged in Western civilization.

Immanuel Kant is an earlier thinker whose thought significantly helped pave the way for religious naturalism. Perhaps he is the first influential philosopher to provide a profound vision affirming the scientific understanding of early modernism while also incorporating moral and religious insights. It should be remembered that in his pre-critical period Kant published scientific works on the rotation of the earth, the nature of fire, and cosmology (he was the first to suggest the Milky Way was a vast collections of suns). The thought of Newton was his standard of explanatory excellence. However, as early as 1773 he wrote that, in contrast to the prevailing metaphysics, he aimed "to give philosophy an enduring new turn that would be far more advantageous to religion and morals" (uncited quotation from Kant in Jaspers 1962: 95). His rationalistic religion supporting morality has little content in common with contemporary religious naturalism. However, it exhibits support for religion within a scientific framework, limits uncritical metaphysical speculation, understands the need for the regulative use of reason in attaining systemic coherence, and cautiously avoids religious dogmatism—all ideas compatible with religious naturalism.

In twentieth-century America, William James, Alfred North Whitehead, John Dewey, Henry Nelson Wieman, and John Herman Randall are among the prominent thinkers whose ideas have much in common with religious naturalism. Dewey's *A Common Faith* is an early statement of essential themes in religious naturalism.

> In the degree in which we cease to depend upon belief in the supernatural, selection is enlightened and choice can be made in behalf of ideals whose inherent relations to conditions and consequences are understood. Were the naturalistic foundations and bearings of religion grasped, the religious element of life would emerge.
>
> (Dewey 1934: 57)

This is religious naturalism in full bloom.

Process theology, based on Whitehead's philosophy, has influenced many thoughtful laypersons and religious leaders. Yet as commonly construed (and indeed supported by Whitehead's postulation of the primordial and consequent natures of God), it still allows for a kind of supernaturalism. And the reach of Whitehead's metaphysics extends considerably beyond the evidence of contemporary science. Hence it seems best to regard process thought as a partial ally of religious naturalism.

A similar comment applies to thinkers associated with the continental theological turn in phenomenology, an expression of postmodern themes in theology. Paul Ricoeur's notion of a second naïveté (Ricoeur 1967: 351, 355) rejects the adequacy of literal understandings of scripture and theology, but in rethinking symbolic meanings it remains open to sophisticated forms of supernaturalism. The title of Richard Kearney's book, *The God Who May Be*, articulates the subjunctive mood taken by many participants in the theological turn. Kearney, a student of Ricoeur, supports what he calls anatheism, "the space where an open theism and an open atheism can come into dialogue" (Kearney and Zimmerman 2016: 80). Openness to the unexpected, to the stranger, can reveal the sacred in the secular and invite one to respond freely with either hospitality or hostility. "In *Anatheism* I divide the wager of hospitality/hostility into three moments: protest, prophecy, and sacrament" (Kearney and Zimmerman 2016: 174). Connection to the sacred without theism is one option in Kearney's thought. Similarly, Jean-Luc Marion (2008) rejects any metaphysical attempt to define God. However, he remains committed to the distant God of love whose revelatory appearance in "saturated" phenomena obliterates our usual categories of thought. By disconnecting thought about God from notions of being and becoming, Marion's phenomenological notion of the divine is an attempt to escape its problematic controlling aspects—but a hermeneutics of supernaturalism remains.

All this leads to a key question, originally raised in relationship to the thought of Charles Hartshorne: "can one be a naturalist in his interpretation of the world and at the same time, within that naturalistic framework, also believe in a deity that is natural?" (Parsons 1964: 533). Dewey makes this move; so does Jerome Stone. "I am convinced that a naturalistic concept of God, particularly when employing both theoretical and devotional language, can provide a naturalistic principle of transcendence" (Stone 2011: 218). This claim might make it easier for religious naturalists to form partnerships with sympathetic theologians, but care would have to be taken to ensure that theological confusion was not the result.

Those taking the theological turn, primarily French thinkers, tend to approach theology via a Catholic background. In the Anglo-American context, progressive Christian theology is primarily Protestant in perspective. The social and historical context of the Bible is the starting point for the theology of such scholars as Marcus Borg, John Dominic Crossan, Diana Butler Bass, Walter Brueggemann, and John Shelby Spong. For some progressive biblical theologians, the faith and actions of Jesus confronting the violence of the Roman Empire is roughly analogous to the spirituality Christians are called to now in order to confront the social, economic, and environmental injustice promulgated by profit-seeking corporate power in alliance with governmental forbearance. More positively, progressive Christians emphasize the importance of offering hospitality to all and establishing justice for all—especially those who are most vulnerable, but also those of other faiths, including agnostics and atheists. Religious progressives question many traditional theological tenets, such as the sacrificial death of Christ as necessary to atone for human sin, or belief in Christ as the only path to salvation (see John 14:6). Most theological progressives, however, continue to use "God-language," although without imputing to the divine such traditional attributes as omnipotence, omniscience, or immutability. In any case, the values of progressive Christians overlap with the values of religious naturalists, even if progressive Christians maintain a more traditional theological vocabulary. Potentially some religious progressives are strong allies.

Ignored voices?

Finally, in considering allies, three authors articulate views that fit well within the scope of religious naturalism, although they have not generally been included in discussions of religious naturalists. Don Cupitt, Lloyd Geering, and Daniel Maguire have written books in which the theme of accepting the moral heritage of Christianity without belief in God is central. "We are stuck now with our own culturally-mediated form of naturalism, and to an outlook therefore whose religion must be ethics-led and purely immanent" (Cupitt 2015: 138). This view, of course, would be radically incoherent to most traditional Christians, and incoherent for at least two reasons, one of which is worth pondering.

First, here is a highly influential argument for the divinity of Jesus, made most famous by C. S. Lewis.

> I am trying here to prevent anyone saying the really foolish thing that people often say about Him: I'm ready to accept Jesus as a great moral teacher, but I don't accept his claim to be God. That is the one thing we must not say. A man who was merely a man and said the sort of things Jesus said would not be a great moral teacher. He would either be a lunatic—on the level with the man who says he is a poached egg—or else he would be the Devil of Hell.
>
> (Lewis 1952: 54–55)

Lewis's way of treating statements from the gospel of John clashes with historical-critical scholarship that recognizes the imaginative, poetic character of the text, written many decades after the death of Jesus. Were one to take literally *all* the gospel statements attributed to Jesus, we would indeed be justified to view his worldview as incoherent. Members of the so-called Jesus Seminar have attempted to distinguish between which of Jesus's statements appear to be authentic, which are possibly attributable to him, and which are best attributed not to him, but perhaps express the needs of the early church community. John Shelby Spong, who appreciates the spiritual insights of the gospel of John, makes this comment in line with typical historical critical scholarship: "There is probably not a single word in the Fourth Gospel that Jesus ever spoke" (Spong 2013: 68).

Second, while Lewis's claim naively violates historical-critical scholarship, it offers a challenge to those who would base an ethical religion on the teachings of Jesus while denying the existence of God. For the thoughts and actions of Jesus, himself a first-century Jew, are not conceivable apart from recognizing his deep faith in the God of his ancestors. On what basis, then, does it make sense to be a follower of Jesus without also believing in God?

There is a human tendency to seek security beyond the chaos of history by relating to that which is seen as transcendent and thereby reliable. Religious naturalists note that all beliefs are influenced by the volatile particularities of the historical circumstances in which they are made. Jesus's faith in a reliable God is rooted in a culture 2,000 years old. That culture's knowledge of the natural world has been superseded, but ethical insights connected to human relationships may remain pertinent for millennia. Maguire indirectly affirms the relevance of Jesus's moral discernment apart from his belief in God: "As for religions, forget their deity and afterlife creations, and recognize that they are at root poetry-rich philosophies that have hit on things that are stunningly relevant, with no authority behind them other than good sense" (Maguire 2014:

130). Reconstructing the relevance today of Jesus's teachings requires a second poetic act of interpretation beyond the poetic interpretation by the author of the gospel of John. It is just as legitimate to take into consideration in a second poetic rendering of Jesus's words the lessons drawn from nearly two millennia of church history and contemporary knowledge as it was to take comparable issues into account in the original poetic rendering of the Fourth Gospel. We can only conjecture whether Jesus would still maintain his first-century faith in God today in light of current knowledge, but his practical insights and actions can still inspire.

It should now be clear that religious naturalists are not appropriately regarded as voices crying in the wilderness. The thinkers and movements listed so far are representative of a large group of persons sympathetic to most aspects of religious naturalism. Some of the persons discussed are active members of Christian denominations. So there are living options for possible amalgamation of religious naturalism into existing institutions. But it is not clear if such a move would have power. Notice the long-term trajectory of claims about the demise of supernaturalism. When Nietzsche first proclaimed the death of God (Nietzsche 1887/1974: 181), his claim was roundly renounced by virtually all Christians. When the death of God was proclaimed in the 1960s by Thomas Altizer, William Hamilton, and Paul Van Buren (and in a less radical form by Bishop John Robinson in *Honest to God*), it made front-page news. But the books of Maguire, Cupitt, and Geering, as well as most books by recognized religious naturalists, have made no comparable public splash. That suggests they may have lost the power to revitalize current institutions.

Anyway, it seems that ideas found in religious naturalism, while not generally embraced, are at least acknowledged as options in mainline theology and broad social consciousness. True, many both inside and outside the mainline churches may see this acceptance as but one more manifestation of Christianity's declining vitality. But some might interpret them in line with Phyllis Tickle's vision (Tickle 2008: 16–17) of the current Great Emergence as a stimulating and necessary adjustment of the church to how social media, globalization, etc. have changed the way we live.[2] Perhaps, but the challenge remains for religious naturalists to demonstrate to ordinary people that their views energize lives and improve society.

Fleshing out religious naturalism as a religion

These considerations suggest that religious naturalism would have the most robust impact if all those excited by its claims resisted being swallowed up in some existing tradition in which the inertia of long-existing incompatible views would likely prevail. What is the prognosis for success if religious naturalists formed their own religious tradition? What might a religious community of naturalists look like?

If I were to choose one book that best qualifies as the "gospel" of religious naturalism, it would be Ursula Goodenough's *The Sacred Depths of Nature*. This work brilliantly combines spiritual sensitivity with biological expertise, and it does so with poetic flair. But just as the gospels do not establish church organization, so Goodenough's book requires elaboration if an organized religion is to be established. The person who has taken the lead in suggesting what such a religion might look like is Donald Crosby. Exposition and critique of his accomplishments may suggest the sorts of issues with which any practical development of religious naturalism must deal.[3]

Crosby helpfully distinguishes his "Religion of Nature" from three other types of religious naturalism.

> A second is naturalistic theism, which rests belief in God on reflections about experience rather than on special revelations and usually regards God as a wholly immanent

> being, presence, or power in the universe. Another is religious humanism, where humanity, rather than nature or God, is the principle focus of religious concern.
>
> (Crosby 2002: 172)

The fourth version is Jerome Stone's understanding of transcendent ideals that establish goodness. Such ideals can be acknowledged as a minimalist notion of the divine.

Crosby understands nature—all of it—to be the ultimate reality and as such to be the sacred object of faith. "Nothing else lies before [nature], behind it, or beyond it as its ultimate ground, source, sustainer, or guide. Nature exists and persists by virtue of its own inherent, self-contained potentialities, principles, and laws" (Crosby 2013: 2). More briefly, "*Reality is nature in all of its manifestations*" (Crosby 2015: 10). That means that experiences humans regard as negative—suffering, evil, natural disasters, violence—are for Crosby to be accepted as sacred. The Religion of Nature "offers no pap, no panaceas, no empty promises" (Crosby 2008: 108). He distinguishes between religious rightness, the sacredness attributable to all that is, and moral rightness, which deals with what humans can control in order to live together harmoniously.

> It is not anticipated that we should reverence nature by imitating it. There is an important disconnection between the object of faith and moral policies, principles, and practices. Nature as the object of faith can provide context and support for moral living but should not be expected to supply its specific precepts.
>
> (Crosby 2008: 85)

The Religion of Nature, it can be seen, represents religious naturalism in its starkest, most unhedged form. If nature cannot supply precepts for moral living, does it embody values that generate prescriptions regarding how it should be treated? Crosby believes so, for ignoring nature's valuative dimension "would be shockingly inadequate and incomplete" (Crosby 2015: 109). "Salvation in the perspective of Religion of Nature ... consists in the continuing protection, betterment, and flourishing of the world as a whole here and now, not in the hope of being transported beyond this finite, fragile, vulnerable world into some other realm" (Crosby 2013: 127). Religious rightness means nature's "eminent fitness for unstinting religious reverence and devotion ... [whereas] religious evil is the absence of such devotion" (Crosby 2015: 125). Disciples of the Religion of Nature ought to care for the well-being of the natural environment in contrast to thoughtless exploitation of it.

Crosby understands that the "well-being" of nature applies chiefly to the various forms of life. The cover of his book, *More than Discourse: Symbolic Expressions of Naturalistic Faith*, displays a brown pelican in flight. The book is dedicated to these pelicans as "Symbols of the Marvel and Vulnerability of Life." Reverent and responsible action, not passive worship, is called for in the Religion of Nature. His plea for responsibility in relation to animal rights (see Crosby 2013: 45–47) has implications for changes in the way animals are treated (or mistreated) in such realms as sport hunting and fishing, the farming of animals (especially factory farming), commercial fishing, zoos, aquariums, circuses, rodeos, endangered species, and animal experimentation. He also calls for rectification of ecological harms caused by various forms of pollution, exponentially increasing human population, and global climate change.

In portraying the free flying pelican as a symbol of life's delights, Crosby is supporting his conviction that non-discursive religious symbols provide indispensable access to the value-laden depths of nature. Crosby believes that religious symbols can make vital and lasting contributions

> to apprehension of religious truths that speak to the whole beings of persons—emotional, volitional, valuational, and practical as well as rational—and that function to awaken and sustain a sense of the daunting, alluring, and healing presence of the sacred in the world.
>
> (Crosby 2014: xi)

Crosby's use of terms such as assurance, demand, empowerment, and even love resonate with affective vitality (see Crosby 2011: 150–153; Crosby 2013: 129). Under the rubric of religious symbols Crosby includes emotionally evocative material from the visual arts, poetry, drama, architecture, ritual, myth, calligraphy, dance—wherever are found "expressions of nondiscursive, non-propositional, nonassertive types of meaning" (Crosby 2014: 4). Religious symbols immerse us in a world of religious meaning in which our behavior matters. Crosby distinguishes between minor, major, and master symbols, which differ in their capacity to orient, inspire, and motivate us. "Christ is the master symbol of the religious ultimate in Christian faith. He is also, of course, the principal exemplar for the Christian path of life" (Crosby 2014: 43).

It is obvious that Christ and the crucifix would function as master symbols in Christianity. It is far less obvious what might assume the role of master symbols in the Religion of Nature. Crosby chooses water. "Let us imagine ourselves seated below a cascading waterfall and adjacent to the stream that rushes away from it and finally enters into a quiet lake" (Crosby 2014: 86). For Crosby, this watery scene captures the wholeness of nature: reliable continuity as well as precipitous destruction, peaceful calm as well as bubbling transition. To symbolize the creative source of all things, Crosby selects the womb.

That master symbols need to be identified points out one of the great challenges facing a Religion of Nature. The passion of Christ, the teachings of the Buddha, the Quran as revealed to Mohammed—these symbols of meaning and obligation grow organically out of the historical (or legendary) beginnings of each of these world religions. Crosby's account of what he terms the salvific and tragic qualities observable in the course of human history (2014: 99–112) is not cloaked with the paradigmatic authority of a sacred history. From the perspective of Crosby's acceptance of all nature as sacred, his distinctions between salvific and tragic or evil events in history seem arbitrary and ungrounded. Similarly, citing John Muir as the exemplary traveler on the saving path of Religion of Nature seems forced and unconvincing when Muir's career is compared with what paradigmatic religious figures of other traditions accomplished, often at the price of great personal sacrifice.

Alas, no master symbols seem obvious for a Religion of Nature. There are powerful connotations associated with water and the womb, but there is also a certain unavoidable arbitrariness to their selection as master symbols. Nature is all-encompassing. What makes some aspects of nature, some symbols, more worthy of affirmation, more valuable than others?

Crosby provides a partial answer to this question by identifying three value domains seated in human experience. Epistemic, artistic, and moral domains set forth their own inherent standards of good and evil. What establishes epistemic and aesthetic excellence? "The goods of discursive knowledge and the goods of art are essential to human well-being" (Crosby 2015: 114). Likewise, moral goods establish personal and social well-being.

Relating epistemic, artistic, and moral values to human well-being raises a question of possible incoherence in the Religion of Nature. Crosby claims humans profane and diminish sacred nature by "centering unduly or exclusively on human beings" (Crosby 2015: 137). Yet, as just noted, he also says rational, artistic, and moral excellence are measured according to what conduces to human flourishing. Perhaps Crosby's view that human morality is a subset of ecological ethics cancels any seeming incoherence.

In addition, the view that all nature is sacred, including what is seen as evil, clashes with the human hope that what is sacred is fundamentally good (although Crosby properly notes that issues of theodicy are problems for all religious ultimates, not just religious naturalism alone).

Crosby justifies his view that even life's problems are to be religiously affirmed by offering his readers a choice (see Crosby 2015: ix–x). Would we prefer to live in a chemically created fantasy world of perfect peace and harmony or in the real world with its risk, pain, and suffering but also its possibilities of real achievement and joy? He bets that most people would opt to live in the real world with all its ambiguities rather than subsist in an illusory state with no challenges and no motivating purposes. An implication of this preference is that we honor reality, and since reality (nature in all its manifestations) is ultimate, and since what is ultimate is generally acknowledged as the sacred horizon of existence (Crosby 2008: 48; Crosby 2014: 41), nature itself (including what is evil from a human perspective) is sacred. Still—can a religious community thrive when such a large gap exists between the cosmic scope of religious rightness and the narrow focus on human behavior in moral rightness?

In a world that is increasingly urban, nature has become an abstraction to many of its inhabitants. What does the claim that nature is sacred mean to city dwellers? Do the symbols suggested for the Religion of Nature have the capacity to appeal broadly and deeply enough to generate widespread religious devotion? Can what is sacred about humanly contrived aspects of nature such as cultural creations, the built environment, and technology be more clearly interpreted within the Religion of Nature? As Crosby notes, religious symbols "choose us" in the sense of "resonating with something deep inside us and arousing in us recognition of profound insights or truths in ways that ordinary metaphors or other figures of speech do not" (Crosby 2014: 139). Crosby is open to expropriating, perhaps in a reinterpreted form, religious symbols that have spoken with power to people in different social settings and in other religious traditions around the world. He mentions some possibly attractive symbols and practices, but more exploration is called for here.

Donald Crosby is to be commended for his pioneering work in suggesting what religious naturalism might look like as an organized religion. Before it can actually function in organized communal form, though, other persons need to join with Crosby in setting up organizational and administrative functions. As suggested earlier, it is rare that a unique religious tradition can be created from scratch. To create a functioning Religion of Nature is a daunting task. So I will now suggest hybrid structures that may have a chance of preserving crucial facets of religious naturalism in case the Religion of Nature or other "pure" versions of religious naturalism falter.

Triadic structures for housing religious naturalism

Crosby's elaboration of his Religion of Nature has three different functional components: nature itself as the basic object of religious reverence, the religious symbols that articulate that reverence, and the spiritual state (thought and action) that results from appropriating reverence to nature through apt religious symbols. I have argued elsewhere that religious states of mind arise through the commitment to basic beliefs about what is of ultimate significance in the world, symbolic expression of those beliefs, and personal transformation affected by assimilating those symbolic expressions. The example I developed was a Christian Religion of Nature (Gulick

2013). I will more briefly sketch a hybrid Christian Religion of Nature as my chief example of hybridization even as I mention the possible use of the triadic structure by other religious traditions. Realization of any hybrid view of religious naturalism would ideally take place in a newly established organization rather than in an existing denomination that might dilute its key notions.

What if one replaced God with Nature as one of three co-equal ultimate objects of sacred attention? As Crosby and others suggest, nature in its ultimacy can evoke feelings of awe for its mysterious fecundity, respect for its power, amazement for its scope, and gratitude for one's existence. Reverent attention to nature as it is configured in all human contexts lifts one out of excessive self-concern. But nature is amoral; it does not tell us how we should live, a primary task of any religion. For that some sort of cherished *symbolic* exemplar is needed. Within Christianity, Jesus is the moral ultimate, the symbol of the fully human being. Through partaking of bread and wine in communion, one participates spiritually in Christ. Through indwelling his teachings, his parables, his acts of healing, one's life can be transformed. As Paul writes, "I have been crucified with Christ; it is no longer I who live, but Christ who lives in me" (Galatians 2:20). The "I" who is crucified is the self-centered sinful self. In its place the third religious ultimate is realized, the internalized Holy Spirit. Being in Christ can be understood as a form of psychological identification that is consistent with religious naturalism. The sanctified self is the product of this identification. In this model, nature is the ultimate object of concern, Jesus is the religious symbol of moral worth, and the Holy Spirit is the experience of personal transformation resulting from identification with Christ.

In this version of a Christian religious naturalism, prayer is an expressive, focusing function, not thought addressed to a higher being. Meditation and mindful attention to the problems and needs found in the wider world are useful spiritual practices. In support of the social nature of human existence, caring attention to others is called for, beginning with those in one's community. And with nature understood to be one of the objects of ultimate concern, devotion to nature's sacred ecological integrity is mandated. Those sections of the Bible that can legitimately be reinterpreted in non-supernatural terms will continue to have a special, but far from exclusive, place in religious thought and practice. In sum, such a Christian form of religious naturalism seems a viable outcome for those serious about creating a community of religious naturalists.

I believe hybrid forms of religious naturalism can be created in combination with most of the world's great religious traditions (see the articles in Part VI). Judaism and Christianity have been so entwined with Western cultural developments, including the rise of science, that non-supernatural forms of religion are plausible options. Many within Reformed Judaism have beliefs largely consistent with religious naturalism. Nature, the Torah, and traditional Jewish cultural identity can be interpreted as Judaism's three ultimates. The Torah, seen historically as based on Mesopotamian antecedents, modified by nomadic egalitarianism, expanded through Talmudic commentary, and interpreted via Enlightenment categories, functions as the central Reformed Jewish moral model. Liberalizing tendencies in Islam, which is strongly based on Allah's self-revelation as expressed in the Quran, have tended more toward mysticism than embrace of a scientific worldview.

In Asia, certain varieties of Daoism, Confucianism, Hinduism, and Buddhism seem quite compatible with religious naturalism, as other essays in this volume make clear. I will offer a few comments on Buddhism. The teachings of the Buddha, as expressed in Theravada Buddhism, suggest a path to individual enlightenment apart from any supernatural belief. Reverence for the creation would seem to come about indirectly in this tradition, not as an initially intended outcome. The immediate issue for the Buddha is dealing with the suffering inherent in the course of life. All material things are seen to be impermanent. Human existence is marked by disease, old age, and death. All that is dear to one decays and slips away. Nature itself, then, is not regarded favorably by unenlightened humans. The Buddha taught a multifaceted way to transcend the negativities of life. The Middle Path he taught was a way between ascetic rejection of life's pleasures and addicted over-indulgence in life's pleasures. Buddhists are enjoined to give up the desires that bind one to impermanence and suffering and follow the disciplined life of the Eightfold Path. Right mindfulness and right concentration guide one following this path. Liberated from egoistic greed and anxious coping, the world is accepted as it is. One is freed to feel compassion (*karuna*) for others and wisely balance intellect and emotion. It is in this transformed state that reverence for nature becomes a possibility. The triadic ultimates of early Buddhism become, then, first the nature of reality (no longer understood through ignorance and illusion); second, the teaching of the Buddha; and third, the transformed understanding of nature/reality concerning the right way to live. It is evident that Buddhist doctrine dominates this hybrid form of religious naturalism, yet it is also clear that the emphases of religious naturalism harmonize well with Buddhist insights.

Conclusion

The future of any program of thought and action is impossible to predict with any accuracy. But based on the foregoing discussion, I will hazard a few predictions regarding the future of religious naturalism.

1. The perspective on life of religious naturalism will continue to become more and more widely accepted unless some social trauma intrudes. During historical disasters, people tend to revert to traditional, conservative beliefs. However, the projected increasing acceptability of religious naturalism is occurring at a time when membership in organizations, including churches, is declining. Therefore, it is likely that persons sympathetic to the views of religious naturalism will either ascribe to their beliefs as discrete individuals or will fit as isolated individuals into existing religious communities. If this individualistic response persists, two possibilities emerge. Either the creative thrust of the movement will be dissipated and the names of its leaders will become entries in obscure historical tomes, or the increasing activity of religious naturalists involved in social media will be a vehicle for maintaining and even increasing its influence.
2. Religious naturalism's future could likely be made more secure if its adherents sought out partnership with some of the creative, largely compatible theological options being forged at present. There is risk: its signal ideas might be diluted if this path leads to institutionalization. Nevertheless, religious naturalism participates in a broad intellectual movement that increasingly focuses on thoughts and practices that provide existential meaning rather than insisting upon specific theological creeds, claims of institutional authority, or other aspects of religious provincialism.
3. It is rare that a movement like religious naturalism evolves into a free-standing organization. Donald Crosby's pioneering effort to describe the rationale for a Religion of Nature that stands on its own is a valiant attempt to beat the odds. But until its notion of nature is recast so potential converts can see more clearly its benefits, its program will struggle to gain long-term influence.
4. Finally, I suggested the advantages of a triadic model in which hybrid amalgamations of religious naturalism with current religious traditions might preserve the insights of both

parties. Indeed, without competing notions of the divine, it is possible that naturalism can serve as the midwife for a truly global religion—one that appreciates such things as Buddhist emptiness for some situations, Christian charity for others, Confucian structures for yet others, and, yes, reverence for nature.

5. It will be fascinating to see which of these options—or some other—comes to pass.

Notes

1 See Dorrien (2006) for a thorough study of the theological milieu out of which religious naturalism arose. On pp. 463–472 he explores the thought of Nancy Frankenberry and Jerome Stone as exemplars of religious naturalism. Hogue (2010) examines the thought of Rue, Crosby, Goodenough, and Stone in his fine exposition of religious naturalism.

2 Tickle sees adulation of science, which she might attribute to religious naturalism, to have had priority in the past 500-year cycle of history. She thinks science has lost its authority as a guide to living. So religious naturalism would not be for her a candidate for the emerging theology in the next historical cycle. Rather she envisions theology as "evolving into something far more Jewish, more paradoxical, more narrative, and more mystical than anything the Church has had for the past seventeen or eighteen hundred years" (Tickle 2008: 162).

3 I offer a fuller exposition and critique of Crosby's work in Gulick (2016). See also Crosby's response (Crosby 2016).

References

Crosby, D. A. (2002) *A Religion of Nature*, Albany: SUNY Press.

—— (2008) *Living with Ambiguity: Religious Naturalism and the Menace of Evil*, Albany, NY: SUNY Press.

—— (2011) *Faith and Reason: Their Roles in Religious and Secular Life*, Albany, NY: SUNY Press.

—— (2013) *The Thou of Nature: Religious Naturalism and Reverence for Sentient Life*, Albany, NY: SUNY Press.

—— (2014) *More than Discourse: Symbolic Expressions of Naturalistic Faith*, Albany, NY: SUNY Press.

—— (2015) *Nature as Sacred Ground: A Metaphysics for Religious Naturalism*, Albany, NY: SUNY Press.

—— (2016) "Response to Walter Gulick's Observations on My Writings about Religion of Nature," *Tradition and Discovery* 42/2, 25–29.

Cupitt, D. (2015) *Creative Faith: Religion as a Way of Worldmaking*, Salem, OR: Polebridge Press.

Dewey, J. (1934) *A Common Faith*, New Haven, CT: Yale University Press.

Dorrien, G. (2006) *The Making of American Liberal Theology: Crisis, Irony, & Postmodernity, 1950–2005*, Louisville, KY: Westminster John Knox Press.

Goodenough, U. (1998) *The Sacred Depths of Nature*, New York: Oxford University Press.

Gulick, W. B. (2013) "Religious Naturalism: A Framework of Interpretation and a Christian Version," *American Journal of Theology & Philosophy* 34/2, 154–174.

—— (2016) "Outlining a Religion of Nature: The Work of Donald Crosby," *Tradition and Discovery* 42/2, 8–24.

Hogue, M. (2010) *The Promise of Religious Naturalism*, Lanham, MD: Rowman & Littlefield.

Jaspers, K. (1962) *Kant*, H. Arendt (ed.), trans. R. Manheim, New York: Harvest Book.

Kearney R. and Zimmermann, J. (eds) (2016) *Reimagining the Sacred: Richard Kearney Debates God*, New York: Columbia University Press.

Lewis, C. S. (1952) *Mere Christianity*, London: Collins.

Maguire, D. C. (2014) *Christianity without God: Moving beyond the Dogmas and Retrieving the Epic Moral Narrative*, Albany, NY: SUNY Press.

Marion, J. L. (2008) *The Visible and the Revealed*, trans. C. M. Gschwandtner and others, New York: Fordham University Press.

Nietzsche, F. (1974) [1887] *The Gay Science*, trans. W. Kaufmann, New York: Vintage Books.

Parsons, H. L. (1964) "Religious Naturalism and the Philosophy of Charles Hartshorne," in W. L. Reese and E. Freeman (eds), *Process and Divinity—The Hartshorne Festschrift*, LaSalle, IL: Open Court.

Ricoeur, P. (1967) *The Symbolism of Evil*, trans. E. Buchanan, Boston: Beacon Press.

Spong, J. S. (2013) *The Fourth Gospel: Tales of a Jewish Mystic*, New York: HarperOne.

Stone, J. A. (2011) "Is a 'Christian Naturalism' Possible? Exploring the Boundaries of a Tradition," *American Journal of Theology & Philosophy* 32/3, 205–220.

Tickle, P. (2008) *The Great Emergence: How Christianity Is Changing and Why*, Grand Rapids, MI: Baker Books.

Whitehouse, H. (2004) *Modes of Religiosity: A Cognitive Theory of Religious Transformation*, Walnut Creek, CA: AltaMira Press.

28

THE SOCIETY OF NATURE AND THE RELIGION OF NATURE

Bruce M. Hannon

I will begin with some personal history. In the midst of graduate studies in the late 1960s, I learned the Army Corps of Engineers was planning a dam and reservoir that would flood a local park, one of the few remaining places of nature relatively untouched by humans in the endless central Illinois corn and soybean fields. I had grown up visiting this park from my nearby small hometown, and I had developed an attachment to it.

I resolved to stop this proposal. I set about doing so in spite of great resistance from the established culture of a major city nearby, strong bureaucratic support for the dam, and the endorsement for the dam from a local congressman. Opposing the dam was a full-time job for eight years, and with the help of many across the state, we stopped the planned dam and legislatively removed all reference to it from the Corps' bureaucratic memory.

There were many times during this exhausting effort that it seemed we would lose the park. There were times when the array of pro-dam forces was so discouraging there was talk of quitting. But an hour or two of walking in the natural beauty of the park restored my will to continue.

It was on those restorative walks that I realized the depth of power the nature in this park held for me. Its deep stillness, its great diversity with all of the interconnections and interrelatedness among its diverse parts, its cycle of the seasons ... all of this contributed to my strong sense of what can only be called a spiritual connection to nature. I was protecting this unique natural area, and in a way, it was including me in its dynamic, mysterious pulse. This was one of my first experiences of religious naturalism.

I developed in late 1960s what is now called the Iron Triangle, the view that these dam plans were the result of a collaboration of three agents: the bureaucratic (Corps of Engineers), the vested interests (construction contractors, local banks, and real estate developers), and the elected officials (Congressmen, Governors). The object of the local environmental organization was to break these connections. It was an excellent teaching device. We expanded our organization to help others across the United States who also had a proposed reservoir impinging on their place. In a few years, all the proposed Corps reservoirs were stopped.

I came to understand a great deal about what is needed to develop and sustain successful organizations. Certainly this includes the ability to identify and resolve environmental threats; however, no group of citizens can succeed in nature discovery, protection, and restoration without competent leaders. Successful volunteer groups are not wholly democratic.

Leadership requires a host of personal characteristics: a sense of fairness, persistence, humor, and humility; the ability to understand natural and social interactions and motivations; the ability to teach others to develop their sense of place; the formulation of an action plan with a strong sense of efficient delegation and management; the ability to inspire others to join the plan of action; the ability to recognize leadership potential in others; the ability to secure and then celebrate successes; the ability to ensure that everyone involved is recognized for their contributions; and the ability to sustain the growth and activities of this community. In the 1970s I proposed a plan for leadership development and called it the Society of Nature. It is the resulting collection of my many environmental organizational efforts and is elaborated as follows.

I made another discovery, and this was also my biggest disappointment. Groups that formed to oppose particular dams or to protect natural areas frequently dissolved once they were successful. I now believe that including the ideas of religious naturalism among our organizing tools would have helped to solve this problem. Religious naturalism would have provided a center of gravity that could have held the group together as a community of environmentalists.

Developing the leadership, forming the Society of Nature

Love and respect for all life and for inanimate nature—past, present, and future—is the basic mission of the Society. Our environmental problem is in our hearts, not in the woods; in our perceptions, not in the air; in our spirit, not in the sea. Through carelessness, we destroy our most fundamental life support in nature. We struggle to prosper, yet we are unwilling to give our descendants the natural resources for their well-being. We do not respect our natural and cultural heritage, and we have provided a poor example, with the result that our children will not respect theirs. Our attempt to avoid the situation either leaves us alone and impotent with shame and guilt, or we seek consumption as a curative as a distraction from facing the reality of the environmental problem.

There is an alternative behavior that promotes happiness based on consistent accomplishment. It requires a gradual withdrawal from unnecessary material consumption, and limiting the population to a truly sustainable one. It requires an unending curiosity leading to a particular sort of education—learning the significance of our natural and cultural heritage and learning how to preserve, enhance, create, and pass on these gifts to an infinite future. It is our responsibility to learn the way all species communicate and then learn to understand and speak that language (Tonino 2016).

The most obvious first step is to begin a process of regular and open communication.

First we learn to talk to each other from a common base of understanding. Therefore, education is a central theme of the Society of Nature. We need to help each other learn environmentally sound ways to carry out a continuing revision of our personal and professional lifestyles. We must learn and teach the connections between our acts of consumption and their environmental and social consequences. Then we should compare the consequences of our lifestyles with those of other lifestyles. Also, we need to study the connections between our actions in the workplace and their environmental consequences and consider alternatives. We should not wear one hat in our personal lives and another conflicting hat in our workplace. A lifestyle that is inconsistent at this fundamental level likely cannot lead to long-term happiness.

Another part of the solution is action. We must continually involve our community in carrying out projects that lead directly and indirectly (locally and globally) to environmental improvement. Indeed, action is also essential to a complete understanding of the connections between consumption and degradation. Action is integral to converting others to necessary lifestyle changes: without such conversion, we may conclude that it is hopeless to sacrifice personally. Conversion of the public is imperative so that individual sacrifices do not lead to increased consumption by others. Thus, the solution requires a community where we learn we are not alone, where we learn together how much consumption is actually enough. The Society of Nature would be committed to changing beliefs in the hope that appropriate behavioral changes will follow.

All of this planning and action should be based on the dynamics of the natural system. We must study nature and derive the basic operating principles of the ecosystem, and then we have to translate this understanding into guidelines for living sustainably. We need to adopt a belief in the natural system and center our lives and community cohesion on that belief.

To help us accomplish all this, we require community organizers, individuals who are totally dedicated to communication, education, and action. The training and education requires broad schooling in the professional and social arts. Modern society is both technically and economically complex, and the organizers should be skilled in these arenas. Economics, engineering, and ecology cannot be overlooked. History, sociology, political science, and psychology are necessary tools of the organizers. A deep understanding of religious naturalism is useful. The educational process should also involve community action projects to bring practical realism to the classroom. Obviously, the preparation of the organizers is long and difficult and will require the most skillful minds and teaching abilities. Attracting and holding the right people in this process has to be a primary aim of the Society.

To begin, the Society must establish a formal training program to attract students in the late high school years and probably by the second year of college. They should live together, bonding for the struggles ahead. The program should help students to find access to the broad range of courses and other activities required by the Society. This process of education can mainly be fashioned out of the course selection of any major university.

Graduates would associate with one of a variety of communities to form local groups dedicated to communication, education, and action. As the Society matures, these communities will become the main source of candidates for the formal training process. To keep the Society vital, graduates would return regularly to their source of education for renewal and reassignment.

The idea of fairness is central to the Society. This includes fairness between members of the current generation and between members of present and future generations. Fairness to the present compels us to include environmental costs in the production processes that produce environmental damage. Fairness in the public sector requires cost sharing in proportion to benefits. Fairness between generations requires that we preserve options for future generations by conserving natural and cultural resources to an ever-increasing degree, well beyond the levels dictated by current economic practice. We must give the future all the necessary technology, population controls, and remaining resources so the next generation can have the same options as we had (Page: 1977). This process requires that we strive to not discount the future and that we do not dictate how future generations will live, though it would be reasonable for us to expect future generations to plan for the generations after them.

Principles for the organizers

The goal of a good environmental organizer is to make all people "combatants" in the struggle for a desirable environment. Good organizers are humble, kind, and clear-headed. They seek to frame the local environmental problem in a way that reflects already established environmental ethics and values. Good organizers respect those in the community group.

There are four stages through which the organizer should progress for a most successful and effective life. The organizer needs to master these levels, and they should form the basis of the educational process. The levels in question range from the most personal to those that involve interaction with the community at large.

The health of your body must be the first priority for environmental improvement. Let us call it the "Spartan Level." You cannot expect others to believe you are genuinely concerned about the quality of the environment if you are not concerned about the health of your body. Also, the body provides a means to prove, in the clearest possible terms, the consequences of a poisoned environment for each of us. Appropriate food, exercise, rest, and daily rhythm are the principal means to a healthy body. One cannot place too much emphasis on this. Repeatedly overstressing the body by lack of sleep (even if caused by work on an environmental action) should be avoided because it reduces the long-term staying power of the organizer. Long-term overstress leads to "burnout," which makes the organizer ineffective. Healthy weight maintenance is essential.

The second level is the life of the mind. Call it the "Scholar Level." By "mind" I mean not only those mental processes that are commonly called "rational" but also those called "spiritual." The organizer should appreciate the importance of the broad educational process already described and the need for some specialized training, for example on the latest in ecological system function. It is more difficult to appreciate the need to constantly reaffirm the intrinsic value (in contrast to mere human value) of the natural world and our cultural heritage. It is this space beyond the human-centered values, which I call *sacred*, upon which our mind must be regularly refocused. The scholar should not only master but fuse the long-standing distinction between theory and practice (Marglin 1996).

Besides the usual educational process, the good organizer must master learning through the cycle of action and reflection, an educational rhythm for which there is no formal schoolroom. Learning what is the best time to intervene requires a lifetime of practice but the rewards can be noticed from the start.

The third level, the "Ecolate Level," is life as a consumer and worker. At this stage, the organizer is prepared for the most difficult of the preparatory tasks. You have to examine in detail the results, both direct and indirect, of your own consumption. Even more difficult is the evaluation of the consequences of your professional activity. For example, if you are employed in production of throwaway beverage containers, is it possible to reconcile this activity with your environmental values? Would someone else just continue to make these containers if you stopped? These are among the most difficult questions we can ask ourselves. They have no clear-cut answers. Organizers answer such questions in their own minds, but the community and fellow ecolates will influence the form of that answer. The answer hinges on which solution builds the community.

The ultimate aim of the ecolate is to reduce disorder created by his or her existence by minimizing his or her own consumption and by helping others learn how to minimize environmental disruption. Deep cultivation of a sense of place should be one of the higher aspirations of the ecolate. Only by establishing a true sense of place can the organizer take action with sufficient seriousness. When our territory is threatened or degraded, we are most aroused. The ecolates fully know how to develop, how to clarify sense of place in the community they are organizing.

The final stage is life with others, the "Instructor Level." This is the pinnacle of the organizer's life. By contributing to the happiness of others through building community in a lasting and consistent way, one achieves a kind of immortality reached by few. Organizers begin as they themselves were educated: first on the personal level and later on the community level. They provide the role model for leadership, and they cultivate leadership in others. Their main duty is to prepare appropriate members of the community for the long struggle on their road to the Instructor Level.

Instructors lead ever-increasing numbers of the local community from the level of mere personal connection to a broad understanding of the more general environmental problem. They seek out those with the strongest sense of place and then aid in the development of that sense in others. They act to ensure that all segments of the community are involved in the organization. Instructors prevent environmental problems from becoming a class struggle, for they recognize that everyone causes such problems. No class is excused; no individual is irredeemable. Instructors realize that people do not always stay in the same "class" and that polluters also may be victims of environmental pollution.

Instructors should establish regular community talks to build habit and expectation. They must understand that community ties are based on love, understanding, communication, common experience, common problems, and a strong sense of place. Instructors recognize that community groups require a series of standards of achievement, beginning with the minimum conditions for entry, the conditions for acceptance of leadership, and the conditions for exclusion. Membership in the organization must bring benefits that are not available to nonmembers. It is important to reduce the "free rider" problem of people who benefit from the actions of others without making a contribution. The problem is a classic one among environmentalists working toward clean air and water.

The organizer's long-range plan is to establish a strong sense of the sacredness of nature in the local group by continuously acquainting people with local nature and identifying ever more details of this nature. Ultimately, the organizer introduces the concepts of religious naturalism and stresses how adopting such awareness is not only compatible with the local group's newfound understanding of nature but can also be shown to be compatible with their existing religious views.

The Society of Nature at work

What keeps humans from more environmentally sound ways of living? There are many obstacles. In a 2005 article, I attempted to summarize the main ones (Hannon 2005). The leaders of the Society of Nature will eventually learn how to overcome these impediments through close work with the local community. The first impediment is the concept of cognitive dissonance. If we have worked long and hard to achieve a certain goal and then discover to our dismay that progress toward that goal has overwhelmingly negative aspects, our tendency is to dismiss these aspects and forget them. We cannot face the fact that our efforts were misdirected. When the Corps of Engineers learned of large opposition to dams it had planned for years to build, it dismissed this opposition as ravings of a few misguided people. It cannot admit that its work and sacrifices were actually harmful and unneeded. When farmers who for generations have worked to improve crop production face criticisms of their intense chemical use, they dismiss the criticisms. They cannot believe that all of that intergenerational thought and work was misdirected. When people are told that climate change is the result of their ever-increasing use of fossil energy, they refuse to admit that their beloved lifestyle has caused global disaster. This response is ingrained behavior. The problem is finding a way around this dissonance. The principal way is to point out that knowing what they knew, their past behavior was intelligently and rightfully pursued. Now, however, the situation has changed. The creativity of everyone must be enlisted to find the appropriate manner of living.

A second obstacle to environmental change is based on the *de minimis* argument. My automobile, one of millions of autos on the road today, uses such a relatively small amount of fuel that I am consuming an infinitesimal amount of nonrenewable resource supply. Were I to stop using this fuel, the price would decline due to a slight drop in demand, causing other consumers to use a bit more ... my former share. So I decide to continue my present use. This obstacle can be overcome by taxing critical resource use to the point where all consumers are motivated to reduce their use. Such resource taxes have been proposed at the highest levels of U. S. government but have been defeated by the energy industries. Someday, when the Society of Nature is mature and widespread, such resource use taxes can be enacted.

A third obstacle revolves around the concept of memetic behavior. We envy the looks of the neighbor's car, a sign of their superior cultural status. We buy, perhaps extending beyond our means, a similar or even better car. This process of conspicuous consumption is rampant in a consumption-based economy and routinely encouraged by advertising. It certainly produces a level of consumption greater than is necessary. Tying consumption to the definition of economic growth builds on this human tendency. The idea is that greater consumption provides jobs somewhere in the economy. But the demand for jobs has everything to do with the level of automation in society, the choice of industrial and commercial technology, and the types of international trade agreements. A reduction in consumption, if not drastic, may have little connection with reducing employment in an economy.

A fourth obstacle is discounting. We favor the present over the future. This is time discounting. To cause people to postpone consumption, we must compensate them. We also discount on the basis of distance. The farther from my home the nuclear power plant is, the less concern I have about it. I discount others by their genetic or income difference. Racism is a form of this behavior. We are quite willing to ignore a landfill or polluting incinerator placed sufficiently far from us, even if it is placed in an area of poor people of another ethnic group. All these forms of discounting are really means of postponement. Active participation in the local community, where we regularly meet and interact with others whose interests we may otherwise be inclined to discount, will diminish the tendency to discount and promote a more rational and sound environmental lifestyle.

A fifth obstacle is our inability to use ostracism as nonviolent, highly efficient means to discourage wanton resource use. By way of example, ostracism in Tokugawa Japan (1600–1860) helped to control the population (Hannon 1985). The Shogun, the feudal lords, and the village mayors exacted a heavy quota of rice and cotton from the peasant population. The resulting scarcity caused the peasants to ostracize any family with more than two children because the extra child's consumption would eventually inhibit that family's contribution to the quota. As a result the population was nearly steady from about 1700 to the opening of Japan by American Commodore Matthew Perry in 1853.

One can imagine this approach being used to reduce gasoline use in the United States. Each state would have a quota, and states would break that quota down for each city, county, and so on. Then imagine the ostracism laid on a family with three cars or with a car that got mileage much less than the community average. Certainly the country could come to equilibrium with this nonrenewable, climate change-causing resource. Ostracism is a powerful force, but because of the way the country is organized, each person deals directly with distant national and state governments for important policies such as taxes. Ostracism can only operate when the main personal types of consumption are mediated through local government where need can be more truly established. The ways in which resources are distributed in a nation could change to bring about actual environmental improvement.

A sixth obstacle to environmental sustainability is what I call the "many hats" problem. Most of us wear several imaginary hats—parent, employee, citizen, etc. We have a different decision-making process for each hat. We may decide our business success depends on not installing the best pollution controls at our factory. As a parent, we realize air and water pollution are harming the family health. Our future action has two possibilities: Put on our civic hat and organize the community to deal with the factory pollution. This would likely cost us our job, the hard way to reduce the number of our hats. Or we could forget the problem and try to move our home to a less polluted zone of the city, likely breaking up any sense of place we might have developed. The development of a Society of Nature community would provide efficient pathways for busy people to become effective agents of change.

Free-rider thinking is a seventh block to environmental improvement. I am depressed as I read of the decline of world fisheries. Local fishermen are going out of business. Deep dragnets are scraping the ocean floor in an ever more drastic effort to maintain catch size. Reading further, I see that scientists and Greenpeace are working on the problem. I become a free rider, assuming that these agencies will take care of the problem. This same process goes on in hundreds of ways every day. I think that I am too busy to become involved; my family, my workplace need all of my time. Again, the development of a sense of community brings about pressure on free riders to become involved.

Real loss versus theoretical gain is the eighth obstacle. Reducing energy resource use in commerce and industry would most likely cause an increase in employment. The decision makers in this issue—the pertinent industry, unions, and government agencies—all realize this is true. But employment growth causes a shift in available jobs. It is easy to point out the names of people who will lose their jobs but not which individuals will get the greater number of new jobs. The decision makers will respond to the identifiable endangered people rather than those who cannot be identified—even if there are more of the latter—and the conservation issue will be scuttled. A strong community would care for those who become unemployed by the change and work to find them new jobs. The shift in product use from a petroleum refinery to making and installing wind turbines and electric mass transit is an example.

The final obstacle example is people's loss of a sense of connection to the ultimate source of the inputs to their lifestyle. Where does electricity really come from? From the two little holes in the outlet plate on the wall? Where does our drinking water come from? The faucet at the sink? Where does our waste go when we flush the toilet and when the trash hauler drives away? Urban and suburban settings severely mask these connections; however, understanding these local connections sets the stage for larger, more complex ones. Consider the difficulty in understanding the complex system of feedback in the socio-ecological system, using climate change as an example. The global temperature rises mainly because of increasing emissions of carbon dioxide (CO2), largely from fossil fuel electric power plants. As the temperature rises, we use more air conditioning, which needs more electricity, thus more fossil fuel use and more CO2 emissions, followed by another rise in global temperature ... a positive feedback. The community developed by the Society of Nature would provide such a continuing education of these connections, particularly for the youth. Just the process of establishing connections reveals the jobs, resource uses, and environmental costs of an individual's lifestyle. Such awareness is the first step toward real behavioral change and action.

This is just a summary of the main obstacles that face the leaders of the Society of Nature. To become effective leaders in the community they must master the ways of overcoming these problems. It is an extraordinarily difficult process, perhaps the most difficult of any occupation. With a spiritual connection to nature, leaders and their followers would have the background strength to persist, to create the solutions that win the hearts and minds of the community.

The sense of place

Ultimately, our sense of the sacredness of nature is inseparably tied to a deep sense of place. We all have such a sense, whether we think of our place as an urban area, a medium-sized suburb, a small town, or a farm. Most of us do not have a deep connection with the prairie, the woods, the mountains, or the ocean's edge because we live where we can earn a living, near to large agencies of commerce or government. My contention is that each of us has an inherent desire to belong to a place, a "home." We can cherish the nearby park, the yard, the farm fields, but none of these is nearly as complex as the ecosystems we find in nature. I contend that it is this complexity we seek and find that can compel our interest and involvement. As we become more familiar with an ecosystem—the more we realize the meaning of providence—a sacredness begins to appear, to have profound meaning for. This is not a pantheistic view of nature but one that celebrates the dynamic wholeness of its ecosystems. This is definition by experience. And the definition deepens as our understanding increases. The experience is so exhilarating that it binds us firmly to preserving nearby nature, to dedicating our lives to nature protection and restoration, and ultimately to helping others gain this same understanding.

Those who stand still, who stay in place—farmers, for example—grow to love their place, though it may be entirely devoid of trees, grass, and other visually tangible signs of deep nature. What is it that binds them to that place? Perhaps they live in the home of their ancestors, maybe they look closely at the soil and see the microorganisms that aid crop growth, or perhaps they have no sense of place but have no alternative place to live (Mullendore et al. 2015). When a farmer advises his son to take up farming, we know he has a true sense of his place.

The study of natural and human history with access to natural areas gives us the basis for continual development of a sense of place within nature. Without such a sense we are incomplete, unfulfilled humans, and we are unable to recognize the ways in which our life habits are undermining the sustainability of nature. Without a sense of place, we have no foundation to support the many protracted battles that are needed to preserve nature. A sense of place provides a fundamental source of patience. It requires us to slow down, to pay attention to what we are doing to nature and ourselves, and to learn why we are doing it. In this way we personally become a part of nature. The leaders of the Society of Nature have a mission to develop a deep and abiding sense of place in a community.

So instead of a materialistic lifestyle we seek work that we can master and devote to purposely acquiring a connection to our place. We can slowly disengage from work we find meaningless and begin this journey by committing to a place and to resolving local environmental problems. Moving from city to city in pursuit of career or even moving about within a city prevents development of a sense of place. Fortunately, in the United States at least, internal migration has been declining since the mid-1970s (Partridge et al. 2010).

We can learn why our ancestors did what they did and if they understood the implications for nature of their actions. Restoring nature has common ground with restoring the works of man. Restoring the works of an earlier people is a show of respect, a tribute to those people. It is therefore a constant reminder to do works in both society and nature that we hope will be continued. By combining a sense of our human history with the natural history of a place, we take the first steps to bring the whole of society into coherence with the rhythms of nature. It is a process of sustainability, of survival.

Developing a sense of place is clearly an ongoing educational process. Learning the natural and social history of a place begins to endear that place to you. It requires you to sharpen your observational abilities, to look carefully for signs of the past. You note the descendants of those who came before. You read archives of the community history. You visit nearby parks and countryside (at sunrise and sunset, in winter and summer), power plants, water treatment plants, city councils, police stations, industries ... a list without a clear end. You never pass someone digging without asking why and noticing the soil layers, asking if they have found any early street car tracks, old piping, sign of landfill, etc. You are constantly building a memory of the place, its people and how it came to be what it is today. You are developing the sense of connection both with past and present and between the living elements before you. This intensive noticing—curiosity really—will slowly lead you toward a full sense of your place. It has much in common with prayer.

Certainly this process has its own rewards. It is gratifying to begin to understand a place and its dynamics. But a greater personal reward awaits those who have the sense of place and then use this gift to protect and restore works of nature and of man. Stopping the migration to a better job, finding a fascination with where you are, stabilizing your lifestyle and of those close to you, gaining time for the learning process, increasingly recognizing the sacredness of nature ... I think of all of these as part of the sense of place that I have in mind.

The religion of nature

Ultimately, we realize that to witness nature is to have gained the key to its salvation and to ours. To ignore nature in the face of what humanity is doing is to kill nature and all of its living elements. To witness nature's dynamic is to acknowledge it, to learn to respect, restore, and protect it. Being a witness paves the intergenerational pathway to humanity's future and puts us on the path to religious naturalism. It is the ultimate achievement in the quest for place.

Yi-Fu Tuan (2009) notes most religions work on dissolving our focus on place. They consider such concern as tribal and direct us toward a mystical heaven. Church buildings break our connection with nature and local community through, for example, the use of stained glass windows, which filter and transform light into a message of faith. This dissolving of a sense of local place ultimately leads to an abandonment of concern for nature in a local and widespread sense. How can someone learn to truly respect nature when he or she believes his or her true home is ethereal, is heaven?

This is a two-way problem. The true home of the most popular religions is not the earth and its biosphere but another imaginary place, painted with a beckoning infinite goodness. Followers depreciate their earthly home by comparison. Some Christians, for example, take the attitude of (so-called) wise-use stewardship of nature and indicate we should quickly use up God-given resources before Christ comes to earth the second time. This attitude, of course, creates the tendency to ignore or even abuse our earthly home while focusing on our heavenly home. It is likely one of the strongest, though often unstated, environmentally destructive forces.

Many religions promise a heavenly home after death. A religion of nature suggests no existence after death, except through the remembrance of our acts of goodness by others, our writings, and the artifacts we made during our lives. Adopting a religion of nature will let us come to peace with thoughts of our own death and endure the grief and loneliness when a loved one dies. We are alive through no choice of our own, and we have about as much reason to be mystified about why we are alive, as have the wolf and deer. The best path for us is to add to human understanding of the world around us and to learn to respect all life on earth. This is our earthly providence, and it is as close to heaven as we can or need to come.

Though science continues to provide greater understanding of nature, it tells us little about how to live sustainably. This we must learn by establishing a belief system based on nature—not the worshiping of nature but its veneration, with deepest respect and ever-widening understanding of how we are an interdependent part of it. To be willing to let nature influence our aesthetics, ethics and religion is to know our place and to act constantly to recognize, restore, and revere this source of all life. Such is an act of faith and thus the adoption of religious naturalism.

The complex and well-defended definition of a religion of nature appears in Donald Crosby's book (2002). He recognizes that nature fulfills all the things people have sought throughout the long history of religious belief in a consistent manner. Religious naturalism cannot make the lame walk nor the blind see. It is not the source of supernatural miracles. But it is the source of restoration, inspiration, and sustainability. Nature records everything that happens. Its natural power to heal itself from the blows of human society is a form of forgiveness and inspiration for knowledgeable and devoted responsibility. Nature is there, through storm and sunshine, through winter and summer, day and night, all the time. It is a source of revelation, genesis, and evolution. Studying it carefully can help us to understand the dynamics of complex systems that involve feedback, resilience, and reliability, and how such systems cope with and require disturbance to maintain diversity. It is a source of how real reincarnation is accomplished, with the end of life being the beginning of new life, a kind of eternal vibrancy. It is the model, the source of personal rejuvenation and of community coherence and strength. It enables the individual to endure and the group to cohere, to grow its numbers and effectiveness. It keeps a comprehensive record of human impacts. It beckons us to join. It will require the local environmental group to recruit, to educate, and to continuously expand its mission to protect and restore nature.

Nature's problem is there are too few environmentalists. There should be more of them, and they need to be trained and effective. We need to cultivate more of them from the great mass of consumers, and we need to know that we have done so. We must not lull ourselves into believing a movie on forest destruction, a newsletter on the effects of the latest chemical pesticide, an editorial in the *New York Times*, or mention on the evening TV news will change people's behavior in any permanent way. We know we are winning the environmental war only when we can look people in the eye, hold onto their hand, and hear them tell us that we have changed their life—nothing else is enough reward to support us through a lifetime of true environmentalism. This is how the Society of Nature would work.

My ideas reflect a long history of successful environmental organizing. For the past 50 years, a group that takes this hands-on approach to environmentalism has existed in the Midwest. Although the group's name changed over the years—from the Committee on Allerton Park, eventually to the Prairie Rivers Network—it has maintained its focus on making environmentalists out of otherwise innocent citizens. The past and present leaders of this group are the founding members of the Society of Nature.

The central problem with the history of such organizing is this: the local organization will dissolve unless it is united by a belief in the appropriateness of direct, continuous connection to nature, and the examination of nature as guide to sustainable living. This is what the religion of nature can provide. Without continuously expanding understanding of nature and growing respect for it, even a successful local group will disappear. Faith in having nature guide our behavior is the only way the group—and all of us humans—will survive. This concept of a nature religion is a path to finding personal restoration, a process that builds a cohesive and ever-expanding community of members who seek to mimic the processes nature has derived for sustainability.

We share this planet with many other forms of life. We humans obviously need and sustain one another, yet the present collective human behavior disrespects other life forms. We consider ourselves so much more important than other life forms, and yet in harming them we harm ourselves. This is why the Society is based on a religion of nature. It insists that the ecosystem, in all of its complex and mysterious beauty, is a necessary and sufficient focus of faith and commitment to build a new structure of consistent belief. Through the Society of Nature we will gather respect for all life, and only then can ecological justice and ecological peace prevail.

Acknowledgement

I thank Jean McDonald for her careful and patient editing and gracious commenting. I am much indebted to Professor Robert McKim, who provided many valuable and insightful suggestions.

References

Crosby, D. A. (2002) *A Religion of Nature*, Albany, NY: SUNY Press.

Hannon, B. (1985) "World Shogun," *Journal of Social and Biological Structures* 8 (4): 329–41.

—— (2005) "Pathways to Environmental Change," *Ecological Economics* 52 (4): 417–20.

Marglin, S. A. (1996) "Farmers, Seedsmen, and Scientists: Systems of Agriculture and Systems of Knowledge," in F. Apffel-Marglin, and S. Marglin (eds.) *Decolonizing Knowledge*, Oxford: Clarendon Press.

Mullendore, N. D., J. D. Ulrich-Schad, and L. Stalker Prokopy (2015) "U. S. Farmers' Sense of Place and Its Relation to Conservation Behavior," *Landscape and Urban Planning* 140, 67–75.

Page, T. (1977) *Conservation and Economic Efficiency: An Approach to Materials Policy*, Baltimore: Johns Hopkins University Press.

Partridge, M. D., D. S. Rickman, M. R. Olfert, and A. Kamar, A. (2010) "Dwindling U. S. Internal Migration: Evidence of Spatial Equilibrium?" MPRA Paper No. 28157. http://mpra.ub.uni-muenchen.de/28157/.

Tonino, L. (2016) "Two Ways of Knowing: Robert Wall Kimmerer on Scientific and Native American Views of the Natural World," *The Sun*. 484: 5–14.

Tuan, Y.-F. (2009) *Religion: From Place to Placelessness*, Chicago: Columbia College.

29
PRACTICES IN RELIGIOUS NATURALISM

Eric Steinhart

Introduction

At present religious naturalism is primarily a philosophical perspective. It involves mainly commitments to abstract doctrines about reality and value. As Crosby once put it, religious naturalism has "no practicing communities, no institutional structures, no duly constituted cadre of leaders, no body of traditional beliefs, no rituals or ceremonies, no revered founders or scriptures, no stories, myths and symbols" (2002: 155). Today the situation remains the same. If religious naturalism ever hopes to be more than merely an intellectual exercise, it needs to define genuinely *religious* ways of living. It needs to develop systems of practices. These practices need to embody the core values of religious naturalism. They also need to be attractive. Practices are costly, and successful religious practices provide benefits to repay those costs.

So far there are two main strategies for developing practices within religious naturalism. The *first strategy* is to naturalize your participation in some established religion. On this way, you could continue to attend a Christian church and to participate in its practices, but you would understand them in a new way that would not commit you to any literal belief in any divine persons. You would regard church life as *live-action role-playing* (*larping*). The *second strategy* involves developing novel natural religious practices. To do this, you start with practices that are already thought of as religious or spiritual but which are open to naturalization. You then work to fully naturalize these practices. To ensure social success, you need to start with attractive practices—practices that return benefits relative to their costs. As they pursue this strategy, religious naturalists have looked at practices in Stoicism, at the practices associated with meditation and yoga, at transformational festivals, at the use of entheogens, and at other practices. If you pursue this second strategy, then you will start to develop a *natural religion* which includes your new practices.

The religions of worship

A *theist* literally affirms the existence of at least one *divine person*. A divine person has superhuman powers and is not essentially embodied in ordinary matter. Divine persons, despite their lack of ordinary physicality, can causally interact with ordinary physical things. Theists literally affirm that human animals can socially interact with divine persons (but these interactions often have unusual aspects). We can *talk* with divine persons (but we can talk with divine persons just by addressing thoughts to them). We can be *observed* by divine persons (but divine persons can observe us at all times and places). We can be *praised* or *blamed* by divine persons (but their standards may be higher than human moral standards). We can be *rewarded* or *punished* by divine persons (but their rewards and punishments may occur after we die).

We can participate in economic relations with divine persons. These are sometimes called *do-ut-des* relations, meaning "I give that you may give." To give a gift to a divine person is to *worship* that person. We worship divine persons by behaving in ways that please them, by praising them, by offering them valuable goods in ritual sacrifices, and so on. Since divine persons are usually thought to be bound by the laws of fair exchange, it is also usually thought that if we give them gifts, then they will give us gifts in return. To ask a divine person for a gift is to *pray* to them for that benefit (it is a *petitionary* prayer). Since we need not ask for benefits we can reliably obtain ourselves, and since divine persons have superhuman powers, we usually pray for benefits which are hard for us to reliably obtain (such as good luck in risky projects, health, wealth, long life, happiness, life after death, and so on). And if we believe that we have received some benefit from a divine person, we thank that person for it.

But religious naturalists are not theists; they do not recognize any divine persons. So they do cannot engage in *do-ut-des* exchange relations with divine persons. They do not worship any gods or goddesses. They neither pray nor give thanks to them. Since *do-ut-des* relations form the practical core of most traditional religions, it might therefore seem that religious naturalists cannot practice in those religions. However, at least one well-known way exists which enables religious naturalists to practice in those religions. This is the first strategy for developing practices in religious naturalism.

According to this strategy, religious naturalists can adopt the practices of theistic religions by naturalizing them. This naturalization generally has three phases. The first phase involves distinguishing between the *logos* of the traditional religion and the *mythos* of that religion. The second phase involves interpreting the *logos* in purely naturalistic terms. This interpretation usually produces a highly abstract naturalistic theology (Peters 2002; Gulick 2013). The third phase involves interpreting the *mythos* in some non-literal way. The *mythos* is treated as a *fiction* valued for its ability to arouse important emotional states, or for its ability to provide moral education, or for its ability to produce prosocial bonds. This strategy is followed by religious naturalists such as Goodenough (1998), Peters (2002), and Raymo (2008). Adopting this strategy allows religious naturalists to participate in Christian church life—they merely interpret that life non-literally. But their way of life remains Christian. Since a large literature already exists on religious fictionalism, this approach will not be discussed further here.

Three types of concern

The second strategy religious naturalists can use to develop religious practices involves turning away from theistic religions to the development of new and entirely nontheistic *natural religions*. Since these natural religions will not involve any divine persons, they will not be religions of worship. As an alternative to religions of worship, many have advocated *religions of self-realization*. A religion of self-realization has practices that aim to move a human animal into some ideal state or to keep it in an ideal state. All practices of self-realization involve techniques of self-modification. They are *technologies of the self* (Foucault 1988). For most religious naturalists, selves are bodies; hence techniques of self-modification are technologies of the body.

An initial classification includes three types of ideal states and therefore three types of self-realization. Each type of self-realization involves a corresponding type of *concern*. Concern for something is a kind of care, which aims at the *eudaimonia* (well-being) of that thing. It aims at the complete flourishing or thriving of the thing. Assuming that things have natures, any concern for a particular thing aims at the full realization of the nature of the thing; if such realization is the *perfection* of a thing, then it aims at the perfection of the thing. Concerns divide into *ontic concerns* and *ontological concern*. Ontic concerns are directed to the well-being of specific things or types of things. They are restricted concerns. Ontological concern is directed to the well-being of existence itself. It is unrestricted or universal concern.

The first type of self-realization is *physiological*. Physiological self-realization aims at ideal physiological states. The practices linked with this type of self-realization either aim to change a human animal from sickness to health or they aim to keep it healthy. Since the mind is part of the body they also aim to restore or to maintain mental health. Physiological self-realization involves techniques that modify any part of the body (e.g. diet, surgery, drugs, exercise). Such practices are discussed outside of religious contexts (e.g. by doctors, athletic trainers, dieticians, and so on). But religious naturalists seek to bring those discussions into their own religious frameworks. Since the body is a thing among things, this type of concern is ontic.

The second type of self-realization is *ethical*. Ethical self-realization aims at ideal ethical states. Its practices either aim to create or maintain an ethically ideal human animal. An ethically ideal human animal is concerned with its own physiological flourishing. So ethical concern includes physiological concern; but it transcends that concern. For the ancient Greeks, an ethically ideal human animal is a *sage*; for the Buddhists, it is an *enlightened* human animal. Since ethical behaviors are rooted entirely in brains, religious naturalists seek technologies that produce virtuous and prosocial brains. They seek technologies for *moral therapy* and *moral enhancement* (Hughes 2015). Such technologies will be grounded in fields like interpersonal neurobiology (Hollingsworth 2008). Ethical concern aims at the well-being of social groups of humans; it ultimately aims at the flourishing of the whole human species. However, since humans are only one type of thing, this concern remains ontic.

The third type of self-realization is *spiritual*. It aims at spiritually ideal states. Its practices either aim to change a human animal into a spiritually ideal state or to maintain it in a spiritually ideal state. A spiritually ideal human animal is concerned with its own physiological flourishing. So spiritual concern includes physiological concern. A spiritually ideal human is also concerned with the well-being of the whole human species. So spiritual concern includes ethical concern. But a spiritually ideal human animal is further concerned with the flourishing of all life on earth, and spiritual concern radiates outwards without any constraint to include all things in this universe and in any others. It aims at the well-being of the whole of nature. What is nature? To welcome diverse approaches to religious naturalism, the concept is left vague. Still, nature obviously exceeds the uncultivated portions of the earth. Nature may include an infinity of universes. Since religious naturalists affirm that nature is all that exists, a human animal in a spiritually ideal state becomes concerned with the totality of existence itself. Its concern transcends the ontic to become ontological.

When your concern expands to embrace all natural things, you become unified with nature. This unification is not merely intellectual; it is existential. Hence the goal of spiritual self-realization is the *existential unification of the self with nature*. As long as the existential depth of this unification is kept in mind, the goal of spiritual self-realization can be stated more simply as *unity with nature*. Since any natural religion aims at spiritual self-realization, its goal is unity with nature. Since spiritual self-realization includes both ethical and physiological self-realization, unity with nature implies health and virtue. What do we want? We want unity with nature. How do we get it? Through spiritual technologies. For Wildman (2011: ch. 7), the main spiritual technologies are meditation and entheogens. The spiritual technologies discussed below include ecological rituals, Stoic practices, mindfulness meditations, transformational festivals, the use of entactogens, and the use of entheogens.

Religious naturalism and ecological rituals

Religious naturalists have proposed practices aimed at extending our concern to the natural world beyond humanity. Some of these practices involve speech. Crosby says that religious naturalists can "express gratitude, trust, and personal resolve in meditations upon nature" (2002: 153). Although petitionary prayers make little sense in religions of nature, Crosby points out that religions of nature can include "something akin to payers of thankfulness, praise, confession, repentance, and endeavoring to live a more worthy life" (2002: 153; 2014: 141–145). Of course, Crosby correctly says that it makes no sense to talk to an impersonal nature (2014: 143). So any verbal practices in religions of nature are either self-talk or talk to other humans. For example, the *Council of All Beings* is a practice in which humans talk to each other (Seed, Macy, and Fleming 2007). The humans in the Council represent different animals or plants, and they speak up for the interests of those non-human life forms. Although these verbal practices involve interesting ideas, they have not proven to be culturally attractive. It does not seem likely that verbal practices will be able to expand human concern beyond human self-concern.

Ecotherapies aim to provide people with physical and mental health benefits by immersing them in wild environments. Thus *shinrun-yoku* involves meditative walking through a forest (Morita et al. 2007). These ecotherapies can produce psychological benefits (Ambra 2007; Bratman et al. 2015). But so far these ecotherapies aim only at physiological self-realization. They do not aim to produce ethical or spiritual self-realization. However, ecotherapies could be designed to expand our concern to include the entire earthly ecosystem. Although such concern would still be ontic, and thus not spiritual, it could be a stepping stone to spiritual concern.

Many religious groups use the four cardinal directions (north, west, south, east) in their rituals. The cardinal directions are used by Native Americans. They are used by Wiccans in their circles (Sabin 2011: ch. 6), by the Catholic Green Sisters in the *Earth Body Prayer* (Taylor 2007: 231–235), and they can be used by religious naturalists. Crosby encourages religious naturalists to develop "rituals orienting to the four points of the compass, suggesting fealty to the whole of the earth and its creatures" (2014: 147). Likewise, many religious groups use the cardinal elements (fire, earth, air, and water) in their rituals. These elements are used by Wiccans (Sabin 2011: ch. 6). Perhaps they can also be used by religious naturalists. Crosby says that "water, fire, air and earth ... can be put to use as religious symbols and, in particular, as symbols of nature as the religious ultimate" (2014: 90). However, while these are interesting ideas, religious naturalists have generally not developed such rituals.

Religious naturalists might perform the *Cosmic Walk* to symbolize the evolution of complexity in our universe (Taylor 2007: 249–252; Crosby 2014: 148). This ritual involves tracing a spiral on the ground. The spiral represents the history of our universe from the Big Bang at its center to the present at its end. Unlit candles can be placed at significant historical events (such as the appearance of matter, the formation of the sun, the beginning of life on earth, and so on). A *Walker* starts in the center of the spiral at the Big Bang. The Walker then walks outwards along the spiral while a *Reader* narrates the history of the universe. As the Walker passes a candle, he or she lights it. This ritual can help increase our awareness of our cosmic environment. But this ritual seems to have only been practiced rarely. It seems to be more of a history lesson than a religious ritual. It does not bind us emotionally to the past history of the cosmos.

Many religious groups hold celebrations on the solstices, the equinoxes, and the four cross-quarter days between them. These seasonal holidays are celebrated by Druids (Greer 2006:

74–82) and by Wiccans (Sabin 2011: ch. 9). They are celebrated by the Catholic Green Sisters, who refer to them as the Earth Holy Days (Taylor 2007: 252–258). More naturalistically, they are celebrated by pantheists (Harrison 1999: 84) and by atheopagans (Halstead 2016). Among religious naturalists, Crosby encourages "rituals recognizing the equinoxes and solstices" (2014: 147). So these holidays can be used to expand human concern to the entire earthly ecology, whose evolution is driven by the cyclical flow of energy from the sun. However, religious naturalists have shown little interest in these practices.

Religious naturalism and Stoic practices

Natural religions can adopt many ancient Stoic practices. The ancient Stoics were concerned with physiological self-realization. They made many recommendations about how to care for the body. They aimed at bodily health through practices such as vigorous exercise and proper diet. Here Stoic practices can be combined with naturalized versions of yoga. Stoics are materialists about human persons: you are strictly identical to your body. Hence physiological self-realization includes psychological self-realization. Many Stoic practices aimed at psychological self-realization. The Stoics aimed to replace unhealthy emotionality with healthy emotionality. To cultivate this replacement, they developed many psychological exercises (Irvine 2009).

Modern Stoics have developed a large system of psycho-physiological practices, described in detail in Robertson (2015). These include the *Morning Meditation*, the *Evening Meditation*, *Acting with a Reserve Clause*, and the *Premeditation of Adversity*. They include exercises for cognitive distancing, decatastrophizing, and decentering. These Stoic exercises inspired modern *cognitive behavioral therapy* as well as *acceptance and commitment therapy*. These exercises are effective against learned helplessness and depression. They can reduce fear and arouse hope in the midst of suffering. They are easily integrated into natural religions.

The Stoics were deeply concerned with ethical self-realization. Their practices aimed to transform an ordinary human animal into an ethically perfected Sage. The Sage is a fully rational and virtuous person. Sages preserve their serenity through all possible adversities (including death). Stoic serenity resembles Buddhist enlightenment, and Sages resemble Buddhas. All Stoic practices aim at ethical self-realization. The Stoic exercise known as the *Circles of Hierocles* involves expanding your concern beyond your body (Robertson 2015: 107–109). You start with your self-concern, expand your concern to include your family, your country, and the whole of humanity. So far this is an ethical exercise that helps you to build an ethically ideal self. But you can continue to expand your circle of concern to include the whole earthly ecosystem. This exercise helps breed compassion for all living things. This outlook is consistent with the religious naturalist valuing of all life on earth. It may inspire ecological activism.

The Stoics were also interested in spiritual self-realization. Stoicism is intensely theological; perhaps it is even religious. But Stoic theology is also highly naturalistic; it is a kind of scientific pantheism. For the Stoics, nature is rationally organized, and the unity of nature is pure reason. As an intellectual anticipation of spiritual self-realization, you can imagine expanding your circle of concern to include all natural things. A related Stoic exercise is the *View from Above* (Robertson 2015: 220–225). This exercise involves adopting a cosmic perspective, wherein you endeavor to cognitively grasp the whole universe. Although your life is only a small part of this great whole, the whole has had enough concern for you to bring your life into existence. You can identify with this cosmic concern, and this can help you with spiritual self-realization.

Stoic practice includes a kind of prayer (Algra 2003: 174–176). For the Stoics, this prayer is a kind of self-talk in which the irrational part of the self talks to the rational part of the self. So Stoics can address petitionary prayers to their higher selves; but your rational self participates in the pure rationality of nature. This cosmic rationality is also pure virtue. Modern Stoics can therefore embrace a version of the Serenity Prayer: "Virtue grant me the courage to change the things I can, the serenity to accept the things I cannot, and the wisdom to know the difference."

Stoicism is currently undergoing a surprisingly strong revival. Ancient Stoic ideas and practices have been translated into modern psychotherapies. Books on modernized Stoicism are widely read. These include *Stoicism and the Art of Happiness* (Robertson 2015), *Philosophy for Life and Other Dangerous Situations* (Evans 2013), and *A Guide to the Good Life* (Irvine 2009). The University of Wyoming runs a week-long *Stoic Camp*. The University of Exeter runs an annual *Stoic Week*, which involves an intensive seven-day course in practical Stoicism. There are annual popular *StoicCons* in London and New York attended by hundreds of people. This revival of Stoicism may provide religious naturalists with an enduring metaphysical and ethical system.

Religious naturalism and mindfulness meditation

Mindfulness meditation is a popular practice. It came to the West from Buddhism; but its current Western incarnation is highly secular and divorced from most Buddhist metaphysics and religion. Meditation can help to facilitate physiological self-realization (Flanagan 2013; Harris 2014). Meditation is effective for stress-reduction. It can help relieve anxiety and help with depression. It is an effective psycho-therapy (Tang et al. 2015). Meditation can change the default mode network in the brain, which is associated with self-reference and the self-concept (Brewer et al. 2011).

Mindfulness meditation can also help to facilitate ethical self-realization. Meditation can also be practiced in order to develop prosocial virtues. These include empathy, altruism, loving-kindness, and compassion (Kristeller and Johnson 2005). More than merely a mind-hack, meditation is a tool for ethical self-transformation. Like the Stoic exercises, it can move the self towards the ethical ideal of enlightenment. It can help you to expand your concern to include the whole of humanity. The scientific study of meditation has led to efforts to precisely formulate the concept of enlightenment in neurological terms (Davis and Vago 2013).

Meditation can help to facilitate spiritual self-realization. It can aim for a state in which consciousness is emptied of all contents (Fasching 2008). When consciousness is emptied of all contents, it ceases to be consciousness of any particular thing. It is no longer the ontic awareness of some being among beings. This empty consciousness is pure awareness. Since this pure awareness is not consciousness of any particular thing, some say it has no intentionality at all; it is entirely self-centered and not directed towards any reality beyond the self. Another interpretation says that it is the ontological awareness of being-itself; but this pure awareness can be thought of as the existential unification of the self with nature. It is spiritual self-realization.

Religious naturalism and transformational festivals

Natural religions may include festivals similar to *raves*. Raves involve dancing to electronic music and computer-generated imagery. Raves can facilitate physiological self-realization and can be therapeutic (Hutson 2000): raving helps overcome anxiety and depression; helps to overcome destructive behaviors; gives hope, confidence, and courage; can facilitate ethical self-realization. It can produce positive social values, expressed in the rave ethic of PLUR (Peace Love Unity Respect). Raving produces an emotional unification with all other people. It generates profound prosocial feelings of love, sympathy, empathy, and compassion.

Raves can facilitate spiritual self-realization. Many raves involved altars, opening and closing ceremonies, and so on (Sylvan 2005; St. John 2009). As they dance, ravers enter *hyper-arousal trances*, in which they often have intense spiritual or mystical experiences. During these trances, ravers often experience a profound energy flowing through their bodies; their egos dissolve; they experience all things as connected and unified; they feel that this same energy flows through all things (Sylvan 2005: ch. 3). Ravers often report *pronoia*, the feeling that nature is ultimately benevolent. Thus raving can lead to an existential unification of the self with nature.

Some ravers used *entactogens* (aka *empathogens*) in order to enhance their prosocial feelings (Saez-Briones and Hernandez 2013). The primary entactogen was MDMA. MDMA may help treat many mental illnesses (Sessa and Nutt 2015); hence it can be used in physiological self-realization. MDMA produces many positive social effects (Wardle and de Wit 2014), including the facilitation of ethical self-realization. Nevertheless, despite its many psychological and social benefits, there is considerable evidence that regular long-term use of MDMA can have serious neurotoxic effects (Parrott 2004, 2013). Natural religions cannot, therefore, endorse the unregulated use of MDMA. It is illegal in many countries; as long as it remains prohibited, the ethical concerns of religious naturalists prevent them from using it or condoning its use. But natural religions need not rule it out entirely. Taking a small number of low doses of MDMA may not produce long-term toxic effects (Morton 2005). Religious naturalists can endorse further research into safe and legal ways to use MDMA and other entactogens.

Classical rave culture flourished during the 1990s and 2000s. It continues in *yoga raves* and in *transformational festivals*. The paradigmatic transformational festival is *Burning Man* (Doherty 2004). The *Ten Principles* of Burning Man include positive ethical and ecological principles, so it can be understood as a festival aiming at ethical self-realization and ecological self-realization. It has been interpreted as a religious festival (Pike, 2001, 2005; Gilmore 2010). Besides Burning Man, there are many other transformational festivals, and there are smaller transformational groups, such as drum circles and fire circles. An example of a drum-fire circle is the *Spark Collective* in San Francisco. It is a member of the *Fire Family* of drum-fire circles. Perhaps surprisingly, the ancient Stoics often wrote approvingly of festivals (such as Saturnalia and the Olympic games). Although they did not endorse the excesses of those festivals, they saw them as models of the cosmic city in which all humans are citizens. So transformational festivals can also be incorporated into Stoic spiritualties.

The ideas and practices associated with classical rave culture continue to have popular appeal. Religious naturalists may use ideas from classical rave culture and established transformational festivals to develop their own technologies of self-realization. There are two ways to do this. One way is for religious naturalists to get involved with existing transformational festivals and to integrate their values into those festivals. Another way is for religious naturalists to develop their own transformational festivals. They will incorporate entactogens if and only if they are safe and legal. These festivals will aim at physiological, ethical, and spiritual self-realization. They will aim at the existential unification of the self with nature.

Religious naturalism and ayahuasca ceremonies

Ayahuasca is a psychedelic tea. For hundreds of years, it has been used in healing ceremonies by shamans in the Amazonian rainforest. Its use has recently exploded outside of its Amazonian context (Tupper 2008) and it is now popular in North America (Harris and Gurel 2012). Its main psychoactive ingredient is DMT. Since it acts strongly on serotonin receptors, DMT is a *serotonergic psychedelic* (like psilocybin, mescaline, and LSD). The effects of serotonergic psychedelics on the brain are slowly becoming understood (Carhart-Harris et al. 2014).

Ayahuasca is used as a sacrament in syncretic religions like the Santo Daime and Unaio de Vegetal churches, which combine Amazonian rainforest shamanism with Christianity. These churches now have branches in the United States. Under the Religious Freedom Restoration Act (RFRA), these churches may legally use ayahuasca in the United States. Unfortunately for religious naturalists, these churches have developed highly supernatural interpretations of the ayahuasca experience (e.g. illness is caused by evil spirits). Others have also offered supernatural interpretations of ayahuasca (e.g. Barnard 2014). But some ayahuasca groups, such as the *Ayahuasca Pantheist Society*, reject supernaturalism. Religious naturalists can accept only naturalistic interpretations of ayahuasca experiences (Shanon 2010). Natural religions will focus on the ways that the use of ayahuasca can facilitate physiological, ethical, and spiritual self-realization.

Ayahuasca can facilitate physiological self-realization. For many hundreds of years, people have used it on a regular long-term basis. The regular long-term use of ayahuasca appears to be safe (Ribeiro Barbosa et al. 2012), and may even be an effective treatment for many mental illnesses (Labate and Cavnar 2013; Dominguez-Clave et al. 2016). It has been used to treat addiction, anxiety, depression, and PTSD. Ayahuasca has also been used in practices of ethical self-realization. It can produce many positive personality changes that facilitate the development of prosocial virtues (Bouso et al., 2012; Harris and Gurel 2012; Bouso et al. 2015; Soler 2016).

But ayahuasca can also be used for spiritual self-realization. Taking it can induce mystical and spiritual experiences, which can be studied using well-established questionnaires (Trichter et al. 2009; Harris and Gurel 2012). Ayahuasca users reported that they gained a deeper "connection to nature, a deep love for living things, belief in a higher power and belief in maintaining a peaceful existence of service to living things" (Trichter et al. 2009: 128). Shanon reports that ayahuasca experiences "usually converge upon a coherent metaphysical outlook, one which is monistic, idealistic, pantheistic, imbued with religiosity and tainted with optimism, joy, and love" (2010: 269). These experiences induce "animism and a platonic realism" (2010: 269).

Religious naturalists can work on developing ayahuasca ceremonies, which move the interpretations of the ayahuasca experience away from supernaturalism and superstition and towards a naturalistic ontological concern. Thus natural religions can use ayahuasca to facilitate existential unification of the self with nature. Here Stoicism may be helpful. Many ancient Stoics praised the Eleusinian Mysteries. These Mysteries involved taking a substance, the *kykeon*, which some have speculated was psychedelic. Whether or not it was psychedelic, Stoic interpretations of the Mysteries may help to place modern psychedelic use into a more naturalistic framework.

Since natural religions seek ethical self-realization, they will not involve any illegal ayahuasca ceremonies. To ensure that their ayahuasca ceremonies are legal in the US under the RFRA, religious naturalists will need approval from the US Drug Enforcement Agency, and will need to show that their natural religions pass the tests used by the US courts. They will need to avoid the often horrific failures of the psychedelic churches from the 1960s (Stuart 2002; Lander 2011). To develop legal ayahuasca ceremonies, religious naturalists can work with the Council on Spiritual Practices, the ICEERS Foundation, the Beckley Foundation, the Multidisciplinary Association for Psychedelic Studies, the Heffter Research Institute, and similar groups.

Religious naturalism and psilocybin ceremonies

Psilocybin, like DMT, is a serotonergic psychedelic. It is an entheogen that has been used in traditional religious ceremonies in the Americas and can be used in practices of physiological self-realization. It has been studied as a treatment for anxiety, addiction, depression, PTSD, OCD, and other illnesses (Burdick and Adinoff 2013; Kraehenmann et al. 2015). It has also been used in practices of ethical self-realization. It can help produce many positive personality changes, which aim at prosocial virtues. Psilocybin has been used to successfully treat end-of-life anxiety in patients with terminal cancer (Grob et al. 2011; Grob et al. 2013). It can remedy existential distress.

Psilocybin can also be used for spiritual self-realization. Here it appears to act more powerfully than DMT. Taking it can induce extremely powerful mystical experiences (Griffiths et al. 2006; Griffiths et al. 2011). Mystical experiences produced by taking psilocybin have been scientifically studied using the *Mystical Experience Questionnaire* (MacLean et al. 2012; Barrett et al., 2015) and the *Altered States of Consciousness* questionnaire (Kometer et al. 2015). Psilocybin often produces mystical experiences in which the subjects experience "a sense of unity without content (pure consciousness) and/or unity of all things" (Griffiths et al. 2006: 277). It may thus facilitate the existential unification of the self with nature. This psilocybin-induced unification is highly correlated with changes in how the sense of self is processed by the brain (Carhart-Harris et al. 2015; Kometer et al. 2015).

Since most religious naturalists are materialists about human persons, all religious experiences are patterns of neural activity in the brain. All ethical and spiritual self-realization depends on physical changes in neural networks in the brain. So religious naturalists should be especially interested in technologies for changing our brains. At the present time, some of the most effective technologies for changing our brains involve the use of neurologically active molecules such as the serotonergic psychedelics. But our knowledge of molecular neurobiology is in its infancy. Religious naturalists will want to pay close attention to further developments in this area.

Conclusion

As a new religious movement, or a new way of being religious, religious naturalism looks to the future. If it wants to be more than just another system of beliefs, then it will need to develop systems of practices. To ensure social survival, these will need to be practices with some motivation. To avoid fracturing into a plurality of conflicting denominations, these practices should not be tied too closely to particular doctrines. They should be such that people can interpret them in many ways. People are naturally driven towards physiological, ethical, and spiritual self-realization; religious naturalists can develop naturalistic interpretations of practices of self-realization.

References

Algra, K. (2003) "Stoic Theology," in B. Inwood (ed.) *The Cambridge Companion to the Stoics.* New York: Cambridge University Press, 153–78.

Ambra, B. (2007) "People and Green Spaces," *Journal of Public Mental Health 6* (3), 24–39.

Barnard, W. (2014) "Entheogens in a Religious Context: The case of the Santo Daime Religious Tradition," *Zygon 49* (3), 666–84.

Barrett, F. et al. (2015) "Validation of the Revised Mystical Experience Questionnaire in Experimental Sessions with Psilocybin," *Journal of Psychopharmacology 29* (11), 1182–90.

Bouso, J. et al. (2012) "Personality, Psychopathology, Life Attitudes and Neuropsychological Performance among Ritual Users of Ayahuasca: A Longitudinal Study," *PLoS One* 7 (8), e42421.

Bouso, J. et al. (2015) "Long-term Use of Psychedelic Drugs Is Associated with Differences in Brain Structure and Personality in Humans," *European Neuropsychopharmacology 25*, 483–92.

Bratman, G. et al. (2015) "Nature Experience Reduces Rumination and Subgenual Prefrontal Cortex Activation," *Proceedings of the National Academy of Science 112* (28), 8567–72.

Brewer, J. et al. (2011) "Meditation Experience Is Associated with Differences in Default Mode Network Activity and Connectivity," *Proceedings of the National Academy of Sciences 108* (50), 20254–9.

Burdick, B. and Adinoff, B. (2013) "A Proposal to Evaluate Mechanistic Efficacy of Hallucinogens in Addiction Treatment," *The American Journal of Drug and Alcohol Abuse 39* (5), 291–7.

Carhart-Harris, R., Kaelen, M. and Nutt, D. (2014) "How do Hallucinogens Work on the Brain?" *The Psychologist 27* (9), 662–5.

Carhart-Harris, R. et al. (2015) "Neural Correlates of the Psychedelic State as Determined by fMRI Studies with Psilocybin," *Proceedings of the National Academy of Sciences 109* (6), 2138–43.

Crosby, D. (2002) *A Religion of Nature*, Albany, NY: SUNY Press.

Crosby, D. (2008) *Living with Ambiguity: Religious Naturalism and the Menace of Evil*, Albany, NY: SUNY Press.

Crosby, D. (2014) *More than Discourse: Symbolic Expressions of Naturalistic Faith*, Albany, NY: SUNY Press.

Davis, J. and Vago, D. (2013) "Can Enlightenment Be Traced to Specific Neural Correlates, Cognition, or Behavior? No, and (a qualified) Yes," *Frontiers in Psychology 4* (Article 870), 1–4.

Doherty, B. (2004) *This Is Burning Man*, New York: Little Brown.

Dominguez-Clave, E. et al. (2016) "Ayahuasca: Pharmacology, Neuroscience and Therapeutic Potential," *Brain Research Bulletin*, 126: 89–101.

Evans, J. (2013) *Philosophy for Life and Other Dangerous Situations*, New York: Random House.

Fasching, W. (2008) "Consciousness, Self-consciousness, and Meditation," *Phenomenology and the Cognitive Sciences* 7, 463–83.

Flanagan, O. (2013) *The Bodhisattva's Brain: Buddhism Naturalized*, Cambridge, MA: MIT Press.

Foucault, M. (1988) "Technologies of the Self," in L. Martin, H. Gutman, and P. Hutton (eds.) *Technologies of the Self: A Seminar with Michel Foucault*, Amherst, MA: University of Massachusetts Press, 16–49.

Gilmore, L. (2010) *Theatre in a Crowded Fire: Ritual and Spirituality at Burning Man*, Berkeley, CA: University of California Press.

Goodenough, U. (1998) *The Sacred Depths of Nature*, New York: Oxford University Press.

Greer, J. (2006) *The Druidry Handbook: Spiritual Practice Rooted in the Living Earth*, Boston, MA: Red Wheel.

Griffiths, R. et al. (2006) "Psilocybin Can Occasion Mystical-Type Experiences Having Substantial and Sustained Personal Meaning and Spiritual Significance," *Psychopharmacology 187*, 268–83.

Griffiths, R. et al. (2011) "Psilocybin Occasioned Mystical-Type Experiences: Immediate and Persisting Dose-Related Effects," *Psychopharmacology 218* (4), 649–65.

Grob, C. et al. (2011) "Pilot Study of Psilocybin Treatment for Anxiety in Patients with Advanced-Stage Cancer," *Archives for General Psychiatry 68* (1), 71–8.

Grob, C. et al. (2013) "Use of the Classic Hallucinogen Psilocybin for Treatment of Existential Distress Associated with Cancer," in B. Carr & J. Steel (eds.), *Psychological Aspects of Cancer*, New York: Springer, 291–308.

Gulick, W. (2013) "Religious Naturalism: A Framework of Interpretation and a Christian Version," *American Journal of Theology & Philosophy 34* (2), 154–74.

Halstead, J. (ed.) (2016) *Godless Paganism*, Lulu.com.

Harris, R. and Gurel, L. (2012) "A Study of Ayahuasca Use in North America," *Journal of Psychoactive Drugs 44* (3), 209–15.

Harris, S. (2014) *Waking Up*, New York: Simon & Schuster.

Harrison, P. (1999) *Pantheism: Understanding the Divinity in Nature and the Universe*, Boston, MA: Element Books.

Hollingsworth, A. (2008) "Implications of Interpersonal Neurobiology for a Spirituality of Compassion," *Zygon 43* (4), 837–60.

Hughes, J. (2015) "Moral Enhancement Requires Multiple Virtues," *Cambridge Quarterly of Healthcare Ethics 24*, 86–95.

Hutson, S. (2000) "The Rave: Spiritual Healing in Modern Western Subculture," *Anthropological Quarterly 73* (1), 35–49.

Irvine, W. (2009) *A Guide to the Good Life: The Ancient Art of Stoic Joy*, New York: Oxford University Press.

Kometer, M. et al. (2015) "Psilocybin-Induced Spiritual Experiences and Insightfulness are Associated with Synchronization of Neuronal Oscillations," *Psychopharmacology 232*, 3663–76.

Kraehenmann, R. et al. (2015) "Psilocybin-Induced Decrease in Amygdala Reactivity Correlates with Enhanced Positive Mood in Healthy Volunteers," *Biological Psychiatry* 78 (8), 57–281.

Kristeller, J. and Johnson, T. (2005) "Cultivating Loving Kindness," *Zygon 40* (2), 391–407.

Labate, B. and Cavnar, C. (eds.) (2013) *The Therapeutic Use of Ayahuasca*, New York: Springer.

Lander, D. (2011) "Start Your Own Religion: New York State's Acid Churches," *Nova Religio 14* (3), 64–80.

MacLean, K. et al. (2012) "Factor Analysis of the Mystical Experience Questionnaire: A Study of Experiences Occasioned by the Hallucinogen Psilocybin," *Journal for the Scientific Study of Religion 51* (4), 721–37.

Morita, E. et al. (2007) "Psychological Effects of Forest Environments on Healthy Adults," *Public Health 121*, 54–63.

Morton, J. (2005) "Ecstasy: Pharmacology and Neurotoxicity," *Current Opinion in Pharmacology 5*, 79–86.

Parrott, A. (2004) "MDMA (3,4-Methylenedioxymethamphetamine) or Ecstasy: The Neuropsychobiological Implications of Taking it at Dances and Raves," *Neuropsychobiology 50*, 329–35.

Parrott, A. (2013) "MDMA, Serotonergic Neurotoxicity, and the Diverse Functional Deficits of Recreational 'Ecstasy' Users," *Neuroscience and Biobehavioral Reviews 37*, 1466–84.

Peters, K. (2002) *Dancing with the Sacred: Evolution, Ecology, and God*, Harrisburg, PA: Trinity Press International.

Pike, S. (2001) "Desert Goddesses and Apocalyptic Art," in E. Mazur and K. McCarthy (eds.) *God in the Details*, New York: Routledge, 155–76.

Pike, S. (2005) "No Novenas for the Dead: Ritual Action and Communal Memory at the Temple of Tears," in L. Gilmore and M. Van Proyen (eds.) *AfterBurn: Reflections on Burning Man*, Albuquerque, NM: University of New Mexico Press, 195–214.

Raymo, C. (2008) *When God Is Gone Everything Is Holy*, Notre Dame, IN: Sorin Books.

Ribeiro Barbosa, P. et al. (2012) "Health Status of Ayahuasca Users," *Drug Testing and Analysis 4*, 601–9.

Richards, W. (2008) "The Phenomenology and Potential Religious Import of States of Consciousness Facilitated by Psilocybin," *Archive for the Psychology of Religion 30*, 189–99.

Robertson, D. (2015) *Stoicism and the Art of Happiness*, New York: McGraw Hill.

Sabin, T. (2011) *Wicca for Beginners: Fundamentals of Philosophy and Practice*, St. Paul, MI: Llewellyn Publications.

Saez-Briones, P. and Hernandez, A. (2013) "MDMA (3,4-Methylenedioxymeth-amphetamine) Analogues as Tools to Characterize MDMA-like Effects: An Approach to Understand Entactogen Pharmacology," *Current Neuropharmacology 11*, 521–34.

Seed, J., Macy, J. and Fleming, P. (2007) *Thinking Like a Mountain: Towards a Council of All Beings*, Gabriola Island, BC: New Catalyst Books.

Sessa, B. and Nutt, D. (2015) "Making a Medicine out of MDMA," *The British Journal of Psychiatry 206*, 4–6.

Shanon, B. (2010) "The Epistemics of Ayahuasca Visions," *Phenomenology and Cognitive Science 9*, 263–80.

Soler, J. et al. (2016) "Exploring the Therapeutic Potential of Ayahuasca: Acute Intake increases Mindfulness-Related Capacities," *Psychopharmacology 233* (5), 823–9.

St. John, G. (2009) *Rave Culture and Religion*, New York: Routledge.

Stuart, R. (2002) "Entheogenic Sects and Psychedelic Religions," *Multidisciplinary Association for Psychedelic Studies Bulletin 12* (1), 17–24.

Sylvan, R. (2005) *Trance Formation*, New York: Routledge.

Tang, Y., Holzel, B., and Posner, M. (2015) "The Neuroscience of Mindfulness Meditation," *Nature Reviews Neuroscience 16*, 213–25.

Taylor, S. (2007) *Green Sisters: A Spiritual Ecology*, Cambridge, MA: Harvard University Press.

Trichter, S., Klimo, J., and Krippner, S. (2009) "Changes in Spirituality among Ayahuasca Ceremony Novice Participants," *Journal of Psychoactive Drugs 41* (2), 121–34.

Tupper, K. (2008) "The Globalization of Ayahuasca: Harm Reduction or Benefit Maximization?" *International Journal of Drug Policy 19*, 297–303.

Wardle, M. and de Wit, H. (2014) "MDMA Alters Emotional Processing and Facilitates Positive Social Interaction," *Psychopharmacology 231*, 4219–29.

Wildman, W. (2011) *Religious and Spiritual Experiences*, New York: Cambridge University Press.

30
NATURALISTIC SPIRITUALITY AS A PRACTICE

Daniel T. Strain

I wake up in the morning and go downstairs. There is a small closet beneath the staircase. I open it and enter. Inside, a string of white Christmas lights illuminates. There are tapestries and decorative sheets on the walls and the low sloped ceiling. One is from Tibet, given to me by a friend. The other I found at a local shop. It has the image of a tree with branches in a complex interlaced design.

I sit down on a cushion on the floor. In front of me is a small wooden box, about nine or ten inches tall. Upon the box is a red cloth, and on top of that are several items, creating an altar. One of the items is a statuette of the Buddha I obtained from a curiosity shop near my home. Another is a small bust of Socrates I purchased on a trip to Athens, but was probably made in China. Other items include a glass sculpture of the Bodhisattva Quan Yin,[1] a glass candle holder with pagan artwork, a rosary I got in the Vatican gift shop, a brass bowl with a wooden striking stick, and a small box of sand with St. Francis of Assisi with a bird engraved upon it. There is also a tiny, smooth oval-shaped stone.

I place my hands together, finger tips pointing up. I bow toward the altar and say, "The way gathers in emptiness alone. Emptiness is the fasting of the mind."[2] Then I open a case and take out some matches and a stick of incense. I strike the match and use it to light the candle. As I light the candle, I say, "This world, no one of gods or men has made, is an ever-living fire. It's kindling exchanges, judging and convicting all things."[3]

Then I hold the stick of incense over the candle flame. Once it catches I gently blow out the flame on the incense so it still glows and smokes. It smells of sandalwood. I then place the stick in the box of sand upright so it continues to burn.

Next, I activate a timer and pick up the striking stick. As I gently tap the bowl, it releases a smooth gong. I return the stick and close my eyes as the gong oscillates and slowly fades. Then I sit without motion in silence for 20 minutes. When I am finished, I blow out the candle and say, "As the death of fire is the birth of air, so too is wantonness extinguished."[4]

I am an atheist.

I am also a skeptic, a Humanist, a Buddhist, a Stoic, and a few other things. I have been an active founder, organizer, and leader in several secular and humanist organizations. I have no supernatural or paranormal beliefs. I am neither confused, discontent, indecisive, wishy-washy, nor a believer wanna-be. Although I fit the definition of *atheist*, I would almost never consider it a fitting term for what's important for who I am. If asked, my first response would be that I am a Spiritual Naturalist.

I have been meditating for over ten years now, along with various other practices. Over that time, I have seen my mental disposition change radically. My very personality has shifted and my life experience has been altered (I would say improved). I am by no means any kind of master or guru, and still have much to learn and explore. But I have seen enough to have become a true believer in something that it took me a long time to really *get*.

Back when I was *only* a secular humanist, I remember seeing some examples of secular people engaging in ritual. There is a close and historical relationship between the Humanist movement and the Unitarians (Wilson 1995: 5). Occasionally attending UU services, I would see people engaged in singing, lighting candles, and so on. The experience always felt hollow to me; like *empty theater*. It seemed to be a kind of "me too-ism," like an extraneous "show" being put on for appearances. I think it likely that some readers who have read about my morning in this essay would find it similarly hollow to perform the same actions.

Yet, for me, everything I have done on the morning described above had a concrete purpose and a specific function particular to the aims and efforts of my practice. Hopefully, I will be able to convey some of how this can work for a naturalist.

About *spirituality*

Religion is a tricky word. Surely, Religious Naturalists mean *religion* in its best sense. That is, a tradition, body of knowledge, community, and practices—those things that promote our current understanding of the world and our place in it, as well as our values, wisdom, meaning, and inspiration. *Spirituality* is another matter. Though spiritual community and communal practices are a healthy part, spirituality is mainly about our *inner* experience and development and practices. Religion cannot manifest in its finest form unless its people have spirituality. Without that, religion becomes an institutional, bureaucratic, or even dogmatic endeavor—even when it is a naturalistic religion. Thus, it is possible for one to be spiritual without being religious. But, if we mean *religion* in its proper sense, it is impossible to be truly religious without being spiritual.

For naturalists, the *spirit* of a thing refers to the *essence* of that thing— its essential nature. For example, "the spirit of the law" is the essential aspect of that law: its central intent. Likewise, spirituality is about focusing intentionally on the essential things of life rather than an unexamined life engrossed in the mundane. The essential aspect of human beings is our consciousness and our experience. Thus, focusing on the essential things of a human life includes our consciousness, wisdom, and awareness.

This meaning of *spirituality*, by the way, is no less true for supernaturalists. It is simply that their worldview leads them to believe supernatural things are of the essence, and so their majority number has created an overly particular popular understanding of what *spirituality* means.

Why practice?

Many Religious Naturalists are inspired and moved by things like the eloquent words of Carl Sagan. He was a true treasure to humanity in many ways (scientifically, educationally, popularly, etc.). Sagan was especially gifted in giving people a glimpse of what it is like to have a religious experience in response to the wonders of the natural universe, perhaps even helping them toward such experiences. There are many other gifted communicators like this whose work is to be respected and admired.

Unfortunately, many of us may not be aware of the adventure of a lifetime to which such inspiring works call us. Many of us stop at the doorstep of awe. But appreciation and awe before the majesty of nature is just the invitation—the beginning, not the end. If we frame the invitation and place it on our wall without responding to it, we miss the chance to move from being "an enthusiast of spiritual and religious naturalism" to becoming "a *practitioner* of it."

Ultimately, the purpose of a spiritual *practice* is a more flourishing and fulfilled life. This is a life of greater equanimity, peace, contentment, joy, and thriving. The ancient Greeks called it *eudaimonia*—or "healthy soul" (Thomson 2016: 161). Today we are too quick to run to surgery, drugs, consumerism, and entertainment to cure what troubles us. But this kind of flourishing is the *natural* result of a character that is compatible with its own nature and the nature of its world.

Practice is about conditioning the character to that end. This is why it is so often described as a *path*. But such a journey cannot be made by the intellectual subset of the human mind alone. All of it must step together. As rational naturalists we cannot abide the *irrational*, but as religious and spiritual naturalists, we must embrace the *a-rational*.

The trap of belief

One of the things I had to learn early on was the very limited usefulness of beliefs as the definer of a tradition or approach to life. In the conservative Christianity of my upbringing, primary importance was placed on whether or not one *believed* a certain set of things. Merely the sincerely held belief and acceptance of these claims were enough to save one's soul. Accepting on faith the supernatural nature of Jesus Christ as one's personal savior would activate one's ticket to heaven. All of the good deeds or ethics of Jesus were seen as something that would become manifest later, as a result of the Holy Spirit entering one's heart when saved.[5]

Years later I would leave that faith and come to Humanism. But as a fish is blind to water, I did not realize that we were all stuck in the same paradigm. We approached Humanism as a list of beliefs. In the Humanist Manifestos we could hold up a list of opinions about things and say, "because I agree with and believe these things, I am a Humanist." Sure, the list was rational, evidence-based, and in every way more reasonable than the list of positions that the Christians I grew up with believed made them Christians. And, to be fair, one does have to live in compatibility with the principles in order that the proclamation of agreement be honest. But both Christians and Humanists identify themselves, like taking up a team flag, by the list of opinions with which they agreed. This approach was so ubiquitous that it never occurred to me to question it.

It was only when my interest in philosophy lead me to Marcus Aurelius, and then to the rest of the Stoics, that I begin to see another approach. Especially after my study led me on to Buddhism and Taoism, I came to see the alternative of *starting with practice*. We call both Buddhism and Christianity *religions*, and that they are. But far more importantly, Buddhism and Stoicism are firstly *pragmatic practices one employs for a purpose.*[6] They are not so much about allegiances to a set of opinions or beliefs. The Stoics value skeptical inquiry, and the Buddhists have a specific teaching (sutra) on not accepting claims on the basis of authority or scripture (Batchelor 2010: 98).

Certainly, we want our beliefs to be rational and true. They inform our path, and irrational beliefs in spiritual traditions have led to countless evils and tragedies. But if we focus on good practice, then wise and true beliefs will tend to flourish from them.

One of those practices is humility, historically associated with the spiritual person for good reason. Humility can let us focus on how we treat others first. It can allow us to stay mindful that none of us in the world has the exclusive keys to Truth. The reality is that we small organisms crawling around this speck of cosmic dust have very little chance of conceiving of the ultimate cause and nature of reality at its root—or even to know if we ever stumbled upon it. What matters first is *how we live* and *how we love* one another. What a fine thing it would be, to have the privilege of quibbling over our little cosmological speculations (and our little rejections of them) endlessly while enjoying tea together in loving friendship and respectful peace. We can get there by beginning first with our own personal practice.

Transformation and subjective experience

Once I was listening to a religious (conservative American Catholic) talk-radio program. There was plenty on this program that I found objectionable, but I try to stay informed about other points of view. There were also some admirable things at times. At one point, an elderly Catholic woman called into the program. She spoke of her nightly routine. She said that every night she knelt beside her bed and prayed. In her prayer, she went through the people in her life and prayed to God for their well-being, happiness, and relief from the problems facing them. She then moved to others in the world and prayed for peace and relief from suffering. The younger Humanist in me would have admired the woman's good intentions but found this an unfortunate waste of time coming from baseless superstition. I might have snipped that she should focus more on helping them directly.

It is true that superstition can be harmful—for example, praying for healing while neglecting rational courses of action. It is also true that I still today have no belief that her messages were reaching any other entity or affecting the realities of the world directly. But now a new kind of thought entered my mind about this.

Her methodical movement from her family, to friends, acquaintances, and others reminded me of the Buddhist practice of loving-kindness meditation, which follows the same process. In these meditations, however, the goal is transforming ourselves to be more compassionate.

So I wondered to myself what profound effect must these kinds of nightly prayers have on this woman's sense of empathy, her kindness, and compassion? As she moves throughout her day and interacts with others, or makes decisions about what kinds of things she spends time on, how must such a regular practice shape her natural reactions and responses? And, in the life of the typical atheist or secular humanist, what could possibly compare to such an intense and regular practice?

Spiritual practice is a process of *spiritual transformation*. That is, the transformation of our essential nature. If you wonder whether such deep transformation is possible, consider that you are already the product of it. No doubt, if you think back to yourself as a child, or other ages between, you will notice something. While you may have many memories of events in your childhood, it is almost as if that was some other person whose memories have been injected into your mind. You are a different person than you were as a child. Your knowledge was different to be sure. But more importantly, *how* you thought, how you weighed things up, how you responded to various stimuli, what kind of things you loved or hated, and how you reacted to those things, were all different. This kind of transformation takes more than mere knowledge. It takes visceral experiences that make certain truths stingingly real to us on a deeper, more intuitive level over time. This is why an older person cannot simply sit a younger one down and tell them what they need to know to be wise.

The natural course of life provides experiences that mold us and shape our character over time. If we are lucky, some of those experiences have led us to wisdom or insight (through no virtue of our own, my wife calls these gifts of the cosmos *grace*). When we undertake a spiritual practice, this is grabbing the reins of our experiences to mold ourselves in a conscious and deliberative way. If life's experiences are the "natural selection" of our character traits, then our spiritual practice is the selective breeding of them. In a practice, we are engineering subjective experiences designed to cultivate certain perspectives and habits of thought.

We have all had the experience of being distressed about something, only to be soothed by a wise passage or bit of advice that reminds us of something we already knew. But if we knew it, why did we not react to it as such in the moment? This is because intellectual assent to a concept is not enough. True knowledge is not simple absorption and assent to data—it is a deep knowing that becomes intuitive. When we are wise, we do not need to be reminded of how we should have felt or reacted to a situation after the fact, by seeking out philosophy, advice, and teachings. Rather, our character has been transformed so that it is consistent with the wise teachings, and our natural, effortless response is in accord with that wisdom.

Because the aim of practice (flourishing/well-being/deep happiness) is a subjective experience, we must accept the importance of subjective experience in building a practice. Spiritual practice, then, involves the engineering of subjective experiences that stimulate many parts of our mind in different ways. If they are appropriately designed and executed, they get us involved and caught up in them emotionally, intellectually, creatively, and so on. This is how the intellectual becomes intuitive and begins to shape who we are and what we can be.

Exploring naturalistic spiritual practice

So where to begin? What constitutes an individual exercise in our practice? From what sources do we pull? Spiritual and Religious Naturalists today are facing a still-new world of access to the histories and teachings of cultures everywhere throughout history. We also have access to more scientific knowledge about human well-being, behavior, psychology, and more. All of this creates wonderful opportunities. Further, we naturalists stand at the beginning of a new movement to rediscover the sacred in the natural. This movement is just starting to make its way in building its own traditions. For practitioners such as us, *the primary spiritual practice is the very building of our personal practice itself.* A big part of what I've learned in my exploration has been how and why I got that initial feeling of empty theater regarding ritual, and how to avoid wasting my time with it. I eventually learned how to build a practice that really did connect with me, engross and move me, and thus have a significant transformative effect.

Many people faced with considering their practice will mention things like how they already enjoy taking a walk on the beach or listening to music. They consider that to be their practice. These things are wonderful and make excellent components to a practice. However, I would invite you to push yourself beyond your comfort level; not for its own sake, but in ways that might be more productive. A spiritual practice is not merely "what we do to blow off steam" or relieve stress. Some parts of our practice should be challenging or even difficult. When we go to the gym to sculpt our bodies, we cannot spend all our time in the hot tub. Likewise, if we want to sculpt our character, we should expect some practices to require great effort.

Some may seek to get very deeply into one tradition, and others may prefer their own personal "cocktail" of features and interpretations. After founding the Spiritual Naturalist Society, I have been fortunate to have met naturalists within many different schools and traditions, including naturalistic pagans, Jews, Buddhists, Stoics, and even Christians.

In order to discover helpful teachings and practices, it is necessary to have a confidence in yourself. Some act as if, were they to be too open-minded, they might forget the principles they already recognize as naturalists (reason, logic, the value of evidence for forming positions, and so on). But your intellect and memory are more resilient than you think. It is possible to be open, refrain from judgment for a time, and have a tolerance for phrasings or interpretations that don't jive with your worldview, without falling off the deep end. As such, you will begin to spot profound ideas about how to think about the world that are compatible with and useful to a naturalist. You can place yourself in other shoes and roles, and return back to your hard empiricism to begin thinking about how to reconcile them in meaningful ways with your beliefs and practices. You will find this kind of exploration to be more rich, interesting, and insightful than trying to reinvent the wheel. While there are many new discoveries in science that should be incorporated into our practice or checked against it, you will find that thousands of years of deep thinking have produced some immensely powerful practices.

The practices

The most prominent practice you are likely to hear about is meditation. There are many kinds of meditation, not all of them compatible with naturalism. The most basic, and yet perhaps the most crucial and foundational, is mindfulness meditation, or breathing meditation. I and others at the Spiritual Naturalist Society have written some helpful guides on this (see "Further reading"). The important point here is that mindfulness meditation is a practice that can help you develop your ability to still the mind, to direct your attention at will with a more concentrated focus, and to stay aware of your inner mental activities as they arise. These skills you will find essential to keeping and maintaining many other spiritual practices. Without them, you will say you want to incorporate certain thinking habits or reactions into your daily life, but will find yourself forgetting or veering. Mindfulness meditation is not closing your eyes and relaxing, however. Meditation is work and requires diligence and effort to build mental discipline, but eventually you will come to see how it plays a role throughout your life.

If you just want to engage in meditation for a while, that is perfectly fine; yet it would be a mistake to settle for "atheism + meditation." There are *many* other practices naturalists will find important and helpful. Another practice I often include is daily journaling about how I am doing in my intended development so as to stay mindful of where I am and the good wisdom I am seeking to habituate. Other practices might include occasional kinds of dance, music, drumming, and the kinds of ritual I described at the beginning of this essay.

If you want to design a ritual, begin with considering what kind of wise concepts you want to incorporate more into your character. Then look for writing, poetry, music, art, or other content that *moves you* and centers on these ideas. Think about what it is in that work that moves you and why it moves you so. Really try to dissect yourself here. Then you can begin to see how you might create a ritual that incorporates moving and meaningful imagery, sounds, readings, or other experiences to make you fully appreciate the truth of these ideas.

If you find yourself going through the motions of a ritual and you aren't *grabbed* by it or fully and un-self-consciously lost in that moment, then stop and think about why that is. It is being moved that deeply instills concepts. This is why artists can sometimes help move society in ways that simple explanations cannot. They can help people to *get* an idea deep down. This is what ritual is about. Of course, you won't be able to maintain the same level of emotional involvement every time, especially if the ritual is performed as part of your daily or regular practice. But if it connects with you personally, then the repeated practice of it serves as a daily reminder to keep you in the mindset you desire as you go about your day.

Remember that not all practices take place in sectioned off time periods and places. Some practices are a commitment to do certain things in your life a different way. Trying to smile more, trying to open up more to people, having a more compassionate demeanor, remembering to be patient, not getting angry—all of these things and more are a part of your practice, and call for the kind of mindfulness developed by meditation.

I mentioned my altar and described the things on it. You do not have to have an altar. But I use it to visually remind myself of the teachings and sources of my practice. It also helps to create a psychological sense of *sacred space*, which helps to put me into a certain mindset conducive to a focused experience (no thinking about laundry while meditating).

There are many ways to create a sense of the sacred in your practice, and this sense is extremely helpful. *Sacred* means "set apart." It is important for conditioning the mind to have times in our practice where our attention is focused and not full of distraction. We set aside certain principles, values, ethics, philosophies, and other teachings as being especially important to us. We also set aside physical space for these. We set aside time to engage in ritual, practice, experience, and so on. All of this setting aside is the act of creating a sense of the sacred in our practice. If you wonder whether this is relevant to you, think about the last time you read a book or saw a film and were moved deeply by something profound. That may give you a clue as to what you find sacred. These are the formative experiences with the potential to mold our mental habits.

Signs of good spirituality

Due to limitations of space, I have not really discussed the content of many of the sources of wisdom themselves that have been so powerful in improving my life and dealing with fear, anxiety, anger, and more. Instead, my focus here is on the approach of building a practice. But the content of one's practice is essential and you will find that the wisdom you encounter in your reading and exploration will also provide a wealth of ideas on building a practice. See the "Further reading" section at the end of this essay for my personal recommendations.

There are certain hallmarks of what I called "good spirituality" in previous writings.[7] Some of the most important of these are:

- The aim of the practice is an inner flourishing apart from circumstance, not a promise of wealth or outward rewards.
- An approach to knowledge that refrains from making unfounded claims and encourages the practitioner to be humble.
- Philosophies that help us to accept impermanence with peace, rather than promising some kind of escape from impermanence.
- Approaches that focus on mastering and changing ourselves before others.
- Ideas that help us to transcend our small egotistical concerns toward a larger view.
- Philosophies that encourage and are founded upon the importance of compassion.

Imperatives for progress

Despite being an extremely good and valuable employee, my wife was laid off recently from a job she had held for many years. She had been in somewhat of a rut, but perhaps not quite enough to move her out of her comfort zone. The layoff now has her looking forward to new opportunities. She was talking to me just today about how she would have handled the news very differently just a few years ago, and feels that she has been making progress. She seems to have incorporated many wise perspectives into her natural reactions over time, and this has been specifically because of her practices, conversations, readings, and so on.

Our practice should have a sense of progress to it (again, it is a *path*). Because of the highly subjective nature of spiritual practice, there are certain attitudes, dispositions, and approaches that will be a hindrance to an effective and transformative practice. We naturalists are adverse to claims that our thoughts, intentions, and hopes can cause changes in the world directly (by prayer, laws of attraction, spells, and so on). But in the case of our own spiritual practice, what we think really does matter because the object of our practice is the very thing doing the thinking.

For the naturalists (Humanists, atheists, freethinkers, skeptics) in particular, there are some things we will need to "get over" if we are going to seriously move forward in Religious Naturalism. Some of these are things individuals will need to take on, and others also need to happen on a group level. I call these our *Imperatives for Progress* in spiritual practice.

1. **We must discard disdain or bias against anything with the taint of traditional religion.**
   This recoil has prevented us from fruitful exploration of the many wisdom streams within the religious and spiritual traditions of our species. Many justify this attitude by thinking themselves to be concerned with confronting the evils of religion. They may also be concerned that others may think they are religious in the supernatural sense. Progress in development of our individual and collective spiritual practices is severely hampered by this attitude. I don't just mean in technical terms of not being willing to approach certain readings; I also mean that the attitude itself is a distraction from the engrossing kind of experience required for ritual and practice to be individually transformative.
2. **We must drop the quotation marks.**
   I never say that I engage in "spiritual" practices or that my "religion" is naturalistic. We do not practice "something like" spirituality or some pseudo-spirituality. This thinking places the supernaturalists as the default or norm, and relegates us to the fringe. The definition of spiritual I gave earlier ***is the*** definition within our tradition. Christians do not speak of "salvation" (in quotation marks) because other traditions have different approaches or definitions for the word.
3. **Discard our concerns over being misunderstood.**
   It is a common concern that, if we use terminology that most people may think otherwise about or may be ignorant of, then we may miscommunicate or be misunderstood. That may seem like a reasonable concern, but it is not. It results in us using over-clinical, dry, bizarre, or formal language when something far more emotional, metaphorical, and moving is required. Metaphorical language that has a moving effect is essential to the kinds of attitudes, dispositions, and experiences necessary for transformative practice. The most important thing in choosing our terms is that they communicate to us, in our own lexicon, what is important and intuitive. Every tradition has its own lexicon, set of definitions, and terms. Even within Christianity, a Catholic and a Baptist have different definitions for *communion*, for example. In Stoicism, the Sage experiences *Joy* but does not experience *Delight*. It is not reasonable to expect to understand exactly what a tradition means by various terms without taking the time to learn about them, and the same should go for our tradition. Will we be misunderstood? Yes, but Christians, Buddhists, Jews, Hindus, and Muslims have

adherents that misunderstand even their own faith. To build religious naturalism and our own spiritual practices, we must push forward boldly and without apology or reservation.

4. **Do not get hung up on semantics.**
   The flip side of the imperative above is that, when we study the many various traditions of the world, we do not get hung up on *their* semantics. One translation of a Taoist text may refer to *heaven*, and yet the passage holds something very valuable for us. Terms such as *heaven, soul, God*, and so on are not as simplistic as they may seem in the mainstream Christian culture many of us emerged from and exist alongside today. As one studies the ancient philosophies, it becomes apparent that these words often had more subtle pre-Christian meanings, some of which approach meanings similar to Spinoza's philosophy. Even within Christian history there can be found innumerable subtle variations of meaning. They may often be more compatible with our views that seems apparent, and even when they are not there may be important wisdom that can be gleaned from a gracious reinterpretation. We must remember that words are only placeholders for concepts and to seek comprehension of the concepts behind the words first and foremost.
5. **Less talking, more doing.**
   Spiritual practice is not the same as intellectual analysis or discourse *about* spirituality, spiritual people, or spiritual practices or history. Being one who studies human spirituality like an anthropologist, or who likes to talk about it as an amateur enthusiast is perfectly fine. I find it an entertaining and informative activity. But we cannot let ourselves think that we are moving along the path or engaging in spiritual practice merely by doing this. I have sometimes presented talks on meditation to rationalists, who listen to the talk then seem to think, "That was interesting! Now I understand that better. What's the next topic?" but never actually meditate. If that is their choice that is fine, but such a community will *never* be able to provide what traditional religion provides the human heart, mind, and society. This is why the Spiritual Naturalist Society has a prohibition in its published articles of essays that make arguments for *why we should be naturalists* or how *naturalistic spirituality is possible*. Instead, we publish on ways and means of proceeding with our spirituality and ritual. We assume our audience are already naturalists and do not need to be convinced of the importance of rationality or evidence. We proceed with being a community of practicing spiritual and religious naturalists. In our eyes, the time for "making the general case" for whether naturalistic spirituality can exist is over. As individuals, the time for practice, and development of the thought behind it, has commenced.
6. **Self-assess our need to de-sacralize.**
   This is an odd behavior I've noticed in popular culture at large, and in many friends and acquaintances. There seems to be a constant urge in our society to de-sacralize all situations, topics, talk, and events. The urge is likely unconscious and goes without notice. By de-sacralize I mean "to make profane" or un-sacred in tone. If someone mentions something that traditionally might have been given some pause or reverence or sincere reflection, there is an overriding kind of effort that is very often made to "lighten the mood" or supposedly "get real." It often takes the form of flippant or obtuse commentary, or perhaps an irreverent joke. The speakers often see themselves as hard-nosed "realists" who are not taken in by nonsense. Or, it may be that our culture has gotten so far away from sacred approaches that many simply have no idea of what these are about. This outlook has likely developed in our culture as a backlash against the many charlatans and disgraced so-called holy men, televangelists, and other authority figures who have let us down. Everything from the U.S. government's handling of the Vietnamese War up to the pedophile cases of Catholic priests

has seemingly played a role. In any case, this cultural trend can cause a hindrance in our spiritual practice. We have been conditioned to think any serious display of spirituality is "taking ourselves too seriously." There is most certainly good cause to have caution about that. But to get back on track, we need to understand that hucksters have no say or power over our own approach to the sacred. We need to reclaim our ability to engage in sacred activity, our tongues fully *out* of our cheeks.

7. **Reconcile inner peace and outward action.**
One of the most common misunderstandings I see in discussing and writing on these topics is the notion that inner acceptance and peace somehow means passivity in the face of injustice or the need to act to do good in the world. We imagine that personal distress is somehow the only incentive to vigorous action. Yet, as Stoicism teaches, virtue can be the imperative and requires no inner distress. We must get past the idea that spiritual equanimity is about sitting in isolated meditation while the world burns, as some kind of escapism. Rather, it is the spiritually nurtured individual who is calm, at peace, and resilient, and who is more suited to effective action in the world over the long haul. Spiritual Practice is not only compatible with action, it is essential to it.

8. **Abandon religious criticism.**
This precept will be hard to swallow and very easily misunderstood. When I say we need to abandon religious criticism, I do not mean for all people in all contexts. People who are abusing others, stunting the growth of knowledge, misleading others, or harming others need to be criticized and exposed. What I mean is that, individually, we need to have a very clear idea of when we are engaged in spiritual activity in a sacred moment or space. In these moments, it is not helpful to get into judging modes of thought or bringing up our grievances, especially given that this often takes place among those who would agree with us anyway. Even when the time comes for working or speaking against others, we carry our spirituality with us in being mindful of compassion for all beings as our motivation—not hatred or a sense of superiority. As members of religious and spiritual naturalist institutions, we need to decide what our purposes are.

There are already skeptical organizations, political organizations fighting for separation of church and state, educational organizations working against religious oppression and persecution. But at some point there must be a spiritual home, a spiritual community, and a safe haven where spiritual and religious naturalists can fellowship in energizing and inspiring practice together. Institutions that take on this mission will find that allowing "a little criticism of other religions on the side" will be very destructive to their intent. It will paint the atmosphere of their events and the *feel* of their very identity. As I have said, the subjective experience is essential in practice and the right *feel* is monumental in its role in cultivating helpful dispositions to transformative practice. When one goes into a Buddhist temple, one does not expect the dharma discussion to be dominated with criticism of Christianity. Buddhists will be too busy talking about Buddhism. Likewise, in Religious and Spiritual Naturalist gatherings, there is a vast body of wisdom we can be talking about that we *do* agree with and believe.

9. **Make rationality about *humility*.**
Coming from a background that cherishes reason, evidence, science, and logic, we naturalists interested in spiritual practice must take on rationality as a *spiritual practice*. We will not have been the first people to have done so. From the Pythagorean Brotherhood, to the Pyrrhonists, to the Socratic schools, and beyond, there have been many such schools who approached logic and reason with a sacred and spiritual tone. As such, rationality should

take on a much different flavor than we have seen in skeptic, Humanist, and atheist organizations. In this context, the rules of reason take on the form of a character virtue; namely, humility. This is the sympathetic, respected, and adored trait of all people who seek wisdom, compassion, and goodness. Humility already has within it a refraining from making statements we cannot substantiate, drawing unsupported conclusions, and thinking ourselves infallible. This is the foundation of reason, and our motive behind it is essentially one of humility. Also, this means that *humility in our approach to knowledge and claims* is a practice *we* have chosen to take on as a *personal* commitment. It is not to be used as a weapon against others or a means to admonish them. This kind of approach will allow us to progress in our spiritual progress as rational naturalists, thus reuniting the natural and the sacred.

10. **Put compassion first.**

    In all of our endeavors—philosophically, spiritually, institutionally, individually—we must commit to putting compassion first. Our entire practice should be predicated on this most foundational of values. Even rationality itself is secondary to compassion. The reason we value rationality is *because* of the good it can do for people, in their lives, and the world. This suggests a more primal and foundational principle. This is the nexus of knowledge, wisdom, and compassion. If this is our continuous motivation, then we will be off to a good start in our practice.

Notes

1 In Buddhism, a Bodhisattva is someone motivated by great compassion to help others attain enlightenment.

2 This is a quote from Confucius as written by the Taoist philosopher, Chuang-Tzu ("In the World of Men"). The words refer to unifying your attention by purifying the mind of the distractions mentioned above—concerns of the ego.

3 This is a paraphrase of the words of Heraclitus, taken from fragments: "This world-order [the same of all] did none of gods or men make, but it always was and shall be: an everlasting fire, kindling in measures and going out in measures" (Fr. 30); "All things are an equal exchange for fire for all things, as goods are for gold and gold for goods" (Fr. 90); and "Fire in its advance will judge and convict all things" (Fr. 66). It helps us to remember what the flame symbolizes: the impermanence and ever-changing and transforming nature of the universe (Kirk et al. 1957: 186, 199).

4 This is the remainder of Heraclitus' statement on the Divine Fire. My phrase comes from a combination of two fragments: 76 and 43 (Kirk et al. 1957: 206, 213). It refers to how everything in the universe is forever transforming, which the Buddhists call impermanence.

5 This is a far cry from the more contemplative Christianity of Thomas Merton I would discover later in life, and certainly the naturalistic Christianity of Michael Dowd after that.

6 To be sure, there are many other traditions that follow this practice format. Buddhism and Stoicism are primary in my experience and liking, however.

7 For my complete article on this, see "Top 10 Signs of Good Spirituality" at: http://spiritualnaturalist-society.org/top-10-signs-of-good-spirituality/.

References

Batchelor, S. (2010) *Confessions of a Buddhist Atheist*, New York: Random House.

Kirk, G., Raven, J., and Schofield, M. (eds.) (1957) *The Presocratic Philosophers*, London: Cambridge University Press.

Thomson, G. (2016) *Thales to Sextus: An Introduction to Ancient Philosophy*, Long Grove, IL: Waveland Press.

Wilson, E. (1995) *The Genesis of a Humanist Manifesto*, Amherst, MA: Humanist Press.

Further reading

Forstater, M. (2000) *The Spiritual Teachings of Marcus Aurelius*, New York: Harper Collins.
Irvine, W. (2009) *A Guide to the Good Life: The Ancient Art of Stoic Joy*, New York: Oxford University Press.
Strain, D. (ed.) (2014) *Exploring Spiritual Naturalism, Year 1*, Spiritual Naturalist Society.
Watson, B. (1964) *Chuang Tzu: Basic Writings*, New York: Columbia University Press.

PART VII

Critical discussions of religious naturalism

31
THE RELIGIOUS AVAILABILITY OF RELIGIOUS NATURALISM

David E. Conner

What aspects of philosophical naturalism might be deemed valuable for reasons that are specifically religious? Or, what specifically *religious* values might be conferred by a naturalistic philosophy? How can we assess the question of philosophical naturalism's religious availability? For that matter, how is it even plausible to connect religious values with *naturalism*, which, many persons would say, entails by its very definition a repudiation of the supernatural, the transcendent, the spiritual—and, therefore, the *religious*?

Criteria for religious availability

Readers of this volume are aware that there are ways of addressing these questions that are not merely cogent, but scholarly. This essay undertakes to organize various expressions of religious naturalism into a schema of five types and then consider briefly the religious availability of each of those types. I explicate the notion of "religious availability" in terms of seven types of religious value: values that are *intellectual, axiological, ethical, esthetic, emotional, interpersonal*, and *institutional.*[1] These types are imprecise and there are areas of overlap, but the typology is nevertheless useful for purposes of analysis.

It is not possible here to present a detailed assessment of each of the five types of naturalism with regard to each of the seven types of value—that method would require 35 instances of analysis—but these seven types of religious value can at least be kept in mind as we broach the overall subject of religious naturalism. We begin by describing these seven criteria for religious availability.

- Related specifically to values that are *intellectual*, we ask whether the type of naturalism under consideration is reasonable. Are its arguments persuasive? Are its claims tenable? Is it compatible with other fields of study? Does it commend itself to thoughtful persons?
- Regarding *axiological* considerations, how does the type of naturalism understand the existence of value? Are values held to be merely epiphenomenal? If values do exist, are they arbitrary and subjective, or are there objective grounds for value? If certain values are commended, do they tend to be viewed as personal choices, or are they more likely to be shared by groups? Are the values that are sustained limited mainly to human interests, or do they transcend humanocentrism?

- Does the type of naturalism encourage *ethical* involvement? If so, what types of commitments and ethical aims are entailed? Is ethical conduct an ancillary consideration, or are ethics and morality such dominant emphases that the religious outlook tends to be subsumed under ethical obligations?
- Pertaining to *esthetic* value, does the type of naturalism foster an appreciation of beauty? There is a well-developed history of collaboration between religion and the fine arts. Does the religious naturalism being considered easily lend itself to expression through music, poetry, visual arts, the use of inspiring language and symbolism, or other forms of esthetic creativity?
- Concerning *emotional* value, does the type of naturalism convey feelings of comfort, security, companionship, and trust? Or does it demand attitudes of self-sufficiency, determination, courage, or stoical resignation? Does the type of naturalism encourage a balance between feelings of serenity with feelings of stimulation and challenge? Perhaps most crucially, are there feelings of redemption, of acceptance and companionship? Is there a conviction of grace—the feeling of a Power that "does for us what we cannot do for ourselves?"
- Regarding *interpersonal* values, we ask whether the naturalistic perspective may comfortably be held simply as a private point of view—or does it invite interaction with others, perhaps even begging to be shared? May it be limited to inner reflection, or does it call for conversation, group activity, and social wholeness?
- The consideration of interpersonal values leads finally to the question of *institutional* value. Values that are institutional rely on the continuing existence of organizations for their creation, preservation, implementation, and development. Because naturalism is not a prevalent perspective in most religious institutions and because some may feel that there is a palpable antagonism between naturalism and institutional religion, it will be helpful here to go into a bit more detail concerning religious values that are specifically institutional.

Whether willingly or unwillingly, virtually all of us interact with institutions. Nowadays it is commonplace especially in Western cultures to encounter attitudes that are disinterested, dismissive, or openly antagonistic towards religious institutions. One imagines that such attitudes are particularly prevalent among the intellectually oriented persons who are most likely to embrace religious naturalism. Most of the reasons for the unpopularity of "organized religion" are well-known, and frequently those reasons are very defensible. Still, there are in fact certain kinds of positive religious value that are not simply interpersonal or social, but *institutional*. When we think of religious values associated with institutions, we may think first of values derived from church-related hospitals, colleges, orphanages, and homeless shelters and from churches that sponsor refugees, receive offerings to alleviate hunger, provide buildings and resources for disaster relief, and so on. These social-ethical benefits are laudable. However, there are less obvious but still momentous institutional values associated with the origination, development, preservation, and interpretation of religious symbols and practices. One thinks of scriptures and other sacred writings, of hymns, anthems, prayers, poems, and sermons; of sacraments both canonical and circumstantial; of the lingering memory of baptisms, weddings, and funerals; of the dissemination of the writings of a host of prophets, saints, reformers, and teachers; of countless study groups focused on moral progress or personal growth; of innumerable shared meals, supportive words, affectionate gestures, and other kind deeds, large and small, all of which are made possible by religious institutions.

Moreover, there is a unique type of *cumulative* value that accrues from sustained involvement in scheduled activities with persons who, over time, may become one's closest friends—and the more so in the setting of a religious institution, where deliberate efforts are made to relate the institution's activities to matters of ultimate concern and self-transcending commitment. It could be objected that this cumulative value is disputable, for many religious institutions have done more harm than good. Nevertheless, the good that has been realized is not simply demolished by the bad, and as we examine the religious availability of philosophical naturalism, the possibility of specifically *institutional* value must not be excluded.

The fact that certain ideas may tend to support certain values is, of course, no sufficient reason for believing that those ideas are *true*. It would not be a sound argument to advocate one type of religious naturalism over others merely because that type is deemed to be well-suited to religious purposes. Of course, "truth" is not a simple attribute; we must bear in mind that there is considerable leeway in establishing religious truth, for it is inherent to religious perspectives to manifest elements of subjectivity. William James ever reminds us that in some situations the will to believe may create the conditions in which the beliefs in question are warranted (James, 1897). In assessing the religious availability of certain perspectives—naturalistic or otherwise—we accept the *pragmatic* or *perspectival* dimension of the beliefs in question.

But subjective interpretation is acceptable only within limits. Though those limits are notoriously daunting to establish, they nevertheless may not be ignored, for religious feelings and motives display a long and lamentable history of leading beyond the realm of reasonable interpretation into the domain of wishful thinking and error. It is at the point of halting the vehicle of religious delusion that we encounter one of the chief advantages of religious naturalism, for among naturalism's most conspicuous goals is the avoidance of beliefs based on special revelation, miracles, and other examples of the opinion, apparently widely held, that the claims that are most illogical and non-evidential are precisely the claims that must be "accepted on faith." We remember the story of the little boy who, when asked what faith is, answered, "Faith is when you believe what you know ain't so."[2] The tension—which should be a constructive one—between objective and subjective elements in the formation of religious beliefs is relevant here because one of religious naturalism's most important roles may be the provision of parameters and new ideas in the reconstruction of traditional religious convictions.

The reconstruction of religious beliefs, however, is primarily intellectual in character. As soon as one moves beyond religious naturalism's intellectual value to the other types of value that we are considering, challenges confront naturalism's ability to function as a perspective that is explicitly religious. These challenges are not as clear-cut concerning axiological and ethical values, since naturalism, or at least certain types of it, has an established (if not entirely affirmative) history related to axiology and ethical responsibility. The most daunting challenges to naturalism seem to be related to esthetic, emotional, interpersonal, and institutional values.

Naturalism's apparent difficulties in attaining religious availability are not a recent development. The stubborn gulf between intellectual philosophy and popular religious conviction was commonly encountered even in antiquity and can be seen in the disparity between the religious thinking of Greek and Roman philosophers, on the one hand, and the popular polytheism of the cultures that surrounded them. Aristotle's idea of God as an abstract philosophical requirement provides an example, as Whitehead reminds us:

> Aristotle found it necessary to complete his metaphysics by the introduction of a Prime Mover—God.... The Greek gods who surrounded Aristotle were subordinate metaphysical entities, well within nature. Accordingly on the subject of his Prime mover, he would have no motive, except to follow his metaphysical train of thought whithersoever it led him. It did not lead him very far towards the production of a

> God available for religious purposes. It may be doubted whether any properly general metaphysics can ever, without the illicit introduction of other considerations, get much further than Aristotle.
>
> (Whitehead, 1925)

A similar assessment has been put forward regarding the theology of one of the foremost advocates of religious naturalism in our own time, Henry Nelson Wieman. Huston Smith writes that Wieman

> uses empiricism to identify what within the world deserves to be regarded as God because it merits man's ultimate commitment. Where he has failed to carry the religious mind with him, we fear, is in his move from "X is the most worthy event in the natural world" to "X can, on a wide scale, evoke religious fervor." Wieman's position carries force for thinkers (a) who accept rational empiricism as the royal road to knowledge, and (b) who, having been nurtured in a religious tradition, feel their lives should be committed to *something*... . How few are the persons who meet the conditions stated in both (a) and (b) ... is evidenced by the smallness of the inroads of Wieman's theology into Christian churches. The moral is: even if something is the best there is, this is not necessarily enough to evoke religious fervor.
>
> (Smith, 1969)

The central question being addressed in this essay is whether there are circumstances in which naturalism may go any further religiously than it did for Aristotle or Wieman. Can a religion that is focused primarily on qualities observed in nature achieve values that are not only intellectual and ethical, but emotional, psychological, and social? It seems clear that if religious naturalism is to move beyond the sphere of intellectual conviction to the arena of widely experienced religious feeling—"fervor" on a "wide scale," in Smith's terms—it can only be because religious naturalism has moved beyond the intellectual and the ethical to include values that are emotional, social, and even institutional.

Types of religious naturalism

We turn now to a consideration of the various types of religious naturalism itself. What, exactly, is *naturalism*? Perhaps most obviously, naturalism may be defined as *a denial of the existence of the supernatural*. There is, naturalism alleges, neither a supernatural realm nor in fact any type of existence other than what is contained within nature. Naturalism so understood is not a uniquely modern idea. Miracles and other outlandish religious claims were doubted even in ancient times by critical-minded people, but wary attitudes toward religion during antiquity have an ad hoc flavor, being based more nearly on objections to specific instances of religious sensationalism and superstition than on a categorical philosophical distinction between the natural and the non-natural. A more precise approach to the issues became possible due to the labors of Thomas Aquinas (1225–1274 CE), who had received the writings of Aristotle—which had been preserved through Middle Ages by Islamic scholars—and thereupon happily baptized them for Christian purposes. It was clear to Thomas that the comprehensive explanation of natural physical causation found in Aristotle's metaphysics made no allowance for the primary doctrines of Christian belief, such as the divinity of Christ, the Holy Trinity, our need for divine grace and the effectiveness of the sacraments as means of that grace, etc. In Thomism it therefore became necessary to make a formal distinction between natural theology and revealed theology, and a concomitant distinction between the natural world and a superior realm beyond nature—the supernatural (*supernaturalis*). It is the existence of this realm and any influence it might be said to have that are denied by naturalism.

The concept of nature took on new preciseness because of the work of Isaac Newton (1642–1726), whose mathematical treatment of objects both terrestrial and astronomical led to the notion that the total causal matrix of the universe can be explained solely in terms of efficient causes. Nature then began to be understood as a vast network of mechanistic physical interactions, an idea that somewhat ironically differs considerably from the Aristotelian naturalism that had originally inspired Thomas. However, Newton's system was not immediately interpreted as requiring an abandonment of the supernatural; the deists of Newton's era held that God in fact made deliberate use of efficient causation to enact the purposes of Divine Providence, and, though his private views were unorthodox, Newton himself remained within Anglicanism.

By the time of the French mathematician Pierre-Simon Laplace (1749–1827), atheism and a complete renunciation of the supernatural were widely recognized as possible consequences of Newtonian mechanism. British geologist Charles Lyell (1797–1875) added impetus to the idea of nature's self-sufficiency with his doctrine of uniformitarianism, the proposition that the same natural laws and processes that we now observe in nature have always operated in the universe and are uniformly applicable everywhere. Charles Darwin (1809–1882) was influenced by Lyell's work, and it was Darwin's theory of evolution that at last led the public at large to recognize a fundamental conflict between scientific naturalism and orthodox Christianity. But even from the time of Newton, continuing during the era of Lyell and Darwin, and enduring through our own day, there have been religious intellectuals who were "liberal" or "scientific-minded" who have sought to reconcile scientific worldviews with religious ones. It is, one dares to say, because of the legacy of these religious intellectuals that we now can speak of "religious naturalism" and produce books such as the present volume.

I offer this capsule version of the history of naturalism as a reminder that, though we may now speak of "naturalism" as if it were a freestanding, self-defined worldview, its roots are inextricably intertwined with a protracted controversy with religion. Therefore the adjective "religious" ought not to be applied to "naturalism" as if the word "religious" were some sort of an optional add-on or intellectual interloper. Whether or not those who now utilize the term "naturalism" are aware of it, the question of "Religious? Or not?" ineluctably lurks at least tacitly in the root meaning of "naturalism" itself. It is instructive to keep this lingering connection with religion in mind as we examine the following typology of the ways in which *naturalism* may be associated with outlooks that are explicitly *religious*.

Naturalism has been expounded in many ways. For present purposes it is useful to analyze naturalism according to the following five-fold typology: (1) purely materialistic systems that tend to be reductionistic and to regard items such as purpose and free will as, at best, epiphenomenal; (2) philosophies that hold the cosmos was primordially material but that things such as purposiveness, consciousness, and free will have evolved spontaneously from matter and energy; (3) philosophies that view physical reality as having an innate tendency to transmit and utilize information—i.e., influence that is not restricted to purely physical content—in order to form novel and increasingly complex wholes, though without positing the action of a single coordinating agency or creator; (4) philosophies that generally resemble type 3 but in which it is argued that the formation of increasingly complex wholes requires the coordinating activity of some transcendent agency that possesses a singleness or unity that is at least functional, if not ontological; and (5) panentheistic philosophies that claim God is not supernatural but is nevertheless somehow greater than the physical universe and that God possesses not merely functional unity but quasi-personal subjectivity. After briefly describing each of these types, I list thinkers whose works provide exemplification of the type being described. It is, of course, not possible here to go into any detail regarding the ideas of any one philosopher or theologian.

In the first type, the total causal matrix of the universe may itself be viewed as an object of religious devotion. This type tends to deny any kind of transcendence other than our *concept* of the totality; in other words, there is a dismissal of any causally effective holism or "downward" causality, at least on the level of the entire universe. The unity of the universe is simply like the unity of gas molecules in a tank; all the molecules contribute to the temperature and pressure of the tank as a whole, but this wholeness hardly seems to point to anything that could be called "transcendent" in a religious sense. Efficient causes are the true basis for the total causal matrix. As an example of this outlook in antiquity, we may look to the Roman philosopher Lucretius (99 BCE–ca. 55), a materialist and atomist, who, like any worthy anti-supernaturalist, denied the efficacy of prayer and the existence miracles and dismissed any hope in life after death (souls, he said, are made of atoms and simply disintegrate). This materialism notwithstanding, the poetry of Lucretius's *De Rerum Natura* expresses eloquent religious devotion to Nature itself as the creative matrix and life-giver (Latham, 1967).

Modern examples of this first type may be found in the thinking of French philosopher Auguste Comte (1798–1857) and contemporary American biologist Ursula Goodenough. Comte developed a philosophy of positivism in which truth claims are derived from the discovery of positive correlations between scientific observations and the laws of nature. Rejecting traditional Protestantism and Catholicism, Comte pushed for the formation of positivist societies that would support an altruistic "religion of humanity" to fulfill the organizing, social functions that had been associated with traditional religions. Goodenough embraces a philosophy of reductionistic materialism (Goodenough, 1998, pp. 28, 45, 109) but nevertheless writes beautifully about the religious inspiration conveyed by nature. Though she sees herself as a humanist rather than a theist (Goodenough, 1998, pp. 60, 139), Goodenough has joined a Christian church and has participated regularly (Goodenough, 1998, p. xi).

How religiously available is this materialistic, reductionistic type of naturalism? We must preface our answer to this question by acknowledging that there is an extent to which each personal expression of religious conviction may ultimately be assessed only by the person who has embraced it. For example, a friend who is a devout pantheist commented, "People say that pantheists can't pray because 'there's nobody up there to pray to,'—but I still pray anyhow." Persons are entitled, we imagine, to formulate religious beliefs and practices however they please, so long as the results are not demonstrably dysfunctional or harmful. A similarly subjective dimension pertains to many non-religious examples of personal choice and commitment, also.

Nevertheless there are objective criteria for religious availability. We have described seven types in this chapter. With reference to those types, then, we might say that the primary strengths of a purely materialist naturalism are *ethical* and *esthetic*. Both Lucretius and Goodenough remind us of the astounding beauty and complexity of nature, and that this beauty and complexity are associated with the sources of life itself. Intuitively, there is a sacredness to the source of life, and this sacredness would entail ethical obligations related to the preservation of nature. But the ontological status of value (related to axiology) in materialistic schemes is dubious; emotional, interpersonal, and institutional values are not accessible in obvious ways; and even intellectual values are open to dispute, since the adequacy of reductionistic materialism as a worldview has for some time been hotly debated by scholars.

The second type of naturalism maintains that the cosmos was primordially material but that life, intentionality, consciousness, free will, and other psychological, social, and moral realities have emerged as products of natural evolution and now make religion a legitimate aspect of any discussion of plausible worldviews. This second type of naturalism supports true holism—the doctrine that wholes truly are more than the sum of their parts—and that causation is not only "upward" (parts influence the nature of wholes) but "downward" (wholes may influence their parts by functioning not only as aggregates but as unitary agents). Though this type of naturalism attaches crucial importance to notions such as novelty, creativity, and emergence, there is an unwillingness to speak decisively about any identifiable ontological structure or single metaphysical functionality that is seen as the cause of these phenomena.

Examples of this type of naturalism may be found in the works of Samuel Alexander (1859–1938), Henry Nelson Wieman (1884–1975), Gordon Kaufman (1925–2011), and contemporary American philosopher Donald Crosby. I am not claiming that the ideas of these four scholars are identical but only that there are significant similarities as we seek to understand various types of religious naturalism. Alexander and Crosby agree that life, mentality, purposes, etc. have emerged spontaneously from matter, which displays an innate tendency to become organized in increasingly complex combinations. Crosby writes that "nature contains within itself the seeds of all of its adventures of becoming" (Crosby, 2016; Crosby, 2002). Alexander utilizes the term *nisus* to refer to the universe's creative tendency towards increasing complexity, but the term *nisus* is simply an allusion to an observed phenomenon, not a metaphysical explanation (Alexander, 1920). Crosby deliberately eschews the word "God," deeming it too laden with anthropomorphic and supernaturalistic connotations to be useful. Though Alexander speaks of "deity," he uses the term simply to refer to the universe's observed tendency to create emergent wholes or to refer to the highest levels of complexity that have emerged so far during natural history. Wieman and Kaufman, both theologians, do describe God at length, but their descriptions are phenomenological or observational rather than ontological or metaphysical. Wieman understands God as "creative good" or "the creative event," which, Wieman argues, is "the Source of Human Good" and therefore worthy of our reverence (Wieman, 1946). Kaufman identifies God as "Serendipitous Creativity." It is indisputable, Kaufman says, that the universe sustains types of creativity that bring the greatest good we know of into being, but Kaufman declines to enter into any philosophical speculation regarding the sources of that Creativity; it is best to describe it simply as "serendipitous" (Kaufman, 2004). Neither Wieman nor Kaufman emphasizes metaphysical materialism, but both may be classified as "materialists" if, with Crosby and Alexander, we enlarge our notion of "matter" to include radical dynamism and spontaneous creativity.

How available for religious purposes is the second type of naturalism? Clearly, along with the first type, the second type urges us to recognize the awe-inspiring complexity of nature. In this awesomeness there are inspiring esthetic elements and also robust incentives to preserve nature and protect the creatures that have evolved within it. In contrast to the first type, the second type recognizes the positive ontological status of values, feelings, thoughts, and goals as emergent realities. This is in closer harmony with traditional religions—though the second type might be critiqued for associating valuation and higher aspiration, as emergent realities, only with more complex life forms, thereby tending towards humanism or humanocentrism. Nevertheless, feelings of kinship or companionship with the universe itself are possible; Crosby stresses this idea, most prominently in *The Thou of Nature* (Crosby, 2013), but elsewhere as well. It seems doubtful, however, that feelings of kinship with nature as a whole are currently accessible to a large proportion of religious seekers. As for interpersonal and institutional values, the burden of proof would remain with the second type to demonstrate that it can move beyond the province of intellectual endorsement to the realm of significant religious community.

The third type of religious naturalism resembles the second, but with the added conviction that it is not enough to attribute the holism present in nature to mystery or serendipity; rather, it is postulated that holism (creative emergence) is caused by a universal interplay between the *physical* and the *informational*. That is, the most primordial tidbits of reality traffic in data—data that inform the behavior of physical objects. This leads to the proposition that reality is *value-laden*, *purposive*, and/or *experiential*, since the notion of "being informed" clearly implies that there are subjects, however primitive, that somehow assess the relevance of abstract content as a goal for imminent actualization. I use the word *informational* here as roughly corresponding to Whitehead's term eternal objects, Plato's forms, or Carl Jung's archetypes. I prefer the word "information" because it focuses on functionality and is less prone (I hope) to reification. One final characteristic of this third type is that, though pan-subjectivism and pan-evaluationism are affirmed, there is a reluctance to speak of any single coordinating principle or guiding source of graded relevance among the data being transmitted; in other words, God-talk is avoided. Examples of this third type of naturalism include William James (1842–1910), John Dewey (1859–1952), Bernard Loomer (1912–1985), and contemporary American philosopher Donald Sherburne.

William James, later in life, advanced what he called a philosophy of "radical empiricism," by which he intended to refocus the attention of philosophy on "things definable in terms drawn from experience." Though James resisted traditional metaphysical attempts to define reality, he clearly regarded *experience*—the reception, utilization, and transmission of what I am here calling "information"—as intrinsic to nature itself (James, 1996). John Dewey largely followed James on these points in his *Experience and Nature* (Dewey, 1925). James was not a traditional theist, and Dewey was a professed humanist. Loomer, a theologian, spoke and wrote extensively about God, but Loomer was a pantheist who identified God with "the totality of the world, with whatever unity the totality possesses" (Loomer, 1987, p. 20). Loomer's God is thus not an *aspect* of reality that serves specifically to guide or interrelate natural processes. Sherburne's work evidences a long-standing commitment to reinterpret the philosophy of A. N. Whitehead so that the idea of God is jettisoned (Sherburne, 1971).

Regarding religious values, the third type of naturalism, like the first two, is primarily intellectual in orientation in which religious feelings find expression most easily in forms of humanism. Humanism, as an expression of religious feeling, has a long history particularly among persons who are scholarly or skeptical-minded. However, it seems improbable that large numbers of non-intellectuals may enter into these gates with thanksgiving or into these courts with praise. Emotional, social, and institutional values are compromised.

The fourth type of naturalism differs from the third chiefly in its willingness to speak explicitly of an identifiable metaphysical or cosmological structure that is viewed as the coordinator of value in the world. Nature is seen as being composed of occasions of experience in which each occasion is a dipolar unity of physical and mental elements, though the term "mental" must be reconceived so that familiar examples of higher brain functioning (consciousness, thinking, step-by-step deliberation, etc.) are not treated as paradigms. *Mentality* is defined simply as the ability to process and act on information, observed in the way that particles in quantum systems spontaneously arrange themselves into patterns that do not depend on physical collision or the transmission of energy. In this fourth type of naturalism, the individual entities of the world do not, on their own, organize themselves; there is an organizing reality or unitary function that fosters holism by interrelating the welter of aims that are possible. This organizing reality is referred to as "God," though God is not viewed as a single being but as an aspect of the ontological structure of every transpiring event. God is seen in a way similar to the way we understand gravity, which is consistent in its effects though not existing as an individual self. Examples of this type of naturalism are found in the writings of German theologian Friedrich Schleiermacher (1768–1834), philosopher Alfred North Whitehead (1864–1947), and American theologian Harvey H. Potthoff (1911–2002).

Schleiermacher, who was influenced by Kant, German idealism, and German romanticism, held that we experience God directly through what Schleiermacher called "the feeling of absolute dependence" (Niebuhr, 1967). This feeling is not merely an experience of awe or weakness when confronted by the Infinite but, rather, the feeling of the groundedness or rootedness of our own feelings, our own preferences and choices, and our own sense of self within the matrix in which all choices and all selfhood arise. This may sound pantheistic, but Schleiermacher's God is not the universe as a whole, and he expressly disavowed pantheism. Whitehead defined God variously, but his common theme is that God is the metaphysical ground or structure (not an individual being) that coordinates the "initial aims" that bring order and wholeness into nature. Potthoff defined God as "the Wholeness Reality" or "the character of reality in its wholeness" (Potthoff, 1969, pp. 185–192). Though he had studied with Whitehead, Potthoff declined to follow Whitehead into the intricacies of constructive metaphysics. Still, Potthoff's phrase "Wholeness Reality" is not simply a descriptive reference to one of nature's traits but an allusion to a specific tendency or integrative functionality that permeates nature. Potthoff elaborated on this concept by describing the functioning of the Wholeness Reality as "Ground, Grace, and Goal"—a deliberate parallel to the traditional Christian understanding of the functioning of the Holy Trinity (Potthoff, 1969, pp. 198–211).

Though they may seem subtle, the differences between the third and fourth types of naturalism are of major importance when assessing religious availability. Even though God is not thought of as the traditional supernatural personal Spirit, the ability within the fourth type to speak of God with definiteness and conviction as the Source of the Sacred is a critical factor. Though Schleiermacher and Potthoff were professors, it happens that both also served with conspicuous effectiveness as pastors who were celebrated not only for their preaching but for their pastoral care.[3] Reference to these two persons does not constitute a statistical sample, but I nevertheless feel prompted to hazard the opinion that it is of inestimable value to be able to approach a religious calling in the context of a theocentric orientation, or devotion to some other transcendent sacred reality. It is hard to avoid the likelihood that some understanding of "God" as an Ultimate Real Other which/who functions as the Source of the Sacred is of decisive value concerning the matter of religious availability.

This brings us to the fifth type of naturalism, which can be identified by the single term *panentheism*. Panentheism may be defined as the belief that God is greater than the universe but includes and interpenetrates it. Whitehead is commonly but incorrectly named as panentheism's foremost proponent; this distinction belongs more properly to philosopher Charles Hartshorne (1897–2000). This assertion cannot be defended here but is developed more fully elsewhere (Conner, 2009; Conner, 1993; Wheeler, 2018, forthcoming). Panentheism usually also includes the assumption that God is not just a functional aspect of nature but a living Being—that is, in Whitehead's terminology, a "personally ordered society of occasions" (Hartshorne, 1948; Cobb,

1965). Distinguished exponents of panentheism include John B. Cobb, Jr., David Griffin, Marjorie Suchocki, Catherine Keller, and Philip Clayton, but there are numerous others. As Keller indicates, the vision originally spawned by the process theologians now comprises "a vast community of authors, teachers, clergy, and activists collectively rethinking the core values and symbols of the West" (Keller, 2008, p. 23; also 53).

Though panentheism has not achieved the status of being a prevalent orthodoxy in the world's monotheistic religions, its relative success as a form of naturalism is extraordinary. A major reason for this success is that panentheism has all the right stuff: it promotes creative dialogue between religion and the natural and social sciences; vigorously supports liberation theology and social justice; is in step with the pluralist and postmodernist attitudes of the twenty-first century—all this while offering the religious availability of traditional personalistic theism: a God who is "the great companion—a fellow sufferer who understands" (Whitehead, 1978, p. 351).

But, perhaps surprisingly, the challenge for panentheism in our present inquiry is whether panentheism truly is a form of *naturalism*, as its proponents typically claim. The challenge to panentheism's naturalism is that, though panentheism alleges that God is thoroughly immanent, God is also said to be "greater than" nature. Panentheists might reply that this is merely a question of stipulative definition; nature may be redefined so as to include spiritual realities, as in the second, third, and fourth types of naturalism. But the objection remains that panentheism's ultimately-greater-than-nature God is by definition stretched beyond the one-order concept of nature that is espoused by naturalism; panentheism's God finally transcends the interrelated system of nature's total causal matrix. A second problem for panentheism is to explain how the God of panentheism—often described as possessing consciousness, a mind, and a will—can simultaneously and uninterruptedly be in touch with and influential upon the entire universe, a universe unimaginably vast and in which, relativity theory assures us, the traditional notion of cosmic temporal simultaneity is meaningless. These issues rapidly become abstruse and cannot be addressed here. I would like simply to acknowledge my own judgment that Philip Clayton is correct in suggesting that panentheism must finally appeal beyond nature in order to support its claims (Clayton, 2004, pp. 172–185). If this be the case, panentheism ironically falls short at the point of *intellectual value*, since attempts to reach beyond nature are, for philosophical naturalism, simply untenable.

Conclusions

In this essay I have not sought—nor would it have been appropriate—to argue for one type of religious naturalism based on its putative religious availability. My hope has been simply to clarify some issues pertaining to a topic that is admittedly complex and ambiguous and in which one's own history and personal commitments play a legitimate role. Based on the foregoing discussion, however, I now offer a few tentative conclusions. It seems quite likely that the religious availability of any form of philosophical naturalism is greatly aided by the presence of the following features, listed in no particular order.

(1) *The presence of a community of faith.* The religious availability of a set of ideas is almost always enhanced by being expressed in a community of belief. This is achieved more readily if the ideas can be adapted to an existing religious tradition. It is not that nature on its own cannot inspire devotion and reverence. It can, and it does; and, in my view, those who are able to embrace a religion based on nature are to be congratulated. But the values that a religion of nature conveys center upon the intellectual and, in a more derivative way, on the ethical,

and the esthetic. The availability of values that are emotional, psychological, interpersonal, and institutional is more tenuous where there is no faith community.

(2) *The presence of durable sacred conventions* (Dean, 1994). Sacred conventions include rituals, symbols, stories, familiar ideas, songs, and other traditions. Sacred conventions accompany communities of faith but community and sacred convention are not identical. Sacred symbols and conventions cannot simply be invented. They seem to come into existence spontaneously and take on a life that often pays little heed to human manipulation.

(3) *The accessibility of a personal dimension pertaining to the sacred.* Whitehead commented that satisfactory religion moves through three stages, "from God the void to God the enemy, and from God the enemy to God the companion" (Whitehead, 1960, p. 16). God need not be thought of as a literal person if there are sufficiently compelling sacred conventions that convey personal imagery.

(4) *The conviction that the beliefs in question relate directly to what is salvific.* That is, the ideas being advanced are believed to pertain to realities that are essential to salvation, redemption, and/or spiritual wholeness.

(5) *The experience involved is not one of merely comprehending the Sacred, but of being summoned, being transformed, and being sent forth by it.* There is an experience of grace—that the Sacred Reality in some way "does for us what we cannot do for ourselves."

On this final point—if I may speak from the perspective of the religious tradition with which I am familiar—it is instructive that Abraham, Sarah, Moses, the prophets, Mary, Peter, Paul, and apostles encountered the Sacred not through acts of deliberate analysis but in experiences that were completely unanticipated and over which they had little or no control. Their experiences stand in contrast to religious naturalism, which, despite elements of intuition and emotion, is mainly a deliberate invention of human intelligence. But our investigation into the question of religious availability reminds us that religion is not primarily a cognitive activity. In Whitehead's words:

> Religion is the vision of something which stands beyond, behind, and within, the passing flux of immediate things; something which is real, and yet waiting to be realised; something which is a remote possibility, and yet the greatest of present facts; something that gives meaning to all that passes, and yet eludes apprehension; something whose possession is the final good, and yet is beyond all reach; something which is the ultimate ideal, and the hopeless quest... . Apart from it, human life is a flash of occasional enjoyments lighting up a mass of pain and misery, a bagatelle of transient experience.
>
> (Whitehead, 1925, p. 275)

Religion centers not upon intellectual acumen, but upon conviction, devotion, and action. Intellectual coherence may be highly valued, but it is not paramount. Vital religion is not merely plausible, but compelling. It is not a matter of grasping certain ideas, but of being grasped.

Notes

1 Intellectual, axiological, emotional, and ethical types of value are utilized as means of assessing religious availability by Charles Milligan in his "Religious Values of Whitehead's God Concept," *The Iliff Review* IX, 3 (Fall 1952).

2 This statement is commonly attributed to a character in one of Mark Twain's stories, but I have not been able to locate the source.

3 The extensive level of public esteem for Schleiermacher is indicated by his funeral procession, which was attended by an estimated 20,000 to 30,000 persons in the streets of Berlin. Potthoff was highly regarded among United Methodists not only regionally but nationally.

References

Alexander, S. (1920) *Space, Time, and Deity*, 2 vols., London: Macmillan.

Clayton, P. (2004) *Mind and Emergence: From Quantum to Consciousness*, New York: Oxford University Press.

Cobb, J. B. (1965) *A Christian Natural Theology*, Philadelphia, PA: Westminster.

Conner, D. E. (1993) "A Functional-Empirical Approach to the Whitehead-without-God Debate," in C. Peden and L. Axel (eds.), *New Essays in Religious Naturalism*, Macon, GA: Mercer University Press.

—— (2009) "Whitehead the Naturalist," *American Journal of Theology and Philosophy*, 30–2.

Crosby, D. (2002) *A Religion of Nature*, Albany, NY: State University of New York Press.

—— (2013) *The Thou of Nature*, Albany, NY: State University of New York Press.

—— (2016) "Probabilism, Emergentism, and Pluralism: A Naturalistic Metaphysics of Radical Materialism," *American Journal of Theology and Philosophy*, 37–3, 217–27.

Dean, W. D. (1994) *The Religious Critic in American Culture*, Albany, NY: State University of New York Press.

Dewey, J. (1925) *Experience and Nature*, New York: G. P. Putnam's Sons.

Goodenough, U. (1998) *The Sacred Depths of Nature*, New York: Oxford University Press.

Hartshorne, C. (1948), *The Divine Relativity*, New Haven, CT: Yale University Press.

James, W. (1897) *The Will to Believe and Other Essays*, New York: Longmans, Green, and Company.

—— (1996) *Essays in Radical Empiricism*, Lincoln, NE: University of Nebraska Press (first published 1912).

Kaufman, G. (2004) *In the Beginning ... Creativity*, Minneapolis, MN: Augsburg-Fortress Press.

Keller, C. (2008) *On the Mystery: Discerning Divinity in Process*, Minneapolis, MN: Fortress Press.

Latham, R. E. (1967) "Lucretius," in *The Encyclopedia of Philosophy*, Vol. 5, New York: The Macmillan Publishing Co., p. 100.

Loomer, B. (1987) *The Size of God*, W. Dean and L. Axel (eds.), Macon, GA: Mercer University Press.

Niebuhr, R. R. (1967) "Schleiermacher, Friedrich Daniel Ernst," in *The Encyclopedia of Philosophy*, Vol. 7, New York: Macmillan, p. 318.

Potthoff, H. H. (1969) *God and the Celebration of Life*, Chicago: Rand McNally & Company.

Sherburne, D. (1971) "Whitehead without God," in R. James, D. Brown, and G. Reeves (eds.), *Process Philosophy and Christian Thought*, Indianapolis, IN: The Bobbs-Merrill Company, pp. 305–328.

Smith, H. (1969) "Empiricism, Scientific and Religious," in B. Meland (ed.), *The Future of Empirical Theology*. Chicago: University of Chicago Press, pp. 131–132.

Wheeler, D. (2018, forthcoming) "Seizing a Whiteheadian Alternative: A Retrieval of the Empirical Option in Process Thought," in D. Wheeler and D. Conner (eds.), *Conceiving an Alternative: Philosophical Resources for an Ecological Civilization*, Claremont, CA: Process Century Press.

Whitehead, A. N. (1925) *Science and the Modern World*, New York: The Macmillan Company.

—— (1960) *Religion in the Making* (first published 1926), New York: World Publishing Co.

—— (1978) *Process and Reality: An Essay in Cosmology*, D. Griffin and D. Sherburne (eds.), New York: The Free Press.

Wieman, H. N. (1946) *The Source of Human Good*, Carbondale, IL: Southern Illinois University Press.

32
HOLY NOSTALGIA
Toward a sympathetic critique of religious naturalism

Michael L. Raposa

Introduction

This essay presupposes that "religious naturalism" is the meaningful label for a broad range of philosophical and theological perspectives, despite their diversity all defending the claims that (1) there is no reality above or beyond "nature," (2) that the scientific method, broadly conceived, is the most reliable source of knowledge about nature, and (3) that what we can know about nature supports belief in the propriety of certain values and attitudes that might be labeled "religious." My critique pursues no single line of analysis but embodies instead a multi-faceted approach to the topic of religious naturalism. I first explore the logic of vagueness that governs the use of some key terms frequently appearing in the articulation of such perspectives, notably, "nature," "religious," "God," "supernaturalism" and "science." In doing so, I argue that the rejection of a sharp nature/culture distinction may be more crucial for the purposes of religious naturalism than the more commonly discussed rejection of the natural/supernatural dichotomy. While agreeing that an awareness of the problem of evil represents the best reason for embracing atheism, I worry out loud about certain varieties of religious naturalism that appear to be primarily motivated by issues of theodicy. Whether they use the word "God" metaphorically or defend a process theism or claim to represent some type of "atheistic religion," I probe the hypothesis that these are all forms of "halting atheism," perhaps not entirely consistent in their naturalism to the extent that they embody a kind of nostalgia for the idea of the holy. I conclude with a sketch of some of Charles Peirce's arguments: his "neglected argument" that belief in God is instinctive and so natural, along with his preference for God-talk over certain philosophical substitutes, more generally, his curious defense of anthropomorphism, not only in the language of theology, but also in scientific discourse.

The fuzzy logic of religious naturalism

The concept of "nature" with which religious naturalists are typically preoccupied is formulated in a great variety of different ways, ranging from very broad and capacious conceptualizations to those that are significantly more narrow and limited. Even assuming an agreement among these thinkers that nature is to be understood as embracing all that there is—so that there is nothing "beyond" or "outside of" nature—nevertheless, there are competing views about how to understand the whole of reality. Despite the ties that otherwise bind them, for example, individuals who portray nature in "ecstatic" terms (Corrington 1992, 1994, 2000) need to be distinguished from those who regard it from a rigorously materialist or physicalist point of view (Hardwick 1996). Spinoza and Emerson developed now classical views of nature that may contrast in certain respects with those of contemporary philosophers whose thinking has been more directly shaped by developments in the natural sciences (Spinoza 1985; Emerson 2009). Correspondingly, different ways of thinking about nature result in alternative strategies for evaluating its potentially religious significance. Any given thinker will be relatively more or less precise about the use of such terms. But if "religious naturalism" is considered as a general phenomenon, both the "religion" to which religious naturalists adhere and the "nature" to which they attend must be recognized as ideas that are typically quite vague.

To be sure, in the process of "losing their religion" (more on this below), many naturalists would opt also to eschew the concept itself, following John Dewey in preferring to speak in terms of a "religious" quality or aspect of experience, rather than talk about religion in substantive terms (Dewey 1934). The noun is problematic insofar as it refers to specific traditions and institutions that many naturalists will no longer regard as viable, or to the belief in a supernatural, personal deity, which most naturalists will reject as implausible. (Hedging a bit with "most" seems prudent here, because while naturalists would be strongly inclined to deny the existence of the supernatural on principle, some of them—William James and certain process theists, for example—will leave open the possibility of a finite, personal deity.) Interestingly, Dewey defended the lingering utility of "God talk," albeit only if properly understood as descriptive of something neither personal nor supernatural, instead, possessing the power and the reality of an *ideal* (Dewey 1934: 42–54). Other religious naturalists talk about "God" for reasons different from the ones that Dewey articulated. For a pantheist like Spinoza, God and Nature are simply synonymous (Spinoza 1985: 420); but since such an idea of God will have been drained of its anthropomorphism, moreover, since there is nothing like a separate supernatural reality posited beyond nature for Spinoza, the logic governing his usage of the word differs dramatically from that shaping its employment among orthodox theists. Equally unorthodox is the claim—one defended by certain ecstatic naturalists—that the "gods" as we know and worship them do not transcend or produce nature, but rather, emerge from its own creative depths (Corrington 1992: 163–65).

It needs to be observed that the distinction between naturalists who continue to make some sense out of talk about "God" and those who clearly profess to be atheists does not entail the conclusion that all religious naturalists are of the former type. In many cases where discourse about a God or gods has been banished, there continues to be reference to some idea of the "sacred" or of "spirit" (Corrington 2009; Goodenough 1999), and even where supernaturalism has been expunged, some lingering appeal to the concept of "transcendence" (Stone 1992). In instances where faith in a Creator God who possesses a determinate nature and personal agency has been ruled out, the belief in a divine "creative act" is still permissible (Neville 1968). In other cases, it is an account of nature's own creative depths, of "nature naturing" (*natura naturans*) that appears to serve many of the same purposes that the idea of God fulfills in classical theism (Corrington 1994; for an explicitly atheistic account see Crosby 2002).

Pantheism seems like the most straightforward and common strategy for embracing naturalism while maintaining belief in God; on such a world view, God and nature are simply coterminous. But certain religious naturalists, especially those indebted to process philosophy, would argue instead for a version of *panentheism*, the belief that nature is a part but not exhaustive of the divine reality. Here God in some ("primordial") sense transcends but is nevertheless continuous with the natural world, so that in the latter respect God is both determinate and subject to change. Now the insistence on such continuity would seem crucial for maintaining a naturalistic perspective. God may transcend the world but not in the sense of being "wholly other," that is, may be greater than but not radically different from the world. Otherwise, one would appear to be defending a notion of transcendence that necessarily entails supernaturalism. Indeed, it has been argued that certain versions of panentheism are perfectly compatible with orthodox Christian thought; these versions can be understood to embody "a demand that ontology undertake thinking out much more profoundly and much more accurately the relation which exists between absolute and finite being" (Rahner and Vorgrimler 1981: 359–60).

Critics of classical theism might protest that no matter how "profoundly" conceived, any account that distinguishes between God and nature as "absolute" and "finite" will necessarily prove to be incoherent. Yet the mere existence of such an account surely complicates the programmatic rejection by religious naturalists of supernaturalism in all of its forms. To the extent that naturalism is to be defined precisely in terms of its exclusion of supernaturalism, it may make a great deal of difference as to which concept of the supernatural is being repudiated. Yet that concept too is exceedingly vague.

Supernaturalism is usually linked to belief in invisible agents, entities or powers standing outside of nature and not subject to the kind of scientific inquiry that the investigation of nature typically involves. God, gods, ghosts, demons and spirits of all kinds represent the most commonly conceived population of the supernatural realm. Naturalists will have no traffic with such a population; these supernatural beings can explain nothing in nature, can have no effect upon the world in which we live, because they simply do not exist. Widespread belief in their existence can be analyzed in various ways: historically in terms of the limits of a pre-modern science that resorted to anthropomorphism and teleology in order to fill in certain theoretical gaps (in modernity, this strategy continues to be embodied in "intelligent design" arguments); sociologically, as the need to invest socially constructed principles and values with a greater weight and significance; psychologically, as a response to intense feelings of helplessness, beginning with those experienced in infancy; or perhaps in terms of evolutionary accounts about how humans have evolved in and adapted to nature in such a way that they might be predisposed to embrace belief in invisible agents.

All, some or none of these explanations may be defended as meritorious. That the possible truth of belief in the supernatural is not listed among these explanations, however, might be regarded as an attempt to "block the road of inquiry," a cardinal error from the perspective of Peirce's pragmaticism (a view that will be evaluated briefly in the final section of this essay). At this juncture, I want simply to remark on the fuzziness of the logic that governs the use of the word "supernatural" in various types of discourse. I want to note the conceptual gap that exists between a belief in Odin or Zeus on the one hand, and on the other hand (to take one useful but hardly unique example) Duns Scotus's idea of God as univocal in being with all of nature, yet distinguished from all things in nature by possessing being in an infinite mode. I note this gap only to suggest that it makes a difference for our understanding of religious naturalism, insofar as the latter is defined as standing in a dialectical relationship with some concept of the supernatural.

"Invisibility" is often identified as a salient feature of those entities designated as supernatural. Yet scientists regularly posit the reality of theoretical entities that can be known only through their effects. Whether talking about gravity or God, quarks or ch'i, invisibility alone is not a criterion that would allow one to banish something to the realm of the supernatural. There would have to be additional characteristics of the thing in question that prevent it from ever becoming the subject of scientific inquiry. Agency might be proposed as one of those characteristics. I can test for and measure the force of gravity in a scientific manner that it might be impossible to duplicate when dealing with an object of inquiry that is also a *subject*, thus, something with its own independent purposes and volition. Invisible agents, indeed, would seem to comprise a category of entities that scientists would have difficulty studying in any straightforward fashion.

Nevertheless, it would be hasty to conclude that such difficulties are insurmountable. Quarks are a bit harder to investigate than water molecules, and human agents may represent a more problematic subject matter for the scientist than inert substances such as metals or minerals. In any given case, the scientific method will have to be adapted appropriately for the purpose of investigating a specific subject matter. So what about invisible agents? Distinguished American philosophers such as William James and Charles Peirce, at the very least, did not rule out the possibility that questions about their existence could be submitted to experiment and scientific scrutiny.

Even if these pragmatists were incorrect, it is important to admit also the vagueness that attaches to concepts like "science" and the "scientific method." Many religious naturalists will regularly delimit "nature" to whatever is accessible, in principle if not immediately, to disciplined scientific inquiry. This seems like a reasonable way to proceed. Indeed, defining naturalism in terms of its subject matter's accessibility to a scientific methodology may be more useful than conceiving of it as a perspective that rejects all belief in the supernatural, especially given the range of opinions about what terms like "nature" and the "supernatural" might be taken to mean. The precise "nature of nature," its content and limits, cannot be established in advance of scientific inquiry, but only in the long run, gradually determined as the result of such investigation. Religious naturalism could be generally characterized, then, as a perspective shaped by confidence in scientific method, chastened by suspicions about knowledge claims not clearly grounded in such inquiry. This is a characterization that is certainly compatible with religious naturalism, at least in most of its modern versions.

Philosophical pragmatism is an intellectual movement in modernity that has been typically linked to scientific developments in the post-Darwinian era. William James and Charles Peirce, already invoked here, were both trained as scientists, firmly grounded in its principles and methods, and to some extent practitioners. Moreover, pragmatism is a philosophy that many contemporary religious naturalists find congenial for their purposes and to which they appeal for support. The observation that James and Peirce were both thinkers for whom the hypothesis about God's reality was a live one is not intended to refute these naturalists but only to complicate their thinking about naturalism. "God" is not a concept in which James or Peirce would have had any interest if they assumed that the reality it might denote could have no observable effects or that it could not become the appropriate subject matter for scientific inquiry. The question is exactly what kind of scientific method would be required for such an inquiry, what level of nuance and sophistication would have to be displayed by the scientist employing such a method. For the classical American pragmatists (including Dewey and others) the scientific method was too important for its application to be limited to use in the laboratories of physical scientists. For these thinkers, everyday human life is a laboratory, inquiry is ubiquitous, and the only path to knowledge is an empirical one. Pragmatism is best typified as being firmly rooted neither in a theory of truth nor of meaning (as it often is), but rather in a distinctive theory of *inquiry;* and the idea of *experience* employed in that theory represents a radical rethinking of the concept as it was deployed by earlier British empiricists such as John Locke and David Hume.

These remarks are intended to suggest that the idea of the supernatural is too vague for the purpose of defining philosophical naturalism simply as any perspective that repudiates supernaturalism. If nature is all that there is then, to be sure, any consistent religious naturalist will be inclined to reject a two-tier view of reality that posits a rigid distinction between the natural and the supernatural. At the same time, eschewing a sharp nature/culture distinction may be more productive for the project of articulating a contemporary religious naturalism. Consistent with classical pragmatism, such a perspective will regard the creation and development of culture—with its symbols, languages, institutions and practices—as precisely the sort of thing that human beings are *naturally* inclined to do. Moreover, in opposition to social constructivism of a radical sort, such naturalists will claim that human experience is not only shaped by social and cultural factors, but by certain biological characteristics that humans share as members of a species—opposable thumbs, eyes in the front rather than on the side of our heads, brains constructed like ours, certain natural drives and inclinations, etc. The brief remarks about Peirce below will supply the case study for such a perspective.

Religious naturalism as theodicy

Religious naturalism, whether it be articulated as some type of process theism or take the form of an explicit atheism, can be portrayed as a response to the traditional problem of theodicy against which classical theists have continuously banged their heads. In the former case, the problem is solved because the deity is conceived as being limited in power, subject to time, thus not responsible for evil and suffering in the world, but simply doing the best that such a deity can. If "ought implies can" then the finite God of the process philosophers ought *not* to be held accountable for existing evils but, rather, should be perceived as one who suffers with and for humanity, struggling along with and perhaps empowering them. The latter response is even more direct and simple; if there is no God then there can be no theological "problem of evil." Indeed, if there is no God, many things regarded as "evil"—that is, those "natural evils" that cannot be linked to human agency—would have to be understood otherwise and responded to with stoicism. When humans observe terrible suffering or themselves suffer, it might raise a practical problem with which they must wrestle, but not a deeply philosophical one.

These "solutions" resolve the classical dilemma (God or evil) far more efficiently and convincingly than traditional theological strategies designed to soften it, such as the "tapestry theory" (whatever seems bad is a part of God's master plan) or the "free will defense" (God chose to create free beings who would thus be capable of doing evil). Nevertheless, it must be recognized that such solutions are as expensive as they are elegant and simple. One horn of the dilemma (God) either is eliminated altogether or dramatically reduced in size and significance. In either case, the traditional belief in God associated with orthodox Judaism, Christianity or Islam must be abandoned. The expense involved here will be problematic, of course, only for someone who attaches value to such belief.

Although it might seem to constitute a peculiar argument from biography, it may be salient to observe that many of those thinkers who would classify themselves as "religious naturalists" (or are typically classified in that fashion) once embraced a much more orthodox perspective. While other motivations might be identified (for example, developments in modern science that made the clinging to an older metaphysics less tenable), the primary reason for moving from classical theism to religious naturalism in many such cases would seem to be the challenge of theodicy. For contemporary thinkers most especially, belief in an infinite, almighty and benevolent deity seems impossible in the wake of two world wars, massive genocide, frequent acts of terrorism and the rising threat of a nuclear and/or environmental apocalypse. In some instances, the arguments of religious naturalists are framed by explicit autobiographical remarks that clearly identify concerns about the problem of evil as a primary catalyst in the decision to abandon classical theism (e.g., Crosby 2002: 8–9; also, Crosby 2008). Yet, even if we refrain from analysis of the motivations that may have shaped the changing minds of individual thinkers, we can observe more generally within Western cultures an increased discomfort with traditional beliefs about God, also the emergence and rising popularity of process and naturalist perspectives, throughout the twentieth and into the twenty-first century.

To be sure, many individuals continue to cling to those traditional beliefs. Moreover, it is a tricky matter to evaluate which factors are most relevant in the evaluation of those cases where such belief has been abandoned. The recent popularity of books about the "New Atheism" represents an interesting example. One could argue that such popularity is more closely tied to a post-9/11 concern about the role that religion plays in engendering acts of violence than to worries about the problem of evil more generally. Nor do the "New Atheists" seem to have any understanding of or interest in preserving religious values or articulating religious ideas in some more acceptable form; their objective is a repudiation of religion pure and simple (although sometimes Buddhism escapes their wrath, at least in its atheistic versions).

More relevant to this discussion may be an observation about the rising number of individuals who claim to be "spiritual but not religious." What motivates the abandonment of organized religion by such persons even as they cling to some form of spirituality? What value does such a spirituality have and what needs does it serve? Moreover, what sort of empirically testable truth claims might one appeal to in support of belief in something "spiritual?"

Another more narrowly focused but potentially relevant observation concerns the tremendous popularity of Harold Kushner's book, *When Bad Things Happen to Good People*, first published in 1981. This book was written by a Rabbi whose journey from Conservative Judaism to belief in a finite God, benevolent but limited in power, was explicitly shaped by his own personal experience of human suffering. The question about this book and its many inspired readers is slightly different: Why jettison belief in the God of traditional theism but maintain belief in a smaller deity, one that simply does the best that it can?

It is possible to argue that any religious naturalism linked to process theology represents a form of halting atheism (Raposa 2012). Such an argument would be rooted in the claim that atheism pure and simple represents a more coherent and honest solution to the problem of evil than any process alternative. It would develop the insight that belief in a "friendly" but finite deity offers the believer very little in the way of tangible resources for dealing with suffering or the struggle with evil. If we are alone in this struggle (practically speaking), then solidarity among ourselves is best achieved by clearly recognizing that fact, expecting nothing in the way of assistance or consolation from mysterious, non-human allies. On a reading of Dostoevsky that takes actions as speaking louder than words, Ivan Karamazov's handing back his ticket to God as witness to the depth of innocent human suffering does in fact represent a rejection of God (Dostoevsky 1950: 291). Why would one tear that ticket in half and keep a piece of it? What would such a gesture represent and how could it best be defended?

Now these questions may seem irrelevant if the earlier discussion of the vagueness of "religious naturalism" is rejected and the claim is pressed that no perspective qualifies for this label unless it is explicitly atheistic. At worst, any God-talk among naturalists should be interpreted as purely metaphorical and at best it should be eschewed altogether. But then questions linger about what makes such a naturalism "religious" in any sense. In the case of world religions that are explicitly atheistic, there are typically certain conceptions—such as "Nirvana," the "Buddha-nature" or the "Dao"—that distinguish such a perspective from a purely non-religious point of view. What conceptions, explicitly, do this work for religious naturalism and what sort of reality do they denote? If the position that one comes to occupy on the other side of an intense wrestling with the problem of evil can be purged of the nostalgia for the God of classical theism, why not eliminate all nostalgia for the supernatural, even in the vague form that concepts like the "religious" or "sacred" may represent?

Proponents of religious naturalism might insist that it is not the positing of any kind of being or reality in or beyond nature that warrants the description of their world view as religious, but, rather, the appropriateness of certain human attitudes in response to nature—such as awe, wonder, piety, humility, compassion and gratitude. Yet it would still be possible to challenge such a view by asking what the value added is by employing the adjective "religious" in the articulation of it. Why must one be even vaguely religious in order to experience wonder, to feel compassion or gratitude? If one need *not* be, then what is religious about the special way that certain naturalists might claim to have such experiences? Moreover, precisely what is it about nature that makes it "spiritually evocative" and thus elicits such responses (Wildman 2009)? If no compelling answer to these questions can be given, then the continuing usage of words like "religious" or "spiritual" might best be explained as nostalgic and the naturalist who values clarity and consistency might be advised to drop them.

The application of a rigorous scientific method to the study of nature should not undermine the conviction that all human experience of nature is always already interpreted experience. (This conviction can be grounded in a number of ways. I prefer to anchor it to a Peircean conception of experience-as-semiosis, but a Wittgensteinean argument that all perception is a matter of aspect seeing or "seeing as" would work equally well.) What is it then about how religious naturalists experience nature, their *interpretation* of it, that causes them to respond with awe or gratitude, to be filled with compassion? If nothing exists beyond nature, nevertheless, religious naturalists tend to discern something sacred in its depths. Is this "something" real in nature or is it merely something peculiar about their way of attending to nature? Perhaps the latter is a prerequisite for discerning the former. Perhaps this way of attending is in fact tinged with a kind of nostalgia, not necessarily for the abandoned God of traditional theism, but for a lost communion with something real in nature that continues to kindle hope and desire.

Was Peirce a religious naturalist?

This is a question that I have explored at greater length elsewhere (Raposa forthcoming). The brief remarks included in this essay are intended to serve as the case study of a perspective that can be juxtaposed to those views typically identified as forms of religious naturalism. Because it can be partially contrasted with those views, Peirce's philosophy of religion may offer a critical vantage point. Yet because it is also largely continuous and compatible with other types of religious naturalism, especially those that take pragmatism seriously, this critique will necessarily be a "gentle" one.

Among the classical American pragmatists, John Dewey is rather easily labeled as a religious naturalist. In his Terry Lectures, he championed the scientific method, clearly jettisoned any concept of the supernatural, challenged organized "religion" with an emphasis on the "religious" dimension of all human experience, and considered the word "God" as a meaningful label only for certain ideals that have the power to shape human endeavor (Dewey 1934). William James seems also to fit into this category, albeit a bit less comfortably. In his Gifford Lectures, he had argued for a "science of religions," distinct from theology and devoted to the empirical study of religion, with an emphasis on examining descriptions of different religious experiences and evaluating them in terms of their pragmatic "fruits" for human life and activity (James 1983). At the same time, James's late radical empiricism tinged with panpsychism, the voluntarism fueling his defense of our "will to believe," and his accommodation of a certain over-belief in something "More" than what science can ascertain to be true—all test the limits of a naturalistic perspective.

At first glance, Charles Peirce would not appear to fit into this category at all. On the one hand, he identified God as *Ens necessarium* and as the Creator of the universe. He also clearly rejected pantheism and challenged the meaningfulness of belief in any kind of finite deity. At times, he even seemed vaguely inclined to defend a rather peculiar version of the Christian doctrine of the Trinity (Raposa 1989). Yet Peirce, like all naturalists, was a philosophical monist. He declined to affirm a dualistic world view that carved up the universe into mind and matter. He came close to accepting materialism, but given the late nineteenth-century scientific resources available to him, Peirce judged materialism to be inadequate for explaining consciousness or understanding the ideality in nature. Instead, Peirce embraced objective idealism in the form of an extreme scholastic realism (inspired by the metaphysics of Duns Scotus). He was convinced that "general principles are really operative in nature," that there is "a reasonableness energizing in the world"; it is the task of science to discover such principles and to discern such a reasonableness (Raposa 1989: 16, 33).

Peirce's insistence that the scientific method is the only reliable means for "fixing" our beliefs clearly suggests an affinity between his philosophy and naturalism. Everything real is potentially knowable and the best way to acquire knowledge is to engage in scientific inquiry. Now Peirce was not excessively optimistic about what any individual inquirer may come to know through inquiry, and he made some notorious comments about how in matters of "vital importance," individuals should not look to reason but rather should lean on human instinct and common sense. Nevertheless, it would be wise to rebuff any interpretation of these remarks as suggesting that Peirce intended to drive a deep wedge between theory and practice or to limit the role played by science in our gradually coming to know more about the world. In his view, the most powerful and important human instincts are those that shape the process of reasoning itself. Instinctively, birds build nests and bees make honey. By comparison, engaging in inquiry is what human beings are instinctively designed to do; even if training and discipline are required for us to do it well, reasoning is something that humans do *naturally*.

The deductive explication of ideas (determining what else would be entailed should any given idea be affirmed as true) and the inductive testing of their validity were both forms of inference that Peirce rooted in "abduction," that special mode of reasoning by means of which new ideas are formulated as hypotheses. It was abductive reasoning in particular that he regarded as instinctive; while more often wrong than right, the hypotheses that we generate to explain natural phenomena are more frequently correct than should be the case if we were not naturally predisposed to learn the truth. This is how Peirce eventually came to understand the criterion of "simplicity," as recommended to the scientist for use in selecting the best available hypothesis. Rather than logical simplicity, as he had previously affirmed, Peirce modified his view so that the simpler hypothesis was to be identified as the one that is more "natural, the one that instinct suggests" (Peirce CP 6.477).

Peirce's theory of instinct is a thoroughly evolutionary account that ought to be congenial to most naturalists. The development of our capacity for human reasoning has occurred in response to the need to adapt to a complex variety of environmental factors. Successful adaptation was linked for Peirce most especially to basic needs associated with human nutrition and reproduction. He admitted that instincts shaping human reasoning were likely to be less reliable the further removed such reasoning became from matters directly associated with these practical needs. At the same time, Peirce insisted that "man is so completely hemmed in by the bounds of his possible practical experience, his mind is so restricted to being the instrument of his needs that he cannot, in the least, *mean* anything that transcends those limits" (Peirce CP 5.536). All human conceptions are to some extent anthropomorphic.

Peirce's anthropomorphism has been largely ignored by scholars and often poorly understood. In correspondence with William James, rather than linking pragmatism to some doctrine of humanism, Peirce indicated his preference for the word "anthropomorphism," the latter being more "expressive of *the scientific opinion.*" Such an anthropomorphism implies theism in Peirce's view, not belief in any kind of finite God, but rather, in a supreme "Ideal" conceived as "a living power." "Moreover, the human mind and the human heart have a filiation to God." Thus portrayed, anthropomorphism represents a "good sound solid strong pragmatism" (Peirce CP 8.262). Elsewhere, Peirce contended that there is no "more adequate way" of conceiving God as the cause or creator of the universe "than as vaguely like a man" (Peirce CP 5.536). Not only did Peirce affirm an "anthropomorphic" God, he also contended that "if we cannot in some measure understand God's mind, all science ... must be a delusion and a snare" (Peirce CP 8.168).

This "strong pragmatism" is Peirce's own rather idiosyncratic version of naturalism. All human reasoning is rooted in instinct, shaped by human practical concerns. "To say ... that a conception is one natural to man," he opined, is "just about the same thing as that it is anthropomorphic"; furthermore, this is "as high a recommendation as one could give to it in the eyes of an Exact Logician." In truth, "'anthropomorphic' is what pretty much all conceptions are at bottom." Once again, this is primarily a claim about the role that instinct plays in abduction, so that one inclined to reject such anthropomorphism would be wise "to remember that every single truth of science is due to the affinity of the human soul to the soul of the universe, imperfect as that affinity no doubt is" (Peirce CP 5.47).

Peirce's "Neglected Argument for the Reality of God," embodies an experiment based on the presupposition that such an "imperfect" affinity exists—a presupposition that Peirce regarded as indubitable given the otherwise inexplicable success of science in so rapidly discovering truths about nature, also given an evolutionary account about how our capacity for reasoning must have developed in continuous adaptation to the natural world in which human beings live and move and have their being (Peirce CP 6.452–91). The reader performs that experiment whenever she engages in the practice of "musement," a kind of playful but disciplined meditation on the origin and nature of the universe. This is a practice in which billions of individuals, from the earliest intelligent hominids and then throughout human history, must have engaged and from which the idea of God will have arisen. The first part of Peirce's essay, a "Humble Argument," invites his readers to enter the "skiff of Musement" in order to test the claim that such an idea will *naturally* tend to suggest itself under these conditions, and then grow in attractiveness and power as the muser continues to contemplate it. Later sections of the essay indicate that it is intended less as an argument for the reality of God (despite its title) than as an argument for the naturalness or instinctiveness of the God-hypothesis. This later material also links musement to Peirce's mature theory of inquiry, so that it will be recognized as the first or abductive stage of any scientific investigation. Peirce offers several proposals, based on the principles of his pragmaticism, about how this hypothesis might be put to the test as an ideal shaping human life and conduct.

Peirce's anthropomorphism should not be read as a claim that God is to be conceived simply as a human being "writ large," nor should it be conflated with anthropocentrism. In Peirce's view, the universe is "perfused with signs" (Peirce CP 5.449, note 1). The whole of nature is a text or "great poem" that can be read as signifying some divine purpose, albeit one shrouded in extraordinary vagueness. While "the nominalists are fond of insisting on the distinction between words and things, between signs and realities," for a realist and an objective idealist (like Peirce) "the very entelechy of reality is of the nature of a sign." To be sure "we ought not to think that what are signs to us are the only signs; but we have to judge signs in general by these" (Peirce 1979: 297). This is not anthropocentrism (ours are not the "only signs"), but an articulation of anthropomorphism in semiotic terms (we can only judge "signs in general" by "what are signs to us"). This "judging of signs" is the task of interpretation. In a world perfused with signs that task is ongoing, but such a world is potentially knowable, at least in the long run, to a community of interpreters operating with sound scientific principles.

How far removed is such a semiotic realism from the perspective of those thinkers most readily identified as religious naturalists? Recall that in his letter to James, Peirce described God as an "ideal" both "supreme" and "living." John Dewey also conceived of God as ideal, more precisely, as an "active relation" between the ideal and the actual. The ideal for Dewey is something that can be known through the imagination but is not itself purely imaginary. It is fashioned from "the hard stuff of the world of physical and social experience" (Dewey 1934: 49). One might propose that such an ideal could even be regarded as "living" in some sense, but in no way did Dewey imagine God as being "vaguely like a man." It is also not clear how much reality Dewey's God possesses independently of the human beings who conceive of ideals and adapt their behavior accordingly. Nevertheless, one could argue for a certain continuity between Peirce's perspective and Dewey's later religious naturalism.

A more intriguing comparison might be one made between Peirce and several contemporary thinkers. Peirce's portrayal of musement resonates with the way that certain naturalists describe the kind of human relationship to nature that best qualifies as "religious": for example, an attentive "transaction" with the book of nature that takes the form of "reading again and again" (Hogue 2010: 64); or the practice of mindfulness as a bio-*semiotic* exercise, that is, as a development of our capacity to discern "something more" in nature, to develop a habit of "really really" seeing it (Goodenough 2003: 107). Now when one "really really" sees nature what is it that one discerns in its depths? Does it display any purpose, possess any value, independently of what one might project upon it? Does it actively elicit awe, gratitude, maybe even a certain kind of love from the one who contemplates it? Does it have the potential to shape dramatically the behavior of one who is continuously mindful, one whose contemplation of it is habitual rather than episodic?

Peirce characterized "it" as being only very vaguely like a person, yet more like a person than like an impersonal thing—something "living." The hypothesis that it *is* personal, in any event, is one that would naturally tend to suggest itself (he conjectured) to any individual who engages in musement for a considerable period of time. To reject such a hypothesis in advance would represent a violation of the principle that one should not do anything to block the road of inquiry. On the other hand, testing the hypothesis would involve continuing to engage in the mindful contemplation of nature, open to whatever reveals itself in the process. It would also involve transforming one's life into a laboratory, testing the values and purposes discerned in contemplation by trying to embody them in conduct, allowing the love and gratitude elicited in musement gradually to shape one's behavior.

Perhaps, as Emerson famously suggested, humans once had "an original relation to the universe" that has dramatically eroded in modern times (Emerson 2009: 1). Such an erosion might be explained in any number of ways, as theories of modernity weave together different narratives: stories about the rise of a "secular age," about the emergence of capitalism and the impact of commodification on human life, or about the formation of enlightenment ideologies and the implication of new technological practices for human thought and consciousness. On any such account, our relationship to nature has significantly changed. If that is the case, then the tendency of some naturalists to cling to the "religious" might represent a nostalgia for what has been lost, an instinctive longing rooted in a natural "filiation" to the universe. Peirce once opined that the facts that "stare us in face" are often *not* the "ones most easily discerned" (Peirce CP 6.162); and so this would not be a foolish nostalgia for something past and irretrievable, but rather an edifying desire to discover—now occluded—the deepest meanings that nature always signifies and continuously renews.

References

Corrington, R. S. (1992) *Nature & Spirit: An Essay in Ecstatic Naturalism*, New York: Fordham University Press.

—— (1994) *Ecstatic Naturalism: Signs of the World*, Bloomington, IN: Indiana University Press.

—— (2000) *A Semiotic Theory of Theology and Philosophy*, Cambridge: Cambridge University Press.

Crosby, D. A. (2002) *A Religion of Nature*, Albany, NY: SUNY Press.

—— (2008) *Living with Ambiguity: Religious Naturalism and the Menace of Evil*, Albany, NY: SUNY Press.

Dewey, J. (1934) *A Common Faith*, New Haven, CT: Yale University Press.

Dostoevsky, F. (1950) *The Brothers Karamazov*, trans. C. Garnett, New York: Random House.

Emerson, R. W. (2009) *Nature and Other Essays*, Mineola, NY: Dover.

Goodenough, U. (1999) *The Sacred Depths of Nature*, Oxford: Oxford University Press.

—— (2003) "Religious Naturalism and Naturalizing Morality," *Zygon* 38: 101–9.

Hardwick, C. D. (1996) *Events of Grace: Naturalism, Existentialism, and Theology*, Cambridge: Cambridge University Press.

Hogue, M. (2010) *The Promise of Religious Naturalism*, Lanham, MD: Rowman & Littlefield.

James, W. (1983) *The Varieties of Religious Experience*, London: Penguin Classics.

Kushner, H. S. (2001) *When Bad Things Happen to Good People*, New York: Schocken Books.

Neville, R. C. (1968) *God the Creator: On the Transcendence and Presence of God*, Chicago, IL: University of Chicago Press.

Peirce, C. S. (1931–58) *Collected Papers*, edited by C. Hartshorne, P. Weiss, and A. Burks, Cambridge, MA: Harvard University Press (cited in text as "CP" with volume and paragraph number).

—— (1976) *The New Elements of Mathematics*, edited by C. Eisele, The Hague: Mouton.

Rahner, K. and Vorgrimler, H. (1981) *Dictionary of Theology*, translated by R. Strachan, New York: Crossroad.

Raposa, M.L. (1989) *Peirce's Philosophy of Religion*, Bloomington, IN: Indiana University Press.

—— (2012) Review of Michael Hogue's *The Promise of Religious Naturalism* in the *International Journal for the Philosophy of Religion*, 72: 59–62.

—— (forthcoming) "Instinct and Inquiry: A Reconsideration of Peirce's Mature Religious Naturalism," in *Pragmatism, Naturalism and Religion*, edited by M. Bagger, New York: Columbia University Press.

Spinoza, B. (1985) *The Collected Works of Spinoza*, volume 1, translated by E. Curley, Princeton, NJ: Princeton University Press.

Stone, J. A. (1992) *The Minimalist View of Transcendence: A Naturalist Philosophy of Religion*, Albany, NY: SUNY Press.

—— (2008) *Religious Naturalism Today: The Rebirth of a Forgotten Alternative*, Albany, NY: SUNY Press.

Wildman, W. (2009) *Science and Religious Anthropology: A Spiritually Evocative Naturalist Interpretation of Human Life*, Burlington, VT: Ashgate.

33

CONCERNING CONSECRATED SCIENCE

The suspect wonder of the new cosmology

Lisa H. Sideris

The astronomer Carl Sagan once predicted that a religion inspired by scientific knowledge of the universe would eventually emerge to rival the traditional faiths. Such a religion, he proposed, "might be able to draw forth reserves of reverence and awe hardly tapped by the conventional faiths" (1994: 50). Sagan's prophecy contains an implicit assumption that science is better positioned than traditional faiths to elicit a powerful response of awe and wonder. Though more elegantly expressed by Sagan than by some of his colleagues, the assumption itself is not unique. It is something of a set piece in the science-religion debate. Richard Dawkins, in characteristically strident tones, often promotes the power and authenticity of scientific wonder over and above religion's vastly inferior offerings. "Science provides the most stupendous sense of wonder at the universe and at life, something that eclipses the meager, puny, paltry, little sense of wonder that any religion has ever managed to muster" (Dawkins, in Burstein and Keijzer 2009: 240). Edward O. Wilson, another outspoken and visionary biologist, exhorts us to look to the sciences for "richer material to work with" because "the real world, the universe—from black holes to the origin of consciousness—offers far more complex and grander themes than does traditional theology" (Wilson, in Barlow 1997: 27).

We might ask how these scientists *know* that science-inspired wonder is more grand and potent than its religious counterpart. Is this claim amenable to empirical investigation? Might there be an instrument or tool—perhaps a handy "grandeur meter"—that can determine objectively and decisively that modern science offers more wonder than its presumed competitors (Gray 1998)? Assuming that scientific forms of wonder are indeed supremely potent, why should we assume that scientific wonder is *morally* superior to similar feelings inspired by religion? Is it better for humanity, for our moral character, or for the well-being of the planet? Curtis White, in *The Science Delusion*, takes scientists like Dawkins to task for their surprising "lack of curiosity about what this feeling of awe means." Scientists frequently claim the feeling and advertise its popular appeal "without thinking that it needs to be 'substantiated statistically,' as everything else they consider is required to be" (White 2013: 22).

Perhaps we, the laypeople, ought to take these claims for scientific wonder on faith, as the verdict of experts far better acquainted with the universe and its wondrous details than we are. On the basis of such faith, we might proceed even further to construct an entire mythic system—a new science-based religion—around the awesome discoveries of modern science, as Sagan suggests. Recast as a mythopoeic enterprise, a universal narrative rich in metaphor and poetry, science might steer us toward an inexhaustible source of wonder, encourage stronger feelings of connection to the universe, and provide a common platform of facts and values from which to bridge cultural divides and galvanize collective action on "wicked" global problems.

Proponents of what I call the "new cosmology," more commonly called the Epic of Evolution, The Universe Story, or Big History, are working to realize Sagan's prophecy. In its various iterations, the new cosmology proffers a grand narrative of cosmogenesis—the unfolding of the universe from the Big Bang to the present—as a sacred story, a common creation myth for the modern world and for all people. Science, in this movement, offers a wondrous new revelation, an updated sacred scripture, a new Genesis. Prominent advocates of the new cosmology include the cultural historian and "geologian" Thomas Berry and his protégé, the mathematical cosmologist Brian Swimme (Swimme and Berry 1992); religion scholars Mary Evelyn Tucker (Tucker and Swimme, 2011), John Grim, and Loyal Rue (2000); astrophysicist and science educator Eric Chaisson (2005); biologist Ursula Goodenough (1998); science writer Connie Barlow (1997); and Christian pastor and popular author, Michael Dowd (2009). As I argue in detail elsewhere (Sideris 2017), these individuals and their projects share an overlapping agenda. All are engaged in a project of mythopoesis or religiopoiesis: the crafting of a new religion from the materials of science. Modern science is here refashioned as an overarching story that outcompetes or perhaps simply subsumes (it is never quite clear) the particular stories of particular peoples, by virtue of its availability to all and its close conformity to fact.

Tucker, Grim, and Swimme are strongly influenced by Thomas Berry's call for a "New Story" to supplement or replace the traditional faiths and long-standing creation myths. Other versions of the new cosmology display therapeutic dimensions, as with self-styled evolutionary evangelists Dowd and Barlow who encourage audiences to "evolutionize" their lives for personal and planetary well-being. Still others, notably Rue and Goodenough, represent a trend toward an atheistic brand of religious naturalism akin to what Bron Taylor (2010) identifies as "dark green religion," that is, nature-oriented and science-oriented spirituality that rejects the supernatural worldviews and values of traditional faiths, notably the Abrahamic traditions.[1]

The movement also has traction beyond of the academy. *Journey of the Universe*, a documentary film written by Brian Swimme and Mary Evelyn Tucker, is widely featured on public television stations, and Tucker, Grim, and Swimme hold screenings and discussions of the film worldwide. Dowd and Barlow, travelling road-show enthusiasts for the Epic of Evolution, introduce the new scientific story to children in religious education classes or summer camps, often in song and storybook form, or in ritual enactments of "evolutionary parables" (Barlow, *Evolutionary Parables*). Rue and Goodenough hope to extend the scientific narrative into the public sphere, urging the adoption in primary and secondary education of a "consilient" curriculum, inspired by the Epic of Evolution (Rue and Goodenough 2009).

The new cosmology can be understood as a fairly broad, interdisciplinary movement, depending on how the parameters are defined. My focus here will be on a subset of new cosmologists—Rue, Goodenough, Dowd, and Barlow—who self-identify as religious naturalists, as evidenced by leadership roles in the Religious Naturalist Association, for example, or in publications and lectures in which they claim the title for themselves. These religious naturalists also derive inspiration for their projects from some of the scientists mentioned above, notably Wilson and Dawkins (among others). Like Wilson and Dawkins, they celebrate the superior forms of wonder made possible by science. For Rue and Goodenough, this celebration of scientific wonder entails advancing Wilson's agenda for consilience—the unity of knowledge—as a key development that makes it possible to tell a coherent "evolutionary epic" of matter, life, consciousness, and the universe. For Dowd and Barlow, the Epic of Evolution is the stuff of *real* magic, in Dawkins's preferred phrase. The Epic stands as the one true myth that reliably informs us of *how things are* in the world and *which things matter.*

Taken as a whole, the new cosmology calls on us to respond with awe and wonder to what is deemed most authentically real. Religious Naturalists who embrace the Epic of Evolution believe that a comprehensive scientific narrative outperforms the mythic and moral functions of traditional religion. Specifically, and importantly, they maintain that a science-based cosmology will inspire an *eco*morality. Awe at the scientific account, they believe, will readily translate into feelings of care and concern for the natural world.

The new cosmologists are invested in a high-stakes competition with the existing religious traditions to determine where we direct our sense of wonder and how we define our ultimate values. As a movement seeking converts to science, however, the new cosmology elicits wonder first and foremost at science and the scientific enterprise, and only secondarily, *at nature.* As it is typically portrayed, the pitched battle between scientific and religious forms of wonder can have only one winner—science, of course, being the projected winner—but it can have more than one loser. Nature itself, I argue, is a neglected category in this competition. Put differently: in their desire to redirect wonder away from religion, and toward science, the new cosmologists tend to divert wonder away from the natural world. The redirection of wonder toward science casts nature as science's poorer cousin—a thing not quite real until explained—while also stripping wonder of its salutary and commendable dimensions.

The quest for a new myth

The idea that a science-based myth offers a headier sense of wonder than the traditional faiths is traceable in part to Wilson, who is generally credited with having coined the phrase "Epic of Evolution" (Wilson 1978) and whose vision of a fully unified body of knowledge ("consilience") inspires some of the new cosmologists (Wilson 1998). His own youthful conversion from Southern Baptist faith to a profound enchantment with evolutionary theory marked him indelibly with a deep sense of science as fulfilling our innate hunger for religion and meaning (Wilson 1998). Like many converts, Wilson feels moved to share his "unification metaphysics" with the wider world (1998: 6). Throughout his career, he has sought a way to "divert the power of religion into the services of the great new enterprise that lays bare the source of that power"—i.e., the scientific enterprise (Wilson 1978: 193). Science, in explaining the sources of religion as an evolutionarily engrained propensity, is positioned to appropriate religion's power for itself in the form of a new mythology.

Rather than write off religion, scientific humanism must confront religion's power if it hopes to turn it to nobler ends, Wilson advises. But where will science find a compelling narrative to rival religious myths? Wilson insists that a consilient unity of the disciplines will eventually knit all knowledge together into a seamless whole. The Epic of Evolution is a mythic, narrative rendering of that whole. "The core of scientific materialism is the evolutionary epic," Wilson explains in *On Human Nature*, and the evolutionary epic "is probably the best myth we will ever have" (1978: 201). *Consilience* makes a similar case: "The true evolutionary epic, retold as poetry, is as intrinsically ennobling as any religious epic" (1998: 289).

Wilson's prophecy regarding the imminent unification of all knowledge entails convergence not just among academic disciplines but between science and religion as well. More accurately, it entails the displacement of religion (but not the religious impulse) with a scientific mythology. Humans' evolutionary predisposition to engage in religion is a valuable "source of energies that can be shifted in new directions when scientific materialism itself is accepted as the more powerful mythology" (Wilson 1978: 207). This appraisal of science as the stuff of a superior myth, issued by Wilson decades ago, resonates strongly with today's champions of the evolutionary epic.

The challenge of amythia

A new myth is urgently needed, the argument usually goes, because "our" culture (commonly, Western culture) is suffering from what Loyal Rue (2004) calls *amythia*—the condition of being without a serviceable, universal myth.[2] The stories we have inherited from the traditional faiths are no longer morally relevant or cognitively plausible. Rue's concern about amythia echoes earlier calls made, for example, by Thomas Berry who anticipated the dawn of a "New Story," a functional religion that takes our modern scientific understanding of the universe as its primary reference point. For both Rue and Berry, the environmental crisis is, at root, a crisis of meaninglessness, of *storylessness*. A new, more accurate story must be found or invented if we are to steer a path back to meaning and purpose, as well as proper concern for nature.

Rue's project of creating a new global myth, or what he calls "everybody's story," includes blueprint for educational reform drawn from the consilient unity of knowledge. "Consilience among scientific disciplines," he argues, "now makes it possible to construct a coherent narrative" (Rue 2005: 615). Like Wilson, he believes that the evolutionary epic gains universality and power to unite from the unity of knowledge itself. Consilience suggests that in *reality* there is not a multiplicity of stories in the universe, as the diverse spectrum of religions would suggest, but just one. The epic of evolution is "the biggest of all pictures ... It is the ultimate account of how things are, and is therefore the essential foundation for discourse about which things matter" (Rue 2000: xii). The disciplines show signs of converging around this ultimate account, moving toward decreased specialization and tighter integration (the current, and highly artificial, state of academic sprawl notwithstanding). This integration suggests that education at all levels should reflect a new core curriculum "focused on the evolution of matter, life and consciousness-culture" (Rue 2000: 130–31). Even children's science education will take the form of narrative instruction because "the brain is a narrative spinning modular system" (Rue 2000: 131).

One story for all

In this effort to construct a consilient curriculum grounded in the evolutionary epic, Rue is joined by Goodenough, whose own work adopts Rue's basic division of reality into How Things Are and Which Things Matter (Rue 2000; Goodenough 1998, 2009). Indeed, she credits Rue with "explain[ing] to me most of what I understand about theology and philosophy" (1998: xi). A cell biologist and author of a widely used genetics textbook, Goodenough attracted broader attention with *The Sacred Depths of Nature* (1998), a scientist's personal and spiritual reflection on nature and natural processes. Rue and Goodenough recognize the affinity of their project of religiopoiesis with Wilson's efforts to mythologize scientific materialism in Epic form. Like Rue, Goodenough regards the Epic as the one narrative capable of fostering the values needed to cope with our current global challenges. Their joint commitment to teaching the Epic as the centerpiece of education draws inspiration from "Wilson's bold vision" of a consilience (2009: 175).

A co-authored essay spells out the authors' rationale for a consilient curriculum. Rue and Goodenough argue, for example, that recent breakthroughs in science make it possible to extend the evolutionary paradigm broadly, in novel directions; doing so will produce a fresh, coherent vision of nature and ourselves that has the potential to transform our lives and culture. "Inherent in this story is a rich and satisfying account of who we are, where we have come from, and how we might become fulfilled," they confidently assert (2009: 181). Culture wars over core curriculum—pitting the Western canon against multiculturalism, and conservatives against liberals—are not only tedious but irrelevant, they contend, because these debates about pedagogy emerged before the watershed publication of *Consilience*. In that less enlightened time, before science emerged as the authority on "human nature," it was popularly assumed that the *humanities* spoke authoritatively about humanity's greatest stories. "Understanding humanity was then the exclusive province of the humanities," Rue and Goodenough note,

> [b]ut this is no longer true, as Wilson's *Consilience* makes abundantly clear. Indeed, it is so far from being true that one might insist (as we do) that any story of human nature not firmly grounded in the sciences does not merit the attention of youthful minds. (2009: 178)

Consilience presents an opportunity for us all "to rethink the issues at stake in the on-going debate over the American college curriculum" (2009: 175).

As we continue to debate which stories young people in our culture should be learning, the evolutionary epic emerges as the one narrative encompassing everything that matters. "Debates over the core curriculum should be focused on how best to tell this story to the next generation"—not *whether* the story should be told as a grand, unifying narrative. Rue and Goodenough go on to argue that a core curriculum is "all about coming to terms with human nature," or what they term "human reality" (2009: 178). "One world calls for one story," they conclude. "The Epic of Evolution is it" (ibid).

Rue and Goodenough invoke the text of *Consilience* much as a fundamentalist might reference the Bible, as self-evident proof that we are now in possession of the greatest story ever told, one that finally clears away doubts about human nature and destiny. Consilience, the unity of knowledge, is treated as a *fait accompli*, rather than a hypothesis—still hotly disputed (Dupré 1998) regarding the possible links among the branches of knowledge. (Significantly, Wilson himself uses the language of faith, conviction, and metaphysics in describing his commitment to consilience.) Also problematic is Epic enthusiasts' naïve positing of a direct relationship between How Things Are and Which Things Matter. They proceed on the assumption that acquainting students—or anyone—with "the" story of the universe will invariably arouse feelings of intimacy and care for the natural world. But science itself—even when gilded with the accoutrements of myth and poetry—is not a sufficient *source* of environmental values or of moral motivation generally. It does not give us adequate grounds for caring.

Building the perfect cult

The conviction that the scientific narrative inspires the "awe," "majesty" and "sense of wonder" (Rue 1997) and that these feelings will in turn motivate environmental concern is taken further in some of Goodenough's work apart from Rue. Epic enthusiasts display a penchant for applying evolutionary frameworks to traditional religions in order to evaluate their adaptiveness and "fitness." Both Rue and Goodenough diagnose these traditions as dysfunctional in the modern context and, consequently, forecast bleak prospects for these faiths.

When Goodenough turns her attention to the existing religious faiths and their stories, she makes short work of them. "I set about analyzing religious systems," she explains, "using the paradigm most familiar to me, the paradigm of biological evolution" (1994: 321). She imposes a simplistic three-fold typology on all the major religions and finds them wanting in terms of their *cognitive* and *universal* appeal and their *environmental* or *ecomoral* potential. Her typology (at which she arrives without consulting any scholarship) consists of ancestor cults, sky cults, and earth cults. Ancestor cults include virtually "all religious systems" insofar as they create continuity with the past by means of art, ritual, and beliefs that the dead "are engaged in bestowing benefit or harm" (1994: 322). Sky cults, which may also function as ancestor cults, pose questions of ultimate origins and destiny, and often display belief in a supernatural creator who is actively involved in day-to-day lives of believers. Together, the first two categories encompass the main established tradition. A scientific worldview has nothing to contribute to these two types of religions—*all* of the major world traditions, according to her classification scheme—because they are essentially closed systems, providing answers to questions about why we are here and where we are going on their own unscientific terms. "After hundreds of years of effort, in thousands of books written by thousands of theologians and physicists, the science/sky-cult dialogue remains a standoff" (1994: 326). Moreover, these traditions tell stories that are too particular to be expanded into a global myth. Because they evolved to fill a particular niche, they cannot tell *everybody's* story.

Goodenough's remaining category of religions, earth cults, initially looks more promising, owing to nature-centered practices and rituals (rain dances, seasonal celebrations) and forms of devotion (earth goddess traditions, contemporary appropriations of Native American rituals). However, they offer little in the way of otherworldly rewards such as an afterlife, and therefore cannot compete successfully with ancestor and sky cults in the evolutionary lottery. Invoking Rue's categories of How Things are and Which Things Matter, she rejects earth cults on the grounds that they lack coherence with modern science, as indeed she rejects all religious narratives crafted prior to "contemporary understandings of How Things Are derived via scientific inquiry" (2009: 373). All such religions produce "all-too-familiar conflicts about which accounts are 'true'" (2009: 373). There are simply too many deficiencies in the existing traditions, even the earth-centered, nature-reverencing varieties. "Therefore," Goodenough concludes, "if we want an earth cult grounded in scientific cosmology, we're going to have to invent one" (1994: 325). This "invented" religion is, of course, the Epic of Evolution.

As Goodenough concedes, many nature-centered religions—what she dubs earth cults—emerged in cultures that knew (or know) nothing of "genes or molecules or plate tectonics" (1998: 328). This lack of knowledge has not prevented them from respecting or even worshiping nature. Nevertheless, Goodenough insists that science ought to function in such religions as something like a main text or canon, like the Bible or Koran. "The earth sciences could be such a text ... a basis preferable to the authority of custom" (1998: 329). But why must an inherently earth-friendly tradition be *reinvented* with science at its core? Does Goodenough believe, for example, that indigenous cultures will develop a more robust environmental ethic following a lesson in genetics or ecosystem ecology? Her commentary suggests that she does. It also points a very narrow vision of what counts as real and true in human relationships to the natural world.

Goodenough contends that "awe" made possible by "understanding how life works" will turn to *affection* for nature (1998: 327). "The more we know about life, the more we can care about it" (ibid). As an example, she offers sociobiological claims about kin selection that present a "calculus" of genetic relatedness. This calculus, she ventures, explains why we care more for organisms with whom we share genes. And since we share genes with all life, sociobiology actually *enables* a broad concern for the whole earth: "Our cognitive understanding of evolution now allows us to take this concept much further: to the extent that the genes are shared throughout all of life, this gives us a lot more to care about" (1998: 328). This is a dubious claim. What is to prevent this "calculus" from justifying a position of caring very little or not at all for organisms far removed from us on the genetic spectrum? The genetic calculus Goodenough defends (and insists earth cults need to adopt) might very well *reinforce* rather than correct existing biases.

The inordinate investment of the new cosmologists in what they assume to be universally true, cognitively compelling, or empirically verifiable, blinds them to other sources of attachment and intimacy with the natural world, including wondrous encounters with nature (spiritual or otherwise) *not* filtered through scientific explanation and up-to-date knowledge of genes, molecules, star formation, or light waves. Mythopoeic science and its narratives suggest that the natural world as humans normally encounter it—without aid of sophisticated instruments or facility with the latest scientific concepts—is neither fully real nor especially valuable. Wonder then becomes the province of the expert who grasps the abstract knowledge that eludes the layperson. Once displaced from lived experience of the world, wonder becomes at best a vicarious experience; the natural world, a derivative reality (Abram 2010). This equation of wonder only with what is scientifically real and verifiable makes additional appearances in devotees of the evolutionary epic who look to Dawkins for inspiration.

"Wonderful because real"

As I have noted, Richard Dawkins' reflections on scientific wonder have deeply impressed some enthusiasts of evolutionary epic. Unlike Wilson, who is explicit about adopting scientific mythology as a surrogate religion, Dawkins denies that his esteem for science amounts to religious-like reverence, and he refrains from presenting science in mythic form. Nevertheless, he understands science to provide spiritual uplift that far outclasses the awe of traditional religion. "Uplift," he writes, "is where science really comes into its own."

> All the great religions have a place for awe, for ecstatic transport at the wonder and beauty of creation. And it's exactly this feeling of spine-shivering, breath-catching awe—almost worship—this flooding of the chest with ecstatic wonder that modern science can provide.... The merest glance through a microscope at the brain of an ant or through a telescope at a long-ago galaxy of a billion worlds is enough to render poky and parochial the very psalms of praise.
>
> (Dawkins 1997: 27)

Dawkins has long championed the superiority of scientifically clarified—"real"—wonder vis-à-vis wonder at perceived mysteries, puzzles, or miracles. In *Unweaving the Rainbow* (1998) he presents science as satisfying the human "appetite for wonder." He chides the Romantic poets who resented Newton for destroying the mystery and poetry of the rainbow by dissecting it into light of different wavelengths. It is true, he concedes, that science banishes mystery and the miraculous, but the *knowledge* it returns is itself a thing of wonder and the stuff of magic. Real wonder, as Dawkins sees it, arises not in the presence of something unexplained but as a payoff for intensely earnest but dispassionate study.

Knowledge, in putting to rest (at least for a time) the questioning impulse, warrants the ultimate prize, the highest form of wonder. Hence, Dawkins disapproves of admiration for nature that does not stem from or culminate in something more scientific. Wonder evoked by such untutored encounters with nature is *fake* wonder, a phenomenon almost as repellant as religion itself: a mystical, muddled response to something unreal and unexplained. He lambasts the poet William Blake who, in gazing with wonder at the world in grain of sand, reveals himself as a lazy mystic "content to bask in the wonder and revel in a mystery that we were not 'meant' to understand." The true scientist "feels the same wonder but is restless, not content; recognizes the mystery as profound, then adds 'But we're working on it'" (1998: 17).

The lofty status Dawkins confers on science and scientific knowledge is further indicated by his horror that science ever be made into "vulgar fun" for laypersons or even *children*. Science, he stresses, is hard work but "worth the struggle" (1998: 23–24). Popular science demonstrations featuring "fun explosions" and "whacky 'personalities'" only "store up trouble for the future," he ominously warns (ibid: 22). Dawkins denounces the "populist whoring that defiles the wonder of science" wherever he finds it (ibid: 23). In the same metaphorical vein, he castigates those who profit from a willingness to "prostitute the language—and the wonder—of science" (2003: 43). Allusions to defilement and debauchery suggest that we that we are indeed in the presence of something sacred.

Dawkins's dedication to ensuring that even children develop demystified and scientifically correct responses to the world around them is one of the peculiar hallmarks of his sense of wonder (Sideris 2015). In 2011, Dawkins published a book for young readers that explains the true, scientific genesis of many wondrous objects, including (once more) the rainbow: *The Magic of Reality: How We Know What's Really True*. He juxtaposes myths and fairytales with lucid scientific explanation in order to discredit longstanding myths that children are likely to learn at a young age. By myths he means everything from fairytales about the rainbow's origin to Judeo-Christian stories such as Noah's ark. A chapter on the sun presents an Aztec myth, an ancient Egyptian myth, and an Aboriginal myth, prior to displacing each and the false wonder they generate, with an account of the sun's true nature. Though he often exhorts children to think for themselves, the book's message is essentially the same as *Unweaving the Rainbow*: what is real and explicable in a scientific sense is most deserving of wonder. Science is not one way of experiencing wonder, but *the* authentic way.

> I want to show you that the real world, as understood scientifically, has a magic of its own ... an inspiring beauty which is all the more magical because it is real and because we can understand how it works ... The magic of reality is—quite simply—wonderful. Wonderful, and real. Wonderful *because* real.
>
> (2011: 31)

A religion of reality

Dawkins' project receives enthusiastic support from evolutionary evangelists Dowd and Barlow who endorse the Epic as a religion of reality and hail Dawkins as its courageous prophet (Dowd 2010). Converts to this religion, Dowd likes to say, are not believers but *knowers*. Together Dowd and Barlow spread the gospel of evolution and host evolutionary revivals for thousands of groups throughout North America. They aspire to attract audiences large enough to rival modern mega-churches, and at times they nearly do. These audiences include children and adults, in both secular and liberal religious venues. Barlow describes her vision of "evolutionary revivals" that preach the message of evolutionary psychology and brain science, particularly to young people:

> Michael and I have been working for more than a year on some cool stuff in evolutionary psychology and evolutionary brain science, that helps us understand WHY we have these challenges, helps us accept our "inherited proclivities," our "unchosen nature." ... We have found that teens and young people especially tune into this part of our programs, as these are their new and frightening struggles. With the help of local liberal churches (talk about re-energizing mainline congregations!), we could pour a

> lot of energy into an amazing event that would be the template for doing more and better "Evolutionary Revivals" all around the country—which would be a new form of participatory concert for college kids, too!
>
> (Atlee 2006)

Dowd and Barlow have also created what can only be called a promotional video for Dawkins' children's book, *The Magic of Reality*. The video opens with the pair affirming Dawkins's commitment to presenting scientific reality as more wondrous than traditional myths and stories: "the truth is more magical ... than any myth or made-up mystery or miracle," in Dawkins's words (Dawkins 2011: 265). The book's message, they note, has "really broad implications for society along the lines that we've been promoting for ten years" (Dowd and Barlow 2011). The two read aloud from sections of the book that meet with their strongest approval, as when Dawkins distinguishes "supernatural" magic (a category that includes religious myths and miracles, Grimm's fairy tales, and the Harry Potter series, among other pernicious sources of fake magic) from *real* "poetic magic" that is the domain of science. Reality—facts as ascertained through science—is magical in the poetic sense, Dawkins contends. Dowd and Barlow also read with approval Dawkins' more controversial claim: "To say that something happened supernaturally," Dawkins observes, "is not just to say 'We don't understand it' but to say 'We will never understand it, so don't even try'" (2011: 23). They estimate that *The Magic of Reality* is suitable for children of approximately fifth grade level but urge viewers to introduce the book to children as early as possible, before they are old enough to understand.

Why so much interest in children and young people? If a new religion is to survive, it must have a fresh supply of young converts. But children are an important part of the ministry for other reasons. Following Dawkins, Dowd and Barlow aim to inoculate children against religious ideas before they are infected. Their goal of battling widespread religious intransigence is not, however, limited to debunking fundamentalist or creationist beliefs. Rather, Dowd and Barlow, like Dawkins, tackle any number of childlike beliefs or fanciful notions, simply because they are not of scientific provenance. Hence, Barlow's insistence in teaching children an enlightened version of "Twinkle, Twinkle Little Star": "now I know just what you are" (Dowd 2009: 91).

Dowd insists that demands for *reality* are coming from children themselves, who are no longer satisfied with stale old myths and fairy tales. Children are rejecting stories that do not meet their evidentiary standards.

> Now kids expect the real deal: magnificent BBC, National Geographic, Discovery Channel, and History Channel productions that enflesh *T. rex* and trilobites, and that spectacularly feature (and animate!) the fresh news delivered by Earth's orbiting population of space telescopes.... Ancient stories that contradict the new stories beloved of modern children (the stories of black holes and fossils).
>
> (2012)

Given Dowd's account of reality-seeking "modern" children, the real mystery is why so many flock to stories of boy wizards, vampires, and mythical beasts.

Nature in the hands of Epic enthusiasts is too often invoked as something in the distant past—extinct dinosaurs—or something far away and fantastical—black holes, Hubble images—rather than a vital, living dimension of our everyday worlds. It is invoked as a *story*, and the story itself is treated as an object of veneration, as indeed it is to the new cosmologists. At other times, nature surfaces merely as a thing to be correctly explained rather than encountered close up and in person, in a state of wonder.

Drawing children into the world of science seems laudable enough. But can it be wise to impose scientific "reality" on children who may not have yet formed attachments to nature in other ways, whether those ways are deeply spiritual or merely sensory and emotional? What does this imposition of reality suggest to children about the trustworthiness of their own impressions of the natural world and its wonder, value, or reality? I would venture that these evolutionary evangelists are less concerned with connecting children to nature than they are in preventing them from adopting religion, in any traditional sense. Appeals to the superior charms of science illustrate that science and religion are cast not simply as competing explanations for the physical world but as competing discourses of wonder. Nature, meanwhile, gets lost in the shuffle.

Wonder, lost and found

What sort of wonder is this, then, that the Epic of Evolution claims? These narratives are demonstrably lacking in what the poet Keats (1899: 277) called negative capability: an ability to dwell in doubt, mystery, and ambiguity and to resist the categorization of all phenomena and experience into a system of knowledge. For all its talk of deep time and space, stars, galaxies, and dinosaurs, Epic science peddles a rather bland yet potentially vicious (I will suggest) form of wonder. True wonder, by contrast, resists and defies the static ordering of a universal narrative and the quest for security that so often impels such narratives, the evolutionary epic included. Oddly, then, the superior wonder, magic, or grandeur said to infuse the new narratives is a function of the systematic *displacement* of abiding mystery and the questioning impulse: *now I know just what you are*. Ambivalence toward uncertainty and mystery—sometimes bordering on hostility by Wilson and Dawkins—runs through these projects. "When we have unified enough certain knowledge," Wilson fervently believes, "we will understand who we are and why we are here" (1998: 7). Paradoxically, these celebrations of wonder seek wonder's eradication at the hands of scientists and their totalizing knowledge. We should think twice about ceding wonder to mythmakers whose only response to wonder and mystery is to start "working on it," as Dawkins advises, as if it were a problem to be solved.

Whether in theology, science, or natural history, wonder has long been valued because of the ethical dispositions it frequently engenders: humility, compassion, generosity, non-exploitative stances, a "concern not to blunder into damaging manipulation of another" (Hepburn 1984: 146). We risk losing wonder's most laudable dimensions when we tether it to such delimited and rigorously patrolled categories of what is real and true. In valorizing certainty and reality (thus construed) Epic science elevates the human mind as the epic "hero"—as Wilson says, the central character in the cosmic drama (1978: 203). Armed with such certain and complete knowledge, we need not proceed with caution, prudence, or humility in our interactions with the world around us, for these dispositions are not necessary and will not flourish in an atmosphere of complete knowledge. Invoking Icarus—a tragic figure—Wilson thus throws caution to the wind: "Let us see how high we can fly before the sun melts the wax in our wings" (1998: 7). This is not the sort of dictum we should heed if it is nature—not merely ourselves and our science—that we hold within our wondering gaze.

Science is a valuable but fallible human enterprise, and as such, remains a dubious candidate as the sole or ultimate object of wonder and reverence. If we seek a new religion that enhances the prestige of science and exalts the human mind, then the Epic of Evolution may fit the bill.

But if the goal of Religious Naturalism is to foster ethical sensibilities that encourage reverent, responsible coexistence with the natural world and enduring wonder for its myriad creatures, then the Epic stands on much shakier ground. Perhaps the lesson to be learned from efforts to consecrate and mythologize science is this: those who advocate strenuously for the superiority of science over religion do not always make the best advocates for nature itself.

Notes

1 Taylor (2010) develops a typology of dark green religions that casts Goodenough's worldview as a form of Gaian Naturalism "whose proponents express awe and wonder when facing the complexity and mysteries of life and the universe, relying on religious language and metaphors of the sacred (sometimes only implicitly and not self-consciously) when confessing their feelings of belonging and connection to the energy and life systems that they inhabit and study" (16).

2 The diagnosis of amythia typically extends primarily to Western cultures or to global industrial society generally, irrespective of particular religious or cultural commitments. Thomas Berry often diagnosed storylessness as a Eurowestern problem. The subtitle of Loyal Rue's book is revealing: *Crisis in the Natural History of Western Culture* (Rue 2004).

References

Abram, D. (2010) *Becoming Animal. An Earthly Cosmology*, New York: Pantheon.

Atlee, T. (2006) "An Emerging Evolutionary Spirituality Project: Evolutionary Revivals!" *Evolutionary Life*, December. www.co-intelligence.org/newsletter/EvolutionaryRevivals1.html.

Barlow, C. (n. d.) *Evolutionary Parables*. http://thegreatstory.org/parables.html. Accessed 28 April 2017.

Barlow, C. (1997) *Green Space, Green Time: The Way of Science*, New York: Springer.

Berry, T. (1978) "The New Story: Comments on the Origin, Identification and Transmission of Values," *Teilhard Studies* 1, New York: American Teilhard Association.

Burstein, D. and Keijzer, A. (2009) "Religion—The Mental Equivalent of a Computer Virus: An Interview with Richard Dawkins," in *Inside Angels and Demons: The Story Behind the International Bestseller*, New York: Vanguard Press: 239–250.

Chaisson, E. (2005) *The Epic of Evolution: Seven Ages of the Cosmos*, New York: Columbia University Press.

Dawkins, R. (1997) "Is Science a Religion?" *The Humanist*, January/February.

Dawkins, R. (1998) *Unweaving the Rainbow: Science, Delusion, and the Appetite for Wonder*, Boston: Houghton Mifflin.

Dawkins, R. (2003) *A Devil's Chaplain: Reflections on Hope, Lies, Science, and Love*, Boston: Houghton Mifflin.

Dawkins, R. (2011) *The Magic of Reality: How We Know What's Really True*, New York: Free Press.

Dowd, M. (2009) *Thank God for Evolution*, New York: Plume/Penguin.

Dowd, M. (2010) "The New Atheists are God's Prophets." http://thankgodforevolution.com/node/2018. Accessed 28 April 2017.

Dowd, M. (2012) "Big History Hits the Big Time," *Huffington Post*, May 8. www.huffingtonpost.com/rev-michael-dowd/big-news-about-big-histor_b_1477137.html. Accessed 28 April 2017.

Dowd, M. and Barlow, C. (2011) "The Magic of Reality—An Inside Look at Richard Dawkins' First Children's Book." www.youtube.com/watch?v=CcOwS8EFSoU. Accessed 28 April 2017.

Dupré, J. (1998) "Unification Not Proved," *Science* 280: 1395.

Goodenough, U. (1994) "What Science Can and Cannot Offer to a Religious Narrative," *Zygon* 29/3: 321–330.

Goodenough, U. (1998) *The Sacred Depths of Nature*, New York: Oxford University Press.

Goodenough, U. (2009) "Ecomorality: Toward an Ethic of Sustainability," in L. A. Mazur (ed.) *Pivotal Moment: Population, Justice, and the Environmental Challenge*, Covelo, CA: Island Press: 372–382.

Goodenough, U. (2010) "My Covenant with Mystery," 13.7, *Cosmos and Culture Blog*, August 27. www.npr.org/blogs/13.7/2010/08/27/129471676/my-covenant-with-mystery.

Gray, M. (1998) "Letter to the Atlantic: Biology and Morality," *The Atlantic Online*, July.

Hepburn, R.W. (1984) *"Wonder" and Other Essays: Eight Studies in Aesthetics and Neighboring Fields*, Edinburgh: University of Edinburgh Press.

Keats, J. (1899). *The Complete Poetical Works and Letters of John Keats*, Cambridge Edition, Boston and New York: Houghton, Mifflin and Company.
Kennard, D. and Northcutt, P. (dir.) (2011) *Journey of the Universe*, InCA Productions. DVD.
Religious Naturalist Association (n.d.) http://religious-naturalist-association.org/. Accessed 1 May 2017.
Rue, L. (1997) "Going Deeper: Spiritual Dimensions of the Epic of Evolution," *Earthlight Magazine* 26 (Summer): 12–13. www.earthlight.org/personal26.html. Accessed 30 April 2017.
Rue, L. (2000) *Everybody's Story: Wising up to the Epic of Evolution*, Albany, NY: State University of New York Press.
Rue, L. (2004) *Amythia: Crisis in the Natural History of Western Culture*, Tuscaloosa, AL: University of Alabama Press.
Rue, L. (2005) "Epic of Evolution," in B. Taylor (ed.) *Encyclopedia of Religion and Nature*, London and New York: Continuum: 612–615.
Rue, L. and U. Goodenough (2009) "A Consilient Curriculum," in Genet, C. et al. (eds.) *The Evolutionary Epic: Science's Story and Humanity's Response*, Santa Margarita, CA: Collins Foundation Press: 175–182.
Sagan, C. (1994) *Pale Blue Dot: A Vision of the Human Future in Space*, New York: Random House.
Sideris, L. (2015) "Contested Wonder: Biological Reductionism and Children's Nature Education," *Journal of Religion and Society*, Supplement Series 11: 193–205.
Sideris, L. (2017) *Consecrating Science: Wonder, Knowledge, and the Natural World*, Berkeley and Los Angeles: University of California Press.
Swimme, B. and Berry, T. (1992) *The Universe Story from the Primordial Flaring Forth to the Ecozoic Era: A Celebration of the Unfolding of the Cosmos*, New York: HarperCollins.
Swimme, B. and Tucker, M.E. (2011) *Journey of the Universe*, New Haven, CT: Yale University Press.
Taylor, B. (2005) *The Encyclopedia of Religion and Nature*. London and New York: Continuum.
Taylor, B. (2010) *Dark Green Religion: Nature Spirituality and the Planetary Future*, Berkeley and Los Angeles: University of California Press.
White, C. (2013) *The Science Delusion: Asking Big Questions in a Culture of Easy Answers*, Brooklyn, NY: Melville House.
Wilson, E.O. (1978) *On Human Nature*, Cambridge, MA: Harvard University Press.
Wilson, E.O. (1998) *Consilience: The Unity of Knowledge*, New York: Vintage.

34

REFLECTING ON RELIGIOUS NATURALISM

Possibilities and critiques

Philip Hefner

My discussion in this chapter consists of an appreciation of religious naturalism's possibilities and also a critique. The appreciation is essential for two reasons: to clarify my own perspective and to express the presuppositions of my critique.[1]

Appreciation of possibilities

My statement of appreciation turns on four basic points:

1. The "turn to nature" is necessary, because nature, broadly conceived, is all there is. In our time, the sciences provide a normative description of nature. Without an understanding of nature, a naturalistic perspective has no specific content.
2. Nature is a text requiring a hermeneutic. If we are to understand nature, we must first engage in a process of interpretation. Part of this process consists of formulating ideas of nature.
3. Since nature is dynamic, frequently changing, we must be open to new ideas of nature. I propose here an "expansive" view of nature.
4. Religions offer such a hermeneutic.

The last 20 years have brought a "turn to naturalism." This turn has occurred in the wake of the sciences, giving us a picture of nature as incredibly varied and comprehensive. As a result, today we look first and foremost for natural explanations of everything. Even those who deny that natural explanations can account for all that we experience, recognize that we must search for answers within nature before we consider other kinds of answers. Naturalistic explanations are the default modes for understanding our world. These explanations are applied not only to the nature outside us, but also, increasingly, to our inner nature—our minds, emotions, our "souls." Those who choose to reject this turn to nature carry a heavy burden of proof. For this reason, it is important that widely accessible naturalistic perspectives be developed. The idea of nature that I propose is offered as such an accessible form.

I endorse an expansive idea of nature. We have merged with the processes of nature and live out our lives within those same processes. Our bodies are occurrences of nature—we *are* nature.

How do we begin to think about nature? When we talk about a turn to nature, what are we turning toward?

Alfred North Whitehead, in his lectures *The Concept of Nature* (Whitehead 1920), argues that even though we are immersed in nature, we know it through our thinking on nature, concepts we construct by means of science. We also think about our thinking in our philosophy of science. While our experience of nature is immediate, our knowledge of nature is once-removed and constructed.

R. G. Collingwood, a British philosopher and historian in the first half of the twentieth century, dealt with this "once-removed" knowledge at some length. He wrote a little book that was published posthumously in 1945 and has become a classic, *The Idea of Nature* (Collingwood 1945). He emphasized that since nature is so rich and various that it is impossible to encompass it in our rational conceptions, we crystallize its multiplicity in our *ideas of nature*. Nature is a kaleidoscope of shapes, colors, sounds, movements—and creatures. A rushing mountain stream is nature, and so are the water striders that we see walking on the calmer pools nearby and the peaks and forests that surround the scene. And of course, the observers of all this—we ourselves, for example—are nature, too. It is impossible to grasp the whole kaleidoscope in our minds at one time, so we bring it all together in a single idea. The idea is an image that helps us put all the natural forms together. It also forms our understanding and guides our investigations.

According to Collingwood, these ideas do more than focus our understanding of nature; they also determine how we view ourselves and how we relate to everything else. For example, our idea of nature influences whether we consider ourselves to be instances of nature and how we think of our mental and spiritual lives—as well as how we think God relates to nature.

Collingwood believed that our ideas of nature have changed over time. The ancient world held to the idea of nature as a living organism, while the Renaissance and Enlightenment viewed nature in analogy to the machine. The twentieth century, under the impact of Darwin's theory of evolution and the quantum revolution in physics, took to the idea of nature as historical process; nature is on a journey, as yet unfinished. Even though he was a historian as well as a philosopher in his study of nature, he was more concerned with the logic that was inherent in these ideas, rather than in the history as such. Collingwood's "idea" is comparable to "ideal type" as used by Ernst Troeltsch (Troeltsch 1912) and Max Weber (Weber 1946; Swedberg and Agevall 2005). Neither idea nor type is amenable to rigorous empirical validation; rather, they are heuristic devices that get us into the material and stimulate interpretation (Allan 2005: 149). In the nearly 70 years since Collingwood wrote, our ideas of nature have continued to develop.

All of these ideas—organism, machine, historical process—are linked, Collingwood suggested, to experience. (1) *Living intelligent organism*. The ancients experienced nature as permeated by mind; nature is an intelligent living animal and as such it orders itself. The animals inhabiting the earth participate in the world's soul and mind just as they participate in its body. Nature was seen as analogous to the individual human being. (2) *Machine and maker*. Since the Renaissance was preoccupied with new and marvelous machines—think of Leonardo Da Vinci—it is not surprising that they viewed nature through this lens. In contrast with ancient views, they considered nature to be devoid of life and intelligence. Nature's order is imposed by outside forces. The analogy at work here is a machine and its maker—nature is God's created machine. This view continued to dominate through the eighteenth century. (3) *Historical process*. The sense that reality is historical process or journey is basic to modern experience—it has been said that the nineteenth century "invented" the worldview or metaphysics of history. Historical studies that placed process, change, and development at the center became the prevailing analogy for an idea of nature. In mid-century, Darwin's evolutionary theories depicted life itself as a developing historical process, and the human species itself had a *natural* history. By the end of the century, Collingwood held, physicists were speaking of the "quantum world," too small to be seen with the naked eye, in which material things are made up of atoms, which in turn are composed of particles, all of which make their own historical journeys. Philosophers began to speak of reality as *process, creativity* (Henri Bergson 1911; Gustav Fechner 1877; Rudolf Lotze 1889).

Collingwood's three-stage interpretation of the idea of nature is useful, but we cannot stop where he did in the mid-twentieth century. We still speak of nature as a living organism (think of the Gaia idea [Lovelock 1979]) and as a machine (we talk about our heart as a pump and also about our internal "plumbing" or "pipes") even though these ideas are really obsolete and misleading. Nature is made up of many organisms; it is not a single super-organism, and there are such things as rocks and volcanoes that are hardly classified as "organic." Viewing our bodies as a machine or an engine may have limited validity, but we soon discover that our gears and pumps and pipes are living things that do not react like the insides of a watch or an automobile engine. It is still valid to look upon nature as a historical process—on a journey, as described by cosmological, biological, and cultural evolution—but the idea of history is not sufficient unless we include that it is a dimension in which surprising new things emerge.

The sciences are changing and expanding our view of nature in ways that defy imagination—this is something to which we often do not give enough attention. How recent our scientific knowledge is! We have increased our knowledge of nature more in the last 50 years than in all of previous human history. Older ideas of nature are deficient.

I suggest three additional elaborations of our idea of nature. (4) *Emergence.* Closely related to the idea of historical process is that of *emergence*—new and unexpected events seem to come forth from what we see before us without any extraordinary or outside causes that we can detect. They come out of the bottle like the genie, so to speak, and they cannot be put back in as they were before. Nature is boundlessly rich and constantly producing new things. This experience of the new and unexpected is so basic to our everyday life that we can well conclude that it is inherent in nature. Little wonder that it is also now a significant scientific research item (Morowitz 2002). Nature has the capability to be self-creating—technically called *autopoeisis* or self-generation.

"Complexity science" is the name that is frequently applied to this research, since novelty seems to emerge in systems composed of interconnected parts that work together in ways that are not predictable if we focus only on the parts individually. The Santa Fe Institute, established in 1984, is an example of a scientific effort that devotes itself entirely to complexity/emergence research. The Institute brings together ideas and principles of many fields—from physics, mathematics, and biology to the social sciences and the humanities—in pursuit of creative insights that improve our world. Their studies include quantum physics, molecular biology, weather, music, and urban traffic patterns—all of them complex systems in which novelty emerges.

(5) *Mystery.* To these ideas we add yet a fifth that is also deeply rooted in our experience: the idea of mystery. Nature is a realm of knowledge, control, and mystery. (a) Our knowledge about nature, chiefly through our scientific research, is mind-boggling. In the last half-century, we have added millions of pieces of information from dozens of specific sciences to our body of knowledge. Scientific exploration has unfolded a picture of nature—human nature and the larger world—that is mind-bending and inexhaustibly rich. (b) This knowledge has led to our control of nature in ways that we scarcely dreamed of just a century ago. Knowledge has spawned technology that is unimaginably complex and successful in bending nature to our will, from

the levels of atoms and molecules to that of computers, earth-moving, and space exploration. (c) At the very same time, our knowledge reveals to us how much more there is to know, and our attempts at control reveal how successfully nature can defy our attempts to tame it. This is not a question of "gaps" in our knowledge or a breakdown in our control functions. Rather, it is precisely the success of our quest for knowledge and control that makes clear that nature is more than we can comprehend and more than we can ever bring under our complete control.

This awareness brings us to the reality of mystery. Mystery is not a matter of ignorance, not a matter of not knowing enough. Mystery is a matter of richness and texture. Mystery is "the endlessly knowable" (a phrase of Paul Crowley's [2011: 9]). Our knowledge about nature continues to grow exponentially, but the more we know about nature, the greater the richness and the deeper the mystery become. This is especially true of our human nature. Theologian Paul Tillich was right when he said that each human being is marked by a mystery, depth, and greatness (Tillich 1948: 159). Mystery is a clue to the meaning of our experience of nature.

Consider two examples. For one, the facts of climate change. Our vast knowledge, even in its present imperfect state, enables us to know that climate change is in fact happening and what some of its physical causes are. At the same time, our knowledge tells us that we can understand and control only a small segment of this change—that in large part we must adapt to it where we cannot hope to control it or roll it back. Recall the quite unexpected onslaughts on the East Coast of the United States within a six-month period in 2011—record snowfall, earthquake, and hurricane. Attempts to adapt, in turn, raise questions of human nature and destiny. To what extent can we adapt or not as a human species? How do we make the decisions as to where or for whom adaptive strategies will be devised? What are our ethical obligations to those regions and peoples that will not be able to adapt successfully? What is the human future to be? Is it to be survival for the wealthiest and for those who live in "safer" places, away from coastlines and drought areas, and extinction for the rest? Our knowledge and control leave the human and planetary future beyond our capacities to determine; they bestow on our decisions the gravity of a wager—leaps of faith into the unfathomable and unmanageable as we try to make our way.

Consider the knowledge and control we gain through cognitive neuroscience, through which we trace in detail the brain processes that correlate to such basics as our thinking about specific things, our emotions, and our interactions with other people. The detail and complexity of our brain's activity is awesome. This research forms the surface of the even more complex work of our brains that science does not explain—how these brain processes bring forth a Beethoven symphony or a Bach chorale, a Shakespearean play or a poem by Emily Dickinson, Darwin's theory of evolution or Einstein's theory of relativity, the proposal of Jeffersonian democracy or the Gandhi/King practice of non-violence. In other words, the scientific charting of our brain's activities overlaps the richness and mystery of the human spirit. Brain scans do not translate into Beethoven's Ninth or into a Marine's act of heroism in falling on an exploding grenade, thus saving the lives of his comrades, even though the achievements of human spirit are fully embedded in our biology—science intensifies our sense of this embeddedness. The more we know about it, the more we realize that we do not comprehend it—it is mystery.

We should be clear about how we are using the word *mystery*. It is not the same as a puzzle. A puzzle can be solved—the more we learn about it and think about it, the closer we come to its solution. Richness and texture characterize mystery. As we dig into it ever more deeply, we come upon ever richer and more profound dimensions of meaning. This endless depth of meaning is not something to be "solved"; it is inexhaustible. So in the example of climate change, we find that over and beyond our increase of knowledge and devising changes in behavior we are engaged with the world of nature in an ongoing quest for understanding our relationship with nature, a quest that in turn brings us face-to-face with questions of human destiny. (a) Just what should we be about in our relations with the natural world? (b) What is the purpose of human behaviors toward nature? (c) What are our obligations to the natural world, and how are we responsible for enabling other humans and other creatures in their relationship to nature? These questions are appropriate to mystery. In this case the mystery of the natural world—despite our enormous amount of knowledge about it—remains a challenge we will never exhaust, provoking us to reflect hard and long about the meaning of human life in this world.

The example from cognitive science proceeds similarly. The more we learn about our brains and how their processes correlate to our mental lives, the deeper the mystery of how our neurobiology can enable the unbelievably rich possibilities that our minds explore every moment, whether it is creative art, problem-solving, devising strategies for living, or deepening our relationships with other people. In these activities, our minds—in their thoroughly natural biological working—do in fact continuously transcend our physical situations. In this mental activity, nature is transcending itself. Nature is a continuously self-transcending realm. When we ponder the meaning of human life in the natural world and when we experience the creativity of a Shakespeare or a Ray Kurzweil, we experience nature itself seeking to go beyond itself. Our encounter with mystery is a signal that we are in the presence of transcendence that challenges us to explore its depths.

We think of Teilhard de Chardin's aphorism, "Humans are evolution become aware of itself" (Teilhard de Chardin 1965: 65). In our very own intellectual and spiritual life, we witness nature going beyond itself and thereby redefining itself. We embody this witness in our experience; it is ours—the witness is us. At the same time, while we possess this self-transcending action since it is our own experience, we are aware that we are not fully in control of its cascading rush; we sense that we are riding a torrent that is larger than we are—we are subject and object at the same time. Rather than possessing this experience, it is more accurate to say that we are possessed by it. Here we are confronted with the basic question: What is the significance of these acts of transcendence? What do they reveal to us about our fundamental human nature and our reason for being? Again we encounter not a lack of knowledge, not a puzzle to be resolved, but the mystery of our very bodily nature—this is Tillich's point: our nature possesses a mystery, a depth, and a greatness that are fully natural.

(6) *Full-bodied/God-intoxicated.* These reflections lead to a sixth idea of nature. We call this idea—that nature is the inexhaustible source from which new things continually emerge—a *full-bodied* idea of nature. As such, nature continually confronts us with and wraps us in *mystery*. Mystery involves us, as we have said, in transcendence, and when some religious people encounter transcendence and mystery, they understand that they are in the presence of God. We must be clear—this engagement with transcendence comes to us, *not* through some supernatural intervention and not by introducing some *unnatural* element. Rather, we are through earthiness bathed in transcendence. Hence, the Christian theological possibility to call this a *full-bodied* and *God-intoxicated* idea of nature.

These ideas of nature, particularly nature as a process of emergence and as full-bodied, bring with them a deep sense of humility. We are much aware of ourselves as being borne upon processes of which we are not the authors or the drivers. We know ourselves to be in possession of freedom, in that we are always faced with decisions that need to be made. Even more, we feel that we are part of something larger than ourselves, and this is the seed of humility in the face of nature—both within us and outside of us. This sense of humility in the face of depth has been called "creature-feeling" (Otto 1923: 10–11); we engage mystery at this point. Mystery surrounds us at every point—when we seek knowledge of nature, when we try to control it, when we ponder its bottomless depth within us—each of these is an avenue that leads us sooner or later to mystery, which in turn may be an opening to transcendence. The poet A. R. Ammons spoke of how "things spiral out from a center" and take shape as they come forth (Ammons 1986: 61). The processes of nature on which we are borne are always spiraling out from mystery and taking shape right before us—and within us.

Interpreting nature—the need for a hermeneutic

Nature is a text that stands in need for interpreting; it is not a self-explicating continuum. To gain knowledge and understanding of nature, we require an interpretive method and concepts—a hermeneutic. Subatomic particles, for example, do not know the laws of physics. We construct those laws in our attempt to understand the behavior of those particles, just as we construct theories to understand the behavior of biological organisms that are equally ignorant of Darwin and his successors. The array of methods and theories that make up the contemporary sciences may constitute the dominant hermeneutic for understanding nature in our time. Throughout the millennia of their history, however, humans have been constructing interpretations of nature—most often perhaps in their religious systems. All of the world's religions show the record of the human effort to understand nature. No religion is de-natured.

The traditions of our seeking understanding of nature form a huge body of knowledge and attempts to cope with nature—thousands of years of trial-and-error and reflection upon the success and failures in that process of coping.

At the outset, we must remind ourselves constantly that both nature and religion are rich and dynamic—so much so that all our concepts will be inadequate. Concepts are necessary, we rely on them, but they must be continually revised.

Are all religions naturalistic? At no point in human history have humans been out of touch with nature. For all of that history, we have struggled with the world of nature in which we live out our lives. We include human nature in the concept of nature. This means that all human thinking, including our religions, is in some sense naturalistic and in that sense a kind of "naturalism." Even world-denying philosophies and religious practices are statements about nature.

Following these assertions, we ask whether all religions are forms of religious naturalism. Buddhist teachings and practices, for example, are responses to nature, and by no means anti-nature. Like all asceticisms, the Buddhist forms are careful strategies for taking nature seriously. At the heart of Christian faith is the belief that in Jesus Christ, divinity and humanity existed in indissoluble unity, each fully its own "nature." Christian sacramental practices are an extension of this belief about Christ. In this respect, I consider the traditional reflections on Christ to be consistent with the expansive idea of nature that I outlined in the first part of this chapter. Several chapters in this book give attention to religions as naturalistic. (See chapters by Oh, Rue, and Tucker in this volume.)

A necessary polarity: science explaining religion versus religion as hermeneutic. It is common practice for later generations to try to update traditional religious accounts of nature by means of explanations derived from current knowledge. Much has been written about the correspondence between traditional Buddhist meditation and contemporary psychology. Stories from Hebrew Scriptures have been explained "scientifically": the account of creation in the first chapter of

Genesis, the parting of the Red Sea, and the sun standing still for Joshua, are prominent examples. The New Testament accounts of Jesus' miracles have received intense scientific scrutiny. These explanations of traditional accounts are useful for those who already accept the tradition; they may have less impact on the critics of traditional religion. Karl Peters' chapter in this book is an excellent example of explaining items of Christian belief in naturalistic terms.

We might say the efforts to explain the religious traditions look through one end of the telescope, the end that focuses on details. This approach applies a science-based hermeneutic to religion. When we look through the other end of the telescope, religion-based hermeneutics come into play. From this end, religious, rather than scientific, issues are dealt with—questions of meaning, purpose, and morality.

The religious end of the telescope involves a sophisticated hermeneutic of its own. Religion includes ritual, symbol, myth, and narrative, among other forms. Each of these forms requires a method of interpretation that is appropriate, methods derived from anthropology, linguistics, literary studies, philosophy, and theology. The scientific and religious approaches use language in different ways—they exist together as a polarity in which each pole must maintain its integrity, neither pole can be collapsed into the other.

Hermeneutic sophistication is required, especially when dealing with the mythic and "supernatural" texts of religious traditions. These texts are frequently dismissed for their literal absurdity. Such dismissal rests on a category mistake. Myths and symbols are not intended literally; their richness is unearthed through patient exploration of layers of possible meanings. The term "God" also falls within this context of sophisticated hermeneutic consideration.

We cannot survey the full hermeneutic agenda here, but one example can clarify my point. A major item on the agenda is the assertion of Christian, Jewish, Muslim, and Hindu faiths that the evolution of the cosmos tends toward fulfillment, not destruction, goodness, not evil. Science cannot adjudicate the truth of this assertion, although it can cleanse it from absurdities (the Christian teaching about the so-called Fall, is, in my opinion, one such absurdity). The assertion of essential goodness and fulfillment is not a simple one—it requires sophisticated interpretation—but it does not violate our scientific understanding of nature.

A related issue is that of natural selection. Arguments have been made that both Buddhist meditation and the Christian teaching of love seek to counter selection (Theissen 1994; Wright 2017). Scientific and religious hermeneutical approaches can work together to explore this idea. By "countering selection," I refer to attempts to go against, even reverse, the effects of natural selection—the chief of which are death and extinction. Humans frequently seek to counter selection. Ironically, we employ science-based medical practice to counter the lethal selection by disease and defect among humans. Gerd Theissen, for example, argues that Jesus proposes an anti-selectionism when he proclaims that God shows equal concern for the poor, when he urges inclusion of eunuchs and love for enemies, and when he subordinates the genetic-related family unit (Theissen 1994: 114–119). Robert Wright makes the case that Buddhist mindfulness counters the biologically programmed desire to seek pleasure, which results in chronic dissatisfaction. Mindfulness meditation, he insists, carries out a rebellion against natural selection (Gross 2017).

A critique for future work

In our time, it is critical that the efforts and insights of religious naturalism be made accessible and persuasive. My special concern is that this task be furthered among traditional religious communities. Most often religious naturalism is rejected in these communities, for several reasons: First, because the concept of nature is often understood solely in reductionistic and one-dimensional terms. I have developed an expansive idea of nature in order to introduce new, more scientifically realistic understandings.

Second, too often religious naturalism ignores the hermeneutic resources that are offered by religious traditions. Religious naturalist thinkers tend too often to look through only one end of the telescope, thus devaluing the richness of tradition. To many adherents of religion, ignoring the tradition results in a flatness and thinness, less rich and less satisfying.

Third, functional approaches tend to dominate the literature of religious naturalism. More acknowledgement is in order of the exploratory and wisdom dimensions of religion. The functional approach is useful and necessary, but religion will not allow itself to be confined to one approach. Religion is polysemic and thus can be understood adequately only by a variety of interpretative methods. Religious naturalism will not succeed if it is in effect a synonym for "scientific" naturalism. If religious naturalism allows itself to be enriched, it can fulfill its historical potential.

Note

1 Portions of this chapter are taken from Hefner, P. (2015), "An Idea of Nature: A Bipolar Proposal." *Zygon: Journal of Religion and Science* 50: 287–303.

References

Allan, K. (2005) *Explorations in Classical Sociological Theory: Seeing the Social World*, Los Angeles, CA: Sage.
Ammons, A. R. (1986) "Poetics," in *The Selected Poems: Expanded Edition*, New York: W. W. Norton.
Bergson, H. (1983 [1911]) *Creative Evolution*, Lanham, MD: University Press of America.
Collingwood, R. G. (1945) *The Idea of Nature*, New Haven, CT: Yale University Press.
Crowley, P. G. (2011) "Tomorrow's Theologians," *America* 204: 9.
Fechner, G. (1877) *In Sachen der Psychophysik*, Leipzig: Breitkopf und Härtel.
Gross, T. (2017) "Can Buddhist Practices Help Us Overcome the Biological Pull of Dissatisfaction?" in http://news.wfsu.org/post/can-buddhist-practices-help-us-overcome-biological-pull-dissatisfaction (accessed 8/15/17).
Lotze, H. (1889) *Microcosmus: An Essay Concerning Man and His Relation to the World*, trans. E. Hamilton and E. E. C. Jones, 4th edition, Edinburgh: T. & T. Clark.
Lovelock, J. (1979) *Gaia: A New Look at Life on Earth*, Oxford: Oxford University Press.
Morowitz, H. J. (2002) *The Emergence of Everything: How the World Became Complex*, Oxford: Oxford University Press.
Otto, R. (1923) *The Idea of the Holy: An Inquiry into the Non-rational Factor in the Idea of the Divine and Its Relation to the Rational*, trans. J. W. Harvey, 2nd edition, New York: Oxford University Press.
Santa Fe Institute (2016) "Our Mission," www.santafe.edu/ (accessed 11/2014).
Swedberg, R. and O. Agevall (2005) *The Max Weber Dictionary: Key Words and Central Concepts*, Palo Alto, CA: Stanford University Press.
Teilhard de Chardin, P. (1965) "My Universe," in *Science and Christ*, trans. Rene Hague, New York: Harper & Row.
Theissen, G. (1994) *Biblical Faith: An Evolutionary Approach*, Minneapolis, MN: Fortress Press.
Tillich, P. (1948) "You Are Accepted," in *The Shaking of the Foundations.* New York: Scribner's.
—— (1957) *Systematic Theology, Volume II*, Chicago, IL: University of Chicago Press.
Troeltsch, E. (1912) *The Social Teaching of the Christian Churches*, 2 vols., trans. O. Wyon, Louisville, KY: Westminster/John Knox Press.
Weber, M. (1946) *From Max Weber: Essays in Sociology.* Eds., H. Gerth and C. W. Mills, New York: Oxford University Press.
Whitehead, A. N. (1920) *The Concept of Nature*, Cambridge: Cambridge University Press.
World Headquarters for Complexity Science (2016) www.santafe.edu/ (accessed 8/19/2017).

Wright, R. (2017) *Why Buddhism Is True: The Science and Philosophy of Meditation and Enlightenment*, New York: Simon and Schuster.

Further reading

Corrington, R. S. (1992) *Nature and Spirit: An Essay in Ecstatic Naturalism*, New York: Fordham University Press. (Creative proposals for understanding deeper dimensions of nature.)

Hopkins, G.M. ([1877] 1967) "God's Grandeur," in *The Poems of Gerard Manley Hopkins*, 4th edition, eds. W. H. Gardner and N. H. MacKenzie, New York: Oxford University Press: 66. (One of the greatest statements of an expansive view of nature in English literature.)

Ricoeur, P. (1967) *The Symbolism of Evil*, Boston, MA: Beacon Press. (An exemplary exercise in biblical hermeneutics.)

Swimme, B. and T. Berry (1992) *The Universe Story: From the Primordial Flaring Forth to the Ecozoic Era*, San Francisco, CA: Harper. (An important text in religious naturalism, synthesizing science, theology, and spirituality.)

Teilhard de Chardin, P. (1963) "The Mass on the World," in *Hymn of the Universe*, trans. S. Bartholomew, New York: Harper & Row. (A classic text in Christian naturalism.)

INDEX

Made in the USA
Monee, IL
13 September 2026

62978961R10240